Interactive TV Standards

191

Interactive TV Standards

Steven Morris
Anthony Smith-Chaigneau

ELSEVIER

AMSTERDAM • BOSTON • HEIDELBERG • LONDON
NEW YORK • OXFORD • PARIS • SAN DIEGO
SAN FRANCISCO • SINGAPORE • SYDNEY • TOKYO
Focal Press in an Imprint of Elsevier

Acquisition Editor: Joanne Tracy/Angelina Ward
Project Manager: Carl M. Soares
Assistant Editor: Becky Golden-Harrell
Marketing Manager: Christine Degon
Design Manager: Cate Barr

Focal Press is an imprint of Elsevier
30 Corporate Drive, Suite 400, Burlington, MA 01803, USA
Linacre House, Jordan Hill, Oxford OX2 8DP, UK

Recognizing the importance of preserving what has been written, Elsevier prints its books on acid-free paper whenever possible.

Library of Congress Cataloging-in-Publication Data
Morris, Steven, 1972-
    Interactive TV standards / Steven Morris, Anthony Smith-Chaigneau.
    p. cm.
    ISBN 0-240-80666-2
  1. Interactive television—Standards.   I. Title: Interactive television standards.
II. Smith-Chaigneau, Anthony, 1959-   II. Title.
    TK6679.3.M67 2005
  621.388'07—dc22
                            2005001355

British Library Cataloguing-in-Publication Data
A catalogue record for this book is available from the British Library.

ISBN: 0-240-80666-2

For information on all Focal Press publications
visit our website at www.books.elsevier.com

05 06 07 08 09 10   10 9 8 7 6 5 4 3 2 1

Printed in the United States of America

To Jasmine and Dylan, for their
patience and support and for just being there;
and to Joan and Emyr Morris

—Steven Morris

To Peter MacAvock, for bringing
me into the world of TV technology;
and my family who think I am
really an international spy!

—Anthony Smith-Chaigneau

Acknowledgments

The authors would like to thank everyone who has given his or her assistance during the writing process, especially Jon Piesing, Paul Bristow, and Martin Svedén, as well as to everyone who has provided illustrations for use in this book. The authors would also like to thank their editor Joanne Tracy as well as Becky Golden-Harrell and Carl Soares for their assistance, guidance, and high professional competence.

Permissions

Tables 4.6, 12.2, 12.5, 12.8–12.12, 12.14–12.17, 15.1, 15.2, 15.7, 16.2, 16.3, and A.5–A.13 and figures 2.3, 4.1, 7.10, 7.15, 12.4, and A.3 are copyright © ETSI 1999–2003. Further use, modification, or redistribution is strictly prohibited. ETSI standards are available from *http://pda.etsi.org/pda* and *www.etsi.org/services_products/freestandard/home.htm*.

Tables B.2, B.4–B.6, B.8, B.9, B.11–B.15 and Figure B.2 are taken from ATSC document number A/65b (Program and System Information Protocol for Terrestrial Broadcast and Cable, Revision B). Tables B.3, B.7, and B.10 are taken from ATSC document number A/81 (ATSC Direct-to-Home Satellite Broadcast Standard). These tables and figures are copyright © Advanced Television Systems Committee, Inc., 2003. Readers are encouraged to check the ATSC web site at *www.atsc.org* for the most recent versions of these and other standards.

Tables 10.1, A.1–A.3 and A.16 are taken from ISO 13818-1:2000 (*Information Technology: Generic Coding of Moving Pictures and Associated Audio Information Systems*; tables 2-30-1, 2-25, 2-28, 2-27, and 2-30, respectively). These tables are reproduced by permission of the International Organization for Standardization, ISO. This standard can be obtained from any ISO member and from the web site of the ISO Central Secretariat at *www.iso.org*. Copyright remains with ISO.

Figures 3.2 and 3.3 are taken from the OpenCable Application Platform 1.0 profile, version I09. Copyright © Cable Television Laboratories, Inc., 2002–2003.

Figures 7.6, 7.7, and 7.8 are taken from the HAVi 1.1 specification. These figures are copyright © HAVi, Inc., 2001, and are used courtesy of the HAVi organization, November 2004.

Table B.17 is reproduced from ANSI//SCTE document number 65 and is copyright © Society of Cable Telecommunications Engineers, Inc., 2002. SCTE standards are available from *www.scte.org*.

Trademarks

CableLabs, DOCSIS, OpenCable, OCAP, and CableCARD are registered trademarks of Cable Television Laboratories, Inc. DVB and MHP are registered trademarks of the DVB Project Office. The HAVi name is a registered trademark of HAVi. Java and all Java-based marks are trademarks or registered trademarks of Sun Microsystems, Inc. in the United States and other countries. All other trademarks are the property of their respective owners.

Contents

Introduction

Millions of people worldwide watch digital TV (DTV) every day, and this number is growing fast as more network operators and governments see the benefits of digital broadcasting. In recent years, interactive TV (ITV) has become the "next big thing" for the broadcasting industry as broadcasters and network operators seek new ways of making money and keeping viewers watching.

Although open standards are nothing new to the broadcasting industry, both public broadcasters and pay-TV operators are starting to use open standards for ITV middleware, to try to bring ITV to a wider audience. Hand in hand with this, governments are requiring the use of open standards for publicly funded DTV systems, and this includes the middleware those systems use.

Around the world, JavaTV and MHP form the core of the open middleware systems that are being built and deployed. Broadcasters, receiver manufacturers, and application developers are all jumping on the MHP bandwagon. In the United States, the OCAP standard (based on MHP) looks poised for a very successful introduction into the marketplace.

Unfortunately, this is still a confusing area for people who are trying to use the technology. This is partly because the market is still young, partly because these standards can have a profound effect on the business models of companies that use them, and partly because the available documentation is spread across many sources and is not always consistent. Both the pro– and anti–open standards camps are pushing their own views of these standards, and impartially researched information is difficult to come by.

The book you are holding is one of the first truly independent discussions of these technologies. Both of the authors have been involved in MHP since the early days. We have been part of the standardization process, both on the technical side and on the commercial side. We have written business cases for MHP and OCAP deployments, we have built middleware implementations, and we have built and deployed applications. We have heard the questions project managers, application developers, and middleware manufacturers are asking, and we hope that this book will answer some of those questions for you.

With this book, we will give you an insight into the background of the MHP and OCAP standards and the issues involved in implementing them. We look at how the different standards fit together, and at how you can use them to build good products and get them to market quickly. This book also acts as a companion to the underlying standards that make up MHP and OCAP. We take an in-depth look at the MHP and OCAP APIs and architecture, at how middleware developers can build efficient and reliable middleware, and at how application developers can exploit these standards to build cool applications. Most importantly, we examine how we can actually make some money from our products once we have built them.

This is not an introduction to DTV. It is not a book on Java programming. We concentrate on the practical issues involved in working with these new middleware technologies and in building products that people want to purchase and use. By looking "under the hood" of these standards, we hope that both new and experienced developers and managers can learn how to exploit these standards to their full potential.

Intended Audience

This book is of interest to anyone who works with MHP and OCAP, building applications, building middleware stacks, or deploying MHP or OCAP in a real network. We assume that you have some knowledge of digital broadcasting, and (for developers) we assume that you have some experience in developing software in Java. We do not assume any familiarity with other middleware standards, however, or with the technical details of how DTV works. At the same time, we cover the material in enough depth that even experienced OCAP or MHP developers will find it useful. In particular, this book is of interest to the entities discussed in the following sections.

Project Managers

If you are responsible for deploying an OCAP or MHP solution (whether it is a receiver, an application, or a complete set of MHP services) you need to make sure you can deliver a successful product on time. Deploying MHP or OCAP is similar to deploying other DTV systems, but this book highlights the differences in the business models and in the way products need to be deployed, and it will help you make sure your products interoperate with others in the marketplace.

Application Developers

You may already be familiar with Java and with programming for DTV systems, and thus this book does more than just cover the basics. It also covers the practical details of how we build an MHP or OCAP application and how we can get the most out of the various APIs. We also examine how you can make your application portable across middleware stacks.

Middleware Developers

The challenges of building an MHP or OCAP middleware stack are very different from those involved in working with other middleware stacks. The design and implementation of the middleware plays a vital role in the success of a project, and thus we examine how you can build the most reliable and efficient middleware stack possible, looking at the design and implementation issues that can affect your entire project. We will also look at the rationale behind some of the design choices in the standard, to help you make more informed decisions about how you should build your software.

Senior Management, Sales and Marketing Staff, and Network Operators

In addition to looking at the technical details of MHP and OCAP, this book examines the commercial aspects of the new crop of open standards. This includes business models for ITV, the advantages and disadvantages of open middleware standards, and the market situations that can affect an MHP or OCAP deployment.

Students

With the growth in DTV systems worldwide, more universities are running courses in DTV technology and application development. This book introduces you to MHP and OCAP, and provides you with practical advice based on many years of experience in the industry. By combining practical examples and real-world experience, this book offers you more than just the theory.

Book Organization

This book consists of four main sections. Starting from a basic introduction to DTV and the issues involved in broadcasting DTV services (both commercial and technical), we then move on to look at the basic features of MHP and OCAP from a technical perspective. This provides a grounding in the essentials of building applications and middleware, after which we look at more advanced topics. Finally, we take a look at the practical issues of building and deploying MHP and OCAP systems, discussing both the technical aspects and looking at how we can actually make money from an open system once we have deployed it. A more detailed breakdown by chapter content follows.

Chapter 1 discusses the current state of the DTV industry and how open systems and proprietary middleware solutions coexist in the marketplace. It looks at the driving forces behind the open middleware systems, and at how the various standards are related.

Chapter 2 introduces the basic technical concepts of DTV and looks at how we get signals from the camera to the receiver. It also discusses the various types of DTV networks and the technical issues that affect them.

Chapter 3 provides an overview of the MHP and OCAP middleware and looks at the different components that make up the middleware stack. We also discuss the high-level decisions that middleware implementers face.

In Chapter 4 we look at a simple MHP and OCAP application, and at the most important things we need to consider when we develop applications for these systems. We also cover the various types of OCAP and MHP applications we may come across, and offer practical tips for application developers.

Chapter 5 is a basic introduction to the concept of services and how they affect the life cycle of MHP and OCAP applications.

Chapter 6 introduces the concept of resource management, and looks at how many of the MHP and OCAP APIs manage resources. The chapter also examines how we make our middleware and applications more resilient to resource contention problems.

Chapter 7 discusses the graphics model in MHP and OCAP, including problems specific to a TV-based display. We discuss how to configure and manage the different parts of the display, at how we can use the user interface widgets provided by MHP and OCAP, and at how we can integrate video and graphics in our application.

In Chapter 8 we look at the basic concepts we need for referring to broadcast content.

Chapter 9 looks at service information, and examines how applications can get information about services and content being broadcast. The chapter also examines how a middleware stack can manage this data most effectively and efficiently.

Chapter 10 discusses how applications can get access to raw data from the broadcast stream using MPEG section filters. We look at the various types of filtering we can perform, the advantages and disadvantages of each type, and the problems the middleware faces in handling these types of filtering.

Chapter 11 looks at the model used by MHP and OCAP for playing video and other media, and discusses the extensions DTV systems use. The chapter also examines the special content formats available to OCAP and MHP applications, and takes a look at how we can control how video is displayed on the screen.

In Chapter 12 we examine data broadcasting and see how we get data from the transmission center to the receiver. You will see the various ways we can send files and other data. This is an area that can make or break a receiver or an application, and thus we also discuss how we can give the best performance possible when loading data from a broadcast stream.

Chapter 13 introduces the MHP and OCAP security model. We cover how the middleware can stop applications from doing things they are not allowed to, and how broadcasters can tell the receiver what an application is allowed to do.

Chapter 14 discusses how applications can communicate with one another, and examines the architectural choices that led to the design of the inter-application communication mechanism in MHP and OCAP. The chapter also looks at one possible implementation of that mechanism.

Chapter 15 looks at how we can use HTML in MHP 1.1 and OCAP 2.0. We look at the HTML application model, and at what has changed from the W3C standards. The chapter also explores how application developers can take advantage of the new HTML and CSS features MHP and OCAP support.

Chapter 16 is an introduction to the new features introduced in MHP 1.1, such as the Internet access API and the API for communicating with smart cards. The chapter also discusses the current state of MHP 1.1 and its place in the market.

Chapter 17 examines some of the advanced features of MHP and OCAP, including advanced techniques for controlling applications, using the return channel to communicate with a remote server, and tuning to a new broadcast stream.

Chapter 18 familiarizes you with the efforts under way to harmonize MHP, OCAP, and the other open middleware standards in use today. We look at the Globally Executable MHP (GEM) specification, and at how middleware developers can design their middleware so that they can reuse as many components as possible between implementations of the different standards. The chapter also explores how GEM affects application developers, and how they can ensure portability between the different GEM-based standards.

Chapter 19 is a discussion of the commercial issues involved in deploying MHP. This covers interoperability and conformance testing, and looks at some potentially successful MHP applications. It also discusses movement toward analog switch-off in various countries, and at how the migration to digital broadcasting is progressing.

Appendix A provides further information on the basic concepts behind DVB service information, one of the most important building concepts in digital broadcasting in Europe and Asia. The appendix provides a technical discussion of DVB-SI for people who are new to DTV systems, and serves as a reference for developers who already know about DVB-SI.

Appendix B covers the ATSC Program and System Information Protocol, the service information format used in North America and parts of Asia. The appendix serves as an introduction to PSIP for beginners and as a reference to the more important components for developers who are familiar with the PSIP standards.

Versions

This book covers the most recent versions of MHP and OCAP at the time of writing. Both MHP 1.0.3 (including errata 2) and MHP 1.1.1 are covered, as are version I13 of the OCAP 1.0 profile and version I01 of the OCAP 2.0 profile.

At the time of writing, most MHP receivers in the market are based on version 1.0.2 of MHP, although they sometimes include minor elements of later MHP versions in order to fix specific problems. OCAP receivers are typically based on a recent version of the OCAP 1.0 profile, but the lack of conformance tests means that some middleware vendors will track new versions of the standard more closely than others.

Shelving Code: Broadcast Technology

Interactive TV Standards by Steven Morris and Anthony Smith-Chaigneau

For any digital TV developer or manager, the maze of standards and specifications related to MHP and OCAP is daunting. You have to patch together pieces from several standards to gather all of the necessary knowledge you need to compete worldwide. The standards themselves can be confusing, and contain many inconsistencies and missing pieces. *Interactive TV Standards* provides a guide for actually deploying these technologies for a broadcaster or product and application developer.

Understanding what the APIs do is essential for your job, but understanding how the APIs work and how they relate to one another at a deeper level helps you do it better, faster, and easier. Learn how to spot when something that looks like a good solution to a problem really is not. Understand how the many standards that make up MHP fit together, and implement them effectively and quickly. Two DVB insiders teach you which elements of the standards are needed for digital TV, highlight those elements that are not needed, and explain the special requirements MHP places on implementations of these standards.

Once you have mastered the basics, you will learn how to develop products for U.S., European, and Asian markets, saving time and money. By detailing how a team can develop products for both the OCAP and MHP markets, *Interactive TV Standards* teaches you how to leverage your experience with one of these standards into the skills and knowledge needed to work with the critical related standards.

Does the team developing a receiver have all of the knowledge they need to succeed, or have they missed important information in an apparently unrelated standard? Does an application developer really know how to write a reliable piece of software that runs on any MHP or OCAP receiver? Does the broadcaster understand the business and technical issues well enough to deploy MHP successfully, or will their project fail? Increase your chances of success the first time with *Interactive TV Standards*.

About the authors:

Steven Morris is an experienced developer in the area of interactive digital television. Formerly of Philips Electronics, one of the major players in the development of MHP, he was heavily involved in the development of the standard, its predecessors, and related standards such as JavaTV and Open-Cable. In addition to work on the standard itself, Steven is the Webmaster and content author for the Interactive TV Web web site (*www.interactivetvweb.org* and *www.mhp-interactive.org*), a key resource for MHP, JavaTV, and OCAP developers.

Anthony Smith-Chaigneau is the former Head of Marketing & Communications for the DVB Consortium. In that role, he created the first MHP website *www.mhp.org* and was responsible for driving the market implementation of this specification. Anthony left the DVB to join Advanced Digital Broadcast, where he helped them bring the first commercial MHP receivers to market. He is still heavily involved in the DVB MHP committees with Osmosys, an MHP and OCAP licensing company, based out of Switzerland.

Related Titles by Focal Press:

The MPEG Handbook by John Watkinson, ISBN: 0-240-51657-6
Digital Television by Herve Benoit, ISBN: 0-240-51695-8

Focal Press
An Imprint of Elsevier
www.focalpress.com

ISBN: 0-240-80666-2

1 The Middleware Market

The introduction of digital TV (DTV) and interactive TV (ITV) is causing huge changes in the television industry. This chapter provides an overview of the current state of the market and examines how open standards for middleware fit into the picture. We will look at the history behind those standards, and take a look at the bodies behind the standards we discuss in this book.

The broadcasting and television industries are in a state of flux on many fronts. The industry has been working toward wooing consumers from a passive role to a more active one, and multimedia and the convergence of the consumer electronics and personal computer worlds play a big part in this change.

The concept of ITV is not new. It commenced with teletext in the 1980s, and unknown to many Warner-Qube was deploying a form of video-on-demand (VOD) in the United States as early as the 1970s. Unfortunately, many of these early attempts were soon shelved due to the cost of having to bring two-way networks into people's homes. Since then, changes in technology and in the markets themselves have made this possible, and broadcasters are looking to differentiate and capitalize on these new technologies, adding new functionality in order to bring a much more active TV experience to consumers.

Proprietary middleware solutions have been available for several years from companies such as OpenTV, NDS, Canal+, PowerTV, and Microsoft, but we are now in an emerging market for middleware based on open standards. The Digital Video Broadcast (DVB) Project's Multimedia Home Platform (MHP) is currently leading the development of this market, with the OpenCable Application Platform (OCAP) and Advanced Common Application Platform (ACAP) standards two to three years behind.

MHP saw Finland go first, with initial deployments starting in 2002. Premiere was to be next, the satellite pay-TV operator fulfilling a German dream with the launch of MHP over satellite to about a million homes. This did not happen for a variety of reasons, although since then many other German broadcasters (such as ARD, RTL, and ZDF) have launched MHP

1

services on both terrestrial and satellite networks. Italy has taken a different approach to DTV migration, and the government has assisted the market by sponsoring the introduction of terrestrial MHP set-top boxes. In Italy, we now have a market of 20 million households that is completely serious about launching MHP terrestrial services. Other countries, especially Austria and Spain, look likely to follow: they are among many countries and network operators presently running trials using MHP.

The growth of these open middleware standards raises many questions for the industry as a whole, and in this book we hope to answer some of those questions. In this chapter, we concentrate on a basic overview of the current ITV landscape, looking at where these standards came from, why they are useful, and how they are changing the industry.

Why Do We Need Open Standards?

We are all exposed, at some time or another in our daily lives, to standards. From reading e-mail over the Internet at work to watching television in the evening, a wide variety of standards exist in daily use. These standards are typically defined through a standardization body.

Consumers in general do not involve themselves in any of the fine details, worrying mainly about usability instead. Customers are happy if a standard does the job it was made for. If it represents something cool and funky, that is even better, especially in today's high-tech culture. From the customer's perspective, the only concern regarding standards is that they should allow the equipment they purchase to perform "just like it says on the box."

On the other hand, standards are of great concern to industry specialists, who bring together various technologies to create new products. Open standards are used because they guarantee that compliant systems will be able to work together, no matter which manufacturer actually provides the equipment. Specialists do not always fully agree on the content of standards, of course, as you will see in this book. There are many instances of competing standards in all technology markets, including the fields of broadcasting and ITV middleware.

Until 2002, DVB specifications concentrated on digital broadcasting and associated technologies. Convergence has led to products that use a mélange of previously unrelated standards, and the work of standards bodies such as DVB is becoming more complex in order to keep control of the technology used in these new devices. DVB offers its specifications for standardization by the relevant international bodies such as the EBU/ETSI/CENELEC Joint Technical Committee and the International Telecommunication Union (ITU-R or ITU-T).

For standards such as MHP, the final documentation for the specification includes such things as commercial requirements, technical guidelines, implementation guidelines, and the technical specification itself. This provides companies with all of the information they need to implement the specification in a consistent way, and many of these documents are available over the Internet either free of charge or for a nominal fee.

Furthermore, specification bodies such as DVB and CableLabs also define a certification (compliance) process. For DVB, this is a self-certification process that includes a process for registering compliant products and that may include payment programs that provide access to test suites and possibly branding for implementations such as permission to use the MHP logo. CableLabs offers a Wave certification program, for which costs are not insignificant. We are still awaiting the definition of the OCAP certification process, and it may well be that this will follow a model similar to that of the DVB self-certification process. A more thorough discussion of the MHP self-certification process can be found in Chapter 19.

From experience in the wider open standards world, we know that standards create a fully competitive and open market, where technologies become more widely implemented because they are available to all players in the industry under the same terms. Standards work!

Driving Forces Behind Open Standard Middleware

In the early days of television, several competing technologies emerged for carrying picture and sound information from the transmitter into viewers' homes. Following many tests and political arguments, three main standards emerged: NTSC, PAL, and SECAM. The fragmentation of the world's television services among these three standards created a complicated scenario for consumer electronics manufacturers, particularly because each standard had a number of variants. This was most evident in Europe, where neighboring countries chose differing variants of PAL or SECAM, leading to a number of market problems such as receivers purchased in one country not necessarily working in another. To illustrate the problems this caused, Brazil chose the PAL-M system for its TV broadcasts, thus isolating itself from the rest of the world and shutting the door to TV import/export opportunities (except perhaps for Laos, the only other PAL-M country).

Standards in DTV

It would have made sense not to replicate this fragmentation in the move to digital, in that the technology was available to unify the digital broadcasting world. However, learning from experience is not always something people are good at. It would have been obvious that by introducing common digital broadcasting systems consumer electronics manufacturers and ultimately their customers would have benefited from massive economies of scale, but this did not happen. There is still fragmentation in the DTV broadcast market—some say on a much-reduced scale, whereas other commentators disagree. Commercial and political issues, as well as the NIH ("not invented here") syndrome created the following competing DTV standards.

- Europe chose the transmission technology COFDM (coded orthogonal frequency division multiplexing) for terrestrial broadcasts, adopting this in the DVB-T (DVB-Terrestrial) specification.

- The United States' ATSC (Advanced Television Systems Committee) chose a system using 8-VSB (vestigial sideband) technology for terrestrial transmission. Canada and South Korea adopted the same system.
- Japan looked at COFDM and the Japanese Digital Broadcasting Experts Group (DIBEG) and then created its own flavor, which included time interleaving, calling it ISDB-T (Integrated Services Digital Broadcasting, Terrestrial).
- Most cable operators use QAM (quadrature amplitude modulation) as the modulation technology for cable systems, although different systems use different transmission parameters and are not always compatible. DVB defined the DVB-C (DVB-Cable) standard, whereas CableLabs defined the OpenCable standard.
- Most satellite operators use QPSK (quadrature phase-shift keying) modulation, although again there are several flavors. DVB defined the DVB-S (DVB-Satellite) standard for satellite broadcasting, whereas the ATSC defined the A/80 and A/81 satellite broadcasting standards for use in North America.
- Brazil recently decided, as has China, it would favor producing alternative "home-grown" broadcasting standards in order to avoid any intellectual property issues outside the country.

As well as choosing different modulation systems, Europe, the United States, and Japan all chose different standards for the service information needed to decode DTV services and for data broadcasting. When these choices are coupled with the continuing use of PAL, NTSC, and SECAM, there are now even more differences among the countries, although some of these are more superficial than others.

Correcting the Fragmented ITV Market

As DTV systems were deployed, network operators wanted to exploit the strengths of these new technologies to provide new revenue streams. This led to an interest in interactive services, which in turn led to a desire for middleware platforms that would enable interactive applications to run on various hardware platforms. Proprietary middleware solutions were by far the most common, with OpenTV, Liberate, NDS, Microsoft, and Canal+ Technologies being among the leading players. Naturally, the services and applications running on proprietary middleware were tightly linked to those platforms, and because operators normally chose different middleware and conditional access technologies this led to the development of vertical markets, the trend away from which is indicated in Figure 1.1.

In a vertical market, the network operator controls the specification of the set-top boxes used on that network and of the applications that run on it. This meant that in a vertical market the set-top box often became the largest financial burden to a network operator, in that the network operator purchases the receiver directly from the set-top box supplier and leases it or gives it away to the viewer as part of a subscription.

With the growth of ITV, television standards bodies around the world decided to create open standards for the middleware needed to run these services. Because these were developed by the same bodies that produced the other DTV standards, the United States, Europe, and

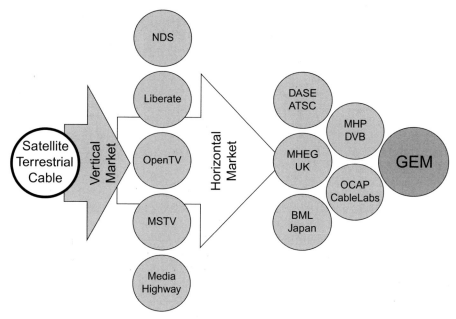

Figure 1.1. The digital TV market is currently migrating from proprietary middleware solutions in vertical markets to open standards in horizontal markets.

Japan all produced different standards for middleware, designed to work with their own service information and data broadcasting standards. These standards are outlined in the following.

- Through DVB, Europe created specifications for all types of DTV networks (DVB-T for terrestrial, DVB-C for cable, and DVB-S for satellite). The common features of these systems led to the development of a common middleware standard (MHP) for all three types of networks.
- ATSC, the standards body for the United States, developed the DASE (Digital TV Applications Software Environment) middleware system based on its DTV standards. This has since been used as the basis for the next-generation ACAP standard. Canada and Korea have also adopted ATSC standards, including ACAP, for their terrestrial transmission services. Cable systems in the United States were largely standardized through CableLabs, which modified the ATSC standards for service information in order to make them more suitable for a cable environment. CableLabs developed the OCAP middleware standard, which has a number of features specific to U.S. cable networks.
- The Japanese DIBEG created the BML (Broadcast Markup Language) markup language for ITV. There are no known users of this system outside Japan, and the Japanese standards body ARIB (Association of Radio Industries and Businesses) is currently developing a more advanced middleware standard that is closer to MHP.

- En route to selection of an open standard, Brazil and the People's Republic of China decided to produce alternative home-grown broadcasting standards for terrestrial transmissions that could lead to yet another challenge for middleware standardization and GEM (Globally Executable MHP).

This is not the end of the story, however. Brazil, China, and Japan all use DVB-C for cable services, and some U.S. satellite operators use the DVB-S standard. According to one middleware supplier, they have already deployed MHP services in China's Shenzhen province. Due to the mix of different transmission systems selected, Korea is the most complicated scenario of all, where OCAP has been chosen for cable networks, MHP for satellite networks, and ACAP for terrestrial networks.

A few open middleware standards were developed before MHP, and these are in use in a small number of public networks (such as for digital terrestrial services in the United Kingdom). Unfortunately, the size of the markets and competition from subscription-based services (which can offer better content) have meant that these markets have not grown very quickly. At the same time, the current crop of middleware standards is a successor to these standards on a technical level and a commercial level, and in later chapters we will see how MHP and OCAP have taken elements from these standards and applied them in regard to the current conditions in the market.

This profusion of middleware standards has had an impact on the market, fragmenting it further so that different operators may deploy different middleware solutions (either proprietary or open) and different transmission standards. Unless the industry as a whole is very careful about how these standards are adopted and used, this level of fragmentation can only introduce more problems for the entire industry.

By painting such a gloomy picture of the current state of digital broadcasting, some people may accuse us of reinforcing the fear, uncertainty, and doubt that have sometimes been spread by opponents of open standards. We need to highlight these effects in order to understand the consequences of these choices, but the overall picture is really not as bad as we may have made it sound. What we have to remember is that despite the differing digital transmission standards the TV market is traditionally a fully horizontal market. Many people forget this in the DTV debate. Televisions and "free-to-air" set-top boxes are available in retail stores across all countries that have made the move to public digital broadcasting, and consumers have the choice of a range of products from low-end set-top boxes to high-end integrated TVs. Open standards for middleware were created to help fix the fragmentation of the ITV market, and these standards are now being taken up around the world.

What Are DVB and CableLabs?

Before we can look at the standards themselves in any detail, we need a clear understanding of the organizations involved. Many groups in the world are developing standards for DTV. Some of these (such as DVB, ATSC, and the ITU) may be familiar to you, whereas others may be less familiar to some readers. A number of bodies have been involved in the development of the middleware standards we will discuss further, but two main organizations have really driven the process and those are the DVB in Europe and CableLabs in the United States.

The Digital Video Broadcasting Project

The DVB Project (also known simply as DVB) is a consortium of companies charged with developing specifications for digital television. Based in Geneva, its original mission was to develop pan-European specifications for digital broadcasting, but DVB specifications are now in use worldwide and its membership extends well beyond Europe.

DVB grew out of the European Launching Group (ELG) of 1991, which consisted of eight mainstream broadcasters and consumer electronics companies. The ELG members understood that the transition from analog to digital broadcasting needed to be carefully regulated and managed in order to establish a common DTV platform across Europe. The ELG eventually drafted a Memorandum of Understanding (MoU) that brought together all of the players who had an interest in the new and emerging digital broadcasting market. The most difficult aspect of this work was gathering competing companies to work in a "precompetitive" situation, and then expecting them to cooperate fully and to share in common goals and joint ambitions. This required a great deal of trust, and this was possible under the neutral umbrella of DVB. In 1993, DVB took up the mantle of the ELG work that had already commenced.

The main success of DVB has been its pure "market-led" approach. DVB works to strict commercial requirements, offering technical specifications that guarantee fair, nondiscriminatory terms and conditions with respect to intellectual property used in the creation of the DVB specifications. This allows those specifications to be freely adopted and used worldwide, even by companies that were not members of DVB, and many countries now use DVB standards for digital broadcasting. More information about DVB is available on the DVB web site at *www.dvb.org*.

Presently the membership of DVB is approximately 300 companies, although this fluctuates to reflect the evolution of the industry as current players merge and new companies join the market. Overall, it can be said that DVB has grown dramatically in recent years and is one of the most influential DTV organizations of the twenty-first century. One of the most difficult DVB specifications created to date was the open standard for DTV middleware known as the MHP, which forms the core of the standards to which this book is dedicated.

DVB-MHP: The Multimedia Home Platform

The MHP specification started life following discussions in 1994 through 1996 in a European community project on platform interoperability in digital television (the DG III—UNITEL project). Later that year, discussion began on the commercial requirements for MHP, and these discussions carried on into 1997. In October of 1997, the DVB Steering Board approved the MHP commercial requirements that form the foundation of the MHP specification. These commercial requirements have been a consistent driving force for the technical aspects of MHP and have provided a way of ensuring that MHP meets the needs of the market. The consistency this approach enforces is illustrated by the fact that the chairperson of the original MHP group is still chairing the DVB-MHP Commercial Module more than 60 meetings later.

This particular DVB work item covered what was always understood to be a very difficult objective: standardizing elements of what we can call the home platform (set-top box, television, and so on), which were seen as key to the success of interactive applications for the DTV domain of the future.

In its simple form, MHP is a middleware layer or application programming interface (API) that allows interactive applications and services to be accessed independently of the hardware platform they run on. This was seen as a natural progression from the pure transmission-related work of DVB and a natural move toward multimedia, ITV software, and the applications that were beginning to bring added value in the transition from analog to DTV. After several more years of hard work, the first version of the specification was released on 23 February 2000 via ETSI.

Since then, the work has expanded to cover not only improvements to the API but to aspects such as the in-home digital network, PVR (personal video recorder), mobile applications, and other technologies as convergence has become more important. The evolving MHP specifications and associated documentation are available from the DVB-MHP web site at *www.mhp.org*.

CableLabs

CableLabs (*www.cablelabs.com*) is an organization based in Denver, Colorado, that has been involved in the cable television industry since 1988. Like DVB, it is driven by common goals, common ambition, and a desire to be involved in the "innovation phase" of new TV technologies. Unlike DVB, however, CableLabs was formed to concentrate purely on cable TV technologies.

From its early work, CableLabs has matured into a group of industry specialists gathering in a noncompetitive framework, who work for the good of its members by providing a form of joint research and development facility. In 1998, the OpenCable initiative was started to exploit the new digital set-top box technologies and innovations that were becoming more and more prevalent in the technology marketplace. One of the aspects of the OpenCable initiatives naturally encompassed the middleware layer, and this eventually led to CableLabs looking into open standards for middleware APIs. To ensure compatibility between different systems and avoid fragmentation of the marketplace, CableLabs has collaborated with DVB in its work on open middleware standards, and this led to the acceptance of the MHP specification as the basis for the OpenCable Applications Platform (OCAP) in January of 2002.

OpenCable Applications Platform (OCAP)

With MHP at its core, OCAP provides a specification for a common middleware layer for cable systems in the United States. This fundamentally delivers what MHP set out to do for "DVB market" consumer electronics vendors, thus helping American network operators move forward with a horizontal set-top box market. OCAP is intended to enable developers of interactive television services and applications to design products for a common North American platform. These services and applications will run successfully on any cable television system running OCAP, independently of set-top or television receiver hardware and low-level software.

CableLabs published the first version of the OCAP 1.0 profile in December of 2001, and the OCAP 2.0 profile followed this in April of 2002. Since the first release, several new versions of the OCAP 1.0 profile have been published. These have moved the OCAP platform closer to the MHP specification (and more recently the GEM specification of MHP), although this has happened at a cost: not all versions of OCAP are completely backward compatible with one another. This has the potential to be a real problem for broadcasters, middleware developers, and application developers. Unless the market standardizes on a single version of the specification or on a specification that uses OCAP (e.g., SCTE standard 90-1, *SCTE Application Platform Standard, OCAP 1.0 Profile*), application developers and network operators will be forced to use only those elements that have not changed among the various versions of OCAP. Given that changes to OCAP have affected such basic functionality as resource management, this may not be easy. At the time of writing, the most recent version of the OCAP 1.0 profile is version I13, whereas OCAP 2.0 remains at version I01.

A History Lesson: The Background of MHP and OCAP

We have already mentioned that MHP was not the first open standard for ITV. In 1997, the ISO (International Standards Organization) Multimedia and Hypermedia Experts Group (MHEG) published the MHEG standard. This offered a declarative approach to building multimedia applications that could be run on any engine that complied with the MHEG standard. The original specification, known as MHEG-1 (MHEG part 1) used ASN.1 notation to define object-based multimedia applications. Conceptually, MHEG set out to do for interactive applications what HTML did for documents; that is, provide a common interchange format that could be run on any receiver.

MHEG-1 included support for objects that contained procedural code, which could extend the basic MHEG-1 model to add decision-making features that were otherwise not possible. The MHEG-3 standard defined a standardized virtual machine and byte code representation that allowed this code to be portable across hardware platforms.

MHEG-1 and MHEG-3 were not very successful, partly because the underlying concepts were very complicated and because the industry was not yet ready for the features offered by these standards.

To remedy this, MHEG defined part 5 of the MHEG standard, known as MHEG-5, which was published in April of 1997. MHEG-5 is a simpler profile of MHEG-1, although in practice it is different enough that most people treat it as a separate standard. Many features are the same, but there are also many differences. Most notably, the U.K. digital terrestrial network uses MHEG-5 for digital teletext and other information services. The U.K. Digital Terrestrial Group (DTG) is the driving force behind the use of MHEG-5 in the United Kingdom. (More information about MHEG is available at *www.dtg.org.uk.*)

MHEG-3 was overtaken by the success of Java, and thus in 1998 MHEG-6 was added to the family of MHEG standards. This took MHEG-5 and added support for using Java to develop script objects, thus mixing the declarative strengths of MHEG with the procedural elements of Java. To do this, it defined a Java application programming interface (API) for MHEG so that Java code could manipulate MHEG objects in its parent application.

Although MHEG-6 was never deployed, it formed the basis of the DAVIC (Digital Audio Visual Council) standard for ITV. This added a set of new Java APIs to MHEG-6, enabling Java to access far more of the capabilities of a DTV receiver. The DAVIC APIs allowed Java objects to access some service information, control the presentation of audio and video content, and handle resource management in the receiver. Although it still was not possible to write a pure Java application for a DAVIC receiver, Java APIs were now able to control far more elements of the receiver than was possible using other standards. DAVIC was published in 1998, shortly after the publication of MHEG-6.

You may have noticed that the publication dates of some of the standards we have discussed seem very close together, or even in the "wrong" order. This is mainly due to the different processes used by the standards bodies for ratifying standards, and some bodies (such as ISO) by their very nature take longer than other bodies to ratify a standard. As a result, subsequent standards may have to wait for another standards body to ratify a standard before they can officially use it. This happened in MHP with the JavaTV standard: MHP was ready for publication before the JavaTV specification was completed, and the DVB could not publish MHP while it referred to a draft version of the JavaTV specification. Once Sun (Sun Microsystems) finalized the JavaTV specification, the publication of MHP could continue very quickly.

Many of the same companies that were involved in DAVIC were also working in the DVB, and thus when the DVB selected Java as the basis for MHP it was natural to reuse many of the DAVIC APIs. MHP was the first open middleware standard based purely on Java, meaning that receivers did not need to implement another technology (such as MHEG) to use it. This was quite a departure from the current wisdom at the time, as Jean-Pierre Evain of the European Broadcasting Union and ex-secretary of MHP, recalls:

You had to be brave or foolhardy to pronounce the acronym "MHP" in the early days, when the only future seemed to be digital pay-TV and vertical markets. DVB's vision was, however, correct—horizontal markets are alive and competition is stronger. MHP could have been MHEG based plus some Java as proposed by DAVIC, but DVB negotiations decided otherwise to the benefit of a now richer far-sighted solution.

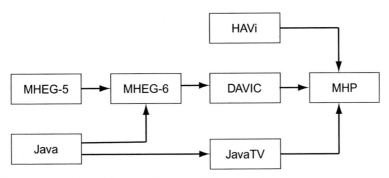

Figure 1.2. MHP and its relationship to earlier open ITV standards.

This shift from declarative approaches such as MHEG toward Java does not mean that the declarative approaches failed, however. For many applications, a technology such as MHEG is perfectly adequate, and may actually be superior to Java. Despite that, Java happened to be in the right place at the right time, and its flexibility made it ideal for exploiting the growing interest in open standards for ITV. Declarative technologies still have their place, and these are included in MHP using HTML and plug-ins for non-Java application formats. More recently, work on the Portable Content Format (PCF) has begun to standardize a representation for the various declarative formats (including MHEG) currently in use. We will look at these topics in more detail elsewhere in this book. Figure 1.2 shows the relationship of MHP to other early open ITV standards.

The MHP Family Tree

The MHP standard defines three separate profiles for MHP receivers, which enable receiver manufacturers and application developers to build different products with different capabilities and costs. Using these profiles, products can be targeted for specific market segments or for specific network operators. The three MHP profiles are outlined in the following.

- *The Enhanced Broadcast Profile (Profile 1)*, defined in ETSI standard ES 201 812 (MHP 1.0): This profile is aimed at low-cost receivers, and is designed to provide the functionality of existing middleware systems and the applications that run on them.
- *The Interactive Broadcast Profile (Profile 2)*, defined in ETSI standard ES 201 812 (MHP 1.0): The main difference between Profile 1 and Profile 2 is that Profile 2 includes standardized support for a return channel. Applications can download classes via the return channel, whereas in the Enhanced Broadcast Profile this is only possible via broadcast streams. This profile also includes APIs for controlling return channel access.
- *The Internet Access Profile (Profile 3)*, defined in ETSI standard TS 102 812 (MHP 1.1): Profile 3 allows for much broader support for Internet applications such as e-mail, web browsing, and other Internet-related activities on the receiver.

In MHP 1.1, profiles 2 and 3 add optional support for DVB-HTML applications. Figure 1.3 depicts the three possible profiles for MHP receivers and applications.

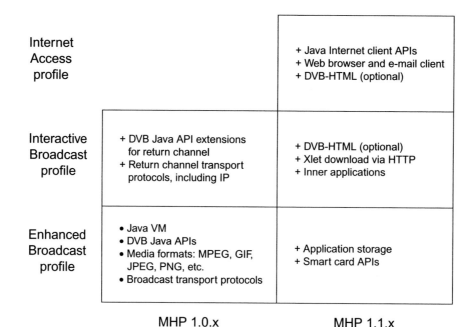

Figure 1.3. MHP defines three possible profiles for MHP receivers and applications.

Following the standardization of MHP, CableLabs decided to use MHP as the basis for the OCAP platform. So far, the following two profiles for OCAP have been defined.

- The OCAP 1.0 profile was first issued in December of 2001. This is based on MHP 1.0.x, taking elements from both profiles. Since then, several new versions of the OCAP 1.0 profile have been published, with version I13 being the most recent at the time of writing. These changes have brought OCAP and MHP closer together, building on the GEM (Globally Executable MHP) standard for the core functionality and making it easier to develop common middleware platforms and applications.
- The OCAP 2.0 profile was issued in April of 2002. This took the OCAP 1.0 profile and added the version of HTML supported by MHP 1.1. To date, only one version of this profile has been defined.

The harmonization of MHP and OCAP led to the GEM process, discussed further in material to follow. Test suites are a significant element in conformance. In June of 2002, the DVB released version 1.0.2 of the MHP test suite, which covered version 1.0.2 of the MHP specification. In December of that year, the DVB released a revised version (version 1.0.2b) that extended the test suite to cover more of the MHP 1.0.2 standard. At the time of this writing, CableLabs has published the first test suite for OCAP, which applies to version I13 of the OCAP 1.0 profile.

Delays are an inherent part of the standards process, especially within the complex and time-consuming process of creating test suites. The issue of conformance testing has proven to be more troublesome than anticipated. DVB Chairman Theo Peek acknowledged during a 2003 conference in Dublin, Ireland, that "MHP is a very complex specification, and DVB under-estimated the effort required to build the MHP test suite."

JavaTV: A Common Standard for DTV

Until the development of MHP, Java was used mainly as a scripting language that extended other platforms such as MHEG. Standards such as DAVIC had already defined many Java APIs for DTV, but these focused on the role of Java as an extension of other technologies.

In March of 1998, Sun announced the JavaTV API, and work began on defining a pure Java platform for DTV applications. Part of this work involved standardizing the use of existing APIs in DTV platforms, such as the use of the Java Media Framework (JMF) for controlling audio and video content as originally specified by DAVIC. At the same time, JavaTV needed to define many new components, including a new application model, APIs to access DTV-specific functionality, and a coherent architecture that would allow all of these elements work well together.

Following on from their cooperation in DAVIC, Sun and a number of consumer electronics companies worked together to develop the JavaTV specification. Although Sun was very much the custodian of the standard, many other companies (such as Philips, Sony, and Matsushita) provided valuable input.

JavaTV was not the only standard under development at this time. As we have already seen, the commercial requirements for MHP had already been agreed upon by the time JavaTV was announced, and a few months after the announcement of JavaTV DVB selected Java as the basis for the MHP platform. Many of the companies involved in the JavaTV initiative were also working on MHP, and in many cases the same people at these companies were involved.

This close relationship between the two standards meant that they were designed almost from the beginning to be complementary, and many of the overlaps in the two standards are a result of JavaTV taking a more platform-neutral approach than MHP. The two standards offered feedback to each other, and in some cases JavaTV took elements from MHP and applied them to all DTV markets.

Unlike standards such as MHP or DAVIC, JavaTV aims to provide a basic set of features to all conformant receivers. It does not define a full DTV platform that offers complete access to every feature in the receiver, because standards such as MHP were designed to do that. Instead, it offers a lightweight set of APIs (which provides the common elements) and a framework in which to use that set.

Although the process was not completely without its problems, Sun published the JavaTV specification at the JavaOne conference in 1999, and shortly afterward Sun and DVB agreed on terms for the use of Java in MHP. Because of this, the inclusion of JavaTV in MHP was assured and version 1.0 of the MHP specification was published shortly thereafter.

Harmonization: Globally Executable MHP

Open standards guarantee that compliant systems will be able to work together, regardless of which manufacturer provided the equipment. With several organizations around the world striving for the same goal in creating open middleware systems, it obviously made sense to seek some form of harmonization. For an industry that is as global as the television industry, this is even more important. The GEM work item in DVB came about after a request from CableLabs to consider the unification of MHP with the original DASE standard from ATSC.

The GEM specification is a subset of MHP that has been designed to take into consideration the various interoperability issues across the various open standard middleware specifications. These issues include the following.

- Technical issues of interoperability arising from previous middleware standards, such as OCAP and DASE
- Issues related to individual transmission systems, such as the choice of modulation systems, delivery mechanisms, and CA systems
- Varied market requirements for network operators

GEM is a framework aimed at allowing varied organizations to create harmony in technical specifications, such as the selection of a single execution engine and (where possible) a common set of APIs. The goal is such that applications and content will be interoperable across all GEM-based platforms. The MHP web site states:

The GEM specification lists those parts of the MHP specification that have been found to be DVB technology or market specific. It allows these to be replaced where necessary as long as the replacement technology is functionally equivalent to the original—so called functional equivalents.

Even though other standards will not be completely compatible with the full MHP specification, GEM ensures that compatibility will be maintained where it is feasible to do so. The set of technologies where functional equivalents are allowed is negotiated as part of the technical dialogue between the DVB and each of the organizations wishing to use GEM. Additionally, the GEM specification contains a list of those other specifications with which it can be used.

GEM 1.0 was published in February of 2003. It contains some guidelines for using GEM with the OCAP specification, which is presently the only entry in the list of specifications that can be used with GEM. Future versions of the GEM specification will probably include similar guidelines for the Japanese ARIB B.23 standard and the DASE and ACAP specifications from ATSC. Figure 1.4 depicts the relationships among MHP, GEM, and other standards.

In addition to standardizing an API on a European level, cooperative effort from several industry consortia led to the adoption of the GEM specification as an ITU recommendation

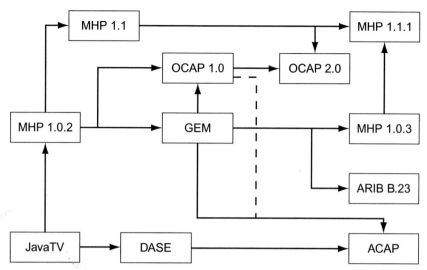

Figure 1.4. The relationships among MHP, GEM, and other standards.

in March of 2003, implying support from industry consortia in the three main DTV markets (i.e., Europe, the United States, and Japan). Agreement was made on a single execution engine, based on the MHP standard (although it should be mentioned that several issues relating to conformance testing and IPR (intellectual property rights) licensing still need to be addressed before deployment of this specification is possible). Effectively, this turns MHP into a worldwide API standard for digital ITV. GEM is discussed in further detail in Chapter 18.

The Difficult Part of Standardization

Although standards offer many benefits to the industry, they also entail a number of potential pitfalls. Most of these are related to the standardization process itself. We often hear the criticism that standards are "created by committee," and we have all heard the joke describing the elephant as a mouse designed by committee. It is true that committees can be very large, are often seen as complicating the effort, and often try to please all stakeholders involved in the process. Reaching a consensus can be the most difficult work for standards organizations: there are many hidden political and commercial agendas lurking at the concept stage, and these will often remain throughout the entire process of developing a standard. With this in mind, standards are often at risk of becoming ineffective or outdated due to the length of time taken to create and publish them.

Standards organizations and standardization work are also commonly attacked when they interfere with the status quo for companies who have implemented and grown their market share using their own proprietary technologies (unless those proprietary technologies form the basis of the new standards). This was a particularly big problem for MHP, and it created a considerable delay in reaching consensus on MHP technologies at the DVB Steering Board. It remains an issue today, with some of the proprietary players (who are also members of the specification bodies) creating a group called the DIF (Digital Interoperability Forum) to push the use of alternative technologies in place of MHP. The European Union in Brussels is considering whether it should mandate a common platform for digital broadcasting in Europe, and thus the DIF is working against efforts to mandate MHP for European DTV. Discussing this properly would take an entire chapter, but let it suffice to say that there are groups for and against a mandated common platform, and the DIF web site will give you some of the arguments against making MHP mandatory.

In a similar move in the United States, Microsoft has recently submitted its NET Common Language Infrastructure (CLI) specification to CableLabs for possible inclusion in future versions of the OCAP standard. Whether this will be successful is not yet clear, but the effects this move will have on OCAP should not be underestimated.

Intellectual Property and Royalties

Another important aspect of open standards is that standards bodies usually try to write the standards in such a way that companies can implement them freely, without the payment of any royalty for using the standard. This is not 100-percent achievable in most cases, and so any patents that affect a specification are often bundled into a patent pool, which is administered either by the standards body or an independent third party. The set of patents used in a standard is then offered on fair, reasonable, and nondiscriminatory terms to companies who are implementing the standard. Companies still have every right to charge for their own implementations of a particular open standard specification, although free implementations may also be available in some cases.

For MHP, the DVB has called for a voluntary joint licensing arrangement for a portfolio of patent rights. In particular, the call for MHP patents was for declarations of intellectual patent rights essential to DVB specifications adopted since May of 1997. This will create a "one-stop-shop" facility for those requiring licenses on a fair, reasonable, and nondiscriminatory basis (similar to the MPEG-LA arrangement for DVB-T). The initial call for a DVB patent pool co-coordinator was made in September of 2001. Currently, the firm of Surghrue Mion PLLC, based in Washington, D.C., is acting as the joint Patent Pool Coordinator for the initial process, covering both MHP and OCAP patents.

One of the concerns regarding the use of patented technologies in standards such as those of the DVB is the licensing terms that will be applied to those standards. There is always a danger that companies will attempt to use the inclusion of patents as a cash cow or as a way of limiting competition. Although this generally has not happened, there is always a risk. The patent pool costs for MHP and OCAP have not yet been defined, and some people believe the problems are not yet over.

During the writing of this book, the patent pool lawyers have completed the patent pool development process, which allows for the next step: choosing a company to gather the royalties on behalf of the companies involved. DVB and OCAP patent holders selected Via Licensing of San Francisco (*www.via-licensing.com*) to form one or more patent pools for MHP and other DVB standards. Via Licensing's business is to develop and manage patent pools in order to bring reasonably priced and convenient licenses to market. Their intention is to work with the DVB and OCAP patent holders to make licenses for MHP, OCAP, GEM, and other standards available as soon as possible. This was announced in the press on June 10, 2004.

This rather unexciting topic highlights one important aspect of open standards; namely, that in an open standards framework one technology is not favored over another. Specifications can contain grouped technologies or a particular technology for a particular function, which consequently provides a cost-effective "technical toolbox" of specifications that results in common functionality and interoperability across consumer devices. This is especially true in the consumer electronics world, in which cost and interoperability are vital elements of a product.

Upon trawling many dictionaries of legal quotations and articles relating to IPR, the following extract aptly conveys some of the concerns raised during the process of developing MHP as an open standard. From *U.S. Supreme Court, Atlantic Works vs. Brady, 1882*:

It was never the object of patent laws to grant a monopoly for every trifling device, every shadow of a shade of an idea, which would naturally and spontaneously occur to any skilled mechanic or operator in the ordinary progress of manufactures. Such an indiscriminate creation of exclusive privileges tends rather to obstruct than to stimulate invention. It creates a class of speculative schemers who make it their business to watch the advancing wave of improvement, and gather its foam in the form of patented monopolies, which enable them to lay a heavy tax on the industry of the country, without contributing anything to the real advancement of the arts. It embarrasses the honest pursuit of business with fears and apprehensions of unknown liability lawsuits and vexatious accounting for profits made in good faith.

Where Do We Go from Here?

Technology improvements have led to set-top boxes and integrated TVs offering better performance and more features, and now it is possible to broadcast applications and other interactive services at a premium. DTV and ITV will become a stronger, more prevalent new business for the broadcast community.

Progress has not always been smooth, however. The dynamics of the industry changed in the late 1990s and on into the year 2000 as the dot-com bubble burst. All technology companies began to suffer in a very weak marketplace. For example, cash-strapped cable and satellite operators suffered tremendously and there were many casualties along the way. Many people assumed that the desire for the horizontal market would have become much stronger, given the cost savings that could be gained. Despite the crisis, however, vertical network operators have had a difficult time letting go of the control associated with vertical markets—not least because of the costs involved in writing off past investments. They have not fully embraced the opportunities put before them, and have looked at cheap solutions to drive receivers even cheaper.

A zapper[1] community has been created with receivers that offer only the most basic features, little or no interactivity, and in many instances (such as Freeview in the United Kingdom) no return channel. Although this may be attractive from a cost perspective, it vastly reduces the opportunities for exploiting ITV and is probably a blind alley for the DTV business. Content is still king, but traditional pay-TV earners such as movies and sports are not the only type of content available to DTV operators.

There is no benchmark open system in broadcasting that would allow for statistical forecasting, for business plans, and for gauging success, so we have no idea how this is all going to progress. Both of the authors have toiled over MHP business plans, market implementation, market statistics, and the technical details of making MHP work in the real world, and the market changes on a monthly basis.

Having followed the market through these trials and tribulations, one thing is clear to us. The industry has seen broadcast equipment manufacturers committed to developing all of the pieces necessary to realize the MHP dream, but for a bright new broadcasting future we need the full commitment of all broadcasters and network operators toward MHP.

What we will say, however, is that this is not about "old middleware" versus "new middleware" or who has the most deployments. This is about fundamentally changing the middleware landscape in order to help a broken and fragmented ITV market.

Open Versus Proprietary Middleware

Proprietary solutions obviously see MHP as a competitor: after all, it came about as an answer to the proprietary solutions that were already installed in a rather fragmented market. Despite the success of MHP at consortium level, companies such as OpenTV, Liberate, and Microsoft still offer their proprietary systems despite showing some signs of support for the new open standards. Behind closed doors, proprietary players have used all manner of tactics to confuse broadcasters and operators, not least of which is an exploitation of fears that MHP is a more expensive solution. As hardware costs have fallen, and middleware capabilities have increased, this has been shown to be false. MHP implementations are now available on hardware platforms that are no more expensive than those used for proprietary middleware solutions, and MHP is often less onerous in terms of IPR than many of those alternative solutions.

Finally seeing that they can unburden themselves entirely from the cost of set-top boxes, broadcasters and network operators are slowly coming to terms with this and are starting to realize the potential of horizontal markets. Many of the latest RFPs (Requests for Proposal)

---

[1] *Zapper* is the common term for a digital receiver that has only channel-changing capabilities. These have a very low price due to the lack of features in the hardware and software.

from broadcasters and network operators ask for middleware-enabled products aimed at a retail market. This has always been the goal of MHP and OCAP, and thus competition from other systems becomes less of a threat as the companies take up the open standards philosophy on a wider global footing.

Proprietary solutions will have their place for a long time to come, however, and it is unlikely that they will ever disappear completely because some markets will always have requirements that open standards cannot satisfy. It may be that proprietary solutions will move closer to today's open standards, or it could be that the market will continue to choose proprietary solutions over open standards. Time and time again, the TV industry has seen the benefits of open standards, but the nature of the industry means that change is slow and thus proprietary middleware and open standards will coexist for awhile, at least.

2 An Introduction to Digital TV

This chapter provides a technical introduction to digital television (DTV), covering the basic technology required to get a signal from a camera in the studio to a DTV receiver in a consumer's home. The chapter explores DTV transmission and other elements of the receiver, such as the return channel and the differences among terrestrial, cable, and satellite networks.

So what is DTV in practical terms? At a simple level, DTV uses digital encoding techniques to carry video and audio information, as well as data signals, to a receiver in the consumer's home. Although the transmissions themselves are still analog, the information contained in those transmissions consists only of digital data modulated on to the analog carrier signal. In addition to other advantages (examined in material to follow), this has a number of quality advantages when compared to analog broadcasts. Analog signals are subject to interference and "ghosting," which can reduce picture quality, and each channel needs a comparatively large part of the frequency spectrum to broadcast at an acceptable quality. Digital signals are less sensitive to interference, although they are not perfect, and the space used by one analog channel can carry several digital channels. As we will see, the technology used in DTV is based on MPEG-2 technology (similar to that used by DVD players), and many of the benefits DVDs have over VHS tapes also apply to DTV when compared to normal broadcasts.

The Consumer Perspective

Technically, DTV offers many new challenges and opportunities, the technical aspects of which are explored later in the chapter. From the consumer's point of view, though, there may not be a radical change in what they see between DTV and analog broadcasting (at least not at first). Today, the changes the viewer is exposed to are evolutionary rather than revolutionary. DTV offers four main advantages.

- More channels
- Better picture quality (up to high-definition TV resolution)

- Higher-quality sound
- Additional services and applications

Although the increased resolution of high-definition TV (HDTV; up to 1920 × 1080 pixels, compared to 720 × 480 as used by standard-definition DTV) is attractive, the new services and applications are probably the most interesting feature to many viewers and people in the broadcasting industry. Digital broadcasting can offer many types of services that are simply not possible with analog broadcasting. This can include extra information carried in the stream to enhance the TV-watching experience, or it can include downloadable applications that let viewers interact with their TV in new ways. These applications can be simple enhancements to existing TV shows such as multilingual subtitles or electronic program guides (EPGs) that show better schedule information. They may be improved information services such as news, or information services tied to specific shows (cast biographies, or statistics in a sports broadcast). They may also be new application areas such as e-commerce, interactive advertisements, or other applications tied to specific shows that enhance the viewer's enjoyment and encourage them to participate. For people working in the industry, the most important thing about many of these types of applications is that customers will pay to use them.

All of these improvements have taken some time to deploy, and thus the uptake of digital has not been as fast as the move to color TV from black-and-white. In that case, sales rose from 1.6 million to 57,000,000 million TV sets in the United States alone over a 10-year period. This was a revolution rather than an evolution, with a significant "wow factor" to drive sales forward. That has not really been the case for digital broadcasting so far, although the introduction of HDTV services and more advanced interactive services is starting to change this.

Many pay-TV cable or satellite services have already moved to digital broadcasting using proprietary middleware systems, and a number of public broadcasters have launched digital terrestrial services with varying degrees of success. As we will see, the nature of these two markets is very different and this affects the success of digital deployments.

We have learned from early players such as the United Kingdom's ONdigital (later ITV Digital) and Spain's QuieroTV that the consumer of today is not expecting to be charmed by the technical aspects of DTV. Both of these ventures failed, partly because of a lack of new and exciting content. Both companies had to battle the incumbent satellite competitors as well as the standard analog services already available, and ultimately they were simply not able to justify any subscription charges viewers had to pay to access anything but the most basic content.

Following the failure of ONdigital/ITV Digital, the BBC, Crown Castle International, and BSkyB jointly launched a free-to-air service called Freeview. Clever use of the word *free* has served its marketing purpose, and the perception that it is free has led to viewer numbers increasing at a significantly faster rate than previous attempts. ONdigital/ITV Digital receivers can receive and decode the new Freeview services, and this has probably helped increase viewer numbers.

In the case of Freeview, the platform for interactive applications is based on the MHEG-5 standard, which was also used by ONdigital/ITV Digital. Because of these legacy receivers and applications, a move to MHP in the United Kingdom will take longer than it would if we were starting from a clean slate. The dominance of proprietary middleware in the United Kingdom's pay-TV networks only makes it more difficult to move toward MHP.

Customizable TV

We have already mentioned that DTV broadcasts are more efficient than analog broadcasts. This means that we can receive more channels, but these channels are also easier to customize. DTV broadcasts can be tailored to suit regional markets much more easily than can analog broadcasts. This is done by replacing specific channels in the digital signal with regional programming, or by inserting region-specific advertisements into national programming. Similarly, interactive services such as news tickers, weather information, and e-commerce can be targeted at specific regions, either within a single country or across more than one country.

Another way to target broadcasts at different regions is by carrying multilingual content. A TV show may have audio tracks and subtitles in several different languages, so that viewers in different countries can watch exactly the same content but still hear or see it in their native language. For satellite broadcasters who operate in several countries, this can be a big advantage.

Managing TV content also becomes more efficient with the move to digital. Traditionally, TV shows and advertisements have all been stored on videotape, which must be placed in a videotape recording (VTR) machine and wound to the correct place before we can see a show or advertisement. If we need to broadcast a show on one tape, and then show ads from another couple of tapes, managing this content is a tricky business that can take a lot of skill. Even managing this tape is difficult. For example, a major broadcaster such as the BBC may need to store thousands or hundreds of thousands of tapes. Some of those tapes will need to be stored securely (working with tapes of Hollywood movies can be a frustrating business, given the security requirements often required by the studios), but all of them must be kept track of and stored in a way that will not damage them.

With digital content, we can store it all on the hard disk of a video server and send it virtually anywhere over a computer network. Backups can also be stored as digital content, and secure content can be encrypted. In addition, it does not matter if an ad is stored on one server and the main show is stored on another—we have almost instantaneous access to all of the content we need.

Digital content storage is nothing new, and many operators that have not yet made the move to digital will store content as MPEG-2. By keeping things digital all the way to the receiver, we can improve quality for the end user, manipulate content more easily, and make storage easier than is otherwise possible.

Understanding DTV Services

DTV services are now being delivered over satellite, cable, and terrestrial networks. Since 1996, the digital set-top box market has enjoyed rapid growth, with satellite broadcasting as the forerunner. According to In-Stat/MDR, set-top box deployments rose from 873,000 to over 14.4 million units in 2001 alone. It is now 2004, and we see the Italian terrestrial TV market moving to digital, which will drive the deployment of many more digital receivers. Analysts have tried to predict the growth of DTV services and have failed in the same way as those who failed us all in the dot-com era: this is a volatile area, and growth and reduction are exponential. No one predicted the rise and fall of QuieroTV, Boxer, and ONdigital, and many analysts doubted the success of MHP until it happened. Digital cable broadcasting has struggled due to financial woes, but we believe that the tide will eventually turn and the evidence is that this is starting to happen.

To the layman, however, DTV can be a little confusing. During a demonstration of DVB-T SDTV at the 2000 International Broadcasting Convention in Amsterdam, visitors asked one of the authors whether the pictures they were looking at were "high definition." From questions like this, it is obvious that their main understanding of DTV was the format and not the technical details. Why the confusion? In this case, the content they were seeing was an SDTV picture being displayed on some of the first 16:9 plasma TVs on the market. Visitors —not even average consumers, but visitors to a broadcasting-related trade show—assumed that this new plasma display meant high-definition signals.

In the consumer world it is still amazing how many people are used to a ghost-ridden, low-quality signal due to bad connections between the aerial and the TV. Many people will put up with a terrible picture provided they get access to their content. We all watch TV, and it is obvious that today people consider access to a TV signal a right rather than an enabling technology. This will not change as we move toward digital broadcasting, and we must be careful to remember that migrating to digital is as much a political and marketing move as a technical one. Without broad consumer acceptance, no DTV deployment will succeed. This is even more the case in free-to-air markets.

How does the layman know about DTV's increased channel capability, the high-quality picture and sound, and the added services? Well, often they do not, and they do not really care. Once they have actually seen it in action and seen the benefits for themselves, they naturally get more excited. Making consumers aware of these features will be one of the keys to the wide deployment of DTV. To summarize, DTV services provide some or all of the following benefits.

- Better picture quality (up to HDTV resolution with no ghosting and interference)
- Better sound quality (including Dolby surround sound in a lot of cases)
- More channels (especially with satellite and cable)
- New services (mobile services and data casting)
- Wider screen format; more DTV content is broadcast in 16:9 (widescreen) format
- Multilingual transmission and multilingual closed captioning and subtitling
- Electronic program guides to make browsing easier
- Interactive applications related to the broadcast show (bound applications)
- Personalized TV (personal video recorders, pay-per-view, VOD)
- Standalone applications and games that are downloaded to the memory of the STB or receiver (unbound applications)

Producing DTV Content

Although this book is largely concerned with ITV middleware, we need to have a decent grounding in the basics before we can really begin to discuss how open standards for middleware will affect the industry. Almost every DTV system deployed today, with the exception of IP-based services, works in the same way (although there are a few differences, examined later in the chapter). Although this does not directly affect our choice of middleware, there is a close relationship between the various parts of the system and many concepts in our middleware layer build on the basic techniques used to transmit DTV signals. These lower-level issues can also have a big impact on the types of applications we can deploy in a given network because of the limitations they impose on the amount of data that can be carried and on receiver hardware itself.

A good comparison is that of wired and wireless networking in the PC world. A wired network connection can carry much more data and is much less susceptible to interference than even the best wireless network connection, simply because of the way data gets from one computer to another. Even though the two types of network connections are the same to an application, applications that need to send and receive large amounts of data over the network will usually work best with a wired connection. Other types of applications will benefit from the advantages of a wireless connection, such as the lack of a cable connecting the PC to the network. Although this analogy is not perfect, it should give you an indication of how the lower-level parts of the system can have a wide effect on other elements of the system.

Moving back to our discussion of DTV systems, there are a number of issues we have to face when transmitting DTV signals over a network. The first of these is getting the data into the right format. Before we can carry video and audio over a digital network, we need to convert the analog information we get from a camera or a VTR into a digital format. This alone is not enough, however, because the amount of data that will result from a simple conversion is too great for most networks to carry. An uncompressed digital signal for standard-definition PAL video may have a data rate in excess of 160 Mbits/second.

To solve this problem, we compress the video and audio information using the MPEG compression scheme. The Moving Picture Experts Group is a group affiliated with ISO that has defined a number of standards for video and audio compression. Most DTV systems today use the MPEG-2 compression scheme, which can encode standard-definition signals at 3 to 15 Mbits/second. DVD players use the same compression system, but encode the content at a slightly higher bit rate than is used for DTV. MPEG is a lossy compression scheme, and thus a lower bit rate usually means worse quality. Typically, video for a DTV broadcast will be encoded at about 4 to 5 Mbits/second. This offers image quality that is about the same as analog broadcasts, but which we can broadcast much more efficiently (as we will see in material to follow).

MPEG content can be coded either with a fixed bit rate (in which the bandwidth required by the stream will always be constant) or with a variable bit rate, in which the bandwidth may vary depending on the complexity of the material being compressed. In the latter case, there is usually an upper bound for the bit rate, just to make it practical for broadcasting. Historically, DTV systems used constant bit rate encoding for video and audio content, but more recently the tide has turned in favor of variable bit rate encoding. The reasons for this are explored later in the chapter. Details of the MPEG compression system are beyond the scope of this book, and thus we will concentrate on the higher-level aspects leading to an understanding of the process of producing and transmitting DTV content.

Elementary Streams

An MPEG encoder will produce a single stream for audio or video content. This is known as an elementary stream (ES), the most basic component we will deal with. We will not look in detail at what an ES contains, because that topic would fill several books on its own. At this point, all we need to know is that each ES contains one piece of video or audio content. In the case of TV content, the video is encoded as one ES and the audio is encoded as another ES containing both stereo channels.

An ES is a continuous stream of information, and can thus be quite difficult to manipulate. For this reason, the content of the stream is often split into packets. Each packet will include a time stamp, a stream ID that identifies the type of content and how it is coded, and some synchronization information. The exact size of these packets can vary from one case to another, although this is typically a few kilobytes. Once an ES has been split into packets, it is known as a packetized elementary stream (PES).

PESs can be used for multiplexing into a form ready for transmission. As mentioned, audio and video content for DTV is broadcast using MPEG-2 transport streams. This is different from the format (known as a program stream) used with DVDs or digital camcorders. Both types of streams use PESs to carry audio and video data, but transport streams will contain additional information and their content will be multiplexed differently.

Transport Streams

Probably the biggest difference between the two types of streams is that a program stream will carry just one MPEG-2 program (hence the name). A program in MPEG terms is a group of video and audio PESs that will be played as a single piece of content. This may include multiple camera angles, or multiple audio languages, but it is still a single piece of content. Again, a DVD is a good example of this.

A transport stream, on the other hand, can carry several programs. For instance, a transport stream will usually carry several DTV channels. In the context of DTV, an MPEG program may also be known as a "service" or a "channel," but all of these terms mean the same thing.

When we multiplex more than one service in a transport stream, what we actually do is multiplex the PESs used by those services. This is simply a case of placing packets from different PESs one after another in the transport stream, with the relative bit rates of the PESs determining the order in which the multiplexer inserts packets into the transport stream.

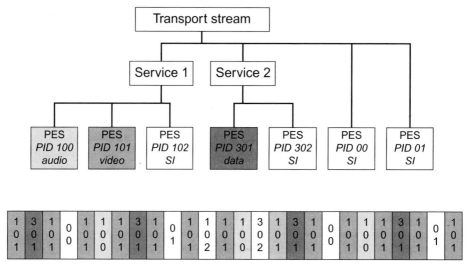

Figure 2.1. How PESs are organized into a transport stream.

Figure 2.1 shows an example of a simple transport stream, with one possible way of multiplexing the packets from the various ESs.

In multiplexing several services, we need some way of telling the receiver which PESs make up each service. There are actually two parts to this problem: identifying a PES in the multiplex and telling the receiver which PESs make up each service.

To identify which packets belong to which PES, each PES packet is labeled with a packet identifier (PID) when it is multiplexed into the transport stream. This is a numeric ID that uniquely identifies that PES within that transport stream.

To tell the receiver how these PESs should be organized into services, we also include some non-audio/visual (AV) information in the transport stream. This is called service information, which is basically a database encoded in MPEG-2 packets and broadcast along with the audio and video data. As well as telling the receiver which streams belong to which services, it tells the receiver the transmission parameters for other transport streams being broadcast and includes information for the viewer about the channels in that transport stream. Although some service information is defined by the MPEG standard, other standards bodies such as ATSC and DVB define the higher-level elements. More details about service information in DVB, ATSC, or OpenCable systems are available in the appendices.

The final difference between a program stream and a transport stream is the way data is split into packets and encoded. Transport streams are used in environments in which errors are likely to happen, whereas program streams are used in environments in which errors are much less common. This means that transport streams include much more robust error correction to ensure that errors do not affect the final signal.

Data in transport streams is split into packets of 188 bytes, which contain some header information (such as the PID of the PES the packet belongs to, as well as time stamp information). In addition to the time stamp information encoded in PES packets, each service in a transport stream will include an extra time stamp called the program clock reference (PCR).

MPEG relies on having a stable clock it can use to decode video and audio packets at the correct time, and for a DTV network we need to make sure the clock in the receiver is synchronized with the clock used to produce the data. If this is not the case, the receiver may decode data earlier or later than it should, and cause glitches in the display. To make sure the receiver's clock is synchronized with the encoder's, the PCR acts as a master clock signal for a given service. The receiver can use this to make sure that its clock is running at the correct rate, and to avoid any problems in the decoding process.

The Multiplexing Process

Multiplexing a stream may seem like a relatively easy thing to do, but in practice it can be a complex task that offers a great deal of feedback to the MPEG-encoding process. At the simplest level, multiplexing involves splitting the various PESs into 188-byte transport stream packets, assigning a PID to each PES, and transmitting the packets so that each PES gets its assigned share of the total bandwidth. There are a few complicating factors (most notably, the multiplexer needs to make sure that audio or video data is transmitted before it is needed by the decoder), but these do not change the basic process.

Recent advances in multiplexing technology and trends in the DTV industry have changed this. Traditionally, a network operator would assign a fixed bit rate to each PES in a multiplex, set the MPEG encoder to generate a constant-bit-rate MPEG stream, and let the multiplexer do its work. This may not make the best use of bandwidth, however, because simple scenes may use less bandwidth than is assigned to them. This means that other streams, which may need more bandwidth to encode complex scenes, cannot use the extra bandwidth that is unused by simpler content.

To solve this, many operators now use a technique called statistical multiplexing, depicted in Figure 2.2. This takes advantage of the fact that only a few streams will contain complex scenes at any given time. The multiplexer uses information about the complexity of the scene to determine how much bandwidth should be allocated to each stream it is multiplexing, and this is then used to set the appropriate bit rates for the MPEG encoders. Thus, streams containing complex scenes can "borrow" bandwidth from streams with less complex scenes.

MPEG encoders can generate this complex information and send it to the multiplexer. This acts like a feedback loop, allowing the multiplexer to dynamically change the bandwidth allocated to various streams as the complexity of scenes changes.

Carrying Transport Streams in the Network

We need to consider a number of factors when preparing a DTV signal for transmission. Some of these relate to the practical issues of radio transmission and some relate to the

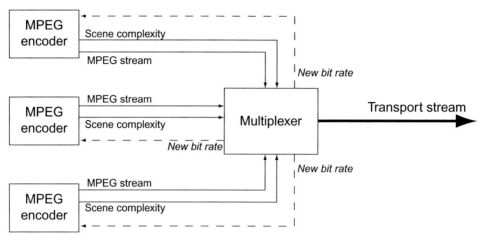

Figure 2.2. Statistical multiplexing uses feedback from the multiplexer to set future encoder bit rates.

problems of generating a signal that can be received correctly. We cannot simply convert the multiplexed transport stream into an analog waveform and transmit it, and in the sections that follow we will look at some of the problems we face and how we overcome them.

Energy Dispersal

The first problem we face is one of signal quality. Signals are not transmitted on single frequency. Typically, they use a range of frequencies, and the power of the signal will vary across the frequency spectrum used to transmit that signal. For a completely random stream of bits, the power would be equal across the entire spectrum, but for nonrandom data such as an MPEG-2 transport stream this is not the case. This can cause problems, because some parts of the spectrum may have very high power (and cause interference with other signals), whereas others may have a very low power (and are thus susceptible to interference themselves). Coupled with this, if one part of the spectrum remains at a high power for a long time it can cause DC current to flow in some parts of the transmitter and receiver, which can cause other problems.

To solve these two problems, the transmitter carries out a process called randomization (energy dispersal). This takes the transport stream and uses a pseudo–random number generator to scramble the stream in a way the receiver can easily descramble. The purpose of this is to avoid patterns in the signal and to spread the energy equally across the entire spectrum used to transmit the signal. DVB systems use the randomization algorithm shown in Figure 2.3.

We cannot apply randomization to the entire signal, because we still need some way of synchronizing the transmitter and the receiver. Each transport packet has a synchronization (or sync) byte at the start to identify the beginning of the packet, and we use these to periodi-

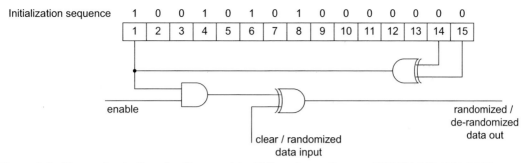

Figure 2.3. The randomization algorithm used by DVB systems. *Source:* ETSI EN 300 744:1999 (DVB-T specification).

cally reset the randomizer (in the case of DVB we do this every eight transport packets). The transmitter will invert the sync byte of the first packet in each group of eight, and when the receiver sees this inverted sync byte it knows that it should reset the randomizer. Other sync bytes are not randomized, and thus the receiver can use this to identify the start of each packet and the start of each group. It can then use this information to reset the derandomizer at the appropriate time.

Terrestrial and cable ATSC systems use different randomization algorithms, but the principle is basically the same. ATSC satellite systems follow the DVB standard.

Error Correction

The second problem we face is error correction. Due to the nature of TV transmission systems, every transmission will contain some errors because of radio-frequency interference. For an analog system, this is usually not a big problem because the quality of the output will reduce gradually, but small errors in a digital signal can have a big effect on the quality of the output. To solve this, we need to include some type of error-correction mechanism.

In an IP network, we can do this using checksums and acknowledgments to retransmit packets if one of them is corrupted. In a broadcast network, this is not practical. There may not be any way for the receiver to tell the transmitter about an error. Even if there were, the transmitter might not be able to resend the data in time. In addition, different physical locations will have different patterns of interference. For instance, electronic equipment may produce interference in one location but not another, or weather patterns may interfere with broadcasts in one parts of a satellite's coverage area but not another. If we were resending data, this would make it necessary to resend different packets for different receivers. In a busy environment, in which a transmitter may serve hundreds of thousands of households, this is simply not possible.

This means that we need to build in a mechanism by which the receiver can correct errors in the data without relying on resending the corrupt packets. We can do this using a technique called forward error correction (FEC), which builds some redundancy into the transmitted data to help detect and correct errors. Most DTV systems use a technique called Reed-Solomon encoding to encode this redundancy and to correct errors where necessary. We will not discuss Reed-Solomon encoding here in any detail, but for now it is enough to know that the version used in DTV systems adds 16 bytes of data to a transport stream packet. This gives a total packet size of 204 bytes for a transport stream that is ready for transmission.

Reed-Solomon encoding alone is not enough to solve all of our data corruption problems, however. Although it does a good job of identifying and correcting small errors caused by noise in the transmitted signal, Reed-Solomon cannot handle larger errors that corrupt several bytes of data. In the case of 16 bytes of redundancy, it can detect but not correct errors affecting more than 8 bytes. These could be caused by a variety of phenomena, such as electrical storms or interference from other electrical equipment.

To solve these problems, we use a process called interleaving, depicted in Figure 2.4. This reorders the data before we transmit it, so that adjacent bytes in the stream are not transmitted together. There are several ways of doing this, but one of the most common approaches is to use a two-dimensional buffer to store the data before transmission. By writing data to the buffer in row order, but reading it in column order, we can easily reorder the content of the stream and reconstruct the original stream at the receiver.

This process can be modified by introducing an offset between the different rows, so that the algorithm will read byte *n* of one row followed by byte *n-1* of the next row. This may use less memory than the more straightforward interleaving previously mentioned.

Interleaving makes the data much more resistant to burst noise. A burst of corruption in the transmission will be spread across several packets in the reconstructed stream. This will usually bring the error rate below the threshold that can be corrected by the FEC algorithm.

The type of coding at the superstructure level is known as outer coding. To prevent noise from reducing the error-correcting capability of interleaving, a layer of coding (called inner coding) is added.

Most inner coding techniques are based on a technique called trellis coding, in which symbols are grouped to form "trellises." For a group of three symbols, a modulation scheme that

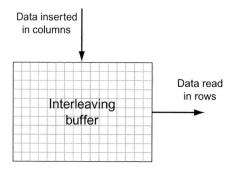

Figure 2.4. Interleaving a bit stream spreads transmission errors across multiple packets.

stores 8 bits per symbol can store 512 separate values. By using a subset of these as valid values, the network operator can introduce some extra redundancy into the signal. The effect of this is that each symbol may carry fewer bits of data, but for every group of three symbols it is possible to correct one erroneous symbol by choosing the value for that symbol that gives a valid trellis. This is the approach used by U.S. digital terrestrial systems.

DVB systems use Viterbi coding instead, which is a slightly different algorithm to finding the trellis that most closely matches the received data. Viterbi coding is an extremely complex topic beyond the scope of this book. By default, a Viterbi encoder will generate two separate output bytes for every byte of input, with each output representing the result of a different parity check on the input. This is known as 1/2-rate encoding, because one symbol of actual data is represented by two symbols in the output.

Because this is rather inefficient, most applications use a technique called puncturing, which involves transmitting one of the output bytes at a time in a well-known ratio. This allows the network operator to trade a reduction in error-correcting power for an improvement in bit rate. Different networks and different applications will use different puncturing rates, depending on the bit rates they need to achieve and the number of uncorrected errors that can be tolerated.

Modulation

Once we have applied inner coding to our bit stream, we are ready to transmit it. The last stage in the process is modulation: converting the digital bit stream into a transmittable analog waveform.

If we used a simple approach to modulate the signal, the amount of bandwidth we would need to transmit each signal would be extremely high. To solve this problem, modulation techniques will normally squeeze several bits of data into each symbol that gets transmitted. This symbol can be coded either by using different amplitude levels to indicate different values or by changing the phase of the signal.

The 8-VSB modulation scheme used by terrestrial systems in the United States uses eight amplitude levels to encode 3 bits of data in each symbol, whereas the QPSK and 8PSK systems use phase modulation to encode either 2 or 3 bits per symbol (by using phases that are 90 degrees apart or 45 degrees apart, respectively). These two approaches can be combined to squeeze even more data into a single symbol. Quadrature amplitude modulation (QAM) is commonly used with cable systems. This modulation combines QPSK modulation with multiple amplitude levels to support up to 8 bits of data in a single symbol.

QAM and QPSK use two different carriers with a 90-degree phase difference between them (one is in quadrature with the other, and hence the name). In both of these schemes, symbols are represented by a combination of both signals. Figure 2.5 shows 16-QAM modulation, in which each symbol can have one of 16 states. This allows each symbol to carry 4 bits of data.

An alternative modulation scheme splits the signal across several carriers simultaneously, using a scheme such as QPSK or QAM to modulate data on to each of these. This scheme,

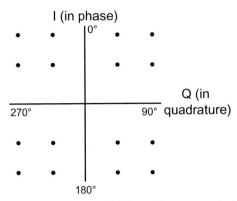

Figure 2.5. QAM modulation carries several bits of information per symbol.

called COFDM, is used by DVB for terrestrial broadcasts. The inner coding of the bit stream is often closely tied to the modulation technique used, and thus these are usually specified together.

Cable, Satellite, and Terrestrial Broadcasting

We have already seen that DTV can be carried over cable, satellite, or terrestrial networks. The differences among these broadcasting systems give each system unique advantages and feature sets, and this in turn affects how open standards can be used and how they change the business model for network operators and receiver manufacturers.

Digital satellite broadcasting allows for the distribution of content to a wide geographical area, with few restrictions on where the signal can be received. This technology now offers return channel capabilities that have been missing until recently. Satellite networks are an extreme case of shared networks, wherein the same satellite may carry transport streams from many different networks. Given the wide coverage of satellite transmissions, these may use different languages or even different video standards. In this case, it is extremely unlikely that much coordination will take place among network operators, although satellite operators may define their own requirements for transport streams carried on their satellites.

Cable networks are able to provide the consumer a large pipe to the home with an automatic (and high-bandwidth) return path for future interactive services. Cable signals will also be aimed at a small geographic region, although some cable operators may band together to broadcast the same content across several cable networks. The disadvantage of cable is the cost of cabling remote areas, or other areas where the geography is not favorable. A cable network is a private network (i.e., only the cable operator will be transmitting signals on that network), and typically there will only be one operator per cable network.

Digital terrestrial TV offers consumers not only the multichannel aspects of the other systems but the possibility of future mobile services such as those provided by TVMobile in

Table 2.1. Bandwidth allocation in a typical transport stream in Italy.

Content	Bit Rate
Advertising	3.9
Interactive	7.1
Launcher	1.0
Margin	0.5
Total for interactive services	**12.5**
Video + SI	11.5
Total	**24 Mbits/s**

Singapore. Reception may be a problem in some areas due to geographical features and the placement of transmitters, but this is less of a problem than for cable systems. Terrestrial networks are typically shared networks. Although each network operator will broadcast to a distinct geographical area (limited by the power and location of their transmitters), these areas may overlap or there may be more than one network operator in the same area. Network operators may or may not cooperate to keep channel numbering the same, or to coordinate other aspects of their services.

There are many considerations regarding bandwidth and interactive services for the various transmission systems. Most notably, the digital terrestrial system offers the lowest bandwidth available, depending on the nature of multiplexes and on channel allocation. In digital terrestrial broadcasting, 8 MHz of bandwidth can carry five or six DTV channels, with a total bit rate of up to 24 Mbits/second. To give you an example of where this bandwidth goes, a broadcaster in Italy typically allocates space within a transport stream as indicated in Table 2.1.

In Finland, up to 15% of the bandwidth in terrestrial services may be allocated to data casting, although cable operators do not have to carry any data services not associated with TV channels or other must-carry content. By doing this, Finland has ensured that there will be enough bandwidth for enhanced services. A typical multiplex in Finland can carry up to 22 Mbits/second, which corresponds to four or five services (each with its own MHP applications).

The comparative lack of bandwidth in terrestrial broadcasts can have its problems. In Spain, for example, one multiplex contains up to five SDTV services from five different broadcasters, and thus there is no bandwidth available for enhancements (e.g., extra audio languages or Dolby 5.1 audio) or for interactivity. Therefore, there is no benefit to broadcasters or consumers over analog services, and this situation inhibits innovative broadcasting and the push for MHP. This will be rectified in due course, and the new Spanish government has already reviewed the previous governments plans and set into motion the required changes and adoptions to put Spain on the route to full interactive digital DTT (digital terrestrial television).

Satellite and cable systems have far more bandwidth available, and up to 10 digital video channels can be carried in the "slot" previously occupied by one analog channel. This significantly expands the total number of channels and programs that can be offered and has better capacity for supplementary services. In a typical DVB-S transponder, a 36-Mbit pipe is available to the broadcaster. This translates into 5×6 Mbit channels, which leaves sufficient capacity for interactive services. Cable networks have similar data rates available.

It would be wise at this point to examine the new DVB-S2 specification, which will likely change the face of digital satellite broadcasting in the future. The increased carrying capacity offered by this new standard is a big step forward when compared to the current DVB-S standard. When DVB-S2 is combined with other technologies such as MPEG-4 part 10 and Windows Media 9 it is not unrealistic to imagine 20 to 25 SDTV services and 4 or 5 HDTV services (all including interactive applications) being carried on a single transponder. This recent adoption by DVB is considered so technologically advanced that the following statement accompanies DVB's documentation: "DVB-S2 is so powerful that in the course of our lifetime we will never need to design another system." In Chapter 19, we will see how these differences between networks can influence the types of services we deploy on them, and the types of business models that best suit each network.

Broadcasting Issues and Business Opportunities

Each of the different types of DTV networks has its own advantages and disadvantages. However, the issue is not simply one of cable versus satellite versus terrestrial. The different modulation schemes used by the three network types (and the different DTV standards used in different countries) play a part as well. In terrestrial systems, for instance, the DVB-T system handles reflected signals much better than the 8-VSB system defined by ATSC.

Consumers in built-up areas are used to seeing ghosting on analog terrestrial services, which is caused by reflected signals arriving at the receiver earlier or later than the main transmission. The receiver can either discard signals arriving early or late or use them to create a single high-quality (SDTV) picture on the screen. Unfortunately for the United States, however, this is not possible in all modulation systems and 8-VSB is one of the systems that suffers because it cannot handle reflections in the same way as some other systems. Consequently, the choice of 8-VSB will seriously restrict digital terrestrial TV in some parts of the United States, especially in cities with large numbers of high-rise buildings. Trials conducted in Taiwan have shown that this can be a big problem for 8-VSB when compared to DVB-T.

One relatively unknown technical aspect of digital terrestrial TV is that the DVB-T allows for mobile services. This was not realized during development of the standard, but was discovered during early implementations. Further investigation into the DVB-T COFDM "reflections" phenomenon showed that DVB-T receivers could make use of the reflections to improve picture quality, and consequently moving the antenna had no effect on the received picture. Further tests including the use of "dual diversity reception" (on the Nurburgring racing circuit in a Ferrari) have seen a full DTV service available at speeds of up to 237 km/h (170 mph). Depending on the parameters chosen for the modulation, broadcasters can choose

Table 2.2. The effect of the modulation scheme on mobile broadcasting.

Modulation Scheme	Approximate Maximum Signal-reception Rate
8K 64-QAM	50 km/h (31 mph)
2K QPSK	400 km/h (248 mph)

to trade data rates for reception at higher speeds. Table 2.2 outlines how choices in the modulation scheme can affect performance.

The resulting market possibility and eventual market reality was an installed mobile DTV service in Singapore called TVMobile. Starting in 2002, some 200 buses (each equipped with a standard receiver and a simple roof antenna) provide entertainment to Singapore's commuters and benefit advertisers. Sound is supplied through an inexpensive FM receiver for those who actually want to look and listen to the service.

This is a perfect example of a network for which a terrestrial system is the only practical choice (although satellite operators are now starting to explore mobile transmissions as well). Of course, for new applications such as these other factors play a part. Receivers may need to be tailored to fit specific market requirements, such as power supply voltages, physical size constraints, and even the means of controlling the receiver.

DVB has not rested on its laurels with respect to transmission standards, and it has continued working to bring the latest technologies to the market. The recently launched DVB-H specification for handheld and mobile devices is a more rugged system that includes a number of features for conserving battery life in portable systems. This opens up yet another market while still providing backward compatibility with DVB-T, and operators could have a mix of DVB-T services and DVB-H services in the same multiplex. If DVB-H is successful, this could mean that future applications will need to be compatible across different devices and screen sizes, which will be yet another challenge for application developers and middleware developers as regards the creation of ITV content. A full discussion of DVB-H is beyond the scope of this book, but more information about DVB-H (including white papers and additional technical information) is available on the DVB web site.

Subscriber Management and Scrambling

So far, we have covered the technical issues involved in getting a signal from a camera or videotape recorder over a transmission system to a receiver. In some cases, this will be enough. Everyone may be able to view our content, but this may not be a problem if we are a public broadcaster or if we are supported purely by advertising revenue. In other cases, we want to restrict that content to specific people—customers who have paid to subscribe to our services. This is called conditional access (CA) or scrambling, which involves encrypting some or all of the transport stream so that only subscribers can access it. The algorithms used in conditional access systems are a closely guarded secret due to the threat of piracy, and many different companies offer CA solutions for different markets.

Access Issues

Most of these solutions work in a similar way, however. When we use the CA system to scramble specific ESs (for instance, to protect some of the services in a transport stream), not all of the data for those ESs is actually scrambled. In this case, PES packet headers are left unscrambled so that the decoder can work out their content and handle them correctly. When we scramble the entire transport stream, however, only the headers of the transport packets are left unencrypted; everything else is scrambled.

Typically, the decryption process in the receiver is controlled using a smart card that uniquely identifies that receiver. The smart card itself may carry out some of the decryption, or it may contain the decryption keys the receiver needs in order to descramble the content.

For every conditional access system used to encrypt content in a transport stream, that transport stream will include one stream of PES packets containing messages for that CA system. These are known as CA messages, and the two most common message types are entitlement control messages (ECMs) and entitlement management messages (EMMs). Together, these control the ability of users or groups of users to watch scrambled content. The scrambling and descrambling process relies on the following three pieces of information.

- Control word
- Service key
- User key

The control word is encrypted using the service key, providing the first level of scrambling. This service key may be common to a group of users, and typically each encrypted service will have one service key. This encrypted control word is broadcast in an ECM approximately once every 2 seconds, and is what the decoder actually needs to descramble a service.

Next, we have to make sure that authorized users (i.e., those who have paid) can decrypt the control word, but that no one else is able to do so. To do this, we encrypt the service key using the user key. Each user key is unique to a single user, and so a copy of the service key must be encrypted with the user key for each user authorized to view the content. Once we have encrypted the service key, we can broadcast it as part of an EMM. Because there is a lot more information we must broadcast in EMMs (the encrypted service keys must be broadcast separately for each user authorized to watch the service), these are transmitted less frequently. Each EMM is broadcast approximately every 10 seconds. Figure 2.6 shows examples of how EMMs and ECMs are used together.

One thing to note is that the encryption algorithms used may not be symmetrical. To make things easier to understand, we are assuming that the same key is used for encryption and decryption in the case of the service and user keys, but this may not be the case. Asymmetric encryption using public-key algorithms is becoming more common.

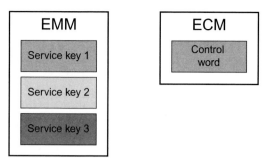

Figure 2.6. Entitlement management messages (EMMs) and entitlement control messages (ECMs).

When the receiver gets a CA message, it is passed to the CA system. In the case of an EMM, the receiver will check whether the EMM is intended for that receiver (usually by checking the CA serial number or smart card number), and if it is the receiver will use its copy of the user key to decrypt the service key.

The receiver then uses the service key to decrypt any ECMs for that service and recover the control word. Once the receiver has the correct control word, it can use this to initialize the descrambling hardware and actually descramble the content.

Some systems use EMMs for other CA-related tasks besides decrypting service keys, such as controlling the pairing of a smart card and an STB so that the smart card will only work in that receiver. Different CA systems will have their own variations on this process (especially between DVB and ATSC systems), and thus it is difficult to provide more than a general overview of this topic here.

DVB and the Society of Cable Telecommunication Engineers (SCTE) both define standard interfaces for CA modules, called the DVB Common Interface (DVB-CI) and Point of Deployment (POD) modules. OpenCable systems use a module based on the POD standard called CableCARD. Although receivers are not required to support these standards, the growth of horizontal markets for DTV means that more receivers will support them in the future. Some CA manufacturers have concerns about security or cost, however, and this may stifle deployment.

An alternative to pluggable CA modules is for all manufacturers to standardize on a single system. This has happened in Finland, with the common adoption of the Conax CA system for pay-TV services. This reduces the cost of the receivers while still ensuring that viewers can use their set-top boxes on all available networks. The only problem with this approach is that it requires a high level of cooperation between the receiver manufacturers and the network operators to ensure that all parties take a consistent approach to their use and interpretation of that CA system.

The Subscriber Management System

To generate the EMMs correctly, the CA system needs some information about which subscribers are entitled to watch which shows. The Subscriber Management System (SMS) sets which channels (or shows) an individual subscriber can watch. Typically, this is a large database of all subscribers that is connected to the billing system and the CA system, and which is used to control the CA system and decide which entitlements should be generated for which users.

The SMS and CA system are usually part of the same package from the CA vendor, and they are closely tied. For this reason, it is difficult to make too many generalizations about how the SMS works.

The Return Channel: Technical and Commercial Considerations

Many debates have taken place concerning what we mean by interactive, and a definition of *interactive* will often vary depending on whom you ask and which industry he or she associated with. Until recently, the phrases "going interactive" and "interactive services" have usually been associated with Internet applications, whereby consumers request data and launch applications with a click of the mouse or the press of a key. Early ITV pushed viewers to log on to certain associated web sites (and in some cases it still does this, especially in the United States). Consumers must leave the TV and go to their PC in order to look up information. This is not what the broadcasters or the advertisers actually want viewers to do, however. Viewer numbers is what makes money in the TV world, not viewers leaving their TV to go do something else.

Today, ITV more often means the use of the remote control to request information over and above the show that is being broadcast, whether the application is associated with or independent from the broadcast channel. This can range from "walled garden" web access, embedded games, or stored applications on the receiver through to information services that are directly associated with a particular TV show.

As we have discussed previously, three types of applications are often grouped together under the term *interactive TV*. Table 2.3 outlines these three groups, and we saw in Chapter 1 that MHP defines profiles that broadly correspond to one of these types.

As we can see from this table, many types of applications need connectivity back to the head-end via some form of communication route, known as a return channel or interaction channel. Return channels come in several different flavors, depending on the cost of the receiver and the type of network the return channel is connected to. Table 2.4 outlines some of the return channel technologies available. Many of these have been tried and tested, although not all are in current use. UMTS, for example, is growing and will probably replace GSM in the future.

Although PSTN modems are a popular choice in many markets, in Finland many households no longer have a fixed PSTN line and thus DSL return channels could be a better choice for

Table 2.3. The basic types of applications that constitute interactive TV.

Enhanced broadcast	• Interaction occurs only between the user and the receiver, either using data embedded in the broadcast or by changing to a different channel. Enhanced TV applications need no communication channel to the broadcaster or other service provider. • Typical applications of this type include information services such as news reports or electronic program guides.
Interactive broadcast	• Interaction relating to specific applications occurs between the user and the broadcaster or another service provider via a return channel. This usually takes the form of two-way applications such as voting, participation in a quiz, chat, or other applications. This may also include some form of "walled garden" web access. • This interaction can take place over a proprietary communication mechanism, although more standard mechanisms such as an IP connection over a PSTN modem are more common.
Internet TV	• Interaction can occur either between the user and the broadcaster or between the user and a server on the Internet. For instance, this may include unrestricted web access, or e-mail via a third-party service provider. • In this case, the receiver needs a standardized IP connection, possibly at a higher speed than is necessary for interactive broadcast applications as defined previously. Depending on the type of application, more support for Internet protocols such as SMTP, HTTP, and HTTPS may also be required.

Table 2.4. Data rates of some possible return channel technologies.

	PSTN	ISDN	GSM	SMS	GPRS	UMTS	DSL	Cable	SATMODE
Downstream (Kbits/s)	56	64–128	14	14	171	2048	256–52000	512–10000	30000
Upstream (Kbits/s)	56	64–128	9.20	160 bits/ packet	171	384	64–3400	64–128	1–64

the return channel in this case despite the increased cost. Figure 2.7 shows how return channel technologies are evolving in today's markets.

Depending on the price point of the receiver, integrating an expensive return channel technology such as DSL may not be a good idea, and in this case external modems may be used. An Ethernet connection on the receiver allows for connection to any type of broadband return channel, such as a DSL or cable modem. Table 2.4 outlines typical data rates for some of the common return channel technologies.

As we can see from the table, a return channel infrastructure may need support at the head-end, and this has consequences for systems engineering, billing systems infrastructure, and the types of services offered. Because of this, receivers sold through retail channels must be equipped with a return channel that is compatible with the infrastructure used by the

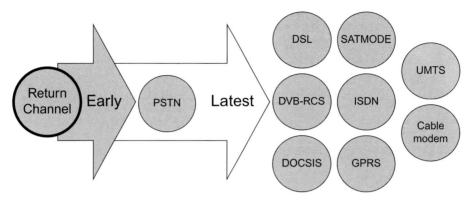

Figure 2.7. Some manufacturers are moving away from PSTN return channels to more advanced technologies.

network operator and/or content provider. If this does not happen, customers may not be able to use the return channel functionality they paid for.

In other cases, such as the market in Finland, the return channel may be completely open. Finnish customers can choose the type of return channel that best suits them, and even choose which ISP to use in that all return channel services to date use a standard Internet connection. To make setup of the return channel easier for nontechnical customers, it is planned to use smart cards for setting up the return channel connections, although this mechanism has not yet been deployed.

Technology is moving fast and receiver manufacturers and network operators may have difficulty keeping up. Partly due to cost and partly for stability reasons, receivers may not support the very latest technology. In satellite networks, for instance, two different technologies are currently emerging that provide a return channel via the satellite: DVB-RCS (return channel via satellite) and SATMODE (short for satellite modem), a joint project among the ESA, SES Global, Thomson, and several other companies. Whereas DVB-RCS is slightly more mature and has more deployments, SATMODE is aiming to provide a lower-cost solution. As well as these, satellite receivers can choose to use any of the other return channel technologies. Until a clear winner emerges from the competing technologies, many receivers will use a PSTN modem or an Ethernet interface for an external broadband return channel.

Decisions about the type of return channel are not just technical ones: they also influence the cost of the receiver and the types of applications that are possible. A receiver that only has a PSTN modem, for instance, will never be able to support VOD over the return channel. For receivers with a broadband connection, VOD applications become much more feasible. Similarly, the penetration of the various technologies into the market is also a significant factor. After all, selling broadband-equipped receivers is no use if only 10% of the market can actually get a broadband connection, or if the cost of a broadband connection is prohibitively high. These types of commercial decisions are important ones, and their impact should not be underestimated.

3 Middleware Architecture

ITV middleware has a number of features that set it apart from other programming environments. This chapter examines these features, and provides a high-level overview of how an MHP or OCAP middleware stack is put together.

Before we start looking in any detail at the various parts of the OCAP or MHP middleware, we need to have a general understanding of what actually happens inside a DTV receiver that is built around these standards. If you are not familiar with the DTV world, you need to understand the constraints and the goals involved, and how they are different from the issues you would encounter working on a PC platform.

We also need a general understanding of how software stacks are put together, and how the various components depend on one another. There are two reasons for this: to understand just how interconnected the different pieces of the puzzle actually are and to see the similarities and differences between the MHP and OCAP middleware stacks.

These two standards are built on the same common platform, as are other up-and-coming standards such as ACAP and the Japanese ARIB B23 system. As middleware developers and application developers, we need to be able to take advantage of these similarities, while knowing how to exploit the differences in order to make a really good product. This is not as easy as it sounds, but it is possible as long as we start with the right knowledge.

MHP and OCAP Are Not Java

Although MHP and OCAP have a very strong base in Java technologies, it is important to remember that Java is not the whole story when it comes to a DTV receiver. Those parts of the middleware that relate to MPEG are equally important, and this is something very easy to forget if you are approaching these standards from the PC world. Many middleware developers underestimate the complexity of the components related to MPEG, and do not realize the length of time needed to complete a good implementation of MHP.

Just because we have ported Java to our platform does not mean we have a complete middleware stack. Integrating the MPEG-related components can be a long and challenging task if you are not prepared for it. Of course, application developers do not need to worry about this, but they will have their own concerns: a much more limited platform, latency issues for data access, and many UI (user interface) concerns.

It is much easier to build a product (whether application, middleware stack, or receiver) if you know in advance that it is not all about Java. Many times we have seen STB developers at various trade shows demonstrating their MHP receiver when all they have is a Java VM (virtual machine) and a web browser, and many times you will not hear much from them again. Until you have added the MPEG-related components, the difficult part is not over.

They Are Not the Web, Either

The same message applies to HTML browsing, although this is a slightly different case. As you know if you are familiar with MHP, OCAP, and ACAP specifications, some versions of both standards support XHTML, CSS level 2, and DOM. In some ways, developing HTML applications is a little easier than developing Java applications, but it also has its pitfalls.

Things are getting easier for web developers as more browsers become compliant with CSS standards, and the DTV situation is even better because problematic legacy browsers such as Netscape 4 are much less common than they are in the PC world. However, developers should not be complacent: many more browsers are available for DTV systems, and no single browser has a stranglehold on the market. Compatibility testing can be even more difficult in this case than for Java systems, and designing a set of pages that looks good and is easy to navigate on a TV is still pretty difficult.

Middleware developers also need to be careful on the HTML front. Although there is much less integration between the browser and the underlying MPEG components, there is still some: a hyperlink in an MHP or OCAP application can refer to a DTV service as well as another web page, and the application model that MHP and OCAP both add to HTML applications must be considered. This application model is examined in more detail in Chapter 15, including how it affects the browser.

Working in the Broadcast World

We must take care not to underestimate the impact of the broadcast model on MHP and OCAP applications: it really does affect every part of your application and middleware. Many of the changes from standard Java and HTML are to cope with the pressures of the broadcast world, and these pressures are probably the most difficult thing to adjust to if you are approaching DTV as a PC developer.

The most important issues are not purely technical ones. Reliability and interoperability play a huge part in the success of any DTV product. Unlike the PC world, people who are watching TV do not expect applications to crash, or to be told to upgrade their browser, or to be told that their STB cannot run an application. Although this may not seem like a big problem, spend some time surfing the Web with a browser other than Internet Explorer or Mozilla and see how many sites have display problems (especially if they use Java or JavaScript). The MHP and OCAP world does not have the equivalent of the "big two" browsers as a middleware provider, and thus application developers cannot simply write to a single platform.

However, much as we would like OCAP or MHP to be completely standard and for every platform to behave the same these things will not and cannot happen. Specifications are ambiguous (sometimes deliberately so), companies make different choices in the capabilities of their receivers, and it is up to developers to make sure their product behaves in a similar way in all situations. For a middleware developer, this means it has to do the same thing as other platforms. For application developers, this means it had better work on all platforms.

These are not the only challenges facing developers, however—things would be too easy if they were. The use of broadcast technology to deliver data to the receiver brings its own problems. If the receiver needs some data, most of the time it cannot simply fetch it from a server. Instead, it has to wait for the broadcaster to transmit that data again. This makes caching data very important, and confronts middleware developers with a completely new set of choices about what information to cache and how much. Application developers do not miss out on the fun, either, because they get to design their applications to reduce latency, and to design how those applications are packaged to make caching easier and to reduce loading time.

It is not all bad news, though. The technologies used in MHP and OCAP do make it easier for application developers to do certain things (such as synchronizing applications and media), and in general the individual elements used have been around for long enough that most of the problems associated with the underlying standards have been worked out.

The Anatomy of an MHP/OCAP Receiver

Now that we have seen some of the challenges MHP and OCAP bring us, let's take a look at how we can design middleware that reduces the problems we face. The MHP software stack is a complex piece of design. A complete middleware stack based on MHP (including the OCAP stack) will be several hundred thousand lines of code in all probability, and this means that getting the architecture right is a major factor in building a reliable and portable middleware implementation.

One of the great things about the components that make up the MHP and OCAP APIs is that many of them can be built on top of other APIs from the two standards. This makes it possible to build the middleware stack in a modular way. Figure 3.1 shows how the components in an MHP stack can be built on top of one another. Each component in the diagram is built on top of those below it, and standardized APIs could be used internally within the software stack as well as by MHP or OCAP applications.

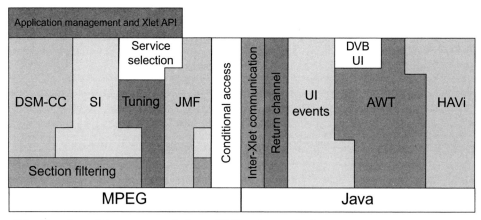

Figure 3.1. Overview of the components in an MHP software stack.

Not every middleware stack will be built this way, but it gives you an idea of how the APIs fit together conceptually, if not in practice. Exactly how closely a particular implementation will follow this depends on a number of factors. Implementations for which more of the code is written in Java will probably follow this more closely than those implementations largely written in C or C++, for instance. Similarly, the operating system and the hardware capabilities of the platform will play a part. Platforms in which most of the work is done in software have a little more freedom in their design, and thus may be more likely to follow this approach.

Although Figure 3.1 only shows one possible way APIs can be built on top of one another, it gives you an idea of the dependencies among the various components. Do not worry if some of these components are not familiar to you. We will take a detailed look at these components, and more, in further chapters.

The APIs that make up MHP and OCAP can be split into two main parts. One part contains the components related to MPEG and MPEG streams. The other part provides services built directly on top of the standard APIs that are part of every Java platform.

At the center of the MPEG-handling APIs sits DAVIC's core MPEG API. This contains just a few classes, but these represent the basic components that describe MPEG services and streams.

Another important API for handling MPEG is the section-filtering component, which is used to filter packets from the MPEG stream. Almost all of the other MPEG-related APIs build on this in some way. The service information component uses it to filter and parse the MPEG sections that contain the SI tables required to build its SI database, which applications can then query using the two available service information APIs. The SI component could use a proprietary API for accessing the section filters it needs, but in some designs it may be equally easy to use the standardized section-filtering API.

The next component we have to consider is the data broadcasting component. This needs to parse the sections that contain broadcast file systems or data streams (using the section-filtering component) in order to decode and provide access to the data they contain, while at the same time using the service information component to find out which streams in a service contain those data streams.

In this case, there may be a benefit to using the standardized SI APIs if possible, although control of section filters may for efficiency be handled at a lower level. It is possible to use the section-filtering API to handle this, but unless you have a very fast platform the performance cost may be too high.

The tuner control component relies on the service information in order to locate the transport stream it should tune to. Once it has the correct frequency and modulation settings (which it may get from scanning the network or from user settings, neither of which has an MHP or OCAP API), it will access the tuner hardware directly in order to tune to the correct transport stream.

The media control component, which is based around Sun's Java Media Framework, needs service information in order to translate a reference to some content into a set of MPEG streams it can decode. It also uses service information to provide the functionality for some of the common JMF features, such as choosing an audio language or subtitles. Once JMF has located the appropriate streams, it will typically access the MPEG decoder hardware directly to decode the stream. In this case, the MPEG decoding hardware or the implementation of the software MPEG decoder will usually demultiplex and decode the appropriate PIDs from the transport stream.

Now that we have seen the low- and mid-level components for MPEG access, we can examine the two higher-level APIs that use them. The JavaTV service selection API uses the service information component to find the service it should select. Once it has done this, it uses the tuning component and JMF to tune to the correct transport stream and display the right service.

In Figure 3.1, the application management component builds on top of the service selection API. In reality, the picture is slightly different and rather more complex. Although application management relies on the service selection API (in that every application is associated with a service), the relationship between the service selection API and the application manager is deeper than this. There is a very close link between service selection and the control of applications, and we will take a detailed look at this in Chapter 4. The application manager also depends on the service information component to get information about available applications.

One of the few other components to directly access the MPEG decoder is the conditional access component. This is largely due to efficiency issues, as with JMF. In that most of the work is carried out in hardware as part of the CA subsystem itself, rather than in the middleware, there is little to be gained from using any standardized API. This accounts for all of the MPEG-based components, with a few minor exceptions. Support for references to broadcast content such as JavaTV locators (see Chapter 8) is not normally based on top of the service information API because the service information available to the receiver may not be enough to decide if a locator is valid. Although this API could be considered part of the core MPEG concepts previously discussed, this is not guaranteed to be the case. Some middleware architectures may have good reasons for keeping them separate.

Turning our attention to the other middleware components, the most obvious place to start is the graphics component. The core of this component is the Java AWT (Abstract Window Toolkit), and although this is very similar to the normal AWT Java developers all know and love it is not completely identical. Differences in the display model between MHP-based platforms and other Java platforms mean that there are a few changes needed here.

The HAVi Level 2 graphical user interface (GUI) API builds on top of AWT to provide the classes needed for a TV user interface, and many of the classes in the HAVi widget set are subclasses of the equivalent AWT widgets. At the same time, there are many elements of the HAVi API that are completely new, including a new model for display devices and a set of new GUI widgets. To complicate matters further, the HAVi API also has links to the media control component to ensure that video and graphics are integrated smoothly. These links may be at a low level, rather than using the standardized APIs, and are not shown in Figure 3.1.

The DVB user interface API also builds on top of AWT. In most cases, this merely extends AWT's functionality in ways that may already be included in the Java platform (depending on the version of Java used). Some elements, especially alpha blending of UI components, may instead use the underlying hardware directly. Alpha blending is a very demanding task the platform may not be able to carry out without some hardware support, and thus this may be carried out in hardware by the graphics processor.

Due to changes in the way user input events are handled by MHP and OCAP receivers, AWT also uses the MHP component for user input. This is used directly by applications as well, of course, but the UI event component will redirect user input events into the normal AWT event-handling process as well as to the parts of the HAVi API that need it.

One component is missing from Figure 3.1. Many components are built on top of the resource manager. This gives the other components in the middleware a framework for sharing scarce resources, and this is exposed to applications and other components via the DAVIC resource notification API (and in the case of OCAP the extended resource management API). This allows all of the APIs that directly use scarce resources to handle these resources in the same way.

The other components are largely independent of one another. The return channel component uses the DAVIC resource notification API, in that most MHP STBs will use a normal PSTN for the return channel if they have one (OCAP receivers will generally use a cable modem instead). In this case, the return channel is a scarce resource, which may not be the case with a cable modem or ADSL return channel.

The inter-Xlet communication API needs to work very closely with the application management component to do its job properly. This is typically built on a private API, however, because none of the MHP APIs really does what is necessary.

The Navigator

One other important component is not shown in Figure 3.1, and indeed this component is not even a formal part of the MHP specification. At the same time, every receiver based on MHP will have one, even if it is known by a different name.

This component is the navigator (also known as the application launcher), which is responsible for letting the viewer actually control the receiver. It displays a list of channels, lets the user watch the content of those channels, and possibly gives the user a simple electronic program guide. In short, it is everything the user needs to watch TV, but that is not the end of the story. It also provides the user with a way of starting and stopping applications, setting user preferences, setting up the receiver, and controlling any other features offered by the receiver.

The navigator may be implemented as an application that uses the standardized APIs, or it may use private APIs to get the information it needs. In many cases, the standardized APIs are enough, although for any nonstandard features (personal video recorder functions, for instance) a private API may also be needed.

Differences in OCAP

If you look at a description of the OCAP middleware stack, you will not usually see any reference to the components discussed previously. They are still there, but OCAP uses a different model to describe its architecture. Figure 3.2 shows what OCAP's vision of the

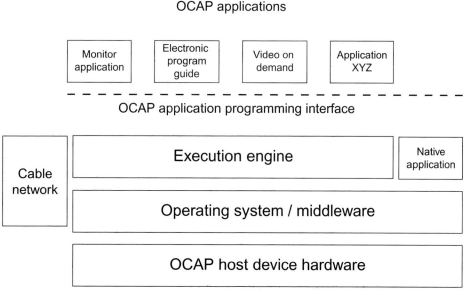

Figure 3.2. Overview of the components in an OCAP software stack. *Source:* OCAP 1.0 profile, version I09.

architecture looks like, and although they are not explicitly mentioned many of the components we saw in the MHP software stack are in this model included in the execution engine.

The execution engine contains most of what we would think of as the middleware. This includes the Java VM (and HTML browser for those receivers that support HTML applications), software related to MPEG handling, and OCAP software libraries.

As you can see, the execution engine contains a number of modules that provide functions to other parts of the receiver. There is some overlap between these and the components we have seen earlier in the chapter in our MHP receiver (some of the modules correspond to one or more components from our MHP software architecture, whereas others are completely new). There are also some of the components from our MHP architecture that are present in OCAP but are not shown in Figure 3.2. An OCAP receiver is more than just the modules described here and shown in Figure 3.2.

One of the reasons for the new modules is the scope of OCAP compared to that of MHP. An OCAP receiver can handle normal analog TV signals as well as digital ones, and for this reason the OCAP middleware has to include functionality for handling these analog signals and doing all of the things a TV normally does. These cannot be left to the TV because there is no guarantee that the output from the receiver will be displayed on a TV. A monitor or projector might be used instead, which will have no support for these functions. Because some functions (such as closed-captioning support) are mandated by the FCC, an OCAP receiver must handle them just like any traditional TV set.

A New Navigator: The Monitor Application

OCAP receivers do not have a navigator such as that we would find in an MHP receiver. Although all of the functionality is still present, it is in different places and the organization of it is a little more complex.

Most of this complexity is introduced because of the nature of the cable TV business in the United States. Network operators traditionally have much more power than their counterparts in Europe, and they exert much more control over some elements of the receiver that are left to the receiver manufacturer in other markets. In particular, OCAP allows the network operator to have a far bigger say in what applications get started, and when (going far beyond the application signaling in the transport stream), as well as in how the receiver responds in certain conditions (such as resource conflicts). Some of the functions in an OCAP receiver are only available after a network operator has downloaded its own software into the receiver.

MHP receivers do not entail these issues, in that the navigator is much more tightly coupled with the firmware. In an OCAP receiver, the navigator functionality and even some of the functionality provided by the MHP middleware is contained in a separate application that must be downloaded from the network operator. The application is then stored in the receiver's persistent memory. This is called the monitor application.

The monitor application is, in effect, just another OCAP application that happens to have been granted special powers by the network operator. It communicates with the rest of the system using standard OCAP APIs, although it can use a few APIs that are not available to other applications. These include APIs such as the system event API added in version I10 of OCAP, which lets the monitor application receive information about uncaught errors in applications or receive notification when the receiver reboots.

Because a monitor application is downloaded to the receiver by the network operator, a newly manufactured receiver does not have a monitor application. It still has to be able to provide all of the basic functions of a DTV receiver, though. The viewer must still be able to connect to a TV network and receive unencrypted services, get basic channel information, change channels, and set up the receiver in the way desired.

When the receiver connects to a DTV network, it will check to see whether the monitor application being broadcast is the same as the one stored in its persistent memory. If it is not (or if there is no monitor application in the receiver already), the receiver will download the new monitor application and replace any it had previously stored. The monitor application will then take over some of the functions previously carried out by the execution engine, such as the navigator and EPG functionality and some of the control over other downloaded applications.

The monitor application can take over many other functions as well. Some of the modules in the execution engine are known as assumable modules, so named because the monitor application can assume their functionality. Later in this chapter we will see what modules are assumable, and how a monitor application can take over their roles.

Assumable modules are useful because they allow the network operator to choose which parts of the functionality they want to customize. Some operators may only want to provide a new EPG, whereas others may want to add customized handling of messages from the CA system or from the network operator. Using assumable modules, the operator can customize most aspects of the user experience. This is a powerful way of strengthening the network operator's brand image, without having to invest time in implementing parts of the system the network operator does not care about.

Modules in the Execution Engine

Now that we have seen the big picture, we can take a look at the execution engine in more detail. Figure 3.3 shows the various modules in the execution engine. The most important function of an OCAP receiver is to allow the user to watch TV, which is normally why they bought it (or subscribed to the network) in the first place. The Watch TV module provides the user with the basic channel-changing functionality that corresponds to the navigator in an MHP receiver. The set of functions offered by this module can vary from the most basic set of controls possible to something much more complex that includes an electronic program guide and a wide range of other features.

Under the right circumstances, the monitor application can take over the functionality of this module. This allows the network operator to provide more sophisticated features (integration with an EPG or with premium services, for instance) or to strengthen their brand image

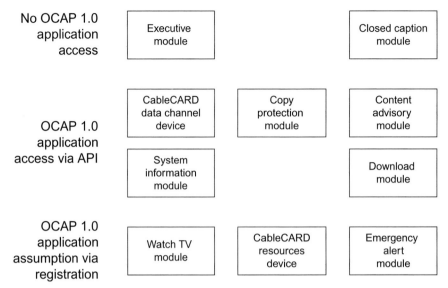

No OCAP 1.0 application access

Executive module		Closed caption module

OCAP 1.0 application access via API

CableCARD data channel device	Copy protection module	Content advisory module
System information module		Download module

OCAP 1.0 application assumption via registration

Watch TV module	CableCARD resources device	Emergency alert module

Figure 3.3. Modules in the OCAP execution engine. *Source:* OCAP 1.0 profile, version l09.

by giving any receiver on that network the same user interface no matter which company manufactured it. As well as the obvious benefit, this has the advantage of reducing support costs: by only having one user interface across every receiver, the network operator's customer support team needs to learn about just one product. In a horizontal market in which many different models of receivers may be present on the same network, this can offer a big advantage.

The executive module is that part of the firmware that boots the receiver and starts the monitor application. It is also responsible for controlling the receiver if there is no monitor application present, or if the receiver is not connected to a digital cable network. This contains the application management component, which is extended slightly from that seen in the MHP software architecture.

The executive module communicates with the monitor application via the standardized OCAP APIs, and once a monitor application has been loaded the executive module delegates many of its responsibilities to it. The monitor application takes over some of the responsibility for application management, although most of the lower-level work (such as parsing the application signaling from the transport stream) will still be done by the executive module. There is far more to the relationship between the monitor application and the executive module than we have seen here, however, and we will look at this relationship in more detail in other chapters.

As its name suggests, the system information module parses and monitors in-band and (if a POD or CableCARD module is inserted) out-of-band service information that complies with the appropriate ATSC or ANSI standard. Applications can then use the JavaTV service information API to access the broadcast SI data, although some SI may have to be read using other APIs. SI related to application signaling is only available via the application management API, for instance.

In some cases, the system information module will also pass on information to other modules in the middleware. For instance, any Emergency Alert System messages will be passed on to the Emergency Alert System (EAS) module (examined in more detail in material to follow).

All analog TV systems in the United States must be able to display closed-caption subtitles because of FCC regulations, and the closed caption module is responsible for handling closed-caption subtitles in an OCAP receiver. Support for this is mandatory, and thus a receiver must be able to display closed-caption information from analog sources even when a monitor application is not running. To support this, the closed-caption module parses any closed-caption signals in the current stream and handles them as necessary. If the receiver has a video output that supports NTSC closed-caption signals, the closed-caption data is added to the VBI on the output. For receivers that only have outputs such as VGA or component outputs, where VBI data cannot be added, the closed-caption data is decoded and overlaid on the video output by the receiver itself.

Two other components are driven by the requirements of FCC regulations. One of these is the content advisory module. This module supports V-chip parental rating functionality for analog signals, and is responsible for decoding any V-chip signals contained in the incoming VBI data. These signals are then forwarded to any other modules in the receiver that need to access them in order to decide whether that content should be blocked.

The EAS module is the other module mandated by the FCC. The Emergency Alert System provides a way for network operators to force all receivers on the network to display messages during local or national emergencies in order to provide information to viewers. If the system information module receives any messages for the Emergency Alert System, it passes these on to the EAS module to be displayed on the screen.

The EAS module is an assumable module, and thus network operators can decide how they want to present Emergency Alert System messages to their viewers. This may involve tuning to a new service, displaying a message, or playing a sound to tell the viewer that an Emergency Alert System message has been received.

The download module implements a common piece of functionality that is also present in many other set-top boxes, including MHP receivers. It allows the network operator to upgrade the firmware of the receiver in the field by downloading it via the same data broadcasting mechanism used for applications. Thus, the network operator can fix bugs in the firmware of receivers that have been deployed. Given the reliability requirements of consumer devices and the problems that would be caused by upgrading receivers any other way, this is an important function.

There are several ways this function can be invoked. The monitor application can use the download module to check for new versions of the firmware, or the network operator can signal the new version of the firmware following the method described in the *OpenCable Common Download Specification*.

In addition to OCAP middleware, other native applications or pieces of firmware could also be upgraded in this way. This method is typically not used for updating OCAP applications stored in the receiver, however. As we will see later, these are updated using the normal method for signaling applications.

The copy protection module is also related to analog functionality. This module controls any copy protection systems used on the outputs of the receiver. These may be analog systems such as Macrovision or a digital system such as the broadcast flag for those receivers that support digital outputs. This module comes into play when you are watching TV, and OCAP receivers that support personal video recorder (PVR) functions can use this module to decide how many times a specific piece of content can be viewed, depending on the copy protection system in use.

The final module in the execution engine is the CableCARD interface resources module, also known as the POD resources module in some earlier versions of the OCAP specification. This handles any messages from the POD hardware that relate to the man-machine interface of the CA system, out-of-band application information, or other data that may be transmitted out-of-band by the network operator.

Although the POD resources module is an assumable module, it is different from other assumable modules we have seen so far. The functionality offered by this module is actually shared as three submodules (called resources), which have different rules about how their functionality can be assumed.

The MMI resource is responsible for presenting MMI messages from the CA system, such as confirmation dialogs, requests for a PIN number, and messages that tell the user why access to a service has been denied. In that the messages from the CA system should be presented in a standard look and feel, to avoid confusing the user the functionality provided by the MMI resource can be delegated to just one application at a time.

The other two resources provided by the POD resources module are the application information resource and specific application resource. These resources allow applications to communicate with a head-end server using the standard functions of the POD module. These can delegate their functionality to several applications simultaneously, allowing more than one application to communicate with a server via the POD.

As you can see, an OCAP receiver has many differences from an MHP receiver. There is no single reason for this, but many factors affect the architecture in different ways. We have already seen how the regulatory environment affects things, and the relative strength of the network operators is an important factor. In addition, the U.S. market has a business environment very different from that of European markets, where DVB was developed. Receivers in the United States may be able to receive several different networks if they can receive terrestrial or satellite signals, and this means that network operators have to work a little more closely than in Europe. If a receiver in France cannot access German pay-TV satellite services it is not a tragedy, but a receiver in southern Oregon should be able to handle signals from a network in northern California.

Technical issues also play their part: OCAP receivers are designed to operate in a cable environment only, and thus the POD module provides an always-on return channel as well as CA functionality. ACAP receivers for cable systems follow OCAP's approach, and thus they too have a permanently connected return channel. The need to include analog functionality also has an effect. Handling closed captions and V-chip functions means that an OCAP receiver has to be able to deal with analog signals as well as digital ones, and this is simply not an issue for MHP receivers.

These are not the only technical issues, however. More mundane issues, such as the differences in service information formats, also change things even though all systems use the same basic format for transporting data. We will return to these issues in later chapters.

Architectural Issues for Implementers

Now that we have seen the various parts of the architecture, any middleware implementers who are reading this may be feeling a little nervous. Relax, it is really not that bad, although there are a few things you need to be aware of. Application developers should not skip this section, however, because by understanding the middleware as well as you can you are in a better position to write applications that take full advantage of the platform.

Choosing a Java VM

Over the last few years, the range of Java VMs available has grown enormously, and we do not just mean the number of companies writing VMs. It used to be simple: you had the Java VM from Sun, and you had a number of companies building VMs that followed the Sun specification. All you needed to do was choose which version of the APIs to use, and the company to purchase them from.

Then Sun introduced PersonalJava, and life started getting more interesting for people working on small devices. pJava provided a subset of the Java 2 platform that was aimed at consumer devices. pJava did not include many of the features required for a desktop implementation that were not useful in a consumer device. Since that time, pJava has been overtaken by J2ME (the Java 2 platform, Micro Edition). This in turn has a number of different configurations (which can use one of two different VM implementations), and thus middleware developers now have a choice in the platform they use. At the time of writing, the following three basic platform choices are available.

- JDK-based systems that provide the full Sun JDK platform (now known as the J2SE platform for versions later than 1.2)
- pJava-based systems
- J2ME-based systems

These platforms have the relationships shown in Figure 3.4.

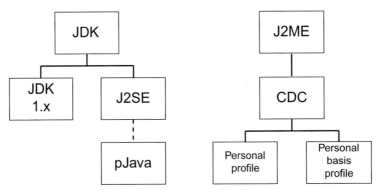

Figure 3.4. Relationships among Java platforms MHP and OCAP can use.

There are two parts to consider when looking at which solution to use: the Java VM itself and the class libraries that go with it. Although J2ME supports two different VMs (depending on the configuration you use), in practice we can ignore this factor. MHP and OCAP both use the Connected Device Configuration (CDC) of J2ME, which uses the same JVM as J2SE and pJava. No matter which platform we choose, they all use the same basic VM.

Now that we have that out of the way, we can turn our attention to the class libraries. J2SE is the latest evolution of the JDK, and is mainly aimed at PCs and workstations. In that a DTV receiver is usually not a PC or workstation, platforms based around a full J2SE implementation are much larger than MHP or OCAP actually requires (especially in the area of AWT). This means that a receiver using J2SE will usually need a lot more memory than a receiver using a different Java platform. Depending on the functionality and the target price point of the receiver, this may not be a big deal, but for lower-cost receivers the extra memory could add substantially to the price.

pJava takes elements of JDK 1.1 and adds a number of useful elements from J2SE, as well as making some classes optional. This is a compromise designed to take the parts of J2SE that are useful to DTV receivers while avoiding the size issues that go with using all of J2SE. Although this is not a bad compromise, it has begun the end-of-life process at Sun. This should not rule it out, however, because many other companies have pJava VMs still in development.

pJava may not be an ideal platform choice for those companies starting to build a middleware implementation, but for companies who already have a middleware implementation based around pJava there is no real reason to be concerned. MHP and OCAP will not for a long time move away from a platform that is substantially compatible with pJava, and there is no compelling reason to change. pJava has several advantages, not least the fact that it is a lot smaller than a full J2SE implementation.

The J2ME CDC is probably the preferred choice of platform at the moment, mainly because it has the size advantages of pJava while still being maintained. At first glance, the multiple profiles of CDC look confusing, but it is actually relatively simple for MHP and OCAP implementers. Both MHP and OCAP make life easy for us by requiring at least the Personal Basis Profile. This profile was designed to provide the features needed by MHP and OCAP set-top boxes, and thus it includes support for Java applications in a way that is most useful to DTV receivers and a simple GUI library that provides only those features we are actually likely to need. This is also Sun's recommended migration path from pJava, and is thus also a possibility for middleware implementers currently working with pJava.

The Personal Basis Profile (PBP) offers more features, and may be a better choice for receivers that offer advanced features that are not standardized in MHP and OCAP. Essentially, the Personal Basis Profile offers the same set of features included in pJava, the Personal Basis Profile is a subset of Personal Profile that provides only those features needed by MHP and OCAP.

Sun has recently announced an initiative as part of the Java Community Process to bring DTV APIs to the Connected Limited Device Configuration (CLDC) of J2ME. This is known as the "On ramp to OCAP" (see Sun JSR 242 for details). We will discuss this more elsewhere in the chapter, but for now it is enough to say that this initiative is not aimed at MHP or OCAP.

So far, we have glossed over the details of the type of VM that is most suitable, apart from saying that the full JVM is needed for both middleware stacks. Regarding the question of clean-room VM implementations, there is no right answer: both clean-room VMs and Sun's own VM have their advantages, although some clean-room VMs have been optimized for limited platforms such as DTV receivers. This may involve improving the memory usage of the VM or of the applications that run on it, or it may involve using just-in-time compilation to improve performance. Many of these techniques can be useful for a VM used in a set-top box, but the choice of vendor depends more on choices made during the design of the hardware platform than any other factor.

Sun's JVM or a Clean-room Implementation?

Making the necessary arrangements for using the Java VM in MHP was not a completely smooth ride. Over a three-year period, Sun Microsystems, the DVB Steering Board, and DVB members negotiated the inclusion of Java into the MHP specification. This was a long and tawdry affair, which had companies such as Microsoft, OpenTV, HP, and Intel continually resisting the acceptance of Java as the sole VM technology, and thus Sun's role as the custodian of a fundamental piece of MHP technology. This anti-Sun, anti-Java lobby within the DVB fought long and hard to ensure that Sun would not have undue influence over implementations of MHP. The outcome of this was an understanding that the use of clean-room Java implementations was an acceptable way of implementing MHP.

As part of this process, a law firm was called in to further investigate and advise on whether the DVB should report this to the European Union's anti-trust watchdog to clarify that the inclusion of Java did not break EU competition law. After much investigation, the lawyers satisfied themselves that the complex arrangements with Sun were legal. The DVB Steering Board was also satisfied that Sun would not monopolize the market because the JVM specification was freely available, and furthermore that clean-room VM implementations already existed and were permissible in MHP implementations.

Following these discussions, the DVB published its licensing and conformance guidelines for MHP (document number A066, the so-called *"Blue Book,"* published in February of 2002). Unfortunately, the story does not end there. After a long battle over the interpretation of the rules described in the *Blue Book* and what is actually permissible in order to build an MHP implementation, many MHP companies are still coming to terms with the issues raised by the Sun licensing scheme. Many interpretations of the *Blue Book* have proven to be incorrect, and these may have important consequences for middleware developers.

At this point, we must offer a word of warning and clarify one of the areas open to misinterpretation. For middleware manufacturers who use Sun's source code, the *Blue Book* rules do not apply because of the nature of Sun's full licensing scheme. Use of Sun's s source code requires the acceptance of a click-through SCSL license that ties you into many additional license conditions beyond those required for MHP. One of the license restrictions that many MHP implementers still disagree with is the automatic supersetting of the JVM. Both the DVB Project (in conjunction with ETSI) and CableLabs have defined their own sets of tests for compliance testing of the JVM. Both of these are a subset of the Sun TCK (Technology Compatibility Kit, the Java test suite) and on their own they are not sufficient for MHP or OCAP implementers who are using the Sun JVM. This means that the JVM must be bigger than actually required for MHP (or OCAP) in order to pass the full Sun TCK, rather than the basic TCK required for MHP conformance. Among other things, there is a cost associated with using the Sun TCK as well as an annual maintenance contract and other terms and conditions.

Because of this, the *Blue Book* rules only apply to clean-room JVM implementations. Some companies have argued this is not what DVB agreed to. As we mentioned earlier, the agreement between Sun and DVB was the result of a long and convoluted set of negotiations in which interpretations were argued over in detail, and thus it is very difficult to track exactly what was and was not agreed to without trawling the minutes of each DVB Commercial Module and Steering Board meeting. Sun have generally played it straight, however, and have not swayed from their own understanding and interpretation of the subject. It has been difficult for DVB to actually reexamine this subject, because it is considered closed by DVB committees involved.

Fundamentally, the problem was that the DVB *Blue Book* did not describe the licenses that would be needed between a middleware manufacturer and a JVM supplier such as Sun Microsystems in order to complete a nonclean-room JVM. This makes it more difficult for MHP implementers to get full details of the licenses they need and to analyze their best course or action without entering into NDAs, agreements, and the Sun licensing process. Sun's SCSL license does not clearly detail the implications of the license terms, and many people feel that DVB should have studied this more carefully so that they could have fully understood the implications and had clear and precise explanations for their members. Even if DVB could not actually publish the associated costs due to nondisclosure agreements and other legal reasons, members would have been able to make a more informed decision whether to build a clean room implementation or use the Sun source code.

Both Sun's JVM and clean-room implementations have their advantages and disadvantages. Sun's s source code may allow manufacturers to go to market earlier, whereas clean-room implementations typically have smaller royalties and less onerous testing requirements for use within MHP. The right choices for you will depend on your plans, but it is important to take into account the different licensing requirements and testing needs of the different approaches.

The Impact of the Java Community Process

Since the introduction of the Java 2 platform (J2SE and J2ME), the user community has participated in shaping the Java platforms via the Java Community Process (JCP). This gives companies using Java a voice in newer versions of the standard, and enables users to add features and functionality that is truly what they need. As part of the JCP, groups of interested parties will produce Java Specification Requests (JSRs) that define new APIs or technologies that can be used with a Java VM.

DTV companies have been an important part of this process, especially when it has been applied to APIs that affect DTV receivers. A number of companies have been involved in the expert groups defining, for example, the J2ME Personal Profile and Personal Basis Profile (both products of JCP), as well as defining a number of lower-level technologies that may not be glamorous but are very important to consumer systems. These include such things as techniques for isolating applications from each other while allowing them to run within a single VM. The JCP has been influential in getting a number of technologies adopted that are extremely useful for DTV receivers, but which otherwise may not have seemed important. Table 3.1 outlines some of these.

Table 3.1. Java Specification Requests (JSRs) important to DTV developers.

JSR Number	Title
JSR 1	Real-time Java
JSR 36	J2ME Connected Device Configuration
JSR 46	J2ME Foundation Profile Specification
JSR 62	J2ME Personal Profile Specification
JSR 121	Application Isolation API Specification
JSR 129	J2ME Personal Basis Profile Specification
JSR 133	Java Memory Model and Thread Specification Revision
JSR 173	Streaming API for XML
JSR 177	Security and Trust Services API for J2ME
JSR 206	Java API for XML Processing (JAXP) 1.3
JSR 209	Advanced Graphics and User Interface Optional Package for the J2ME Platform
JSR 242	Digital Set Top Box Profile—"On Ramp to OCAP"

The big advantage offered by this process is simply one of getting the right people involved: Sun is not a CE company, and they cannot know all of the technologies useful or important to people building CE products. From the perspective of DTV developers, more involvement in the JCP can only help move the platform toward the functionality we need to improve products and applications. The more companies involved the more likely we are to arrive at solutions that work on our platforms and solve problems that affect our industry.

The last JSR in Table 3.1 is one that may be of interest to some developers. This is designed to provide a way for low-end STBs available in U.S. markets (such as the General Instrument/Motorola DCT 2000 receiver, for instance) to run JavaTV. These receivers are not capable of supporting an implementation of the Personal Basis Profile of CDC, and thus JSR 242 is aimed at the CLDC of J2ME to provide a basic set of features needed for simple ITV applications. Whereas MHP and OCAP receivers may support JSR 242, not every JSR 242-compliant receiver will support MHP of OCAP.

Portability

One of the obvious issues when choosing a Java VM is how portable it is to new hardware and software platforms, but the significance of this does not stop there. MHP has been implemented on a wide range of platforms, from Linux running on an x86 PC, to the STMicroelectronics 551x and NEC EMMA2 platforms, to DSPs such as the Philips Trimedia processor. Given the range of chip sets and operating systems in use in the DTV business, it makes sense for middleware developers to make their software stack as portable as possible.

We will not get into a discussion of portability layers here, because far too much has already been written on that topic in other places. All we will say here is that portability layers are generally a good idea, because all developers know that eventually they will get asked to port their software.

Instead, we will look at some elements that may not be immediately obvious. The platforms available to DTV systems have a wide range of hardware and software features, and these can complicate the porting process in ways people new to the DTV world may not realize. DTV systems typically rely on a number of hardware features not available in other platforms. Many of these features are related to MPEG demultiplexing and decoding, or to the graphics model. DTV platforms often do not have the general-purpose computing power required to perform all of these tasks in software, and thus the hardware capabilities of the platform play an important part.

MHP and OCAP are both demanding software stacks, simply because they are current-generation technology. It is possible to build a DTV software stack that needs fewer resources, but it will not have as many capabilities as MHP or OCAP. The reason for mentioning this is very simple: not all hardware platforms are created equal. Given the hardware needs of MHP and OCAP, the single most important thing developers can do to make their software portable is to use resources wisely. We must never forget that both MHP and OCAP define minimum platform capabilities that have to be available to applications. These do not define the computing power that is available, but instead define the number of section filters available, the number of timers and threads available to each application simultaneously, and the amount of memory available. Middleware stacks that use these resources wisely make themselves easier to port.

The balance between using resources wisely and being too conservative has to be carefully managed. Limiting the number of section filters used by the middleware can lead to reduced performance in parsing service information or broadcast data streams, and this could affect the applications more than simply having one less section filter available. Striking the right balance is something that can be done only on a platform-by-platform basis.

Performance Issues

One of the problems facing DTV middleware developers is that a middleware stack must be reliable and fast, while at the same time handling any behavior buggy or malicious applications can throw at it. The reliability part is something we will look at in other places, but there is a trade-off between reliability and speed that has to be considered when the architecture is developed.

This trade-off includes issues such as the mapping of applications (or groups of applications) to VMs. Middleware designers can choose to have all applications running in the same VM (which can pose some challenges in making sure that applications are truly independent from one another), or at the other extreme every application may have its own VM. In this case, middleware developers have to coordinate the behavior of the different VMs and make them communicate with each other where necessary (for instance, when sharing scarce resources or when caching data from service information or from broadcast file systems).

There is a middle ground, wherein applications that are part of the same service may execute in the same VM, but even this poses some challenges middleware architects need to answer. In each case, there is the same trade-off between reliability and performance, in that more VMs will almost certainly have an effect on performance. Another approach to this is the application isolation API defined in JSR 121, although this must be supported by the underlying Java VM.

Another example of the trade-off lies in choosing the correct number of threads to use when dispatching events to applications is an issue. One is too low, but how many should be assigned? Use too many, and it can affect the performance of the rest of the system. We will look at this issue again in later chapters, and maybe provide some insight into how you can make the best choice for your middleware implementation and hardware platform.

The trade-offs involved may not always be obvious to architects who are more familiar with the PC world. Should you use a VM that supports just-in-time compiling? This can increase the performance of an application, but it will not make it load faster and will probably use more memory. Instead, that extra memory may best be spent caching objects from the broadcast file system in order to speed up loading times.

As we mentioned earlier in this chapter, your middleware architecture may use some of the standardized OCAP and MHP APIs for communicating between middleware components.

In some cases this is a good thing, but in other cases the performance impact may simply be too high. The MHP section-filtering and SI APIs use asynchronous calls to retrieve information from a transport stream, and this makes perfect sense for applications. For middleware, however, this may result in a high number of threads being used for nothing except sending data from one component to another.

Given that you can trust your middleware components more than you can trust downloaded applications, this may be a prime target for optimization. Passing data through shared buffers without using so many Java events may cut the processing time for important parts of your code. This is especially true for section filtering, in which every MPEG-related API relies on the use of section filters either directly or indirectly. By using a private API to access sections that have been filtered, it may be possible to improve dramatically the performance and memory usage of your middleware.

Platform choice is also an important factor in how efficient your middleware will be. We have already mentioned the range of platforms that support MHP and OCAP, and in each of these cases there are different architectural challenges. It is important that developers, especially those coming from the PC world, do not underestimate those challenges.

Linux is an increasingly popular platform, but a number of PC-centric companies have tried to use Linux as a basis for an MHP system with no real understanding of the impact the MPEG-related elements of MHP have on the overall architecture, or the costs involved in developing those components. Architects should ask themselves, "Does my platform give me all of the major components I need (namely Java, MPEG decoding, and MPEG processing components); and if not, how easily can I get product-quality components that meet my needs?" Like every other platform, Linux has its strengths and weaknesses (a number of companies have built extremely successful Linux-based MHP and OCAP stacks), but you need to understand those strengths and weaknesses before you can design a middleware stack that suits it.

Some components you get with your platform, some you can buy, and some you have to build. Make sure you know the needs and the risks of the components you have to build, as well as how you can put these components together most efficiently.

4 Applications and Application Management

MHP and OCAP receivers support several types of interactive applications. In this chapter we will look at these types of applications and see how they differ from one another. We will also look at a simple example application and at what developers should and should not do to make their applications as reliable as possible.

OCAP and MHP can support applications written in either Java or HTML, although support for HTML applications is optional in both standards and thus Java is currently the most common choice by far. To emphasize that they are not just normal Java applications, Java applications for MHP receivers are called DVB-J applications, whereas Java applications written for OCAP receivers are called OCAP-J applications. Similarly, HTML applications are known as DVB-HTML or OCAP-HTML applications. This is not an either/or naming system, though; applications can be both DVB-J and OCAP-J applications at the same time, and the same obviously applies to HTML applications. The same application may have different names in different circumstances to indicate that it is compatible with a particular middleware stack. In the ACAP environment, for instance, they are called ACAP-J and ACAP-X applications (X in this case stands for XHTML).

Although many applications in an MHP or OCAP system will be downloaded from the MPEG transport streams being broadcast, these may not be the only types of applications available. MHP 1.1 adds support for loading applications over an IP connection, using HTTP or another protocol. OCAP also supports this for some applications, and defines several other types of applications that can be used. Although they all have a great deal in common, and look almost identical from an application developer's perspective, there are some important differences among these applications.

Service-bound applications are downloaded applications tied to a specific service or set of services. These are downloaded every time the user wants to run them, and can either be started by the user or be started automatically by the receiver if the network operator chooses. Typically, service-bound applications are associated with a specific TV channel (such as a news or stock ticker application for a news channel) or event (for instance, an application that lets viewers play along with a quiz show or that gives biography information about the cast of the current drama show).

Unbound applications are downloaded applications that are not associated with a service. Like service-bound applications, these can either be started automatically by the middleware or started by the user. Unbound applications are typically used by the network operator to offer value-adding services such as an EPG or a pay-per-view system. They could also be used for home shopping, online banking, games, and a range of other applications. Unlike service-bound applications, unbound applications are not a part of MHP and thus cannot be used in MHP systems. A similar effect can be achieved, however, by signaling an application on every service on the network. Only unbound applications can be loaded over a two-way IP connection such as a cable modem.

Downloading an application every time the user wants to run it can be fairly time consuming, especially for complex applications. To help avoid this, both OCAP and MHP support stored applications. These are downloaded applications that are stored locally in the receiver, and which are then loaded from local storage when they are needed. This allows much faster loading, and is especially popular for commonly used applications such as an EPG. OCAP and MHP handle stored applications in different ways. In OCAP, the network operator can tell the receiver that unbound applications can be stored (assuming enough storage space is available on the receiver), whereas in MHP systems the user has a little more control over the process and can reject a request to store applications. Applications are not stored forever, however. If there is no space to store an application, any applications with a lower priority may be removed to make room for the application. Similarly, all stored applications will be removed when a receiver is moved to a different network.

MHP 1.0.x does not support stored applications, and thus such applications are not currently widely used in MHP systems. Stored applications in MHP can be either broadcast-related applications bound to a specific service and controlled by application signaling information in the broadcast stream or standalone applications that execute independently of the current broadcast content. OCAP does not have this distinction, and all stored applications must be unbound applications.

OCAP's monitor application is a class of application all by itself. Although it uses the same APIs as other applications, it has some special capabilities that make it rather distinct. Given the level of control a monitor application can exert over an OCAP receiver, it is much more than just another downloaded application. The monitor application is stored on the receiver the first time the receiver is connected to a given network, and is automatically started every time the receiver starts until it is upgraded by the network operator or until the receiver is connected to a different network. In this case, a new monitor application will probably be downloaded.

Built-in applications are applications provided by the receiver manufacturer. In OCAP, these are applications stored by the receiver manufacturer in the same way as other stored applications, except that they are stored when the receiver is first manufactured. MHP does not distinguish these from other applications, and thus the navigator will typically add them automatically to the list of available applications. The user can then choose to start them from that list if he or she wishes to.

System applications are a subset of built-in applications. These provide specific functionality that is not necessarily tied to the experience of watching TV but is needed for other reasons. The Emergency Alert System may be a system application, for instance. Unlike other built-in applications, these applications may not be shown in the list of available applications and it may not be possible for users to invoke them directly.

OCAP also supports native applications. These are applications written in a language other than Java or HTML, and which are built-in by the receiver manufacturer. Native applications still need to be controlled by the OCAP application manager, and thus must have a Java wrapper that provides the basic methods required to control the life cycle of the application. This means that they must also be written to follow the OCAP application life-cycle model, in that they are in effect OCAP applications that happen to have been written in native code. Broadly speaking, these different types of applications can be grouped as the following four main types.

- Service-bound applications
- Unbound applications
- Stored applications
- Native applications

Of these, only OCAP supports unbound applications and native applications, whereas stored applications are only supported in MHP 1.1. As you would expect, having so many different types of applications affects the architecture of the middleware. Having the ability to store applications means that an OCAP system needs more persistent storage than a typical MHP receiver, as well as the ability to store those parts of the application signaling that refer to those stored applications so that it knows how to start them and what they are called. Native applications do not affect the big picture very much, in that they are forbidden in MHP (or at least they are not considered to be MHP applications), and we have already seen that in OCAP they must have a Java wrapper that provides the same interface as interoperable applications.

An Introduction to Xlets

Now that we have seen the different types of applications we can find in an MHP or OCAP receiver, let's look at how we actually build one of our own. If we are using Java to develop applications for DTV systems, we run into a few problems if we try to use the normal life-cycle model for Java applications.

Xlet Basics

The normal Java application model makes a number of assumptions about the environment that are not compatible with a consumer product. In particular, it assumes that only one application is running in the Java VM and that when the application stops running so does the VM. On a PC, this is not a problem, but it causes headaches in a system wherein you cannot make these assumptions. The normal life-cycle model also assumes that an application will be loaded, start running immediately, and then terminate, which is another assumption that does not work very well in a consumer environment.

The life cycle of Java applets from the Web is far more appropriate: the web browser loads a Java applet into a JVM, initializes it, and executes it. If a page contains two applets, they can both run in the same VM without interfering with each other. When an applet terminates, it is removed from the VM without affecting anything else running in the same VM.

Because the applet model still has a lot of functionality tied to the Web, or is not appropriate for all cases, it was generalized into something more suitable for consumer systems. The result is the Xlet. This forms the basis for all systems based around JavaTV, including MHP and OCAP. Like applets, the `Xlet` interface allows an external source (the application manager in the case of an MHP or MHP receiver) to control the life cycle of an application, and provides the application with a way of communicating with its environment. The `Xlet` interface, shown below, is found in the `javax.tv.xlet` package along with some related classes.

```
public interface Xlet {

   public void initXlet (XletContext ctx)
      throws XletstateChangeException;

   public void startXlet ()
      throws XletstateChangeException;

   public void pauseXlet ();
   public void destroyXlet (boolean unconditional)
      throws XletstateChangeException;
}
```

Although there are some similarities between an Xlet and an applet, there are also a number of differences. The biggest of these is that the execution of an Xlet can be paused and resumed. The reason for this is very simple: in an environment such as a DTV receiver several applications may be running at the same time, and yet hardware restrictions mean that only one of those applications may be visible to the user. Applications that are not actually being used may need to be paused in order to keep resources free for the application currently being used.

An Xlet is also much simpler than an applet. Given the importance of reliability and robustness in DTV systems, an Xlet has many more security restrictions imposed on it than an applet does. Many of the functions supported by the `Applet` class are also supported by

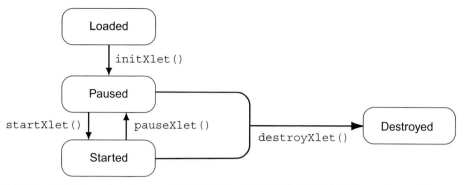

Figure 4.1. The Xlet life cycle. *Source:* ETSI TS 101 812:2003 (MHP 1.0.3 specification).

Xlets, but they are provided through other APIs in which better security checking can take place.

An Xlet has five main states: *Not Loaded*, *Loaded*, *Paused*, *Started*, and *Destroyed*. An application may also be in the *Invalid* state, when it cannot be run on the current service but the Xlet object has not yet been garbage collected. If we examine the life cycle of an Xlet, shown in Figure 4.1, we can see where these states fit into the overall picture.

When an Xlet is run, the steps outlined in Table 4.1 are taken.

When an application moves between states, or when a state transition fails, the middleware will send an AppStateChangeEvent to any listeners that have been registered for those events. This tells the listener which state the application is currently in, and the state it was in previously. We will look at the AppState ChangeEvent in a little more detail in Chapter 17.

It is important to remember that an Xlet is not a standard Java application. There may be more than one Xlet running at any one time, which means that Xlets should not perform any actions that will affect the global state of the Java VM, and indeed most of these actions are explicitly disallowed by the OCAP and MHP specifications.

For instance, an Xlet should never, *ever* call the System.exit() method. We have seen some early applications that do this, and it is highly frustrating when an application simply shuts down the entire Java VM when it terminates. Other dos and don'ts are discussed in the sections that follow.

Xlet Contexts

Like applets, JavaTV Xlets run in a context. For applets, this is represented by the java.appplet.AppletContext class, whereas Xlets use the javax.tv.xlet.Xlet-Context class. In both cases, the context serves to isolate the application from the rest of the VM while still enabling it to interact with the environment it is running in. This interaction

Table 4.1. Processing steps in the life cycle of an Xlet.

Step	Description
Loading	When the receiver first receives information about an application, that application is in the *Not Loaded* state. At some point after this, usually when the Xlet is started, the application manager in the receiver may load the Xlet's main class file (as signaled by the broadcaster) and create an instance of the Xlet by calling the default constructor. Once this has happened, the Xlet is in the *Loaded* state.
Initialization	When the user chooses to start the Xlet (or the network operator tells the receiver the Xlet should start automatically), the application manager in the receiver calls the `initXlet()` method, passing in a newly created `XletContext` object for the Xlet.
	The Xlet may use this `XletContext` to initialize itself, and to preload any large assets (such as images) that may require some time to load from the object carousel. When the initialization is complete, the Xlet is in the *Paused* state and is ready to start immediately.
Execution	Once the `initXlet()` method returns, the application manager can call the `startXlet()` method. This will move the Xlet from the *Paused* state into the *Started* state, and the Xlet will be able to interact with the user.
Pausing and resuming	During the execution of the Xlet, the application manager may call the `pauseXlet()` method. This will cause the application to move from the *Started* state back to the *Paused* state. The application will later be moved back to the *Started* state by calling the `startXlet()` method again. This may happen several times during the execution of the Xlet.
Termination	When the user chooses to kill the Xlet, or when the network operator tells the receiver the Xlet should be terminated, the application manager will call the `destroyXlet()` method. This causes the Xlet to move into the *Destroyed* state and free all of its resources. After this point, this instance of the Xlet cannot be started again.
	Versions of MHP prior to MHP 1.0.3 and 1.1.1 have a problem whereby applications that have been destroyed remain in the *Destroyed* state and cannot be restarted. MHP 1.0.3 and 1.1.1 change this, so that the application is only temporarily in the *Destroyed* state before moving back to the *Not Loaded* state. In practice, many MHP 1.0.2 implementations have already fixed this problem using the same approach.

can involve reading information from the environment by means of system properties, or it can involve notifying the middleware about changes in the state of the Xlet.

This isolation is important in a DTV system because the environment each Xlet operates in may be different. Each Xlet may have different values for the properties that make up its environment, and applications should not be able to find out about the environment of any other applications that may be running. The `XletContext` interface looks as follows.

```
public interface XletContext {

    public static final String   ARGS = "javax.tv.xlet.args";
```

```
    public void notifyDestroyed();
    public void notifyPaused();
    public void resumeRequest();

    public Object getXletProperty(String key);
}
```

The `notifyDestroyed()` and `notifyPaused()` methods allow an Xlet to notify the receiver that it is about to terminate or pause itself. Through these, the receiver can know the state of every application and can take appropriate action. These methods should be called immediately before the Xlet enters the *Paused* or *Destroyed* state, because the receiver may need to carry out some housekeeping operations when it is notified about the Xlet's change of state. If the Xlet is not prepared for this (for example, it has not finished disposing of resources it is using, or is still writing some data to persistent storage), it may get a nasty surprise that can cause problems for the rest of the middleware.

An application can request that it be moved from the *Paused* state back to the *Started* state using the `resumeRequest()` method. This allows an application to pause itself for a while, and then resume when a certain event is received or when a certain time is reached. For instance, an EPG application may let the user set reminders for a specific show, and then pause itself. At the start of a show for which the user has requested a reminder, the EPG can request to be resumed in order to tell the user the show is starting.

As with many elements of the application management process in an MHP or OCAP receiver, the application can only request that this happen; there is no guarantee the request will be honored. If an application with a higher priority is active, the application may not get to resume.

The `getXletProperty()` method allows the Xlet to access information about its environment. These properties are defined by the network operator in the SI that tells the receiver about the application. Table 4.2 outlines which properties are defined by the various standards.

Not all of the information about the Xlet's environment is defined by the broadcaster. Some of it is specific to the middleware in the receiver, and Xlets can access this using the

Table 4.2. Xlet properties.

JavaTV	OCAP	MHP
`javax.tv.xlet.args` (specified by the field `javax.tv.xlet.XletContext.ARGS`)		
	`ocap.profile`	
	`ocap.version`	
		`dvb.app.id`
		`dvb.org.id`
	`dvb.caller.parameters`	

Table 4.3. System properties available to OCAP and MHP applications.

Java/JavaTV	OCAP	MHP
	`path.separator`	
	`dvb.persistent.root`	
	`dvb.returnchannel.timeout`	
	`mhp.profile.enhanced_broadcast`	
	`mhp.profile.interactive_broadcast`	
		`mhp.profile.internet_access`
		`mhp.eb.version.major`
		`mhp.eb.version.minor`
		`mhp.eb.version.micro`
		`mhp.ib.version.major`
		`mhp.ib.version.minor`
		`mhp.ib.version.micro`
		`mhp.ia.version.major`
		`mhp.ia.version.minor`
		`mhp.ia.version.micro`
		`mhp.option.ip.multicast`
		`mhp.option.dsmcc.uu`
		`mhp.option.dvb.html`
	`havi.specification.vendor`	
	`havi.specification.name`	
	`havi.specification.version`	
	`havi.implementation.vendor`	
	`havi.implementation.version`	
	`havi.implementation.name`	

`System.getProperty()` method. This is a standard Java method for finding out about the system settings for a Java platform. The nature of a DTV receiver means that only a few of the system properties found in a desktop Java implementation are available, but MHP and OCAP define a few properties of their own. Table 4.3 outlines the system properties that can be used by a JavaTV, MHP, or OCAP application.

Writing Your First Xlet

Now that we have seen what an Xlet looks like, let's actually write one. In the grand tradition of first programs, we will start with a "Hello world" application. This example should work on all JavaTV, MHP, or OCAP implementations. The following is a very simple application that includes only the most basic elements required to produce a running Xlet.

```
// Import the standard JavaTV Xlet classes.
import javax.tv.xlet.*;

// The main class of every Xlet must implement this
// interface—if it doesn't do this, the middleware
```

```
// can't run it.
public class MyFirstXlet implements javax.tv.xlet.Xlet
{
  // Every Xlet has an Xlet context, just like the
  // Applet context that applets in a web page
  // are given.
  private javax.tv.xlet.XletContext context;

  // A private field to hold the current state. This
  // is needed because the startXlet() method is called
  // both to start the Xlet for the first time and also
  // to make the Xlet resume from the paused state.
  // This field lets us keep track of whether we're
  // starting for the first time.
  private boolean hasBeenStarted;

  /**
   * Every Xlet should have a default constructor that
   * takes no arguments. No other constructor will
   * get called.
   */
  public MyFirstXlet()
  {

    // The constructor should contain nothing. Any
    // initialization should be done in the
    // initXlet() method, or in the startXlet() method if
    // it's time- or resource-intensive. That way, the
    // middleware can control when the initialization
    // happens in a much more predictable way
  }

  /**
   *   Initialize the Xlet. The context for this Xlet
   *   will be passed in to this method, and a reference
   *   to it should be stored in case it's needed later.
   *   This is the place where any initialization should
   *   be done, unless it takes a lot of time or resources.
   *   If something goes wrong, then an
   *   XletStateChangeException should be thrown to let
   *   the runtime system know that the Xlet can't be
   *   initialized.
   */
  public void initXlet(javax.tv.xlet.XletContext context) throws
    javax.tv.xlet.XletStateChangeException
  {
```

```
    // store a reference to the Xlet context that the
    // xletXlet is executing in
    this.context = context;

    // The Xlet has not yet been started for the first
    // time, so set this variable field to false.
    hasBeenStarted = false;

    // Since this is a simple Xlet, we'll just print a
    // message to the debug output (assuming we have one)
    System.out.println(
        "The initXlet() method has been called." +
        "Our Xlet context is " + context);
}

/**
  * Start the Xlet. At this point, the Xlet can display
  * itself on the screen and start interacting with the
  * user, or do any resource-intensive tasks. These
  * kinds of functions should be kept in startXlet(),
  * and should not be done in initXlet().
  *
  * As with initXlet(), if there is any problem this
  * method should throw an XletStateChangeException to
  * tell the runtime system that it can't start.
  *
  * One of the common pitfalls is that the startXlet()
  * method must return to its caller. This means that
  * the main functions of the Xlet should be done in
  * another thread. The startXlet()method should really
  * just create that thread and start it, then return.
  */
public void startXlet()
    throws javax.tv.xlet.XletStateChangeException
{
    // Again, we print a message on the debug output to
    // tell the user that something is happening. In
    // this case, what we print depends on whether the
    // Xlet is starting for the first time, or whether
    // it's been paused and is resuming

    // have we been started?
    if(hasBeenStarted) {
        System.out.println(
            "The startXlet() method has been called to " +
            "resume the Xlet after it's been paused." +
```

```
        "Hello again, world!");
    }
    else {
      System.out.println(
        "The startXlet() method has been called to " +
        "start the Xlet for the first time. Hello, " +
        "world!");

      // set the variable that tells us we have actually
      // been started
      hasBeenStarted = true;
    }
}

/**
 * Pause the Xlet. Unfortunately, it's not clear to
 * anyone (including the folks who wrote the JavaTV,
 * MHP, or OCAP specifications) what this means.
 * Generally, it means that the Xlet should free any
 * scarce resources that it's using, stop any
 * unnecessary threads, and remove itself from the
 * screen.
 *
 * Unlike the other methods, pauseXlet() can't throw an
 * exception to indicate a problem with changing state.
 * When the Xlet is told to pause itself, it must do
 * that.
 */
public void pauseXlet()
{
  // Since we have nothing to pause, we will tell the
  // user that we are pausing by printing a message on
  // the debug output.
  System.out.println(
    "The pauseXlet() method has been called.";
}

/**
 * Stop the Xlet. The boolean parameter tells the
 * method whether the Xlet has to obey this request.
 * If the value of the parameter is 'true,' the Xlet
 * must terminate and the middleware will assume that
 * when the method returns, the Xlet has terminated.
 * If the value of the parameter is 'false,' the Xlet
 * can request that it not be killed, by throwing an
```

```
 * XletStateChangeException.
 *
 * If the middleware still wants to kill the Xlet, it
 * should call destroyXlet() again with the parameter
 * set to true.
 */
public void destroyXlet(boolean unconditional)
   throws javax.tv.xlet.XletStateChangeException
{
  if(unconditional) {
    // We have been ordered to terminate, so we obey
    // the order politely and release any scarce
    // resources that we are holding.
    System.out.println(
      "The destroyXlet() method has been called " +
      "telling the Xlet to stop unconditionally. " +
      "Goodbye, cruel world!");
  }
  else {
    // We have had a polite request to die, so we can
    // refuse this request if we want.
    System.out.println(
      "The destroyXlet() method has been called " +
      "requesting that the application stops, but " +
      "giving it the choice. Therefore, I'll " +
      "decide not to stop.");

    // Throwing an XletStateChangeException tells the
    // middleware that the application would like to
    // keep running if it is allowed to.
    throw new XletStateChangeException(
      "Please don't kill me!");
  }
}
```

As you can see from this code, it simply prints a different message to the debug output when each method is called. This is about the simplest Xlet you will find, but it does enough to let you see what is going on.

You will notice that our Xlet has an empty constructor. This is deliberate. When the middleware starts an application, it first needs to create an instance of the main class. Doing this will invoke the default constructor (if it exists), and any code in the constructor will be executed. However, the Xlet has another method that should be used for most initialization tasks (the initXlet() method) and thus the constructor should not be used for initialization.

Doing this work in the `initXlet()` method gives the middleware more control over when this happens, and means that it only gets done when the Xlet is actually initialized. In short, do not provide a default constructor for your Xlet. Do all of the initialization work in the `initXlet()` method, or in the `startXlet()` method if the initialization uses a lot of resources.

Dos and Don'ts for Application Developers

As we have already seen, applications do not have the virtual machine to themselves. There are a number of things an application can do to be a "good citizen," and there are many things an application should not do. Some of these are formalized in the MHP and OCAP specifications, but the following are those that may be less obvious to developers who are coming from the PC or web development world.

- Applications should not do anything in their constructor. They should definitely not claim any scarce resources.
- Any initialization should be done in the `initXlet()` method. At this point the application should initialize any data structures it needs (unless they are extremely large), but it should not claim any scarce resources it does not need to run. `initXlet()` is a good place to preload images, but it is not a good place to reserve the modem, for instance.
- Applications should wait until `startXlet()` has been called before they create any very large data structures or reserve any scarce resources. An Xlet may be initialized without ever being run, and thus it is best to wait until it is actually running before claiming a large amount of system resources.
- The `startXlet()` method should create a new thread to carry out all of its work. `startXlet()` will be called by a thread that is part of the middleware implementation, and thus `startXlet()` must return in a reasonable time. By creating a separate thread and using that to do all of the things it needs to do, the Xlet can maintain a high degree of separation from the middleware stack.
- All calls to methods that control the Xlet's life cycle should return in a timely manner. We have already seen that `startXlet()` should create a new thread for any tasks it wants to carry out. No other life-cycle methods should start a new thread, especially not `initXlet()` or `destroyXlet()`.
- Resource management is especially important in a DTV environment, as we will see later. An application should cooperate with the middleware and with other Xlets when it is using scarce resources, and it should not keep a resource longer than it needs to.
- When a resource can be shared among several applications (e.g., display devices), an application should minimize the changes it makes to the configuration of that resource. If it does not care about some elements of the configuration, it should not change them to avoid causing problems for other applications.
- Applications should release as many resources as possible when the `pauseXlet()` method is called. All scarce resources should be released, and the receiver should ideally free as much memory as it can. When an application is paused, should hide any user interface elements it is displaying, and should not draw anything on the screen until it is

resumed. The application should free as much memory and resources as possible for the applications that are not paused. Paused applications will be given a lower priority when they ask for resources. Applications that do try to draw to the screen when they are paused may be regarded as hostile by the middleware, and the middleware may choose to kill the application in this case.

- Calling `System.exit()` is never a good idea in a DTV receiver, and both MHP and OCAP specifications say that applications should not do this under any circumstances. A number of other methods from the `java.lang.System` and `java.lang.Runtime` classes are also not available, and application developers should not use these.
- The `destroyXlet()` method should kill all application threads and cancel any existing asynchronous requests that are currently outstanding in the SI and section-filtering APIs.
- The `destroyXlet()`, and ideally the `pauseXlet()`, method should also free any graphics contexts the application has created. The middleware will maintain references to these unless they are disposed of properly with a call to `java.awt.Graphics.dispose()`. This can use a lot of memory, and graphics contexts may not be freed properly unless the application specifically destroys them.
- The application should remember that it may be paused or destroyed at any time, and it should make sure it can always clean up after itself.
- Applications should never catch the `ThreadDeath` exception. This is generated when a thread is terminated, and may be used by malicious applications to prevent them from being killed. Because everyone reading this book is interested in developing well-behaved applications that make life as easy as possible for all members of the MHP and OCAP communities, your application should never do this. By doing all of the processing in the `destroyXlet()` method, there is no need to catch `ThreadDeath` exceptions.
- Finalizers may not be run when an application exits. Application developers should not rely on finalizers to free resources.
- Applications may not always be able to load classes from the broadcast file system when their `destroyXlet()` method is called. For this reason, applications should not rely on being able to load extra application classes when they are being destroyed.
- Exceptions are thrown for a reason. An application should catch any exceptions that can be generated by API methods it calls, and it should be able to recover in those cases where an exception is thrown. This is extremely important for application reliability.

Application Signaling

For a receiver to run an application, the network operator needs to tell the receiver about that application: where to find its files, what the application is called, and whether the receiver should automatically start the application. For service-bound applications, MHP and OCAP both use an extra SI table called an application information table (or AIT) to do this. Each service that has an application associated with it must contain an AIT, and each AIT contains a description of every application that can run while that service is being shown. For every application, this description carries some or all of the following information.

- The name of the application
- The version of the application
- The application's priority
- The ID of the application and the organization it is associated with
- The status of the application (auto-start, startable by the user, to be killed, or another state)
- The type of application (Java or HTML)
- The location of the stream containing the application's classes and data files
- The base directory of the application within the broadcast file system
- The name of the application's main class (or HTML file)
- The location of an icon for the application

A receiver can only run the applications that are described in the AIT associated with the service it is currently showing. This applies to all applications, although the situation for applications that are not service bound is a little more complex. We will examine this in more detail later in the chapter.

Each application has an ID number that must be unique when the application is being broadcast, and the organization that produced that application also has a unique ID. This may be a broadcaster, a network operator, or a production company. For instance, CNN may have an organization ID that is used for all applications associated with CNN shows, such a news ticker. A network operator such as BSkyB may also have an organization ID, used for applications such as their EPG. A receiver manufacturer will also have an organization ID, used for built-in applications.

In extreme cases, satellite operators such as Astra may also have an organization ID, to provide a guide to all free-to-air channels that are broadcast on their satellites. This allows a receiver to identify the different groups that may be responsible for applications delivered to it. The combination of application ID and organization ID is unique to every application currently being broadcast, and this allows a receiver to distinguish one application from another even if they have the same name and base directory. It is possible to reuse application IDs when an application is no longer being used, but it is not recommended that you do this unless you have to.

The priority of an application lets the receiver make decisions about allocating resources and the order in which application requests should be processed. Applications that have a low priority will be the first to be removed from persistent storage when space is short, the last to get access to scarce resources, and may not be run at all if the receiver cannot start all of the signaled applications.

Each application also has a version number so that the receiver can make sure it loads the correct version of the application. Receivers that have cached some of the application's files should check the version before they run the application, in case those files have become outdated. For stored applications, this is even more important and the receiver should check the version of any stored applications and store newer versions of any applications that are available.

Table 4.4. Application status values used in the AIT.

Code	Used in MHP	Used in OCAP	Meaning
AUTOSTART (0x01)	Yes	Yes	The application will start automatically when the service is selected. If killed by the user, it will not be restarted automatically, but the user may start it again.
PRESENT (0x02)	Yes	Yes	Startable by the user, but will not start automatically. If killed by the user, the user may choose to restart it.
DESTROY (0x03)	Yes	Yes	When the control code changes from AUTOSTART or PRESENT to DESTROY, a Java application will be conditionally terminated (i.e., destroyXlet() will be called with the value *false*). In MHP 1.1 or OCAP 2.0, an HTML application will move into the *Destroyed* state.
KILL (0x04)	Yes	Yes	When the control code changes from AUTOSTART or PRESENT to KILL, a Java application will be unconditionally terminated (i.e., destroyXlet() will be called with the value *true*). In MHP 1.1 or OCAP 2.0, when the control code changes from AUTOSTART, PRESENT, or DESTROYED to KILL, an HTML application will move to the *Killed* state.
PREFETCH (0x05)	Yes (DVB-HTML only)	Yes (OCAP-HTML only)	The HTML application entry point is loaded and the HTML engine is prepared. When all elements are initialized, the application waits for a trigger before moving to the *Active* state.
REMOTE (0x06)	Yes	Yes	The application is not available on this service, and will only be available following service selection.

One of the most important values signaled in the AIT is the status of the application. In MHP and OCAP systems, this can take one of the values outlined in Table 4.4.

These allow the network operator to control when an application starts and stops, and to tell the receiver whether it should start automatically. To see why this level of control is useful, examine the possible applications outlined in Table 4.5.

As you can see, by not tying the start or end of the application tightly to a particular event we gain a lot of flexibility in how applications behave. In some cases, it is useful to let the user finish interacting with an application before we kill it, although there are also problems with this.

Interactive ads are one of the more complicated cases, because having an application that runs for a very short time may be frustrating to the user. It may not even have loaded before it has to be killed. At the same time, the application cannot run for too long, because then

Table 4.5. Example start and stop times for various types of application.

Application	Starts	Ends
News ticker	Does not matter	Does not matter
Sports statistics application	When the sports show starts	When the show ends
Game associated with a quiz show	When the show starts	Up to 5 minutes after the show ends (long enough to let the viewer complete their game)
Stand-alone game	Does not matter	Does not matter
Interactive ad	When the ad starts	Up to 5 minutes after the ad ends (so that the user has a chance to use the application a little bit more)

other advertisers with interactive ads may miss out. Balancing this is not easy, and the network operator needs to think about how long each application will be available. Many networks already use interactive ads and they can be very effective. Running interactive ads at the end of an advertising slot gives the user more time to use the application, if there is no interference from applications associated with the show following the ad break.

This is not just an issue for interactive ads. For games or for e-commerce applications associated with a specific show, network operators must think about what happens at the end of a show. If the user is in the middle of a game, or in the middle of a transaction, simply killing the application at the end of the show will frustrate the user and give him or her less reason to use that interactive application in the future. On the other hand, this has to be balanced with the needs of any show that follows because you do not want the user to be concentrating too hard on the application and missing whatever is currently showing.

Setting the priority of an application appropriately, along with setting sensible limits on when the application can run, becomes an important issue when scheduling a number of interactive shows close together. Having many interactive shows is only good for users if they get enough time to use the applications they are interested in.

In material to follow we will see some more of the values contained in each AIT entry. First, though, we will take a closer look at the format of the AIT itself, which is outlined in Table 4.6.

The AIT is broadcast with a table ID of 0x74. Each AIT contains two top-level loops. The first of these is the common loop, which contains a set of descriptors that apply to every application signaled in that particular AIT or that apply to the service as a whole. The other top-level loop is the application loop, which describes the applications signaled by that AIT instance. Each iteration of the application loop describes one application (i.e., it represents one row in the table of signaled applications). Among this information is the application control code, which will take one of the values shown in Table 4.4.

Table 4.6. Format of an MPEG-2 section containing the AIT.

Syntax	No. of Bits	Identifier
application_information_section() {		
table_id	8	uimsbf
section_syntax_indicator	1	bslbf
reserved_future_use	1	bslbf
reserved	2	bslbf
section_length	12	uimsbf
test_application_flag	1	bslbf
application_type	15	uimsbf
reserved	2	bslbf
version_number	5	uimsbf
current_next_indicator	1	bslbf
section_number	8	uimsbf
last_section_number	8	uimsbf
reserved_future_use	4	bslbf
common_descriptors_length	12	uimsbf
for(i=0; i<N; i++) {		
descriptor()		
}		
reserved_future_use	4	bslbf
application_loop_length	12	uimsbf
for(i=0; i<N; i++) {		
application_identifier()		
application_control_code	8	uimsbf
reserved_future_use	4	bslbf
application_descriptors_loop_length	12	uimsbf
for(j=0; j<N; j++) {		
descriptor()		
}		
}		
CRC_32	32	rpchof
}		

Source: ETSI TS 101 812:2003 (MHP 1.0.3 specification).

The most useful of the descriptors in the common loop is the external application authorization descriptor. This lists applications not signaled as part of this service, but which may continue to run if they are already running when this service is selected. This gives slightly different semantics from anything we could achieve using control codes, because it in effect says "allow this application to keep running if it is already doing so, but do not let the user start a new copy of it."

Another useful descriptor that can be used in the common descriptor loop is the transport protocol descriptor. This tells the receiver which protocol should be used to load an application, and where it can be found. It tells the receiver if the application is available as part of a broadcast DSM-CC object carousel, and the service (and stream within that service) that contains the object carousel. If the application is available via a multicast IP stream carried in an MPEG stream, it tells the receiver the service and stream containing the IP data, and the URL of the application within that data. This descriptor can be contained in either the common loop or in the application loop (see material following), or in both.

Two other descriptors can also be included in this descriptor loop. The IP signaling descriptor tells the receiver which multicast IP streams contain application data used by the applications signaled in that AIT. By telling the receiver the platform IDs of any streams that may be used, this descriptor lets the receiver determine which IP notification tables should be loaded from the SI in order to access those streams. Only MHP 1.0.3 and later will use this descriptor, because the handling of multicast IP data was changed slightly between MHP 1.0.2/OCAP and MHP 1.0.3.

MHP 1.0.2 and OCAP receivers will use IP routing descriptors instead. These tell the receiver which multicast addresses and port numbers will be used by applications signaled in the AIT. These are described as address masks and port numbers, and thus more than one IP-routing descriptor may be needed to define all of the addresses used by applications. Two versions of the IP-routing descriptor may be used: one version describes IPv4 addresses, and the other supports IPv6 addresses.

The final descriptor we might encounter in this descriptor loop is the DII location descriptor. DSM-CC `DownloadInfoIndication` (DII) messages tell the receiver which DSM-CC modules make up the carousel. This descriptor tells the receiver which DII messages it should pay attention to in addition to the default DII messages. We will look at DII messages in Chapter 12.

Each iteration of the application loop has its own nested descriptor loop, which contains descriptors specific to that application. Table 4.7 outlines the descriptors that can be placed in the application descriptor loop.

As you can see, AIT descriptors fall into two basic groups. The first group is application related and gives the receiver any details it needs about a given application. Most of these are mandatory, with the exception of the application icons descriptor. In an OCAP middleware stack, any instances of `DVB` in the descriptor names previously outlined should be replaced with `OCAP`. The content of the descriptors stays the same; only the names are different between the two standards.

The second group of descriptors contains additional transport information that is specific to a given application. This can extend any transport information given in the common descriptor loop, although it is also possible to put all transport information in the application loop and not include any in the common loop.

All descriptors in the second group are optional, although an AIT must contain at least one transport protocol descriptor for every application in the AIT. If one is not present in the common descriptor loop, every iteration of the application loop must include a transport protocol descriptor.

Table 4.7. Descriptors found in the application descriptor loop of the AIT.

Descriptor	Application Type	Purpose
Application name descriptor	Both	Gives the name of the application.
Application icons descriptor	Both	Gives a reference to an icon that may be used to represent the application.
DVB-J application descriptor	Java	Gives basic information about the application priority and the receiver profile and version needed to run the application.
DVB-J application location descriptor	Java	Gives the base directory, initial class name, and class path extension for the application.
DVB-HTML application descriptor	HTML	Gives a set of parameters that should be passed to the application upon start-up.
DVB-HTML application location descriptor	HTML	Gives the initial page for the application.
DVB-HTML application boundary descriptor	HTML	Describes the set of HTML pages that should be considered as part of the application.
Transport protocol descriptor	Both	Describes the protocol that should be used to load the application (object carousel, IP in multiprotocol encapsulation, or IP over the return channel), and gives some information about where to find the data.
IP signaling descriptor	Both (MHP 1.0.3 only)	Used to identify the organization that is providing IP multicast streams used by the application. This lets the receiver determine which INT tables to load. (Valid for applications transported via IP using DSM-CC multiprotocol encapsulation only.)
Prefetch descriptor	Both	Provides hints to the receiver about which DSM-CC modules should be cached in order to improve application start-up times. (Valid for applications transported on DSM-CC object carousels only.)
DII location descriptor	Both	Used with the prefetch descriptor to tell the receiver which DSM-CC modules should be cached. (Valid for applications transported on DSM-CC object carousels only.)

When the AIT is updated and applications are added or removed, the middleware will check the entries that have changed and carry out the behavior signaled for them. New applications signaled as AUTOSTART will be automatically started, as will applications previously signaled with another control code (although if these are already running a new copy will not be started). If an application is signaled as KILL or DESTROY, the middleware will take the appropriate steps. Applications that are no longer signaled will be killed, just as if they were signaled with a control code of DESTROY.

To tell the receiver where to find the AIT, the PMT (program map table) for the service must include an entry for the ES that contains the AIT. This ES will have the stream type 0×05, and the entry for that stream in the PMT's ES loop must include an AIT application signaling descriptor. This descriptor tells the middleware which version of the AIT is currently being broadcast, so that a receiver can monitor changes in the AIT without having to monitor the AIT itself. In some receivers with limited hardware capabilities this is an important feature. Similarly, the ES carrying the AIT must not be scrambled unless the entire transport stream is scrambled, so that receivers can monitor it more easily.

Although this information is necessary for a receiver to detect an MHP application, other SI may be needed before it can actually be started. For instance, any broadcast file systems an application needs must also be signaled in the SI so that the receiver can connect to them and load any files needed.

Extending the AIT

Although the AIT works well for applications bound to a given service, it does not cope very well with applications that may be stored, or that are not bound to a service. OCAP and MHP deal with this in slightly different ways. MHP's approach applies to stored applications only, and we will examine this in more detail in Chapter 16.

OCAP takes the normal AIT and uses it as the basis for the extra information needed. This table is known as the Extended Application Information Table (XAIT), which is sent to the receiver as an out-of-band signal via the extended channel of the POD module (the channel in the POD module that supports out-of band communication between the head-end and the receiver). The extended channel can carry MPEG sections as well as other data formats, and the XAIT is carried in MPEG sections on PID number 0x1FFC.

Recent versions of the OCAP specification, starting at version I09, transmit the XAIT in the same binary form used for the AIT in broadcast streams. The MPEG table ID value used for an XAIT is 0×74, which is the same as that used for a normal AIT. This does not matter to the receiver, in that the receiver can tell it is dealing with an XAIT because the XAIT is sent via an extended channel.

Apart from the transmission mechanism, the XAIT has very few differences from the AIT. XAIT entries will usually include a few OCAP-specific descriptors in order to support the special requirements of the applications signaled in the XAIT. This extra information helps the receiver decide the relative priority with which they should be stored, or how unbound applications can be run without interfering with one another. Table 4.8 outlines the extra descriptors that can be found in a typical XAIT.

These only apply to Java applications. OCAP profile 2.0, which adds support for HTML applications, has not been updated since the adoption of this XAIT format. It is thus not clear which descriptors are added for OCAP-HTML applications that may be signaled in the XAIT.

Table 4.8. OCAP-specific descriptors that may be used in the XAIT.

Descriptor	Purpose
Abstract service descriptor	Describes an abstract service, which will be used to control the life cycle of one or more unbound applications.
Unbound application descriptor	Gives the version number of an unbound application, and identifies the abstract service to which it belongs.
Application storage descriptor	Describes the priority on an application that should be stored in the receiver's persistent storage.
Privileged certificate descriptor	Gives a hash value of the certificate used for signing the monitor application. See Chapter 13 for more information about signing applications.

Previous versions of OCAP used an XML-based file format. Because this format is now depreciated, we will not look at it in too much detail, but the following example gives you an idea of what this file looks like.

```xml
<?xml version="1.0"?>
<!DOCTYPE xait SYSTEM "xait.dtd">

<xait>
  <abstract_service>
    <service_type> OCAP_ABSTRACT_SERVICE </service_type>
    <organisation_id> 0x0000002A </organisation_id>
    <service_id> 1 </service_id>
    <service_name> "Local News" </service_name>
    <auto_select> "false" </auto_select>
  </abstract_service>

  <application>
    <application_type> OCAP-J </application_type>

    <application_name>
      "newsTicker"
    </application_name>

    <application_version> 0 </application_version>

    <application_control_code>
      PRESENT
    </application_control_code>

    <application_identifier>
      0x000000001234
    </application_identifier>
```

```
<service_id> 1 </service_id>
<visibility> VISIBLE </visibility>
<priority> 128 </priority>
<storage_priority> 128 </storage_priority>
<launchOrder> 0 </launchOrder>

<platform_version_major> 1 </platform_version_major>
<platform_version_minor> 0 </platform_version_minor>
<platform_version_micro> 0 </platform_version_micro>

<ocap_j_application>
  <base_directory>
    "/apps/news_ticker"
</base_directory>

<classpath_extension>
  "/apps/common_classes"
</classpath_extension>

<initial_class_name>
  "newsTicker"
</initial_class_name>
</ocap_j_application>

<transport_protocol>
  <transport_via_OC>
    <OCAPlocator>
      "ocap://news_service"
    </OCAPlocator>
  </transport_via_OC>
</transport_protocol>

    </application>

  </xait>
```

More details of the file format are available in older versions of the OCAP specification, if you can find them. Pre-I09 versions of the OCAP standard allowed applications (in particular, the monitor application) to pass fragments of the XAIT to the application manager via the application management API found in the `org.ocap.application` package. This makes things a little more complex for older versions of the middleware, in that they need to parse fragments of XML from a number of sources, not just from the main XAIT. It also means that they need to parse two separate formats for application signaling, which can get rather painful for middleware developers. This is just one of the things that make it worthwhile for middleware developers to move to at least version I09 of OCAP.

Adopting a binary format for the XAIT makes it less feasible for network operators to produce XAIT fragments, and thus version I09 of OCAP changes the OCAP application management API so that the values for the XAIT entry can be passed to the application manager as separate arguments in a method call. We will look at this in a bit more detail in Chapter 17.

Because the XAIT describes applications that are not bound to a given service, the same XAIT is sent regardless of the service the receiver is currently presenting. This makes it easier for the receiver to monitor the XAIT, but it still has to check whether the XAIT has been updated. In I09 and later versions, the receiver simply needs to check the version number of the XAIT.

Older versions of the standard have a different method of signaling the presence and location of the XAIT. I08 and older versions of OCAP used an XAIT descriptor in the common loop of the Virtual Channel Table to tell the receiver where to find the XAIT. In these versions, the XAIT could be sent via the POD module's extended channel, or it could be downloaded from a server via HTTP or even embedded in the descriptor itself. For these older versions of OCAP, the receiver just had to monitor the version number of the XAIT descriptor in order to discover if the XAIT had been updated.

Controlling Xlets

As we have already seen, MHP and OCAP middleware stacks both include an application manager component that is responsible for starting and stopping applications based on the AIT and XAIT. In MHP systems, this application manager is solely responsible for deciding which applications get started, and when. All other parts of the system (including the user and other applications) can only request that applications are started or stopped. What this means is that the receiver manufacturer is in complete control of which applications can run at any given time. This makes network operators in the United States a little nervous, and thus OCAP gives the network operators some more control over when an application will run. Although the executive module in an OCAP receiver is mainly responsible for deciding which applications are running, the monitor application can also have a say in which applications get started.

The monitor application can use the `org.ocap.application.AppManager` Proxy class to access some of the features of the application manager. Among these features is the ability to set a filter that decides which applications can be run. The `setAppFilter()` method takes an instance of the `AppsDatabaseFilter` class as an argument. By implementing a subclass of `AppsDatabaseFilter`, the monitor application can decide which applications are allowed to run. We will take a more detailed look at how `AppsDatabaseFilters` can be used in Chapter 17.

Registering Unbound Applications

The monitor application has another way of influencing which applications can run. It is the monitor application's responsibility to register any stored applications with the application manager, so that the application manager can see and control them. Any applications that are not registered by the monitor application will not be available to the user.

It is pretty unlikely that system applications or other stored applications will not be registered, though, simply because this would remove functionality from the receiver. As a rule, the monitor application should register every stored application. The only exception to this is manufacturer-specific applications. These will be included in the applications database automatically by the manufacturer, and thus do not need to be explicitly registered.

It is not just the monitor application that can register new unbound applications, however. Any application that has been given the right permissions can do so. Although this is normally only given to the monitor application, that does not have to be the case.

Applications are registered using the `appManagerProxy.registerUnbound App()` method. To know which applications should be registered, this method takes an `Input-Stream` that refers to a binary representation of the XAIT. This XAIT will be parsed and loaded into the application database, just like any other XAIT data, and thus it must include section headers and all of the other information the application manager would expect. For those XAITs that are larger than a single MPEG section, the XAIT data should contain all of the necessary sections concatenated together. The following code sample shows how unbound applications are registered.

```
AppManagerProxy managerProxy;
managerProxy = AppManagerProxy.getInstance();

try {

  // Create a FileInputStream that points to some system-
  // specific applications
  FileInputStream xaitInputStream;
  xaitInputStream =
    new FileInputStream("/system/apps/xaitdata.dat");

  // Register the applications that this XAIT refers to
  managerProxy.registerUnboundApp(xaitInputStream);
}
catch (IOException e) {
  // do nothing
}
```

In this example, we use a `FileInputStream` to load the data, but in practice we could use any type of `InputStream`. We are not just limited to loading one set of XAIT data, either: we can load as many as we like. These XAIT fragments will be treated like any other XAIT, and thus if the information in one fragment conflicts with information that is already loaded the new data will overwrite the old data.

If we want to remove an application previously registered, we can use the `AppManager-Proxy.unregisterUnboundApp()` method to remove an unbound application from the database. Allowing any application to remove previously registered applications is rather dangerous, and thus only the application that registered an unbound application (using the

`AppManagerProxy.register UnboundApp()` method) can unregister it. Applications registered via an XAIT from the network may not be unregistered in this way.

The ability to update the XAIT in this way can cause problems, and thus the monitor application can register a signal handler with the application manager by calling the `AppManagerProxy.setAppSignalHandler()` method. This takes an `org.ocap.application.AppSignalHandler` object as an argument, and once this signal handler is registered it will receive notification of any attempts to update the XAIT, either by the monitor application or any other application with permission to update the XAIT, or by a new XAIT from the network. This notification includes a list of the applications contained in the new XAIT fragment, and the signal handler can use this to choose whether to allow the update or not.

This is an all-or-nothing operation. That is, either the new XAIT is accepted or it is rejected. It is not possible to allow a subset of the applications in the new XAIT to be added while rejecting the rest, and for this reason it is better not to update the XAIT from multiple sources. In general, there is no reason for any application other than the monitor application to register unbound applications, and it is best to signal them in the XAIT. Another reason for using the XAIT in preference to registering or unregistering them in another way is the fact that changes to applications signaled in the broadcast XAIT can only come from a new version of the broadcast XAIT. Any attempt to make change to these applications via the `AppManagerProxy` will fail.

Making Applications Coexist Reliably

The AIT and XAIT give network operators a way of improving the reliability of their applications by limiting which applications can run simultaneously. Of course, this is only useful if network operators test those applications together, and test them on a number of different receivers. This type of interoperability testing is a standard part of deploying applications in closed systems, and thus using an open standard does not change this except that we also need to test for compatibility with any built-in or stored applications.

This does not let application developers off the hook, of course—it is the people who write the applications that have the most power to make applications work reliably together. The big message here is very simple: be flexible, and do not be too greedy. Design applications that can degrade gracefully if they do not get all of the resources they need, can be flexible in their use of screen real-estate, and only change the configurations of resources when they absolutely have to. There is no magic formula for doing this, unfortunately. Developers should ask themselves, "Do I really, really need to use this resource? What should my application do if it can't get it?"

Sometimes there is no good answer, and the application cannot run without that particular resource, but in other cases the application may be able to work with reduced functionality or with only some of the assets it needs. This may be a good reason for building your applications as a set of components, with some components being optional rather than required.

That way, even if a component cannot be used because of a lack of resources the rest of the application may still be able to provide some value to the user.

This flexibility is tough to achieve, and thus network operators still need to test applications to make sure they can work together. There are some applications that simply cannot work reliably together, and in these cases the network operator has to make a choice to run that application alone or to ask the application developer to change it.

OCAP's use of unbound applications does not change the picture much, although it does make it more difficult for testers to know what applications may be running at any time. They have to coordinate the version of the XAIT that will be used with the version of the AIT that will be transmitted, and make sure that all application combinations are tested. Only having the AIT makes it slightly easier for those testers dealing with MHP systems. Although the sets of applications may actually be identical in both cases, there is only one source of application signaling testers must check.

Pitfalls for Middleware Developers

Managing applications is one of the most important tasks middleware needs to carry out. If this is not handled well, the quality of the entire middleware suffers. It may not matter how well applications run when they are finally launched if the user has problems launching them in the first place.

For applications written in Java, there are some special pitfalls for middleware developers. The biggest of these is the lengths they have to go to in order to kill an application. There are times when applications will need killing, such as at the end of a show or when an application misbehaves, and this can be extremely difficult in Java. Stopping the threads belonging to an application is not easy in Sun's normal Java VM, for several reasons. The middleware needs to know which threads actually belong to the application, and this is a good reason for giving every application its own ThreadGroup (a Java concept that groups threads together) and making every thread for an application be a member of that Thread-Group. At least that way an application's threads can be easily identified.

Even when we can identify all of the threads in an application, actually killing those threads may still be difficult. Calling the stop() method on a Java Thread object may not be enough to kill it. There may be a delay while the thread is actually terminated, and applications may catch the ThreadDeath error that is thrown when a thread is forcibly stopped. Although applications are supposed to re-throw this error if they catch it, a malicious application may use this to stop its death by simply continuing its processing in the exception handler. Because of this possibility, the middleware needs to take steps to prevent this from happening, either by not throwing the ThreadDeath exception at all or by not relying on this to terminate the thread.

There is also the question of what happens if a middleware thread (e.g., one used for dispatching events) is executing application code when the application is killed. If the middleware waits for the event handler to finish executing, the thread may be trapped if the event handler does not return in a timely manner. For instance, it may be waiting for notification from an application thread that has already been killed. In this case, it may be necessary for the middleware to kill that event-dispatching thread and start a new one. We will look at this particular case in later chapters, when we discuss this type of event dispatching in more detail.

Even if it does manage to terminate all of the application's threads, the middleware's troubles are not over. It still needs to make sure that any resources used by the killed application are freed correctly, including graphics contexts, file handles, and network sockets. Applications may not always do this correctly, and if these are not freed they can cause resource leaks that may eventually exhaust the resources of the receiver. This task is not easy and it may need some changes to Java system classes in order to provide the means for the middleware to know exactly what resources were used by which applications. This is not always complex, but it does add to the task of middleware developers.

The Java system classes also need to be changed to take account of some behavior that is specified in the MHP specification relating to application termination. In particular, finalizers that are part of application classes will not be executed, and thus application developers need to be aware of this and middleware developers need to make sure they are never called. There are a number of other changes needed by the MHP and OCAP specifications, and this is just another change that should be added to that list.

In short, middleware developers have to assume that every application is hostile, because even the best application developers will miss things that cause their Xlets to hang, or to fail to be destroyed properly. Genuinely malicious Xlets will be very uncommon, but we all know that bugs are always present in any piece of software.

5 The JavaTV Service Model

The concept of services, representing individual TV channels, is a central concept in a DTV system. This chapter discusses how the service model used in MHP and OCAP affects the life cycle of applications, as well as looking at how applications can change to a new service.

For any DTV receiver, the basic unit of content is the service. To most people, this most often means a TV channel or a digital radio channel. Conceptually, a DTV receiver decodes and displays services, rather than transport streams or particular events, just like a normal TV. It shows one TV channel at a time, and if you want to see something else you have to switch to a new channel.

In analog systems and in basic DTV networks, a service consists of audio, video, and some data (such as subtitles or teletext data, and service information in the case of digital systems). Many interactive systems extend this to include applications, and JavaTV follows this model. This relationship forms one of the foundations of the software architecture in MHP and OCAP. Because a receiver may be able to present more than one service at a time (using picture-in-picture, for instance), it needs some way of linking the various elements of a service while keeping them separate from elements of any other services currently being shown.

The answer to this is the service context. A service context acts like a container for all of the pieces of a service the receiver is currently showing, including media and applications. The middleware can use this to tell which applications or media are part of a given service, and applications can use this to present a new service.

Any service the receiver presents will have its own service context, as shown in Figure 5.1. This means that a receiver will have one service context for every service it can display simultaneously. A receiver that can show two services at the same time will have two service contexts. In practice, things are a little more complicated than this, but we can ignore this for now.

Figure 5.1. Service contexts act as a container for all parts of a service, including applications and media.

Service contexts and the classes that support them are defined in the `javax.tv.service.selection` package. Using this package, applications can get a service context and use it to find information about the current service or choose a new service. Every service context in an MHP or OCAP receiver will implement the `ServiceContext` interface, shown below.

```
public interface ServiceContext {

    public void select(javax.tv.locator.Locator[] components)
        throws InvalidLocatorException,
               InvalidServiceComponentException,
               SecurityException;

    public void select(javax.tv.service.Service selection)
        throws SecurityException;

    public void stop() throws SecurityException;

    public void destroy() throws SecurityException;

    public javax.tv.service.Service getService();

    public ServiceContentHandler[]
        getServiceContentHandlers()
        throws SecurityException;

    public void addListener(ServiceContextListener listener);
    public void removeListener(
        ServiceContextListener listener);
}
```

Applications can refer to a service using a set of `Locator` objects to identify individual streams within the service, or a `Service` object to refer to the default streams within the service. We will not look at these in detail now (locators are covered in more depth in Chapter 8, and the `Service` object is examined in Chapter 9). All we will say for now is that only some types of streams can be selected. In particular, an MHP application is only guaranteed to be able to select audio, video, or subtitle components when using `Locator` objects to select a service. OCAP does not say which types of streams can be selected, and thus an OCAP receiver may or may not allow applications to select other stream types. If the receiver can select other stream types, the effect of this will not be the same across all receivers and may be different depending on the type of stream you select. Generally, you should be careful doing this, and in most cases applications will not need to select streams containing any other type of data (other APIs are available for accessing other data streams).

Any service context can be in one of four states: *Not Presenting, Presentation Pending, Presenting*, and *Destroyed*. At first, a service context will be in the *Not Presenting* state. When an application or the middleware selects a service for that service context, it enters the *Presentation Pending* state, and then moves to the *Presenting* state when it is actually displaying the content from that service. To stop presenting content, an application can call the stop() method, which moves the service context back to the *Not Presenting* state. A service context can go through these transitions many times over the course of its lifetime.

When an application has finished with a service context, it calls the destroy() method. This stops any content that is being presented, and moves the service context to the *Destroyed* state.

At first, this may look like a media player API, but this is not the case. Do not forget that content in this case means applications as well as media. Selecting a new service may start some applications, whereas stopping a service context will kill any applications that were running as part of the service being presented in that service context. This applies to our own service context as well, and thus an application that calls the stop() method on its own service context will be killed along with any others running in that service context. Figure 5.2 shows possible state transitions for a service context.

State transitions in the service context may involve a number of complex operations, and thus all of these transitions are asynchronous. Applications and the middleware are notified about changes in the state of a service context via a ServiceContext Listener object. This receives the ServiceContextEvent, which will inform any interested components

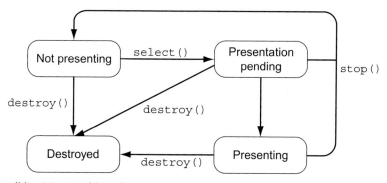

Figure 5.2. Possible state transitions for a service context.

Table 5.1. Events generated by state transitions in a service context.

Event	Description
PresentationChangedEvent	This event may be generated at any time when a service context is presenting content. It indicates that the content being presented by a service context has changed, either because a new service has been selected or because some of the content that was being presented is no longer available. The service context is in the *Presenting* state.
SelectionFailedEvent	The middleware will generate this event following a service selection operation to indicate that the service context could not present a new service that was selected. The service context will be in the state it was in before select() was called (either the *Presenting* state or the *Not Presenting* state, depending on whether a service had previously been selected successfully).
PresentationTerminatedEvent	This event will be generated after the stop() method is called, or after the destroy() method if the service context was not stopped first. It indicates that the service context has stopped presenting content and is now in the *Not Presenting* state.
ServiceContextDestroyedEvent	The service context has been destroyed by a call to the destroy() method, and is now in the *Destroyed* state.

about changes in the state of a service context. A service context listener can receive any of the events outlined in Table 5.1.

The service context does not present any content itself. Instead, this task is given to one or more service content handlers. These come in two basic flavors: service media handlers (implemented by the ServiceMediaHandler class) that present the media components of the service and service content handlers for applications. To give some flexibility to implementations, it is not specified how a middleware stack should implement service content handlers for applications (however, they must implement the ServiceContentHandler interface like any other service content handler). This can be implemented by the Xlet context, or another way that makes more sense in a given implementation. Figure 5.3 depicts service content handlers and their relationship to the service context.

MHP and OCAP specify that service media handlers will always be instances of a Java Media Framework player. Each service media handler will present any element of the media stream that uses the same clock, which for MPEG streams means that components will share the same PCR. This will typically include the video and audio streams, plus any subtitles, and thus in most cases there will only be one ServiceMediaHandler object for the entire service.

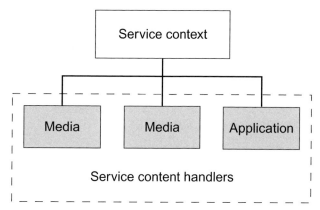

Figure 5.3. Service content handlers and their relationship to the service context.

The `getServiceContentHandler()` method lets an application get the service content handlers for the current service. This will include the content handlers for media and for applications (no distinction is made between them at this point). Depending on the service, it may include only service media handlers, only service content handlers for applications, or both. In every case, the behavior of the service context is the same.

Because a receiver can usually only present a small number of services simultaneously, we can only have a small number of service contexts active at any time. We have already seen that we can destroy a service context using the `destroy()` method, but so far we have no way of creating a new service context.

Service contexts are managed by the `ServiceContextFactory` class, which has the following interface. This lets us create new service contexts and discover how services and applications are connected.

```
public abstract class ServiceContextFactory
    extends java.lang.Object {

  public static ServiceContextFactory getInstance();
  public abstract ServiceContext createServiceContext();
  public abstract ServiceContext getServiceContext( XletContext ctx);
  public abstract ServiceContext[] getServiceContexts();
}
```

The `ServiceContextFactory` is a singleton object, and thus we can get a reference to it using the `getInstance()` method. Once we have an instance of it, we can create a new service context by calling `createServiceContext()`. If a service context can be created, this method will return a reference to a new service context (or a null reference if all of the service contexts are being used).

Sometimes, however, we do not want to create a new service context. We may want to get a reference to the service context that our application is executing in, or to a service context that is presenting another service. The `getServiceContext()` method lets us get the service context associated with a particular Xlet, or we can use `getServiceContexts()` to get references to all active service contexts in the receiver.

What Happens During Service Selection?

When a new service is selected in a given service context, the service context will stop any content that is not part of the new service, and start presenting the content associated with the new service. Any content present in both services will continue to be presented, although there may be some disruption to audio or video if the receiver has to tune to a new transport stream to present the new service.

This also applies to applications, and thus any applications running in that service context will be killed unless they are also signaled in the new service. Service selection is about far more than this, however.

An application can tell the receiver to present different media content, even content outside the current service, using the interfaces examined in Chapter 11. It can tell the receiver to tune to a new transport stream, as we will see in Chapter 17. These change the media content and data streams available to the application, but they do not have an effect on the application itself. They will not cause any changes to which applications are running, or to which application-signaling the receiver listens to.

Selecting a new service will have an effect on the applications running in the service context that is used. When a new service is selected, the receiver will take the following steps.

1. The service context will change its state to the *Presentation Pending* state, possibly via the *Not Presenting* state.
2. Tune to the transport stream containing the new service, if necessary.
3. Use the available service information (and user preferences such as language settings) to decide which streams from the new service should be presented.
4. Start presenting those streams in the service context.
5. The service context will move to the *Presenting* state and generate a `Presentation-ChangedEvent`.
6. Read the AIT for the new service and launch any applications that are signaled as auto-start in the AIT and that are not already running. Applications should be launched in priority order, so that applications with the highest priority are given a better chance of running.
7. Kill any applications running in the service context that are not signaled in the AIT for the new service (either as AIT entries or in external application authorization descriptors).

Apart from the obvious effects, selecting a new service may tell the receiver to flush any cached data it held about that service, such as cached files, cached service information, or any other data used by applications in the old service. Although this may not be necessary, service selection can be a useful point for identifying that this type of information can be freed.

When we call `select()` with an array of locators, the middleware takes another couple of steps during the service selection process. It will first ensure that all of the locators are part of the same service. If this is not the case, the service selection operation will throw an exception. If all locators belong to the same service, the middleware will then check that all of the locators can be presented at the same time. This involves several different checks, as follows.

- Is there only one audio and video stream in the list of locators?
- If any of the locators specifies event IDs, are those event IDs currently valid?
- Is one of the locators a subset of another in the list (e.g., a locator that only specifies the service, and one that specifies an event ID as well)?

Although not all of these are fatal errors, the behavior in some cases (especially the last two previously listed) is undefined. Developers should take care to avoid this type of problem from arising by making sure that the locators they specify are compatible with one another.

Any applications that survive service selection may find that their broadcast file system is no longer available. In this case, applications may not be able to load any new files from broadcast file systems that were only available in the old service. If an application is using a broadcast file system that was present in both the old service and the new service, the middleware will automatically remount the file system after service selection has completed. The question then arises whether two file systems are identical. In this case, the middleware uses the service information for the new service to decide this (discussed further in Chapter 12).

Abstract Services

The model we have seen so far works very well for applications associated with a broadcast service, but not as well when we start dealing with unbound applications or with built-in applications. In practice, this model works pretty well with a couple of adjustments. The biggest of these is a concept called abstract services. These are just like other DTV services except that they contain only applications, and they do not exist as part of a broadcast stream. Abstract services are defined by the broadcaster (or in some cases by the middleware standard or by the receiver manufacturer), and provide a way of grouping applications that can run together.

To run an application that is part of an abstract service, that service must be selected just like a normal broadcast service. This can happen either in a service context that is currently showing a broadcast service or in a new service context.

Earlier in the chapter, we mentioned that a receiver has one service context for every service it can display simultaneously. Usually, this means that a receiver with two tuners and two MPEG decoders will only support two service contexts, but receivers may support more service contexts than this minimum. Many receivers that support unbound applications (e.g., OCAP receivers and some MHP 1.1 receivers) will support additional service contexts so that unbound applications can run while the receiver is presenting broadcast services.

Abstract services are most commonly used in OCAP receivers, although the Internet Access profile of MHP 1.1 uses something similar to provide support for built-in Internet clients. In the case of MHP, these abstract services are represented by API-defined specific subclasses of the `javax.tv.service.Service` object. These can be used to select the abstract service for a particular client.

OCAP supports abstract services in a different way, in that unbound applications are more common in networks that use OCAP. In this case, the network operator (and possibly the receiver manufacturer) defines which abstract services are available. In the rest of this chapter, we will see how OCAP networks use abstract services.

Managing Abstract Services in OCAP

Before an abstract service can be started in an OCAP receiver, the executive module needs to know about that service, and in fact the executive module needs to create that service using the OCAP service the API defined in the `org.ocap.services` package.

Abstract services are not created at the whim of the receiver; something has to tell the receiver what abstract services it should create and what applications are associated with it. This information is contained within the XAIT. To define an abstract service, the broadcaster inserts an abstract service descriptor in the common descriptor loop of an XAIT. They must then define at least some of the applications in that XAIT as belonging to that abstract service, by specifying the XAIT's service ID in the unbound application descriptor of the application in question (see Chapter 4 for more information about this descriptor). An abstract service descriptor takes the format shown in Table 5.2, using the value 0xAE for the descriptor tag.

Abstract services defined by the network operator must have a service ID in the range 0x020000 to 0xFFFFFF. In an OCAP receiver, abstract services are represented by an instance of the `AbstractService` class, as in the following. This extends the basic JavaTV `Service` class to add a way of finding which applications are included in that service.

Table 5.2. Format of the OCAP abstract service descriptor.

Syntax	No. of Bits	Identifier
`abstract_service_descriptor() {`		
`  descriptor_tag`	8	uimsbf
`  descriptor_length`	8	uimsbf
`  service_id`	24	uimsbf
`  reserved_for_future_use`	7	uimsbf
`  auto_select`	1	bslbf
`  for (i=0; i<N; i++) {`		
`    service_name_byte`	8	uimsbf
`  }`		
`}`		

```
public interface AbstractService
   extends javax.tv.service.Service {

   public java.util.Enumeration getAppAttributes();
   public java.util.Enumeration getAppIDs();
}
```

Registering Applications

Defining an abstract service is just the first thing we need to do. Next, we must register any applications associated with that service. These are defined in the XAIT just like other applications, and thus from the network operator's perspective there are no big surprises here. When the OCAP receiver gets a new XAIT, it needs to insert any applications defined there into the application database. Exactly how this is done is not defined in OCAP, but there are a couple of possible solutions.

OCAP does specify a way for an application to register unbound applications with the middleware, so that receiver manufacturers can register any built-in applications. The OCAP middleware can use this same approach to register XAITs received over the network, although there are some minor security differences between the two cases that need to be considered.

Applications can register new abstract services and unbound applications using the `App-ManagerProxy` class from the OCAP application API, defined in the `org.ocap.application` package. This class has two methods, as shown in the following, we are interested in at the moment.

```
public class AppManagerProxy {

   public static AppManagerProxy getInstance();

   public void registerUnboundApp(java.io.InputStream xait);
   public void unregisterUnboundApp(int serviceId, org.dvb.applica
      tion.AppID appid);
}
```

The `registerUnboundApp()` method registers an unbound application, as you would expect. Because the XAIT specifies which applications are associated with which abstract services, this method takes an `InputStream` pointing to a complete XAIT as an argument. This could be an XAIT from the network, or an XAIT from a local file that defines the applications included by the receiver manufacturer. The name `registerUnboundApp()` is a little misleading, because it actually registers every unbound application contained in an XAIT and creates any abstract services specified in that XAIT.

To avoid conflicts with services defined by the network operator (be they abstract services or broadcast services), any abstract services defined by the receiver manufacturer must have a service ID in the range 0x010000 to 0x01FFFF. The middleware must take care to distinguish between XAITs from the network and XAITs registered by other applications. An application (even the monitor application or one defined by the receiver manufacturer) cannot modify any abstract services or application entries contained in the broadcast XAIT.

`unregisterUnboundApp()` does the opposite of `registerUnboundApp()` and removes a previously registered application from all abstract services in the applications databases that refer to it. When all applications for an abstract service that are signaled as either auto-start or present have been removed, the receiver treats that service as if it is no longer available and removes it from the list of available services.

Selecting Abstract Services

Once an application or abstract service has been registered, the middleware treats it just like any other application or service. When a user wants to start an application that is part of an abstract service, that service must be selected in a service context just like any other service.

The monitor application or executive module will use the JavaTV service selection API to select that service, either in an existing service context or in a new one. Thus, starting an abstract service may mean that the receiver cannot display the service the viewer is currently watching. As we mentioned earlier, however, a receiver may support additional service contexts so that abstract services and broadcast services can be presented at the same time.

When an abstract service is selected, all of the applications associated with that service are added to the application database, just as they would be if they were signaled using real AIT signaling. Similarly, any applications signaled as auto-start for that service will be started, following the same rules that apply to broadcast services. Applications in an abstract service have exactly the same life cycle and application model as those applications that are part of a broadcast service.

One slight difference appears when we start asking if applications in an abstract service are service bound. Even though they are not bound to any broadcast service, they are bound to a specific abstract service (or services). What this means is that applications in an abstract service follow the same rules as service-bound applications in any other service: if a new service is selected (a broadcast service or another abstract service) that does not have a given application signaled as present, that application will be killed. Therefore, unbound applications associated with an abstract service are actually bound to that particular abstract service.

6 Resource Management Issues

Managing resources is an important part of a DTV receiver. In this chapter we will see how the receiver manages resources and how applications can reliably handle the situation in which resources are taken away from them. We will also look at the various strategies for resource management a receiver can use.

As you will have noticed from earlier chapters, resource management is a vital part of the tasks a middleware stack must perform. If a receiver does not manage its resources effectively, it may not be able to run as many applications as would otherwise be possible, and applications may not be able to use all of their functionality. This makes resource management a major issue for receiver manufacturers and application developers, and it is something you will encounter fairly frequently in the MHP and OCAP APIs.

Resource management is not a new problem, and the solution used by MHP was defined as part of the DAVIC standardization process. Many of the APIs we will see in later chapters are based on the DAVIC resource notification API contained in the `org.davic.resources` package. Note the choice of words there; this is not a resource *management* API. Resource management is carried out purely by the receiver middleware as it sees fit, and an application has very little say in whether it gets to keep a scarce resource that has been requested by another application. Resource *notification* simply tells the application when another application needs access to a resource, or when the receiver has revoked access to a resource.

The resource notification API consists of three main classes, and is not intended to be a complete API in its own right. Instead, it is designed to be used by other APIs in a way that best suits them, although it does define some common concepts. To make this clearer in the following description, any references to a "using" API mean an MHP or OCAP API that implements the resource notification API as part of its own specification.

Whereas MHP follows this model accurately in most cases, OCAP makes a few changes. Early versions of the OCAP specification did include a formal resource manager, and although this has been removed in more recent versions a few traces of it can still be seen. The approach taken by OCAP implies that a central resource manager is present, but it is possible to implement the OCAP middleware without it.

In MHP, on the other hand, individual API implementations can implement their own resource management strategies, and they may not take account of how other middleware components manage their resources. This is only one way of doing it, however. An MHP implementation can also have a central resource manager that handles every scarce resource in the receiver, but this is not required by the MHP standard. We will take a closer look at the OCAP resource manager later in this chapter, and see how it affects the APIs that use it.

Introducing the Resource Notification API

The resource notification API is based on a client-server model, but it is not quite the same model you may be used to. DAVIC introduced a few changes in order to improve security and resilience to misbehaving applications, although conceptually it is more or less the same. The API itself consists of the following three main elements.

- *Resource server:* The part of the using API that processes requests for resources
- *Resource client:* The part of the application or middleware component that actually uses the resource
- *Resource proxy:* An intermediate object that stores settings for the resource and enforces the middleware's security policy

Every API that uses the resource notification API must implement at least one resource server to process requests for the resources that API exposes. As well as processing requests for scarce resources, the resource server manages how those scarce resources are allocated to the various entities that request them. In practice, the resource server may process those requests itself or it may pass them on to a central resource manager.

The resources controlled by a resource server may be either software (e.g., a software section filter) or hardware resources (e.g., a modem or MPEG decoder). The resource notification API does not care, and MHP and OCAP use it for both situations. The class that implements the resource server does so by implementing the `ResourceServer` interface, as follows.

```
public interface ResourceServer
{
  public abstract void addResourceStatusEventListener(
    ResourceStatusListener listener);

  public abstract void removeResourceStatusEventListener(
    ResourceStatusListener listener);

}
```

As you can see, the only methods this interface defines allow an application to register as a listener for events indicating the change in status of a resource. APIs that use this interface can provide events to show how specific groups of resources change their status. Registering a listener for resource status events for a given API means that resource status events for any resource managed by that resource server will be sent to the listener.

Many people are confused when they first encounter this API, because it has no standardized methods for reserving and releasing resources. We must stress once again that the DAVIC resource notification API is just that—a resource notification API rather than a resource management API.

This is an important distinction, because by not standardizing an interface for resource management we allow APIs that use this API to provide this functionality in a way that fits best with that particular API. The way we reserve a modem in a telephony API will very likely be different from the way we reserve a section filter in a section filter API, for instance. Table 6.1 shows the methods provided by MHP that use the resource notification API to reserve and release resources.

Table 6.1. Methods for reserving resources in MHP.

Class	Method
org.havi.ui.HScreenDevice	reserveDevice()
org.havi.ui.HVideoDevice	reserveDevice()
org.havi.ui.HBackgroundDevice	reserveDevice()
org.havi.ui.HGraphicsDevice	reserveDevice()
org.havi.ui.HEmulatedGraphicsDevice	reserveDevice()
org.davic.mpeg.sections.Section FilterGroup	attach()
org.dvb.net.rc.ConnectionRCInterface	reserve()
org.dvb.event.EventManager	addExclusiveAccessToAWTEvent()
org.dvb.event.EventManager	addUserEventListener()
org.ocap.event.EventManager	addExclusiveAccessToAWTEvent()
org.ocap.media.VBIFilterGroup	attach()
org.davic.net.tuning.NetworkInterface Controller	reserve()
org.davic.net.tuning.NetworkInterface Controller	reserveFor()
org.davic.net.ca.DescramblerProxy	startDescrambling()

The `ResourceClient` interface is, not surprisingly, the client side of the API. This interface, as follows, is implemented by a class in the application or component that uses the resource, and it gives the middleware a way to revoke the resource or to query whether the resource is no longer needed. Maybe now it becomes clearer why this is called a resource notification API.

```
public interface ResourceClient
{
   public abstract boolean requestRelease(
      ResourceProxy proxy, Object requestData);

   public abstract void release(
      ResourceProxy proxy);

   public abstract void notifyReleased(
      ResourceProxy proxy);
}
```

As you can see, the `ResourceClient` interface contains three methods the middleware uses to tell the application, with varying degrees of politeness, that another application needs the resource. The `requestRelease()` method is the most polite of these, requesting that an application give up a scarce resource so that it can be used by something else in the receiver. A client can decide to ignore this request, and if this method returns the value `false` the middleware knows that the client still wants the resource. If the client does not need the resource anymore, it can free the resource and return `true`. This tells the middleware that the client has finished with the resource, and that it can be reused by another application or middleware component. In this case, the middleware may reallocate the resource immediately, and thus the client should finish any housekeeping that needs the resource before returning `true`.

`requestRelease()` implies a level of cooperation between applications regarding resource usage, and the `requestData` parameter that is passed to `requestRelease()` takes this one step further than may be obvious at first. Some APIs that use the resource notification API may take the value of this parameter from the call to the using API that tries to reserve the resource. If the `requestData` argument is anything except a null reference, it must implement the `java.rmi.Remote` interface. This provides a way for the two components or applications to communicate, and possibly to reach a decision between themselves about which one gets the resource. The two components may actually be part of the same application (or they may both be part of the middleware), and thus there is no requirement that remote method invocation be used to communicate between the two components if they are part of the same application. If they are not, they must use the inter-Xlet communication API for any communication between them. We will look at this API further in Chapter 14 and see why this is necessary.

The `release()` method is the next level of impoliteness, and it tells the client that it must give up the scarce resource. In this case, the client has no choice—it can only carry out any housekeeping it needs to in order to continue, and then release the resource. When this method returns, the receiver will assume that the resource is available for other parts of the system to use and will almost certainly reallocate it immediately.

This assumes that a client is going to be cooperative, of course. The client could simply not return from the call to release() and it would keep access to the resource forever. For this reason, the resource notification API can be even less polite. If the release() method does not return within an appropriate (and implementation dependent) timeout period, the middleware will assume that the application is crashed or malicious and will reclaim the resource anyway.

If this happens, the middleware will call the notifyReleased() method from the ResourceClient interface, which tells the client that the resource has been taken away from it and that it should clean up as best it can. This is potentially a rather brutal operation from the point of view of the client, and thus it is only saved for those cases in which the application really is not cooperating. It also shows why applications should be polite and return from a call to release() or notifyReleased() in a reasonable time.

If a call to notifyReleased() does not return in a reasonable time, the resource notification API does not really care because it has already reclaimed the resource. Other parts of the middleware may care, however, and may decide to kill the thread that is not responding (or even the entire application). Thus, an application that tries to make trouble by not giving up resources may find itself being killed.

Now that we have seen the client and the server, let's look at where the resource proxy fits into this picture. A class implementing the ResourceProxy interface sits between the resource client and the actual resource. This serves an important security function, because the middleware does not have to give an untrusted application direct access to a scarce resource. The resource proxy will forward commands from the application to the resource, but because it is a trusted class it will only forward commands when the middleware tells it that it can. This makes it easy to take a resource away from an application when it becomes necessary to do so.

Although this is the main reason for having a resource proxy in the API design, there are a couple of other benefits from having it. The first of these is to provide an application a simple means of setting up the resource. Some resources, such as a modem, have many possible parameters to set. Application developers do not really want to set parameters on a resource every time they reserve it, because it can be time consuming and may introduce silly errors in the application. The design of the resource notification API lets you set parameters on the resource proxy, which can then be downloaded to the real resource when you successfully reserve it. Because an application can use the same resource proxy across multiple request/release cycles, this can make life significantly easier when dealing with complex resources.

A second benefit of having the resource proxy is that in most using APIs instances are created by the client and have no link to a real resource until they are attached to it. This enables the resource client to create multiple resource proxies, each with different settings for any parameters the resource may have. When the application actually reserves the resource it can choose which set of parameters it wants to apply. This means that the application can create these once as part of its initialization process and then simply use them throughout the rest of its life.

Now let's take a more detailed look at the security benefits of the resource proxy. As we have already mentioned, by using the resource proxy as an indirection mechanism we prevent the application from getting a direct reference to the Java object that actually controls the scarce resource. This makes it much easier for the receiver to take a resource away from a misbehaving application.

If the application has a direct link to the object controlling the resource, there is no way it can be forced to break that link. However, consider the situation in which we have the resource proxy acting as an indirection mechanism. In this case, the following steps happen when the application requests access to a resource.

1. The resource client tells the resource server for a given API that it would like to reserve a resource. This is done by calling a method on the resource server that is defined by the using API. This method will take an instance of a `ResourceProxy` as one of its arguments.
2. If access is granted, the resource server calls a private method on the resource proxy. This tells the resource proxy that it is now valid, and establishes a connection between the resource proxy and the resource itself.
3. The resource client calls standardized methods on the resource proxy to manipulate the resource. The resource proxy forwards these requests to the real resource.
4. When the resource client wishes to give up the resource, it calls a method on the resource server that is defined by the using API to release the resource. This method takes the `ResourceProxy` as an argument.
5. The resource server calls a private method on the resource proxy. This tells the resource proxy that it is no longer valid and that it should not forward any more requests to the underlying resource. The resource server updates its internal table of the state of the resources and marks the resource as free again. Figure 6.1 depicts the reservation of a resource.

This is all fairly clear, and it is not very different from what you would expect. Now let's take a look at what happens when the middleware revokes an application's access to a resource (see also Figure 6.2).

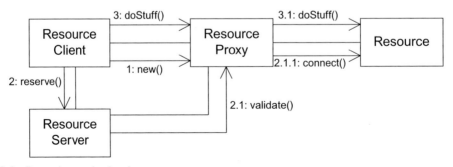

Figure 6.1. Reserving and releasing resources.

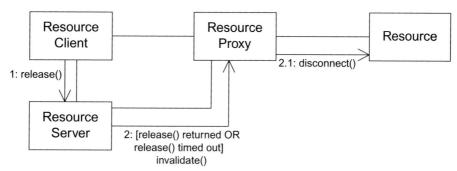

Figure 6.2. Invalidating a resource proxy prevents the application from using resources it is not allowed to.

1. The resource client fails to return from a call to `release()`.
2. After a timeout period defined by the middleware, the resource server calls a private method on the resource proxy that has access to the scarce resource. This tells the resource proxy that it is no longer valid.
3. The resource proxy breaks the link between it and the scarce resource, and resets its internal state to take account of this.
4. The receiver calls the `notifyReleased()` method on the class in the application that implements the `ResourceClient` interface. This informs the application that it no longer has access to the resource and that it should do any housekeeping necessary.
5. Any further attempts by the application to access the resource have no effect on the scarce resource.

The application can do nothing to interfere with the communication between the middleware and the resource proxy. The resource proxy classes are implemented by the receiver manufacturer, and thus can be trusted by the rest of the middleware. This means that unlike the application a resource proxy will always give up the resource when the middleware tells it to.

Similarly, because the methods used to validate or invalidate the resource proxy are not defined by the resource notification API (and should ideally be declared as `protected`, if possible), a malicious application can do nothing to spoof a message from the resource server granting access to the resource. It cannot know which method to call to do this, and the middleware implementation can make it extremely difficult for an application to find out.

This may seem a little extreme, but it is the only way to ensure reliability and security in a way that satisfies both the application and the middleware. In practice, this adds very little overhead to the process of reserving and releasing resources. The following code shows how a middleware stack may choose to implement this functionality in a resource proxy.

```
public class MySimpleResourceProxy
    implements org.davic.resources.ResourceProxy {
```

```
   private boolean valid;
   privateObject resource;

   private ResourceClient client;

   public MySimpleResourceProxy(ResourceClient c) {
      client = c;
      valid = false;
   }

   protected void validate(Object res) {
      valid = true;
      resource = res;
   }

   protected void invalidate() {
      valid = false;
      resource = null;
   }

   // Illustrate how we could use the proxy to securely
   // pass on requests to the resource
   public void SomeMethod(int myParam) {

      // Only pass on the request if the proxy is valid.  N
      // other cases, it will be ignored.
      if (valid) {
         resource.someMethod(myParam);
      }
   }

   public ResourceClient getClient() {
      return client;
   }
}
```

Using the Resource Notification API

In those APIs that use the resource notification API, resources are typically accessed as follows.

1. The application creates an instance of a `ResourceProxy` object.
2. The application sets some parameters on the `ResourceProxy` that will be passed to the real resource when it is acquired.
3. The application calls a method on the `ResourceServer` defined by the using API in order to reserve the real resource. This takes the `ResourceProxy` object as an argument.
4. The application uses the resource as it wants to.

5. When it has finished, the application calls a method on the `ResourceServer` to release the resource for use by another application. This method is also defined by the using API.

As you can see, there is nothing too complex about this. Knowing the philosophy behind this approach makes it much easier to see what actually happens in the various APIs that rely on the DAVIC resource notification API.

There are some things an application developer should consider when writing an application, however. You must be careful when you are using a scarce resource, and not use it more than you have to (your application may be stopping others from using the resource). Your application must also be able to run if it does not have access to the resource, because if another application with a higher priority wants the resource you are using your access to it may be revoked at any time.

Graceful handling of situations like these is the type of thing that makes for a reliable application. Situations like this will happen more commonly than you would wish for, and thus the application must be able to handle them and still run where it possibly can. Although your application does not have to implement any sensible precautions in the methods provided by the `ResourceClient` interface, it is obviously much better if it does. On a limited platform such as a DTV receiver, every application should take care to reserve only those resources it actually uses, for only as long as it needs them.

Handling Resource Contention

The whole point of scarce resources is that they are scarce, and this means that more than one application will often need access to the same resource at the same time. This may be less of a problem in MHP because the set of applications that can run together is pretty well defined, which makes it easier to actually test the applications running together. For OCAP, however, the situation is more complicated. Among service-bound applications, unbound applications, and stored applications, an OCAP receiver may have to run a bigger and less predictable set of applications at the same time.

For both MHP and OCAP, there is another concern that needs to be addressed. Given the use of Java as a programming language, some applications that will run on OCAP or MHP receivers may not be developed specifically for DTV (applications that were originally used on web sites or on other platforms may be re-purposed as digital TV content). Applications that were not originally written for DTV systems will pay less attention to issues such as resource management and may not play nicely with other applications. Although easy migration of content between the Web and the broadcast world has a number of advantages, it also has disadvantages such as this.

All of these things mean that MHP and OCAP middleware stacks need to handle contention for resources. If several applications are competing for the same resource, which one gets the resource? What happens if another application with a higher priority comes along? These are all issues the middleware has to worry about. Fortunately, the folks who standardized DAVIC, MHP, and OCAP gave us a few guidelines, which are outlined in Table 6.2.

Table 6.2. Guidelines for resolving resource contention in MHP and OCAP.

Permission	If the application does not have permission to access a resource, the request is denied even if the requested resource is available. There is a close link between the security status of an application and the resources it may access. An application that has not been signed by the network operator is granted access to very few resources by default, whereas applications that have been verified and signed by the network operator may be granted access to more resources. Control over this is pretty fine-grained, and a network operator can specify in detail which resources an application may access. This topic is covered further in Chapter 13, in discussion of the MHP and OCAP security model.
Request from other applications	If no resource is available, the middleware will call the `requestRelease()` method of every `ResourceClient` using an instance of the resource. These requests will be made in order of application priority, from lowest to highest. When an application gives up a resource by returning `true` from a call to `requestRelease()`, that resource is granted to the application making the current request for a resource.
The monitor application	For OCAP receivers, the monitor application may decide (under some circumstances) which application should get the resource. Exactly what strategy is used is up to the monitor application, and the monitor application will give the middleware a list of applications in the order that they may reserve the resource. Depending on the number of applications in this list, and the total number of resources, the application making the current request may or may not get access to the resource. The priority order returned by the monitor application does not have to reflect the actual priority of the applications as signaled by the network operator. The monitor application can choose to ignore this if it wants and impose its own priorities.
Application priority	For MHP receivers, or for OCAP receivers when the monitor application does not resolve resource conflicts, the request for a resource is ultimately resolved by the priority of the application given in the application signaling. The application with the lowest priority will have its access to the resource removed, and access will be granted to the application making the current request. If the application making the request is the one with the lowest priority, its request is denied.

These guidelines make it easy for us to tell which application should get access to a resource in any given situation. The only confusion arises when two applications with the same priority request access to the resource. In this case, neither OCAP nor MHP defines what the resource manager should do, and thus the request may be granted or denied depending on the choice made by the middleware implementer.

The lesson here for broadcasters is to set application priorities sensibly, and make sure that they reflect the real priorities of the applications. By doing this, you can help the middleware to make the right decision about how resources are allocated to the various applications.

Resource Management in OCAP

As mentioned at the start of the chapter, there are some differences between how the resource management API works in MHP and how it works in OCAP. For application developers, there are few obvious changes. Middleware implementers need to take some other changes into account.

OCAP's resource management API is contained in the `org.ocap.resources` package. This API has undergone a number of changes since early versions of the standard. Originally, the OCAP API featured a complete resource manager that was visible to the applications, and in this respect it is not very different from some early proposals for the DAVIC API. As with DAVIC, this was removed and the API has followed the DAVIC model fairly closely since then. Traces of this resource manager remain, however, as we will see in material to follow.

Despite the similarity between the two APIs, there are a number of philosophical differences. Not least of these is the fact that the task of managing resources may be shared with the monitor application. This allows the network operator to have a voice in how resources are to be shared between the applications that run on the receiver.

The importance of this should not be overlooked: it allows the network operator to impose a completely standard behavior with respect to resource management across all receivers on the network. No matter which manufacturer built a receiver, on that network it will behave the same as any other receiver. This is a marked difference from MHP, wherein the receiver manufacturer is in overall control of how resources are shared.

The OCAP resource management API defines the `ResourceContentionHandler` interface, which provides applications with a way of resolving resource conflicts. Only one `ResourceContentionHandler` can be registered with the middleware at any time, and this is typically registered by the monitor application. This is optional, and if the monitor application does not want to implement its own resource management strategy it can simply not register a `ResourceContention Handler`. In this case, the middleware will follow the resource management strategy we have already seen.

To register a `ResourceContentionHandler`, the application calls the `setResourceContentionHandler()` method on the `ResourceContentionManager` class. This is a singleton class, and is the only thing that remains of the standardized OCAP resource manager. We may only want to handle requests from certain applications using the `ResourceContentionHandler`, and let other requests be handled by the default mechanism (see Figure 6.3). To do this, we call the `ResourceContentionManager.set ResourceFilter()` method. This takes an `org.dvb.application.Apps DatabaseFilter` instance as its argument, which specifies the applications whose resource requests should be handled by the `ResourceContentionHandler`. The `AppsDatabaseFilter` class is part of the application control API (discussed in more detail in Chapter 17).

If the network operator wants to standardize resource-handling behavior across all receivers in the network, a class in the monitor application should implement the `ResourceContentionHandler` interface. Once this is registered with the middleware, anytime a resource

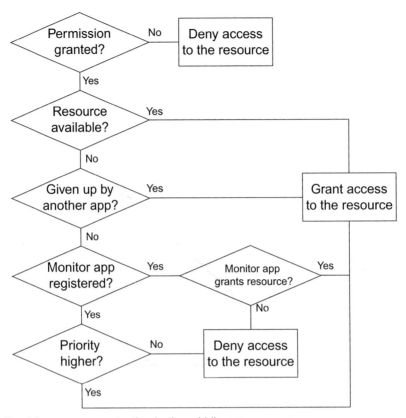

Figure 6.3. Resolving resource contention in the middleware.

conflict occurs (i.e., anytime an application needs a resource that is not available, and when no other application is willing to give up its access to that resource) the middleware will call the resolveResource Contention() method belonging to this class. There are two versions of this method: one used by versions of OCAP prior to version I12 and one used by version I12 and later.

Resource Contention Before Version I12

For earlier versions of OCAP, the full method signature for resolveResource Contention() looks as follows.

```
public org.dvb.application.AppID []
    resolveResourceContention(
        org.dvb.application.AppID requester,
        org.dvb.application.AppID[] owners,
        java.lang.String resourceProxy);
```

This takes as arguments the application ID of the application requesting the resource, the application IDs of those applications that already hold the resource, and the class name of the resource proxy being reserved. This class name should be the name of a concrete class such as `org.davic.mpeg.sections.SectionFilterGroup` rather than one such as `org.davic.resources.ResourceProxy`.

When this method returns, it returns a list of the application IDs granted access to the resource. This list is ordered in the priority they should get the resource, and thus those applications with higher-priority access to the resource come earliest in the list. This way, if there are not enough resources to go around the applications at the end of the list may not be granted access.

Some readers may have noticed a problem with this approach. Calling `resolveResourceContention()` works fine if all of the resources are being used by applications, but what if one or more of them is being used by the middleware?

This is partly solved by the way application IDs are inherited as part of the resource reservation process. If an application tries to reserve a resource indirectly through a middleware call (for instance, when a call to the service selection API needs to reserve the tuner to tune to the correct service), the middleware uses the application ID of that application to make the reservation request. This is not explicitly used anywhere else in the API, although there are other links between the application signaling and the resource reservation process.

For cases in which the middleware is making the request on its own, with no application responsible for doing it (e.g., when the user presses the Channel Up button on the remote and the middleware needs to tune to the new service), a null reference is used instead of the application ID. Although this may not seem like an ideal solution, the monitor application can be sure that only the middleware will use a null reference as its application ID.

So far, so good, but we still have another problem. What happens when the same application (either a real application or the middleware) tries to reserve two separate instances of a resource at once, via two separate resource clients? This case is actually handled by a mechanism we have already seen but have discussed only briefly.

We have already seen that the `resolveResourceContention()` method takes a list of application IDs representing the applications that already have access to the resource. It is not specified whether this list can have duplicate entries, as when, for instance, one application has reserved two instances of the resource. Because of this, it is also not specified what the monitor application should return if the list does contain duplicate entries, or if the application ID of the resource making the new request is the same as one already holding a resource. In this case, a sensible set of behavior parameters is as follows.

- It should always return a list containing all of the application IDs in the list of resource holders and the application ID of the resource making the request. The only time it should not do this is when it explicitly denies an application access to a resource.

- If the list contains duplicate entries, the entries should be placed next to one another in the list. In other words, instances of the same application ID should have priorities as close to one another as possible.
- Applications, the OCAP middleware, and the `ResourceContentionHandler` must all recognize that there is no way for the `ResourceContentionHandler` to define exactly how resources may be shared. It can only determine the order in which applications get access to resources, and if high-priority applications reserve more resources than is available for all applications then low priority applications will be unlucky.
- Applications must recognize that if they make another request for a resource they are already holding the middleware may choose to take that resource away from the part of the application currently holding the resource to give it to another part. Alternatively, it may decide to take the resource away from the application altogether.

This is our interpretation of the OCAP specification and it is not formally specified anywhere, and thus our interpretation may not be correct. Version I12 of OCAP changed this and clarified things, as discussed in material to follow.

Resource Contention in Later Versions

Before we look at this specific case, though, let's look at the wider changes made to resource management in version I12 of the OCAP specification. In this version, the method signature for `resolveResourceContention()` is as follows.

```
public org.ocap.resources.ResourceUsage[]
  resolveResourceContention(
    org.ocap.resources.ResourceUsage newRequest,
    org.ocap.resources.ResourceUsage currentReservations);
```

The `ResourceUsage` class represents a group of resources needed by a single operation. In many situations, only one resource will be needed, but other cases may require more than one resource to be reserved implicitly. For instance, a service selection operation may implicitly reserve a tuner and an MPEG decoder. The interface to `ResourceUsage` follows.

```
public interface ResourceUsage {

  public org.dvb.application.AppID getAppID();

  public org.davic.resources.ResourceProxy
    getResource(java.lang.String resourceName);

  public java.lang.String[] getResourceNames();

}
```

As you can see, each `ResourceUsage` instance includes the `AppID` of the application that owns the resource, or in the case of resources reserved by the middleware the `AppID` of the application that triggered that reservation. It also contains a list of the fully qualified names of the resource proxy classes needed (e.g., `org.havi.ui.HVideoDevice`). In that this describes a single request, there will be one instance of `ResourceUsage` for every request that results in resource contention.

The `getResource()` method lets the monitor application find out a little more about the current state of a request. This returns a reference to the `ResourceProxy` object the application is using to communicate with the underlying resource. If no resource of the specified type is currently reserved by that request, this method will return a null reference. The basic process for handling resource contention, as follows, is very similar to the one we have already seen.

1. The middleware implementation will create a `ResourceUsage` object representing the current request.
2. It will call the `resolveResourceContention()` method of the `ResourceContentionHandler` registered by the monitor application, passing the newly created `ResourceUsage` object as one argument and an array of `ResourceUsage` objects representing any conflicting reservations.
3. The monitor application will return an array of `ResourceUsage` objects representing the requests granted access to resources. As with the earlier version of the method, these are listed in priority order, and thus applications whose requests are listed later in the list may not get access to a resource.

The OCAP specification says that the middleware will use this approach to handle any contention for resources reserved by the methods outlined in Table 6.3. One of the effects of this is that the behavior for multiple requests is now fully specified, in that resource contention is explicitly handled on a per-request basis.

Table 6.3. Resources that will use `ResourceUsage` objects to resolve contention.

Class	Method
`org.havi.ui.HScreenDevice`	`reserveDevice()`
`org.havi.ui.HVideoDevice`	`reserveDevice()`
`org.havi.ui.HBackgroundDevice`	`reserveDevice()`
`org.havi.ui.HGraphicsDevice`	`reserveDevice()`
`org.davic.mpeg.sections.SectionFilterGroup FilterGroup`	`attach()`
`org.dvb.event.EventManager`	`addExclusiveAccessToAWTEvent()`
`org.ocap.media.VBIFilterGroup`	`attach()`
`org.davic.net.tuning.NetworkInterface Controller`	`reserve()`
`org.davic.net.tuning.NetworkInterface Controller`	`reserveFor()`

This approach means that it is easier for the middleware implementation to create a `ResourceUsage` object whenever an application calls one of these methods. This is a slightly simplistic approach, however, and it does not completely solve our problem.

As mentioned previously, some operations such as service selection may not use the DAVIC resource management API, but they will implicitly reserve one or more resources. In the case of service selection and some other operations, more than one resource may be reserved. Service selection will need to reserve both a tuner (represented by the `org.davic.net.tuning.NetworkInterfaceController` class) and an MPEG decoder (represented by the `org.havi.ui.HVideoDevice` class). This means that we may need to create our `ResourceUsage` object at a level above the individual APIs, in order to create a single `ResourceUsage` object that represents both resources.

One of the implications of this is that resource management in OCAP is rather more centralized than it is in MHP. Although each API in an MHP implementation may choose to manage its own resources, it is far more likely that an OCAP implementation will have a central resource manager that is at least responsible for creating and tracking `ResourceUsage` objects.

Common Features of Resource Contention Handling

For both versions of `resolveResourceContention()`, the monitor application can delegate the decision back to the middleware should it choose to do so. By returning a null reference, the application tells the middleware it should handle this particular case of resource contention instead, following the rules we have already seen. Thus, the monitor application does not have to reimplement the standard rules for cases in which those rules apply.

Alternatively, the middleware can exclude one or more applications from access to the resource. Any applications not listed in the array returned by `resolveResourceContention()` do not get access at all, even if there are resources available for them. If `resolveResourceContention()` returns a zero-length array, no application gets access to the resource at that time. This is another example of how the monitor application can control resource allocation.

An Example of a Possible Resource Contention Solution

Now that we have seen how the `ResourceContentionHandler` should deal with the various situations in which resources need to be shared, let's take a look at one possible implementation. This implementation follows the behavior described here, but it also gives higher priority for applications from the network operator and applications that have been stored in the receiver's memory.

```
public AppID[] resolveResourceContention(
  AppID requester,
  AppID owners[],
  java.lang.String resourceProxy) {
```

```
AppID newPriorities[] = null;

// Applications from the network operator or the
// receiver manufacturer get priority, and so they get
// first choice of any resources that they need
if ((requester.getOID() == NETWORK_OPERATOR) ||
    (requester.getOID() == RECEIVER_MANUFACTURER)) {

  // insert the requester at the first element in the
  // new array.
  newPriorities = new AppID[owners.length + 1];
  newPriorities[0] = requester;
  for(int i=0; i<owners.length; i++) {
    newPriorities [i+1] = owners[i];
  }
}
else if (requester.getAID() == UNIMPORTANT_APP) {

  // This application is unimportant no matter what its
  // application priority is, and so it never gets
  // access to resources. In this case, we just copy
  // the original array.
  newPriorities = new AppID[owners.length];
  System.arraycopy(
    owners, 0, newPriorities, 0, owners.length);
}

return(newPriorities);
}
```

Another implementation is shown in Annex L of the OCAP specification. This implementation is not identical to the one shown there, although there are a number of similarities between the two. By comparing and contrasting the two approaches, you may find something that is more suitable to your needs than either of the two examples on its own.

To help with managing priorities, version I10 of OCAP adds the `setApplicationPriority()` method to the `org.ocap.application.OcapAppAttributes` interface. This lets the monitor application change the priority of an application, to override the value specified in the AIT or XAIT. By changing the priority of the application, any resource requests from that application will happen at the new priority. Using this, the resource contention handler can affect the strategy of the middleware's resource manager toward that application.

Resource Management Strategies in OCAP

The ability of the monitor application to resolve resource conflicts means that the network operator has almost complete control of how resources are allocated within the receiver. If a network operator chooses to implement their own resource management strategy, they

Table 6.4. Possible approaches to resolving resource contention.

First come, first served	Applications will not be granted access to a resource unless that resource is already available, and an application will never be pre-empted. This is a purely cooperative model (in that the middleware will always ask applications holding resources to give up those resources), and is only practical in very limited circumstances. Typically, this means that new applications may not be able to start properly if they need access to scarce resources.
Last come, first served	The opposite of first-come/first-served, this strategy grants the resource to the last application or component to request it. This has one big advantage, in that the application that most recently requests a resource is likely to be the one the user is currently using. It does have some disadvantages, however. Auto-start applications will always be granted access to resources, even if the user is busy interacting with another application that was started earlier. For instance, an application associated with a quiz show may take resources away from an EPG. In cases in which applications are not well implemented, this may lead to a tug-of-war between applications as an application that has just lost access to a resource tries to reclaim it.
Application priority	This is the default strategy used by the middleware, and thus it makes very little sense for a monitor application to use this approach. Do not waste space in your monitor application implementing this.
Weighted priority	This can take one of a number of forms. The most common of these is the form in which the application priority as signaled by the network operator is weighted based on a factor such as the organization ID of the application, or on the type of application. For instance, applications with an organization ID matching that of the network operator may have a higher weighting than those with other organization IDs. Alternatively, built-in applications may have a higher priority than downloaded applications, with system applications having a priority that is higher still. This approach must be followed at least in part, in that the application or middleware component that implements the Emergency Alert System functionality must always have the highest priority.

should consider carefully which strategy to employ. Each possible strategy has its own advantages and disadvantages. Table 6.4 outlines potential approaches.

Note: As we have mentioned, the application or middleware component that implements the Emergency Alert System functionality in the receiver will always get the highest priority and must always be granted access to the resources it requests.

Despite all of these strategies, there may still be times when resources become too short for any but the most drastic action. To support this, versions of OCAP later than I10 add the system event API in the `org.ocap.system.event` package. This includes the

`SystemEventListener` interface the monitor application can use to receive notification from the middleware about certain system-level events. At the moment, the most interesting event is the `ResourceDepletionEvent`. This tells the monitor application that the middleware is running so low on resources that killing an application is the only option. By listening for this, the monitor application can make its own choice about which application to kill, and avoid leaving the choice up to the middleware.

Merging OCAP and MHP Resource Management

As well as being the last remnant of the standardized resource manager, the `Resource-ContentionManager` and `ResourceUsage` classes tell us a little bit about the underlying assumptions made by the designers of OCAP. Although there does not have to be a central resource manager, these classes strongly imply that such an entity is present, and thus using a central resource manager may make life easier for OCAP middleware developers.

It is equally possible for MHP middleware stacks to use a central resource manager, and thus this approach provides a good way for middleware architects to use a common design across the two systems. The OCAP resource management component is almost identical to the design of the MHP resource management component, and if no `ResourceContention-Handler` is registered the two components are identical.

Both middleware stacks can use a central resource manager by simply passing on requests to reserve and release resources from the APIs handling scarce resources to the resource manager, and having this resource manager make the decision about whether a request should be granted. All of the details about validating or invalidating resource proxies are left up to the using APIs, and the resource manager is only concerned with keeping track of which application is using which resources, and with resolving conflicts between applications.

From the point of view of each API, there may not seem much point in using a central resource manager. It does have one big advantage over a more decentralized approach, however. In the case in which an application is killed (either because the user or application signaling says it should be killed, or because the middleware has decided it is misbehaving), having a single entity that keeps track of all resources makes it easier to ensure that any resources used by the newly killed application have been freed correctly. Without this central resource manager to do this, the middleware would have to notify each API separately, which would then have to free the resources themselves.

If the middleware does not free resources correctly when it kills the application, this can have serious consequences. Any resource clients the application has registered will not be deleted because the resource management component (or components) will still have references to them for resource notification and these will not be garbage collected. On top of this, not freeing the resources correctly obviously implies that there will be a resource leak within the middleware. This can cause huge problems in supposedly reliable systems, and thus middleware developers must avoid this type of problem. Ensuring that the middleware always frees resources correctly may not be easy but it is vital.

7 Graphics APIs

Almost every ITV application will need to draw some type of user interface (UI) on the screen. In this chapter we will show you how the MHP and OCAP graphics model works, and how to configure the receiver's graphics device to meet your needs. We will also look at higher-level issues such as integrating graphics and video, the UI widgets that are available, and how we can handle input from a remote control or keyboard.

One of the major differences between developing applications for a PC and developing applications for a TV is the way the platform handles graphics. Developers in DTV environments have to be aware of the different graphics models, the issues involved in configuring the display devices and integrating video into their applications, and the need to design a UI that works well given the limitations of a TV screen.

The Java Abstract Window Toolkit (AWT) is designed specifically for a PC environment, and it does not cope well with many of the differences previously mentioned. This is not entirely surprising, in that the AWT was never designed to handle the problems that come with developing consumer products.

To solve these problems, MHP and OCAP have borrowed a great deal from other standards, and in particular they have borrowed heavily from the HAVi APIs for graphical interfaces. The HAVi (Home Audio/Video Interoperability) specification defines a way of streaming audio and video between consumer devices. More interestingly to us, it also defines a set of Java UI classes for use on consumer devices.

DVB does not believe in inventing a new solution to a problem if a suitable solution already exists, and thus DVB adopted the HAVi UI API as the UI solution for MHP. Technically, the UI classes used in MHP are the HAVi Level 2 GUI to distinguish them from any HAVi GUI components that use native code.

Before we look at how HAVi solves the problems that come with developing DTV applications, we need to take a closer look at what these problems are. The biggest single cause of problems is cost: DTV receivers are very cost sensitive, and thus typically do not have a high specification. This has a big effect on the way graphics work in a DTV receiver. We will look at this in more detail later, but we need to consider many other things, including the following, when we develop software for a TV display.

- *Pixel aspect ratios:* Video and TV applications typically use non-square pixels, whereas computer graphics APIs usually assume that pixels are square. This is one of the reasons TV and computer displays have different resolutions, even when they have the same aspect ratio.

  To make it worse, some receivers may have a separate graphics controller, and the graphics and video layers may not be combined until they are about to be drawn on the screen. If the two layers use different pixel shapes (which is possible; see Figure 7.1), an application trying to overlay graphics on video to pixel-perfect precision could be in for a very difficult time. This simply may not be possible on some platforms.

- *Aspect ratio changes:* As if the different pixel aspect ratios were not enough, the aspect ratio of the TV signal itself (the display aspect ratio) may be either 4:3 (standard) or 16:9 (widescreen). It does not even stop there, in that an image in one aspect ratio can be displayed on a TV of another aspect ratio in one of several ways. The aspect ratio of the display, and the way the picture is mapped, may be changed either by the TV or by the user. In either case, the effect on graphics and images not designed for that aspect ratio can be pretty unpleasant from an application developer's point of view as an image is stretched or squashed. Figure 7.2 shows some of the approaches that can be used for aspect ratio conversion.

- *Translucency and transparency:* How can we make graphics transparent so that the viewer can see what is beneath them? If we were using the Java 2 platform, we could use the transparency support included in the Java 2 version of AWT. Unfortunately, we cannot guarantee that we will be using a Java 2 platform.

- *Color space issues:* How do we map the RGB color space used by Java on to the YUV color space used by TV signals? This is more complex than it may appear, given that the two color spaces do not overlap completely.

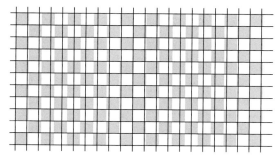

Figure 7.1. Pixel aspect ratios can be different between the video layer (gray squares) and the graphics layer (black grid).

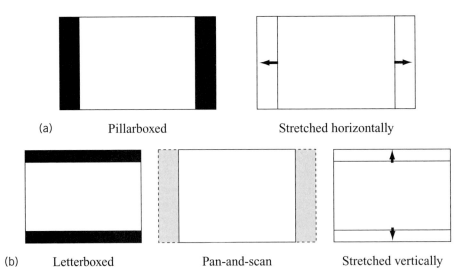

(a) Pillarboxed Stretched horizontally

(b) Letterboxed Pan-and-scan Stretched vertically

Figure 7.2. Devices can handle aspect ratio conversion in many different ways; for example, (a) displaying 4 : 3 images on a 16 : 9 display and (b) displaying 16 : 9 images on a 4 : 3 display.

- *No window manager:* Window managers are too complex for many DTV receivers, and thus an application needs some other way of getting some screen real estate it can draw on. This also means that an application will have to coexist with other applications in a way different from standard Java or PC applications. Different middleware standards take different approaches to this problem, and we will examine this in more detail later in the chapter.
- *UI differences:* A low-resolution display such as a TV, and the difference in the type of applications, means that conventional computer-style UIs do not work well. The application may also need a new UI metaphor, if the user only has a TV remote to interact with it.
- *No free-moving cursor:* Many receivers do not support a pointing device such as a mouse, and thus any navigation must be carried out with the arrow keys and a few other buttons. This means that complex navigation models do not work well.

Given all of these differences, it is probably not surprising that developers need to think about many graphics-related issues. Before we discuss all of the differences, however, we will look at the areas of similarity between desktop Java and Java in DTV applications.

Many of the lessons learned from developing AWT applications will still apply here, although typically you will not want the standard PC-like look and feel for any applications. MHP uses a subset of AWT as defined by the Java 1.1.8 API, with the addition of the HAVi UI classes and some elements from PersonalJava 1.2a. There are a few differences caused by the constraints of a TV environment, but most of the time anyone who can develop an AWT application will be happy working with the graphics APIs in MHP or OCAP.

The differences come in the extensions MHP has added. The HAVi UI classes are contained in the `org.havi.ui` package, and these relate to many of the graphics-related tasks a receiver will need to carry out. These fall into the following four basic categories, which we will look at in more detail later in this chapter.

- A UI widget set for TV displays
- Classes for manipulating the different elements of the display and modeling the display stack
- Classes for handling resource sharing between different applications
- User input extensions for supporting input from a remote control unit

The Display Model in a DTV Receiver

Before we can really understand the changes in other areas, we need to understand how the graphics model works in an MHP or OCAP system. This is slightly different from what you may be used to in the PC world, because we do not have enough processing power to do all of the work in software. Because of this, the capabilities of the hardware play an important part in the structure of the graphics model.

We can split the display in a DTV receiver into three logical layers. From back to front, these are the background layer, the video layer, and the graphics layer.

The background layer lets us show a simple color behind any graphics on the screen, and this will fill in any areas not covered by video (for instance, if we scale the video so that it is displayed in only part of the screen). If we are lucky, this layer will also let us display a still image, although this usually has a number of limitations (examined in material to follow).

Next is the video layer. As its name suggests, this is the layer at which any video content is shown. This video may not cover the entire screen, in that a DTV receiver can usually do some limited scaling and repositioning of the incoming video. MHP and OCAP only say that a receiver should be able to display video at full-screen resolution or at quarter-screen resolution in any of the four quarters of the screen. More advanced receivers, however, can perform arbitrary scaling and positioning operations on the video.

The top layer of the display stack is the graphics layer. This layer contains any AWT widgets an application wants to display. Given the nature of a typical set-top box, you cannot expect anything too fancy from this graphics layer. MHP receivers are only required to support a resolution of 720×576 pixels, whereas OCAP receivers will support a minimum of 640×480 pixels. In both cases, the receivers have to support a minimum of 256 colors. These are minimums, however, and thus receivers can support higher resolutions and color depths. This has its own problems, however, in that at higher resolutions a pixel in the graphics layer will not map directly onto a pixel in the video layer.

Some receivers will actually have more than one layer making up the graphics layer. There may be a separate layer for a cursor, or there may be two graphics layers with some trans-

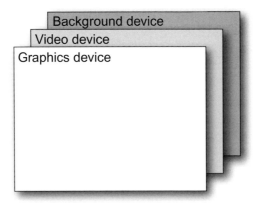

Figure 7.3. Devices in the MHP display stack.

parency between them. To MHP and OCAP application developers, this does not matter. The MHP and OCAP middleware only have one logical graphics layer, and it is up to the middleware to decide how this gets mapped to the underlying hardware. Figure 7.3 shows various devices in the MHP display stack.

Each of these layers can be configured separately, in that their content may be produced by completely separate parts of the hardware. In practice, however, there are usually some interdependencies between two or more of the layers. The cost pressure on DTV receivers normally means that a component will carry out more than one task, if possible, and this does have an impact on the capabilities of the receiver. Because these differences are hardware dependent, we cannot even assume they will be the same from one platform to another.

The lesson to be learned here is a pretty simple one: change only those parts of the graphics configuration you really have to. An application must be flexible when dealing with graphics and video configuration, and it may not always get what it needs. At the same time, it must not disturb other applications that are already running unless this is necessary.

Although layers are logically organized like this, different hardware platforms may support a different physical representation of these layers. High-end platforms may carry to video decoding in software, for instance, and use software compositing to present background, video, and graphics in a single physical graphics layer. Other platforms may support more graphics layers, or even more video layers. In each case, the logical organization of these layers will follow what we have discussed here.

HScreens and HScreenDevices

To help solve the problems of configuring a display device and finding out what its capabilities are, HAVi defines a number of classes that model the various parts of the display stack. The first of these is the `HScreen` class. This represents a physical display device, and every MHP or OCAP receiver will have one `HScreen` instance for every physical display device connected to it. In most cases, a receiver will only have one `HScreen` object active at any time.

Every `HScreen` has a number of `HScreenDevice` objects associated with it. These represent the various layers in the display. `HScreenDevice` is only a base class, and we are more interested in its subclasses, as follows.

- `HBackgroundDevice` represents the background layer.
- `HVideoDevice` represents the video layer.
- `HGraphicsDevice` represents the graphics layer.
- `HEmulatedGraphicsDevice` represents a graphics device being emulated by one with a different "real" configuration.

Each instance of an `HScreen` object will be associated with at least one `HVideoDevice` instance, at least one `HGraphicsDevice` instance, and (if it is an MHP receiver) one `HBackgroundDevice` instance. As we can see from the `HScreen` class definition that follows, we can get references to the devices associated with a given `HScreen` instance by calling the appropriate methods on the `HScreen` object.

```
public class HScreen {

  public static HScreen[] getHScreens();
  public static HScreen getDefaultHScreen();

  public HVideoDevice[] getHVideoDevices();
  public HGraphicsDevice[] getHGraphicsDevices();

  public HVideoDevice getDefaultHVideoDevice();
  public HGraphicsDevice getDefaultHGraphicsDevice();
  public HBackgroundDevice getDefaultHBackgroundDevice();

  public HScreenConfiguration[]
    getCoherentScreenConfigurations();

  public boolean setCoherentScreenConfigurations(
    HScreenConfiguration[] hsca);
}
```

For each `HScreen`, we can get references to the default instances of the three devices. The instances returned by these methods will usually be the general-purpose devices for this `HScreen`, such as the general-purpose graphics plane or the main video plane. In the case of the background device, the default device is the only one we will get because we can only have one background device.

Support for the `HBackgroundDevice` is optional for OCAP receivers, and thus any OCAP applications that want to use the background layer should check that it is present before they try to use it. Otherwise, the user may see some ugly display glitches when the application

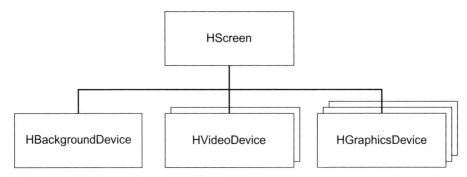

Figure 7.4. `HScreenDevices` in an MHP/OCAP receiver.

developer assumes that a background layer is present when it is not. On an OCAP receiver, the `getDefaultHBackgroundDevice()` method will return a null reference if no background device is present.

For the video and graphics layers of the display stack, we may get more than one `HVideoDevice` or `HGraphicsDevice` instance, respectively. If the device has two tuners and MPEG decoders (for instance, to support picture-in-picture), we may have two video devices. Similarly, we may have several graphics devices in some receivers. Figure 7.4 shows the use of `HScreenDevices` in an MHP/OCAP receiver.

Configuring Screen Devices

Once we have a reference to a device, we may want to change its configuration in order to set some specific options we need. Alternatively, we may want to query a device to see if it can support a certain configuration, or to see how close we can get to our desired configuration. In both of these cases, we use the configuration class associated with each type of device.

Each screen device class has an associated configuration class. This class has a name that matches the device it applies to. For example, for an `HVideoDevice` the configuration class is the `HVideoConfiguration` class, `HGraphicsDevices` are configured using the `HGraphicsConfiguration` class, and the `HBackgroundDevice` uses the `HBackgroundConfiguration` class. All of these are subclasses of the `HScreenConfiguration` base class.

These classes let us set a number of parameters, such as the pixel aspect ratio, the resolution, and other settings specific to each device. If you are familiar with the Java 2 platform, this may seem familiar. Java 2 uses a similar pattern of graphics devices and graphics configurations. The similarity is deliberate. After all, there is no point reinventing a design pattern when a perfectly good one already exists. These similarities only go so far, however. In the Java 2 platform, configuration classes represent the different configurations (resolutions and color depths) of the display device. In a DTV receiver, these configurations are more complex

Table 7.1. Screen devices and their associated configuration and configuration template classes.

Screen Device	Configuration Class	Configuration Template
HBackground Device	HBackgroundConfiguration HStillImageBackground Configuration	HBackgroundConfig Template
HGraphicsDevice	HGraphicsConfiguration	HGraphicsConfigTemplate
HEmulatedGraphics Device	HEmulatedGraphicsConfiguration	
HVideoDevice	HVideoConfiguration	HVideoConfigTemplate

because of the interaction between the various layers of the display. For this reason, we can modify the settings in a configuration without having to get a new configuration object.

Each configuration is defined by a number of parameters, and applications can define a set of parameters for a device using the appropriate subclass of the HScreenConfigTemplate class. As with HScreenConfiguration, there is one subclass for every type of HScreen-Device we have. The relationships among the various classes are outlined in Table 7.1.

The HBackgroundDevice has two separate configuration classes. HBackgroundConfiguration defines a configuration in which the background is set to a single color. If we want to display an image in the background layer, we should use an HStillImageBackgroundConfiguration instead. This is a subclass of HBackgroundConfiguration, and thus it can be used anywhere a normal HBackgroundConfiguration can be used.

We have several ways of finding a suitable configuration for a given device. Each screen device class has a getDefaultConfiguration() method, which returns the default configuration for that device. If we want to use a different configuration, the getConfigurations() method will return all of the possible configurations supported by that device. For every configuration, we can use the getConfigTemplate() method to find the device settings that match that configuration. For each configuration class, this will return a matching instance of a subclass of HScreenConfigTemplate. The corresponding template classes are shown in Table 7.1.

An alternative approach is to create an HScreenConfigTemplate that matches the configuration we would like to set, and then check to see how closely the middleware can match our preferences. Since the screen is a shared resource, we may have more than one application that would like to change the configuration of a device, and thus we need some way of managing this process in order to make sure that applications get the configuration that is best for all of them.

Once we have a configuration template, we can use that to change the settings for a device. Each configuration template has a number of parameters that can be set, and these parame-

Table 7.2. Configuration parameters and the devices they apply to.

Parameter	HBackgroundDevice	HGraphicsDevice	HVideoDevice
PIXEL_ASPECT_RATIO	✓	✓	✓
PIXEL_RESOLUTION	✓	✓	✓
INTERLACED_DISPLAY	✓	✓	✓
FLICKER_FILTERING	✓	✓	✓
VIDEO_GRAPHICS_PIXEL_ALIGNED		✓	✓
SCREEN_RECTANGLE	✓	✓	✓
ZERO_BACKGROUND_IMPACT	✓	✓	✓
ZERO_GRAPHICS_IMPACT	✓	✓	✓
ZERO_VIDEO_IMPACT	✓	✓	✓
CHANGEABLE_SINGLE_COLOR	✓		
STILL_IMAGE	✓		
IMAGE_SCALING_SUPPORT		✓	
MATTE_SUPPORT		✓	
VIDEO_MIXING		✓	
GRAPHICS_MIXING			✓

ters map onto the underlying device settings. Due to the interaction between the various graphics layers, we cannot simply set these values without thinking about the rest of the system or other applications that may be running. When we want to apply a new configuration to a screen device, it is best to think of these parameters as a set of constraints that are applied to the configuration. An application tells the middleware what constraints it would like applied, and the middleware does its best to apply these constraints without violating any other constraints that have already been applied by other applications, by the middleware, or by the graphics hardware itself.

Each parameter that can be set is represented by a constant value in `HScreenConfigTemplate` or one of its subclasses. Table 7.2 outlines which parameters can be set for every screen device.

We will describe just a few of these parameters here. The ZERO_GRAPHICS_IMPACT, ZERO_BACKGROUND_IMPACT, and ZERO_VIDEO_IMPACT parameters are used to specify that any configuration should not have an affect on already-running graphical applications or on currently playing video. These are useful if the application wants to be very polite and not disturb other applications.

The VIDEO_GRAPHICS_PIXEL_ALIGNED parameter indicates whether the pixels in the video and graphics layers should be perfectly aligned (e.g., if the application wants to use pixel-perfect graphics overlays onto video). This is useful if the application wants to overlay specific parts of the video. For instance, a quiz show may show the current question (and

Table 7.3. Priorities for device configuration parameters.

Priority	Meaning
REQUIRED	This preference must be met.
PREFERRED	This preference should be met, but may be ignored if necessary.
UNNECESSARY	The application has no preferred value for this preference.
PREFERRED_NOT	This preference should not take the specified value, but may if necessary.
REQUIRED_NOT	This preference must not take the specified value.

the possible answers) on the screen as part of the video stream, but an application associated with that show could place its own graphics over this to let the user play along at home.

Now that we have seen the parameters that can be set, let's take a closer look at how we can set them. Each parameter has two parts: the desired value of the parameter and a priority. The desired value is self-explanatory, and the type of this value depends on the parameter we want to set. The priority tells the middleware how important a particular setting is to the application. This can take one of the values outlined in Table 7.3.

This allows the application a reasonable degree of freedom in specifying the configuration it wants, while still allowing the receiver to be flexible in matching the desires of the application with those of other applications and the constraints imposed by other screen device configurations.

Once we have created an instance of the appropriate subclass of HScreenConfigTemplate and have set its parameters to match the configuration we want, we can see whether the screen device we want to configure can actually support that configuration. Each subclass of HScreenDevice has a getBestConfiguration() method that takes a matching subclass of HScreenConfigTemplate as a parameter.

When an application calls this method, the middleware checks the preferences specified in the configuration template and attempts to find a valid configuration. The best configuration is one that matches

- all of the parameters specified as REQUIRED,
- none of the parameters specified as REQUIRED_NOT,
- as many as possible of the parameters specified as PREFERRED, and
- as few as possible of the parameters specified as PREFERRED_NOT.

If the middleware cannot satisfy all of the REQUIRED or REQUIRED_NOT parameters, getBestConfiguration() will return a null reference. Otherwise, it will return a reference to a configuration that comes closest to matching what we requested.

A variant of this method can take an array of configuration templates. In this case, the middleware will try to meet the constraints specified in all of the templates, and return one configuration that comes closest to matching all of them. Again, if no configuration can satisfy all of the REQUIRED and REQUIRED_NOT parameters a null reference will get returned. This method allows a very polite application to find one configuration that best fits the set of configurations it may need to apply, and then simply use that as a best compromise in order to avoid changing the configuration too often.

As we saw earlier, selecting an appropriate configuration is very much an exercise in constraint modeling, and this is not just due to the interactions between devices in the graphics hierarchy. The various parameters set by other applications, combined with the capabilities of the various screen devices in the system, define a set of constraints that must be applied to a new configuration. The middleware has to solve these constraints in order to produce a new HScreenConfigTemplate that matches the constraints supplied by our template.

For instance, let's assume an application has specified a value of REQUIRED for the ZERO_GRAPHICS_IMPACT parameter. If our application then sets a value of REQUIRED for the VIDEO_GRAPHICS_PIXEL_ALIGNED parameter, the receiver can only satisfy both of these constraints by changing the configuration of the video device to match the graphics device. If the video device does not support the same resolution and pixel aspect ratio of the current graphics configuration, the middleware will not be able to define a configuration that meets our needs. If this happens, any attempts by our application to get a configuration template will return a null reference.

For graphics devices, the application may choose to meet the constraints in a different way. In some cases, the middleware can emulate the configuration the user wants while using a different "real" configuration for the device. For instance, the receiver may emulate a display with an aspect ratio of 4:3 on a 16:9 display. Any emulated graphics device will be an instance of the HEmulatedGraphicsDevice class, a subclass of HGraphicsDevice. Although using an emulated graphics device may be much slower than using a real graphics device, from an application's perspective there is no difference in behavior.

Any HEmulatedGraphicsDevice will use an HEmulatedGraphicsConfiguration instead of a normal HGraphicsConfiguration. This provides all of the information you can get from an HGraphicsConfiguration, but it also extends it to give some information about the mapping between the emulated device and the real device the receiver uses to perform the emulation. An application can call the getImplementation() method on an HEmulatedGraphicsDevice to get an HGraphicsConfigTemplate object representing the configuration of the underlying graphics device. The application can use this to compare the emulated and real devices, and then take any necessary actions to optimize performance if necessary.

One real HGraphicsDevice might emulate several HEmulatedGraphicsDevices at the same time, depending on the configurations the underlying hardware supports. Of course, supporting many emulated configurations may result in an unacceptable drop in performance, and thus middleware implementers need to take this in to account when deciding whether to support emulated graphics devices. Being able support several emulated devices does not always mean this is a good idea.

To make sure that designers and application developers can make some basic assumptions about the design of their application's UI, MHP and OCAP specify some minimum resolu-

Table 7.4. Minimum device resolutions for an MHP receiver.

Device	Horizontal Resolution	Vertical Resolution
Background	720	576
Video	720	576
Graphics	720	576

Table 7.5. Minimum device resolutions for an OCAP receiver.

Device	Horizontal Resolution	Vertical Resolution
Background	640	480
Video	640	480
Graphics	640	480

tions for the various devices in a receiver. For an MHP receiver, Table 7.4 outlines the minimum supported resolutions.

These assume a PAL display, and each receiver must support at least one device of each type capable of full-screen display. In the case of background devices, there will be only one device of that type.

The graphics layer is required to support only non-square pixels. Optionally, an MHP receiver can also support graphics devices with square pixels at resolutions of 768×576 (for 4:3 displays) and 1024×576 (for 16:9 displays). Receivers also need to support an emulated graphics device with an aspect ratio of 14:9 in order to provide acceptable output on either 4:3 or 16:9 displays.

In practice, this does not help images much, but use of the 14:9 aspect ratio means that text will be the same width (in pixels) on both 16:9 and 4:3 displays because of the pixel-to-point ratios MHP specifies for text at the different aspect ratios. This useful design trick avoids any problems with text rendering interfering with a graphical layout. Because the text will be the same pixel width no matter what the display aspect ratio, life becomes slightly easier for designers. In that OCAP is primarily aimed at U.S. markets, it uses the NTSC display standard. OCAP receivers must provide the minimum resolutions outlined in Table 7.5.

As of version I11 of OCAP, receivers that support high-definition video output must also support a graphics resolution of 960×540. Unlike MHP, OCAP specifies that these resolutions are based on square pixels, and thus OCAP developers do not have to worry about pixel aspect ratios. An OCAP receiver must support both 16:9 and 4:3 display aspect ratios, however.

OCAP receivers can optionally support other resolutions for graphics devices, such as 1280 × 720 (supporting a full-screen 16:9 display with square pixels) or 704 × 480 (a full-screen 4:3 display with non-square pixels), but only 640 × 480 has to be supported. OCAP receivers also do not have to support any emulated graphics devices. The application can find the size of the current graphics device by calling the `java.awt.Toolkit.getScreenSize()` method. This will return the pixel resolution of the default graphics device's current configuration. Because this configuration may change over time, the screen size returned by this method may also change. As we have already seen, support for the `HBackgroundDevice` is also optional in OCAP receivers and thus applications should check that an `HBackgroundDevice` is available before they use it.

Screen Devices and Resource Management

Screen devices are scarce resources, and thus an application must reserve the right to change the configuration of a device. By allowing only one application to change the configuration at any time, the middleware can make sure that any changes are compatible with the configurations of other screen devices in the system. If an application tries to set the configuration when it has not reserved the device, the middleware will throw an `HPermissionDeniedException`.

Like many of the other APIs in an MHP or OCAP receiver, the resource reservation model in HAVi uses the DAVIC resource notification API we met in the last chapter. To reserve a device, the application must call `HScreenDevice.reserveDevice()` on the instance of the screen device it wishes to reserve, passing an object implementing `org.davic.resources.ResourceClient` as an argument. If no other application has reserved that device, the method returns `true` and only that application can change the configuration of the device. Calling the `HScreenDevice.releaseDevice()` method on the same screen device releases the resource for other applications.

As a good citizen, the application should only reserve a device for as long as it needs to query and set the configuration it wants, and it should only reserve the devices it absolutely must. Although reserving it for longer will prevent any other applications from changing its configuration, this is not a very friendly thing to do, and thus applications should avoid doing this unless it is vital that the configuration not change. Offhand, we cannot think of any reason an application would ever need to do this. In most situations, the application should release the device as soon as it is finished setting the configuration.

Unfortunately, things are not quite as straightforward as we have made them sound. Earlier in the chapter we mentioned that the configurations of different layers in the display stack might not be completely independent. Cost pressures mean that some hardware components may be shared between layers in the display stack. This means that changes to the configuration of one screen device may require changes in the configuration of another screen device.

The most obvious example of this is the relationship between the background layer and the video layer. Still images in the background layer will normally be MPEG I-frames, and in order to display these images the receiver will often reuse the hardware MPEG decoder.

Doing this means that the MPEG decoder cannot be used for decoding a broadcast video stream, and so decoded video may stop while a background image is being displayed. Depending on the hardware platform, this may mean just a small glitch in the video play-back while the I-frame is decoded, or it may mean disruption to the video for the entire time the background image is displayed.

What this means when we are configuring a device is that if applying a new configuration to one device will cause changes to the configuration of another device we must reserve that device too. If your application gets HPermissionDeniedExceptions when it attempts to configure a device, check which devices you are reserving. It is probably the case that your configuration means that other devices need to be reconfigured, and you have not reserved the necessary devices. Reserving other devices of the same type as the one you are recon-figuring (e.g., reserving all HGraphicsDevices if you are configuring an HGraphicsDe-vice) or reserving devices that use the same underlying resources (for example, reserving the HBackgroundDevice when you want to reconfigure an HVideoDevice) will probably solve this problem.

In practice, reserving only the devices you need to change may not be easy. We have seen that there may be dependencies between the different devices, but there is no easy way of telling what these dependencies are for any given platform. An application can only guess at what they are, and try to reserve the smallest number of devices it thinks will satisfy those dependencies.

A Practical Example of Device Configuration

Let's examine how this works, taking the graphics device as an example. The HGraphics-Device class has the following interface.

```
public class HGraphicsDevice extends HScreenDevice {

  public HGraphicsConfiguration[]
    getConfigurations();

  public HGraphicsConfiguration
    getDefaultConfiguration();

  public HGraphicsConfiguration
    getCurrentConfiguration();

  public boolean setGraphicsConfiguration(
    HGraphicsConfiguration hgc)
    throws SecurityException,
        HPermissionDeniedException,
        HConfigurationException;

  public HGraphicsConfiguration getBestConfiguration(
    HGraphicsConfigTemplate hgct);
```

```
    public HGraphicsConfiguration getBestConfiguration(
       HGraphicsConfigTemplate hgcta[]);

}
```

We have already seen most of these methods earlier in the chapter, and thus we will not describe them here in any detail. The `setGraphicsConfiguration()` method lets us set the configuration of the device. This can throw a number of exceptions. If the network operator has not granted permission for this application to change the configuration of a screen device, it will throw a `SecurityException`. If we have not reserved the device, or if setting the configuration of this device will affect another device we have not reserved, this method will throw an `HPermissionDeniedException`. Finally, if the screen device cannot provide the functionality we need (i.e., if `getBestConfiguration()` would return a null reference when called with that configuration template), an `HConfigurationException` will get thrown. Now that we have seen the interface to the graphics device, let's look at how we configure it. Consider the following configuration code.

```
// First, get the HSCreen we want to use (the default
// one in this case)
HScreen screen = HScreen.getDefaultHScreen();

// Now get the HGraphicsDevice we want to use (again,
// we use the default).
HGraphicsDevice device;
device = screen.getDefaultHGraphicsDevice();

// Create a new template for the graphics configuration
// and start setting preferences
HGraphicsConfigTemplate template;
template = new HGraphicsConfigTemplate();

template.setPreference(
   template.IMAGE_SCALING_SUPPORT,
   template.PREFERRED);

template.setPreference(
   template.ZERO_VIDEO_IMPACT,
   template.REQUIRED);

// Now get a device configuration that matches our
// preferences
HGraphicsConfiguration configuration;
configuration = device.getBestConfiguration(template);

// Finally, we can actually set the configuration. Before
// doing this, we need to check that our configuration is
// not null (to make sure that our preferences could
// actually be met).
if (configuration != null) {
```

```
// we must reserve the device before we can set a
// configuration, and we should release it as soon as
// we are done. We need to pass an
// org.davic.resources.ResourceClient when we reserve
// the device
device.reserveDevice(myResourceClient);

// Now set the configuration
try {
   device.setGraphicsConfiguration(configuration);
}
catch (HPermissionDeniedException hpde) {
   // do nothing for now
}
catch (HConfigurationException hce) {
   // do nothing for now
}

// release the device as soon as possible to avoid
// inconveniencing other apps
device.releaseDevice();
}
```

HScenes and HSceneTemplates

The changes to the display model are only a few of the changes MHP and OCAP developers will have to deal with. We also need to consider a number of changes to the way AWT works.

Some of these (examined in material to follow), such as the changes to the widget set, are straightforward. There are also some philosophical changes, however, and we need to understand these properly if we are going to develop high-quality applications.

Normally, if we wanted to display something on the screen our application would create one or more instances of the java.awt.Frame class and then add and remove graphical components to this. In essence, the Frame represents the window the application is drawing in, and this is in fact how it is described in the API documentation: "A Frame is a top-level window with a title and a border." A Frame can be moved around the screen, it can overlap other Frames, and it may even be partially off the screen.

On a PC or a workstation, this is fine because we have a full window manager that supports this. DTV receivers often do not have a window manager, partly because a window manager may be too big and partly because we only have a remote to control everything. If you have ever tried running Windows at 640×480 without a mouse, you know how painful this can be. Now imagine controlling it with only a few of the keys on the keyboard. The concept of a "ten-foot UI" (one designed to be used from ten feet away) affects much more than just the widget set used for the UI.

If the middleware does not use a window manager, using a Frame as our top-level component causes a few headaches for middleware developers. First, many of the methods on the Frame class assume that we have a full GUI that supports multiple overlapping windows, and these methods are not very useful if we do not. Second, using the Frame or Window class for our top-level component means that an application can create more than one window, and that it can put it anywhere on the screen at any size. If we are running a full window manager, this is no problem, but on a TV it quickly becomes very difficult for both the user and for the middleware.

To solve this problem, HAVi introduced a new concept: the HScene. On systems that use the HAVi GUI classes (including OCAP and MHP), the HScene becomes the top-level graphical component for an application. Conceptually, an HScene is similar to an AWT Window or Frame, although there is no need for a full window manager. HAVi supports a number of possibilities for displaying these top-level components, including the following.

- A full windowing system in which top-level components can overlap and can be resized without any restrictions.
- A paned windowing system in which components cannot overlap and may be resized when a new top-level component is created. Similarly, there may be restrictions on how they can be moved and resized.
- A full-screen approach in which only one top-level component is visible at a time but without restrictions on its size or position. The component will be hidden, however, when another application gets focus.

What HScenes do is give us a way of defining an area of the screen that applications can draw into, without saying anything about how those areas interact with each other. In essence, it makes applications reserve screen real estate as if it were a scarce resource. On some platforms (depending on the choices made for managing how applications share the screen) it may actually be a scarce resource. This is not the case in an OCAP receiver, however, because receivers that support versions I10 and later of OCAP must support a full windowing system. Despite this, some restrictions may still apply.

Applications can have only one HScene at any time, which makes things simpler for both the middleware and the user. It means that each application has one top-level component, and that it must draw all of its UI elements in that. Restricting the application like this makes navigation far easier for the user, but it also means that the middleware can take a simpler approach to managing how HScenes are displayed, such as using any of the schemes we saw previously or a different scheme entirely. The middleware has the freedom to choose the implementation strategy that best suits its needs and the capabilities of the underlying platform, while still giving applications a consistent model of the display.

To applications, the HScene is just like any other AWT Container object (although creating it is a bit different, as we will see in material to follow). An application can add and remove components to the HScene as if it were any other container. As we have already

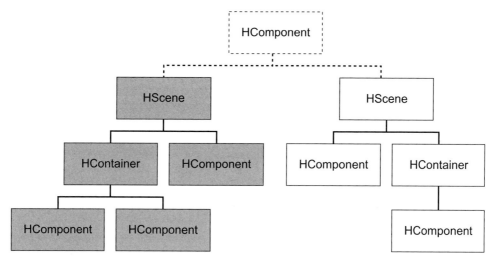

Figure 7.5. Components and HScenes in the display hierarchy.

seen, however, an HScene is the application's top-level component, and thus different platforms may have different values for the HScene's parent object in the graphics hierarchy. For this reason, applications should not call getParent() on an HScene, because the results are not interoperable.

Figure 7.5 shows the effect this has on applications. The application that owns the white components cannot manipulate the shaded components, in that these belong to other applications. Indeed, the application may not even know that these components exist.

When the application terminates, or when it wants to create another HScene, it must use the HScene.dispose() method to free the screen resources it is currently using and destroy any existing HScene. This tells the middleware that the area of the screen used by that HScene can be reallocated, and lets the middleware free any other resources the HScene used.

Creating an HScene

Given the differences between an HScene and a Frame, we cannot create an HScene in the same way we would create a Frame. The middleware needs to make sure that the HScene meets any constraints that must apply to it, and thus we use the org.havi.ui.HScene-Factory class to create an HScene for us.

Using a factory class allows the middleware to make sure that the HScene we get is as close to our requirements as it could be, while at the same time meeting all limitations imposed by the platform and other applications. We tell the HSceneFactory what type of HScene we want using almost the same technique used for screen devices, and the HSceneFactory tells us which type of HScene it can create that comes closest to what we want.

```
public class HSceneFactory{

  public static HSceneFactory getInstance();

  public HSceneTemplate getBestSceneTemplate(
    HSceneTemplate hst);

  public HScene getFullScreenScene(
    HGraphicsDevice device,
    Dimension resolution);

  public HScene getBestScene(HSceneTemplate hst);

  public void dispose(HScene scene);

  public HSceneTemplate
    getBestSceneTemplate(HSceneTemplate hst);

  public HSceneTemplate resizeScene(
    HScene hs,
    HSceneTemplate hst)
    throws java.lang.IllegalStateException;

}
```

As with screen devices, we use a template object to do this. The HSceneTemplate allows us to specify a set of constraints on our HScene (such as size, location on screen, and a variety of other factors; see the following interface), and lets us specify how important those constraints are to us. For instance, we may need our HScene to be a certain size so that we can fit all of our UI elements into it, but we may not care so much where on screen it appears.

To handle this, each property of an HScene has a priority: REQUIRED, PREFERRED, or UNNECESSARY. This lets us specify the relative importance of the various properties, just as we can for screen devices. The middleware will try to create a scene that matches all of the REQUIRED parameters, while meeting as many of the PREFERRED parameters as possible. If the middleware cannot satisfy all of the REQUIRED parameters, it will not create a new HScene. The interface to the HSceneTemplate class follows, which indicates the preferences and priorities that may be set.

```
public class HSceneTemplate{

  // priorities
  public static final int REQUIRED;
  public static final int PREFERRED;
  public static final int UNNECESSARY;

  // possible preferences that can be set
  public static final Dimension LARGEST_DIMENSION;
  public static final int GRAPHICS_CONFIGURATION;
  public static final int SCENE_PIXEL_RESOLUTION;
```

```
public static final int SCENE_PIXEL_RECTANGLE;
public static final int SCENE_SCREEN_RECTANGLE;

// methods to set and get priorities
public void setPreference(
    int preference, object object, int priority);

public Object getPreferenceObject(int preference);
public int getPreferencePriority(int preference);
}
```

The `HSceneFactory.getBestScene()` method takes an `HSceneTemplate` as an argument, and will attempt to allocate an `HScene` that satisfies the constraints described by the `HSceneTemplate`. Properties that are UNNECESSARY will be ignored when the `HSceneFactory` attempts to create the `HScene`, and PREFERRED properties may be ignored if that is the only way to create an `HScene` that fits the other constraints. REQUIRED properties will never be ignored, however. If the `HSceneFactory` cannot create an `HScene` that meets all REQUIRED properties specified in the template while still meeting all other constraints that may exist, `HSceneFactory.getBestScene()` will return a null reference.

If we only want to query the middleware about the type of scene we can get, we can call the `HSceneFactory.getBestSceneTemplate()` method instead. This returns an `HSceneTemplate` representing the `HScene` we would get if we used the specified template. By doing this, we can check whether an `HScene` meets our needs before we actually allocate it, and if necessary we can change the parameters on our `HSceneTemplate` based on the template returned.

There are a couple of shortcuts available to us, if we are not too fussy about the type of `HScene` we want. The `HSceneFactory.getFullScreenScene()` method returns an `HScene` that covers the entire area supported by the specified `HGraphicsDevice`. As we saw earlier, however, a receiver may not support overlapping `HScenes`, and in that case this method will return a null reference if another `HScene` is also using that graphics device.

Alternatively, we can use the JavaTV graphics API to get an `HScene`. The `javax.tv.graphics.TVContainer` class defines the `getRootContainer()` method. This takes the Xlet context of our Xlet as an argument, and returns an `HScene` for that Xlet. If we have already created an `HScene` for that Xlet, we will get a reference to that `HScene`, but otherwise this method will create a new `HScene` (if possible) and return it to us.

Developing Applications Using HScenes

Use of an `HScene` instead of a `Frame` for the top-level component means that applications developers need to take a little more care than usual. Applications should not assume that they can get the area of the screen they want. If another `HScene` is using that area, they may not be able to. Instead, it is better for the application to set the size of the `HScene` it wants and let the middleware position it.

This approach may not always be practical, however. For instance, if the application wants to place graphics over a specific part of the video the HScene needs to occupy a specific area of the screen. This is still possible even if another application already has an HScene covering that area of the screen, because the middleware can choose to move HScenes (and even resize them if necessary). Of course, this is only possible if the applications owning those HScenes have not requested a specific position on the screen (or size, in the case of a resizing operation). This is another reason it is generally a good idea not to set a REQUIRED position or size in any HSceneTemplates you create unless it is absolutely necessary.

An application can register a java.awt.ComponentListener with the HScene so that it is notified when its HScene has been moved, resized, hidden, or shown. Applications can also try to move or resize their HScenes by calling the HSceneFactory.resizeScene() method, as follows.

```
public HSceneTemplate resizeScene(
    HScene hs,
    HSceneTemplate hst)
    throws java.lang.IllegalStateException;
```

This method takes the HScene to be resized and an HSceneTemplate that indicates the desired size and position of the HScene. The HSceneFactory will try to meet these requirements, just as when an HScene is created. If the HScene can be resized or repositioned, the method will move and/or resize it, and then return a new HSceneTemplate that gives the new position and size. If the HSceneFactory cannot carry out the desired operations because of restrictions in the middleware (for instance, because the operation would result in two HScenes overlapping when the platform does not support that), this method will return a null reference and the size and position of the HScene will not be changed.

It is worth remembering that because HScenes are top-level components they will not be visible by default. An application should call the setVisible() method to display the HScene when it is ready. Typically, an HScene will have no visible components itself, and its main purpose is to clip and control the display of any components it contains. By default it will not receive input focus either, and thus applications need to request focus explicitly should they want it. An application will only have input focus if the component that has input focus is a child component of that application's HScene.

For many parts of an application, developers will see very few changes from other Java-based environments when dealing with HScenes. An HScene is just another AWT Container object. Components can be added to or removed from an HScene just like any other Container, and most Java developers will feel right at home here (apart from the changes in the widget set, which we will explore in material to follow).

HScenes do add one useful piece of functionality to our top-level container, however. One of the more common things we want to do in a DTV application is to display some UI elements on the screen while other areas are transparent (so that the background video is visible). When we are using HScenes, we can do this by setting the background mode of our HScene. The setBackgoundMode() method lets us choose whether the background of the HScene is transparent (HScene.NO_BACKGROUND_FILL) or filled (HScene.BACK-GROUND_FILL). When the background is filled, it will typically be filled with the background color, like any other AWT component.

As well as this ability to show transparent backgrounds, HScenes have another useful trick up their sleeves. By calling the setBackgroundImage(), an application can set a background image for the HScene instead of a single color. The setRenderingMode() method allows an application to control how this background image is rendered. It can be stretched to fit the HScene, tiled, centered, or not displayed at all. Remember that this is different from any background image displayed in the background layer of the graphics hierarchy. This image is only visible in the background of the HScene, and will be clipped by the HScene's bounding box.

It is sometimes useful to know whether or not any parts of our HScene are transparent, and thus we can use the isOpaque() method to find out whether our HScene contains any transparent elements or not. In most cases, this is controlled by the HScene's background mode, although subclasses of HScene may choose to override this if they can guarantee that the entire content of the HScene's bounding box will be opaque.

When doing this, however, we must be careful not to cover those elements of the screen that may be used for other purposes, such as subtitles. Unfortunately, we have no way of knowing exactly which areas will be used for subtitles, and thus the network operator and the application developer must agree which areas of the screen will be available to applications.

The HAVi Widget Set

We have already mentioned that there are many changes in the widget set between a standard AWT implementation and an MHP or OCAP middleware stack. Again, one major reason for this is that there may be no underlying window manager on a DTV receiver, and without this it is very difficult to use any heavyweight components.

Most of the components in AWT are heavyweight components, in that they have a peer class and will map almost directly onto an element in the host platform's GUI. HAVi gives an alternative to these by providing lightweight versions of a number of the common GUI components, and by adding features to these components that are very useful in a consumer environment.

We could have used the Java Swing API, but this is not part of pJava or J2ME. Although widgets can be implemented as lightweight components (either Swing components or a proprietary widget set), an MHP or OCAP middleware stack can choose to implement them as heavyweight components, provided that HScenes are also implemented as heavyweight components and that implementations make sure they follow the correct Z-ordering rules. This is a fairly well-known problem in Java and thus we will not describe it here. It means that middleware implementers need to take care when building their GUI libraries. In general, it is better to avoid mixing heavyweight and lightweight components if possible.

Table 7.6. Java AWT classes supported by MHP and OCAP.

Adjustable	AWTError	AWTEvent
AWTEventMulticaster	AWTException	BorderLayout
CardLayout	Color	Component
Container	Cursor	Dimension
EventQueue	FlowLayout	Font
FontMetrics	Graphics	GridBagConstraints
GridBagLayout	GridLayout	IllegalComponentStateException
Image	Insets	ItemSelectable
LayoutManager	LayoutManager2	MediaTracker
Point	Polygon	Rectangle
Shape	Toolkit (partial support, including some methods from pJava)	

Table 7.7. HAVi UI classes included in MHP and OCAP.

HActionable	HAdjustableLook	HAnimateEffect
HLook	HMatte	HMatteLayer
HNavigable	HOrientable	HState
HSwitchable	HTextLayoutManager	HAnimateLook
HAnimation	HComponent	HContainer
HDefaultTextLayoutManager	HFlatEffectMatte	HFlatMatte
HGraphicButton	HGraphicLook	HIcon
HImageEffectMatte	HImageMatte	HListElement
HListGroup	HListGroupLook	HMultilineEntry
HMultilineEntryLook	HRange	HRangeLook
HRangeValue	HScene	HScreenDimension
HScreenPoint	HScreenRectangle	HSinglelineEntry
HSinglelineEntryLook	HSound	HStaticAnimation
HStaticIcon	HStaticRange	HStaticText
HText	HTextButton	HTextLook
HToggleButton	HToggleGroup	HVideoComponent
HVisible		

From the standard `java.awt` libraries, we can use the classes outlined in Table 7.6, as well as the `java.awt.event` package.

The HAVi GUI API adds the classes and interfaces outlined in Table 7.7 for supporting the various UI elements.

These are not the only classes in the HAVi UI API, of course. These are just the ones related to drawing GUI widgets in the screen. As you can see, this gives us a fairly comprehensive set of widgets, including buttons, text entry boxes, radio buttons, icons, and many others. We also get more advanced features such as mattes that let us change the appearance of a component, and the `HLook` interface that lets us define a completely new graphical look for our components.

HAVi defines its own `HComponent` and `HContainer` classes, which are used by the API instead of the normal Java `Container` and `Component` classes. These are both subclasses of the relevant Java class, and thus can be used anywhere that the normal Java `Container` or `Component` can be used. They also provide some extra features. The middleware will use the `HComponent` and `HContainer` classes anywhere the AWT classes should be used, including internally. Any instances of `Container` or `Component` returned by the MHP or OCAP middleware will actually be instances of `HContainer` and `HComponent`.

One of the changes between HAVi components and containers and their AWT equivalents is that the HAVi specification explicitly allows `Components` to overlap. For this reason, the `HContainer` class adds improved support for Z-ordering among the components in a container via the `HComponentOrdering` interface, which follows. This allows applications to add components to a container in front of or behind other components, and to change the order of components that have already been added to the container. Applications can use this together with transparency and alpha blending to provide a more attractive UI.

```
public interface HComponentOrdering {
   public java.awt.Component addBefore(
      java.awt.Component component,
      java.awt.Component behind);

   public java.awt.Component addAfter(
      java.awt.Component component,
      java.awt.Component front);

   public boolean popToFront(java.awt.Component component);
   public boolean pushToBack(java.awt.Component component);
   public boolean pop(java.awt.Component component);
   public boolean push(java.awt.Component component);

   public boolean popInFrontOf(
      java.awt.Component move, java.awt.Component behind);

   public boolean pushBehind(
      java.awt.Component move, java.awt.Component front);
}
```

HAVi also adds support for mattes by providing the `HMatteLayer` interface. This is implemented by the `HComponent` class, and it enables alpha blending between the components contained within a container and any components under the container. We will take a more detailed look at mattes and how we can use them later in the chapter.

Changing the Look of Your Application

Unlike a Windows PC or a Macintosh, wherein every application has a similar look and all GUI elements have a similar design, the CE world is very different. Every CE manufacturer will have its own UI style (or sometimes more than one), but different devices and different applications may have different UI needs. A stock-trading application will usually have a graphical design that is different from an application associated with a pop music show, for instance.

Although the middleware manufacturer will define a default look and feel for the GUI widgets, this will not be suitable for all cases, and thus application developers may want to change the look of the UI to suit their target audience. If we were using the standard Java AWT classes, doing this would be pretty painful. We would have to subclass every Component we used, and then override the paint() method and any related methods to draw our new UI look. Luckily, this is not something we want to do very often in a desktop Java implementation.

For consumer devices this is a problem, and thus HAVi adds support for "pluggable" UI looks. These separate the look of a component from its functionality, and thus an application developer can build the basic UI from the standard HAVi GUI components and then later implement a new look for them without having to subclass those components.

These looks are implemented by the HLook interface, which follows. Classes that implement this provide a showLook() method that replaces the paint() method of the component it is associated with, as well as a couple of other methods related to the appearance of the component.

```
public interface HLook
    extends java.lang.Cloneable {

    public void showLook(
        java.awt.Graphics g, HVisible visible, int state);

    public void widgetChanged(
        HVisible visible, HChangeData[] changes);

    public Dimension getMinimumSize(HVisible hvisible);
    public Dimension getPreferredSize(HVisible hvisible);
    public Dimension getMaximumSize(HVisible hvisible);

    public boolean isOpaque(HVisible visible);
    public java.awt.Insets getInsets(HVisible visible);
}
```

An object implementing HLook can be associated with an HVisible object (which is a subclass of HComponent and represents the base class for any visible components) using the HVisible.setLook() method. Once this has been done, the paint() method belonging to that HVisible will call the showLook() method on the associated HLook instead of drawing the component itself. Thus, the `showLook()` method should implement the same functionality that would be included in the `paint()` method for the component.

In cases in which the content of the widget has changed, but nothing else, the `widgetChanged()` method will be called instead. This gives the `HLook` a set of hints about what has changed in order to make redrawing the widget more efficient. Using these hints, the `HLook` can choose to only redraw those parts of the widget that have actually changed. This is similar to the `update()` method, but the hints that may be provided allow a more efficient way of computing what elements of the component need to be redrawn. In a limited environment such as a set-top box, this may be more efficient when drawing complex components.

Other methods in the original component are also overridden, especially those that get the size of the object. Any method that gets the maximum, minimum, or preferred size of an `HVisible` will call the appropriate method on the `HLook` instead. Applications can get the `HLook` associated with an `HVisible` by calling the `HVisible.getLook()` method. This will return the `HLook` object currently associated with that `HVisible`, or a null reference if the object is using its own `paint()` method.

Having to set an `HLook` for every instance of an `HVisible` we use would be even worse than having to subclass the `Components` we want to change, and thus many subclasses of `HVisible` let us set a default look using the `setDefaultLook()` method. Doing this means that the middleware will use the given `HLook` for all instances of those classes, thus avoiding the need to explicitly set the `HLook` every time we create an instance of a particular class. As you can probably see, this gives us a great deal of flexibility in how we approach this. We can set a default look for all of our components, but we can still set a specific look for one or two components should we need to. For instance, we may set a default look for all of the buttons in our application, but we want to give the Exit button a different appearance so that the user knows it is special. Using looks, we can set a default look for all of our buttons, and then set a separate look for our Exit button.

The advantage of doing this is that we do not need different subclasses of the button. We can simply implement a new look and associate it with the button in question. If we later want to apply the same look to other buttons, or change that look, we do not need to change the button itself. Even better, we can actually change the look of the button at runtime, simply by calling `setLook()` with a new `HLook`. This may not be a good idea (as with any UI, consistency is generally best), but for those times when you do need it the alternatives are much less flexible than this strategy.

If you have ever tried implementing a set of GUI widgets, you know that there are a lot of details that need to be considered, such as how the appearance changes when the user interacts with the components, and how that interaction takes place. To make life easier for developers, and to ensure that the behavior of the UI stays consistent, HAVi defines a number of subclasses of `HLook` to help us implement looks for specific components. Table 7.8 outlines which subclass of `HLook` we must use as the base class for a given component's default look.

Table 7.8. HAVi UI classes and their associated `HLook` classes.

Component	Look
HAnimation	HAnimateLook
HGraphicButton	HGraphicLook
HIcon	HGraphicLook
HListGroup	HListGroupLook
HMultilineEntry	HMultilineEntryLook
HOrientable	HAdjustableLook
HRange	HRangeLook
HRangeValue	HRangeLook
HSinglelineEntry	HSinglelineEntryLook
HStaticAnimation	HAnimateLook
HStaticIcon	HGraphicLook
HStaticRange	HRangeLook
HStaticText	HTextLook
HText	HTextLook
HTextButton	HTextLook
HToggleButton	HGraphicLook

Some widgets may change their behavior based on the representation used to draw them. To cope with this, subclasses of `HLook` may define methods to provide additional information that may be used by the associated widget. For instance, the `HRangeLook` includes a `getValue()` method that returns the value the component should adopt when the user clicks on a specific point in the component. Because only the `HLook` knows how the component is drawn, only it can tell how this should behave. For example, clicking on a point on a slider bar usually has a different effect from clicking on a different representation of a range.

Whereas the `HVisible.setLook()` method takes an `HLook` as its argument, subclasses of `HVisible` may impose some restrictions on the subclass of `HLook` that they support. Setting a graphical look for a textual component is rather silly, and thus in cases such as this `setLook()` can throw an `HInvalidLookException` to indicate that the application has tried to set an incompatible look.

HLooks in Practice

We have already discussed the basics of how `HLooks` work in practice, but we glossed over some of the details. In particular, a component may need to be redrawn differently depending on specific circumstances.

Everyone knows the order in which the `paint()` method gets called on various components in a container. This is well defined in the Sun specifications, and apart from a few inconsistencies (especially when dealing with a mixture of heavyweight and lightweight components) it is pretty easy to predict the order in which drawing operations will happen. So how does this change when we start associating `HLook`s with some of our components?

In some cases, there is no change. The order of drawing operations remains the same as usual (assuming that all `HVisible` objects are treated as lightweight components). The only thing that does change is what happens when the `paint()` method gets called.

For those components that do not have an associated `HLook`, this is pretty straightforward: the `paint()` or `repaint()` method does what it normally does and draws the component. If the component does have an `HLook` associated with it, one of two things will happen. In those cases for which the entire component needs redrawing, the `HLook.showLook()` method will be called (as we have already seen). This could happen because a component has just been added to a container and made visible, or because it was partly obscured by another component but is now fully visible.

Sometimes, however, there is no need to redraw the entire component. If the component's state has changed, but nothing else, we may only need to draw the changed parts of the component. For instance, when the user moves a slider bar we will redraw the slider itself, but we may not need to redraw the borders and the outline of the component. In this case, the `HLook.widgetChanged()` method will get called. This is a replacement for `repaint()`, and thus calls to `HVisible.repaint()` should call `HLook.widgetChanged()` if an `HLook` is associated with that component.

This takes an array of `HChangeData` objects as an argument, and each of these provides a hint about what has changed. Each hint consists of an integer key (representing the type of change) and some data that contains the changed values. These values are always the values before the change. If the `HLook` needs to know the new values it can always get these from its associated `HVisible`.

We will not look at all of the possible hints here, but it is important to know that an `HLook` does not have to use this information to redraw the component. It can simply choose to repaint the entire component if it wishes. This approach is probably the most common approach used at the time of writing, simply because the large number of hints the middleware can provide can make the implementation more complicated if the application tries to pay attention to all of them. In cases in which performance is not a problem, repainting the entire component is probably the simplest approach.

This is not the end of the story, however: some subclasses of `HVisible` may be in one of several states, and this state will have an impact on how that component is drawn. We will look more at the state model of these components later, but for now we will just look at how these states affect the drawing of a component.

There are eight states a component can be in, and we must support a basic set of these. Other states are combinations of these basic states, and their appearance can be emulated using the same style as some of these basic states. Table 7.9 summarizes the states a component can be in. We only provide an overview of them here, but we will take a closer look at how components use these states later in the chapter.

Table 7.9. The look of a component will often change depending on its state.

State	Content if State Not Handled	Example Style
NORMAL_STATE	None	Drawn normally
FOCUSED_STATE	As for NORMAL_STATE	Highlighted (either using a frame or by changing the color scheme of the object)
ACTIONED_STATE	As for FOCUSED_STATE	Pushed in
ACTIONED_FOCUSED_STATE	As for FOCUSED_STATE	Highlighted and pushed in
DISABLED_STATE	As for NORMAL_STATE	Grayed out
DISABLED_FOCUSED_STATE	As for DISABLED_STATE	Grayed out and highlighted
DISABLED_ACTIONED_STATE	As for ACTIONED_STATE	Grayed out and pushed in
DISABLED_ACTIONED_FOCUSED_STATE	As for DISABLED_STATE	Grayed out, highlighted, and pushed in

Using this information, an application developer can design a set of styles that handle the states that must be supported, without complicating the HLook implementation any more than is necessary.

The Behavior of Components in MHP and OCAP

Simply drawing components is not usually enough. The whole point of a UI is that a user will interact with it, and the behavior of a component usually reflects this. We have just seen how the appearance of a component may change to provide feedback when the user interacts with it. For instance, pressing a button may change its color, or make it look like it has been pressed in. The component may be in a number of different states, and the behavior of a component and the states it can be in are an important part of the feel of an MHP or OCAP application.

Five classes determine the basic behavior model of HAVi components. We have already met the simplest of these: HVisible. This acts as the core of the HAVi widget set, in that every visible component is a subclass of this. Although the behavior of the HVisible class is very simple, it defines the basic model used by other interfaces to implement different behaviors. In conjunction with the HState class, HVisible defines a number of states a component can be in. There are four main states for HAVi components, outlined in Table 7.10.

Combinations of these states are also possible, as we saw earlier. All of these states (and valid combinations of them) are represented by constants in the HState class. Table 7.10 outlines some of these constants for the basic states.

Table 7.10. The main states of an HAVi component.

State	Meaning	Constant for This State
Normal	The component is not currently being interacted with.	NORMAL_STATE
Disabled	The user cannot interact with this component.	DISABLED_STATE
Focused	The component has input focus.	FOCUSED_STATE
Actioned	The user has performed an action on that component (e.g., clicked on a button).	ACTIONED_STATE

Although the HVisible class defines the state model, not all of these are valid states for an instance of that class. Subclasses of HVisible that do not implement one of the other behavior-related interfaces can only be in the normal or disabled states.

This is not a very useful behavior for constructing UIs, and thus HAVi defines the HNavigable interface. This lets components receive input focus so that the user can interact with them. Classes that implement the HNavigable interface can receive input focus, and will exhibit certain behaviors when focus is gained or lost. In particular, these classes will generate events to tell other components that those components have gained or lost focus. These classes may also change their appearance or play a sound to indicate this to the user.

Any object implementing HNavigable can be in one of four states: normal, focused, disabled, or disabled and focused. Figure 7.6 shows the possible state transitions for HNavigable objects. Although a disabled object can be focused, the user cannot interact with it while it is disabled.

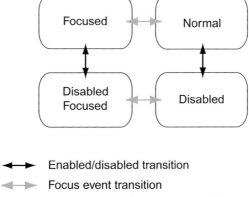

Figure 7.6. State diagram for the HNavigable interface. *Source:* HAVi 1.1 API specification, courtesy of HAVi Organization, November 2004.

147

An important part of being able to navigate between components is to define how that navigation happens. This is especially important when our only input device is a remote control unit. We cannot simply click on a new UI component to give it focus, and thus defining a sensible navigation path through our components is vital. The `HNavigable.setMove()` and `HNavigable.setFocusTraversal()` methods let us define the order components will gain focus. `setMove()` takes two parameters: a key code and an `HVisible` component that should gain focus when that key is pressed. Any key can be set to switch the focus if that key is available on the remote.

Most of the time, we will use the standard up/down/left/right keys for moving the input focus, and the `setFocusTraversal()` method gives us a shortcut to set the navigation targets for each of these four keys at the same time. Instead of taking a single `HNavigable` parameter, `setFocusTraversal()` takes four, representing the navigation targets for the up, down, left, and right keys. If any of these is null (or if the navigation target in a call to `setMove()` is null), no navigation target will be applied to that key.

We cannot stress enough the importance of having a simple and logical navigation model for your application. If a user cannot navigate easily between components, they are less likely to use that application. A good navigation model can make the difference between a successful application and an unsuccessful one.

Being able to navigate between components is useful, but it is even more useful to be able to interact with them. The `HActionable` interface extends the `HNavigable` state model by adding support for user interaction with a UI component. This takes the form of another state (the actioned state) the component enters when a user interacts with it. This is not used for all types of interactions. Of all GUI components defined by HAVi, only the button classes actually implement this interface. Other elements, such as text boxes and slider bars, do not follow this model because these components have a different interaction model. The state diagram shown in Figure 7.7 indicates how the actioned state fits into the state model we have already seen.

The `HSwitchable` interface extends this model further, to provide a model that allows a toggle based on the action. Unlike the normal `HActionable` interface, an action on an object that implements `HSwitchable` causes it to remain in the actioned state until another action is carried out on that object (at which point it moves back to its normal state). Even enabling or disabling an object may not affect this. An `HToggleButton` may remain toggled, for instance, even after it has been disabled. In this case, the `ACTIONED_STATE_BIT` field of an `HSwitchable` object may remain set until another action event is generated by that object. Figure 7.8 shows a state diagram for the `HSwitchable` interface.

By explicitly defining this state model, HAVi defines the behavior for each component that implements the interfaces we have just seen. The advantage of this is that the definition of the behavior of a component is separate from the component itself, which is in turn kept completely separate from its look.

The last interfaces that affect the behavior of a component are the `HTextValue`, `HAdjust-mentValue`, and `HItemValue` interfaces. Although these do not implement a state in the

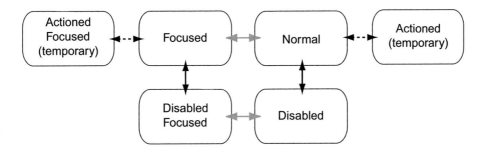

Enabled/disabled transition

Focus event transition

Action event transition

Figure 7.7. State diagram for the `HActionable` interface. *Source:* HAVi 1.1 API specification, courtesy of HAVi Organization, November 2004.

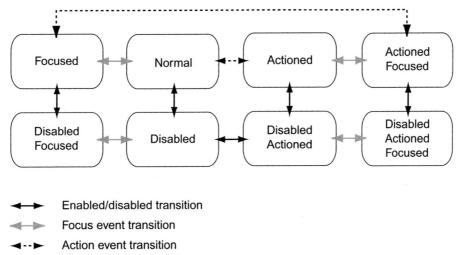

Enabled/disabled transition

Focus event transition

Action event transition

Figure 7.8. State diagram for the `HSwitchable` interface. *Source:* HAVi 1.1 API specification, courtesy of HAVi Organization, November 2004.

way the others do, they do contain state-related information. In particular, these interfaces add a standard way of working with the values associated with some types of UI widgets (for instance, a slider bar or a text input field). By defining this functionality through a separate well-known interface, the behavior it implements is independent of any specific component, thus ensuring consistency across all components that implement that behavior.

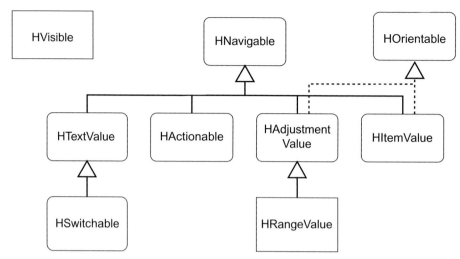

Figure 7.9. The class hierarchy of HAVi behaviors.

Although these three interfaces have different methods, the functionality they provide is very similar: they let applications register listeners to find out when a value changes, or to set some of the behavior related to the changing of the value. For instance, this could tell the component to play a sound when the value changes, or to set the value by which a slider increments.

Any UI widget that is not completely static will implement at least one of the interfaces we have just seen. In some cases there will be two classes for presenting certain types of content. Images, text, and animation can be completely static (as graphical or textual labels, or simply as eye candy), or the user can navigate to a particular component and possibly interact with it. Although this may not seem very useful for content such as text labels or images, it does provide a simple way for applications to implement functionality such as tool tips and rollovers. Figure 7.9 shows the class relationships among these various behavior interfaces, which build on one another.

To see how these interfaces are used in practice, Table 7.11 outlines how each UI widget uses these interfaces to define its behavior under different circumstances. Note that the interfaces here may not be the main super-interface of a given component. Instead, this table shows the most specific interface out of the previous set that a given class implements. In other words, if a component implements both `HVisible` and `HActionable`, `HActionable` will be the interface listed in Table 7.11.

The only GUI component that does not appear in the table (and that does not implement any of these interfaces) is the `HVideoComponent` class. There is a good reason for this: the `HVideoComponent` is not just any GUI component. It is designed to be a place for the receiver to render decoded video in the AWT hierarchy, and is not really an object that users

Table 7.11. The primary behavior for each HAVi UI widget.

Component	Interface
HAnimation	HNavigable
HGraphicButton	HActionable
HIcon	HNavigable
HListGroup	HItemValue
HMultilineEntry	HTextValue
HRange	HNavigable
HRangeValue	HAdjustmentValue
HSinglelineEntry	HTextValue
HStaticAnimation	HVisible
HStaticIcon	HVisible
HStaticRange	HVisible
HStaticText	HVisible
HText	HNavigable
HTextButton	HActionable
HToggleButton	HSwitchable

will interact with. Given the limitations that may be imposed by the receiver hardware, we do not want to complicate this class any more than is necessary. An HVideoComponent is not an interactive element in a GUI, and it may not even obey all of the rules laid down in the HVisible interface. This component is treated as a special case because, in short, it is a special case. It is a shortcut for developers who want an easy way of manipulating the position and size of decoded video and who want to logically place it in the AWT hierarchy.

Interacting with Components

The basic behavior of a component will remain the same for many applications, just because the components will usually do what we expect them to: when we press a button, we want the button to behave as if it has been pressed, by changing its appearance and maybe by playing a sound. Objects implementing the HLook interface will take care of how the appearance of a component changes when the user interacts with it, but other behavior is controlled by the interfaces we have just seen.

The HNavigable interface lets us set a particular sound that should get played when the component gains or loses user input focus using the setGainFocusSound() and setLoseFocusSound() methods. Similarly, we can register a class implementing the HFocusListener interface that will be notified when the component gains or loses focus. HActionable extends this so that we can set a sound when the user carries out an action on the component (e.g., pressing a button), and register an HActionListener that is notified when this action is carried out.

For `HItemValue`, `HTextValue`, and `HRangeValue` interfaces, we can register listeners that tell us when the value changes. In the case of the `HTextValue`, two different types of listeners can be registered: an `HKeyListener` that is notified whenever a key is pressed or an `HTextListener` that is notified when the value of the text changes or when the text cursor is moved. Different types of events let the application respond to different types of interactions.

All of these different listeners have the same basic function: to tell an application that the user has done something and to give the application a chance to react. Some feedback (such as playing a sound) may be automatic, whereas real user interaction (rather than just navigation between components) usually needs a more sophisticated response from the application. Most components that implement `HActionable` should have at least one listener registered to handle user interaction. Objects implementing `HItemValue`, `HTextValue`, or `HRangeValue` may choose to do this, although it is less important. Each of these objects contains a way of getting the current value of the component, and thus an application may just need to read this information in response to another event. For instance, an application in which the user must enter a PIN number may wait until the user presses the OK button before it reads the value. On the other hand, if the user is choosing a product from a list in an e-commerce application, the application may want to display some details about that product without the user having to press another button.

Coordinate Schemes

So far, we have mentioned nothing about the coordinate systems used by an MHP or OCAP receiver. There is a good reason for this: the issue of coordinate schemes is complicated enough that you need a decent grounding in the basics before you approach it. It is not actually difficult to understand, but it is not exactly straightforward either.

In a normal PC implementation, there are three coordinate spaces we have to worry about. First, we have the coordinate space of the screen, which is defined by the resolution of our display. Second, we have the coordinate space of our root window, which starts from (0, 0) in the top left-hand corner of the window and ends at (w, h) in the bottom right-hand corner. Here, w and h are the width and height of our root window, respectively. Finally, we have the coordinate space of an individual component, which starts at (0, 0) in the top left-hand corner of the component and ends at (c, d), where c and d are the width and height of the component, respectively.

Table 7.12 indicates how these are used.

MHP and OCAP use a similar approach, but some changes are necessary because of the hardware limitations. The screen coordinate space is used for positioning video and `HScenes` on the screen. This uses a normalized coordinate scheme defined by the HAVi specification.

Normalized coordinates are independent from the resolution of the display device, and use a pair of floating-point numbers between 0.0 and 1.0 to define a point on the screen. Using normalized coordinates makes it easier for applications to position objects absolutely on the

Table 7.12. Coordinate spaces and their functions in Java.

Coordinate Space	Purpose
Screen	For positioning root window
Root window	For positioning components in root window
Component	For positioning elements when drawing an individual component

screen without having to worry about the device resolution, and they are used by many parts of the HAVi UI API that refer to components outside the AWT component hierarchy.

The `HScreenPoint` class represents a point in the screen using normalized coordinates, whereas the `HScreenDimension` class uses normalized coordinates to define the dimensions of a rectangle. `HScreenRectangle` uses both of these to define an area of the screen. These classes are used by the `HSceneTemplate` class to define the size and dimensions of an `HScene`, or by the `HScreenConfiguration` interface to define the area of the screen that is accessible using a given device configuration. The `org.dvb.media.VideoTransformation` class (which we will see in more detail in Chapter 11) uses normalized coordinates to position decoded video on the screen.

Because the video and graphics layers may be implemented using different hardware components, a graphics device may not cover the entire screen (general-purpose video devices usually will, although a video device for picture-in-picture may not). It is also possible for a graphics device to support more than one resolution, depending on what graphics configurations are supported for that device. By using normalized components to define the on-screen position and size of the graphics device, applications do not have to worry about this issue. The receiver will take care of any mapping onto real device coordinates, and thus life gets simpler for developers.

Of course, normalized coordinates are not always useful, and thus we can get "virtual" coordinates representing the position of a component in terms of pixels measured from the top left-hand corner of the screen. When using virtual coordinates, applications should be very careful to compensate for any offset on the origin of the graphics device. These are not used very much in practice, but you may see them being used from time to time. Figure 7.10 shows various coordinate spaces used by the HAVi UI classes.

Several methods let us convert coordinates between the different coordinate systems we have seen. The first of these is the `HGraphicsConfiguration.getComponentHScreenRectangle()` method, which converts as follows from the coordinate space of the graphics device in that particular configuration to normalized components.

```
public HScreenRectangle getComponentHScreenRectangle(
    java.awt.Component component);
```

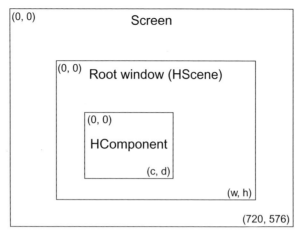

Figure 7.10. Coordinate spaces in the HAVi UI classes. *Source:* ETSI TS 101 812:2003 (MHP 1.0.3 specification).

This method returns an `HScreenRectangle` corresponding to the area on screen defined by the component that is passed in. The `HScene` class implements a variation of this method, which returns as follows an `HScreenRectangle` representing a component in that `HScene`.

```
public HScreenRectangle
  getPixelCoordinatesHScreenRectangle(
    java.awt.Rectangle r);
```

In this case, the rectangle defines an area in the `HScene`'s coordinate space. We can also convert in the other direction, using the `HGraphicsConfiguration.get PixelCoordinatesHScreenRectangle()` method as follows.

```
public java.awt.Rectangle
  getPixelCoordinatesHScreenRectangle(
    HScreenRectangle sr,
    java.awt.Container cont);
```

This returns an AWT `Rectangle` object that corresponds to the `HScreenRectangle` passed in as a parameter. The location of this rectangle is relative to the `Container`, also passed as an argument.

Table 7.13 summarizes how the coordinate spaces in HAVi are used. The names of the coordinate spaces show how they map onto the standard Java coordinate spaces.

Integrating Graphics and Video

One of the more important uses of a DTV receiver is to show TV programs, and many applications will be used to enhance a TV show rather than replacing it. This means that an application will often want to display graphics while still allowing the user to see the video being

Table 7.13. Coordinate spaces in MHP and OCAP.

Coordinate Space	Purpose
Normalized	Used for resolution-independent positioning of the root window and decoded video Defines on-screen location of the graphics device
Virtual	Used for pixel-based positioning of the root window and decoded video Overlaps with normalized coordinate space
Screen	Pixel-based coordinates within the area of screen supported by the graphics device Generally not used by applications
Root window	Used by applications for positioning components in the `HScene`
Component	Used for positioning elements when drawing an individual component

broadcast. Not surprisingly, this is another area in which there are a number of differences between Java on a PC and Java in a broadcast environment. Mixing video and graphics like this is not something a PC needs to do very often, and it usually has far more processing power available when it does.

We have already seen that the display model separates the video and the graphics in most MHP and OCAP implementations, and thus developers need to think carefully about what they are trying to achieve with an application. Integrating graphics and video is not very difficult, but it does take some planning to do it well. Coordinate conversion is just one of the problems we face. A number of other issues may not be as obvious, and we will look at how to address these later in the chapter.

Transparency

We have seen how we can use the coordinate conversion methods to position graphical elements on the screen relative to the content of the video layer, but we also need to be able to see through the graphics layer to the video beneath. `HScenes` are the first part of the solution to this. As we have already seen, an `HScene` is different from other components in that it does not draw its background by default. This means that any part of the `HScene` that does not contain another component will be transparent.

This is useful, but what if we want to draw a component that has transparent colors? Again, HAVi helps us. Each graphics device has a color the middleware will treat as transparent. The `HGraphicsConfiguration` class defines the `getPunchThroughToBackground-Color()` method, which lets us get an AWT `Color` object that represents a transparent or partially transparent color. There are two basic versions of this method, which provide us with transparency to different layers of the graphics hierarchy. Both versions take an integer percentage as a parameter, which defines the amount of transparency the application wants: a value of 0% indicates full transparency, whereas a value of 100% will be fully opaque.

The first version of this method returns a color that is transparent to the background layer. In effect, pixels of this color act as a window through both the graphics layer and the video layer to display the content of the background layer. The second version of this method takes an `HVideoDevice` object as a second parameter. In this case, the returned color will be transparent only through to the specified video device. Any video that would be displayed on screen at that location will be displayed, but background pixels of the content of any other `HVideoDevice` will not be displayed.

Each of these two basic versions has a variant that can also take an AWT `Color` object as an argument. These return a color with the requested level of transparency that keeps the same RGB value as the specified color.

This gives us a great deal of flexibility in how we display the various elements of the graphics hierarchy. Of course, there are still limitations, and because the graphics hardware in a receiver may not support full alpha blending only a limited range of transparency values may be available to us. The `Color` object returned by `getPunchThroughToBackground-Color()` will have the closest supported transparency value, but this may not be very close. Receivers only have to support transparency values of 0% (transparent), 100% (completely opaque), and a value close to 30%. The exact value is not specified, and thus exactly how close is acceptable is not very clear.

Other transparency values are rounded up to the closest supported value, with two exceptions: values greater than 10% will not be rounded to 0%, and values smaller than 90% will not be rounded up to 100%. Therefore, assuming that the receiver supports 0%, 30%, and 100% transparency levels an application requesting a transparency level between 10 and 90% will always get a `Color` object with 30% transparency. Table 7.14 outlines this breakdown.

MHP and OCAP also add support for alpha blending using the `org.dvb.ui.DvbColor` class. This extends the AWT `Color` class to add support for alpha values, and implementations will use `DvbColor` in place of the `java.awt.Color` class where possible. Of course,

Table 7.14. Some transparency levels and their closest approximations.

Requested Transparency (%)	Returned Transparency (%)
50	30
90	30
91	100
10	30
9	0

applications can still use the `java.awt.Color` object by creating a new instance of it, but any `Color` objects returned by API calls or used internally by the API will be instances of `DvbColor`. Any calls to `getPunchThroughToBackgroundColor()` will always return a `DvbColor` object, for instance.

The hardware capabilities of MHP and OCAP receivers mean that they may not always support 24-bit graphics, and thus the color that is actually displayed may not be the color that is actually defined by a `Color` object. Those receivers that only support CLUT-based graphics will map the color specified by a `Color` object to the closest supported color before it is displayed. The same applies to alpha values: the actual alpha value of a `DvbColor` will be mapped to the closest supported value (following the rules described previously) before the color is actually rendered.

The `HVideoComponent` class does not behave like the other components with respect to transparency and translucency. Because the computing power needed to support translucent video is simply too high, instances of `HVideoComponent` will always have an alpha value of 1. This means that when the `SRC` or `SRC_OVER` compositing rule is used with an `HVideo-Component` the video will completely replace any components below it in the Z-order. Similarly, translucent components placed above the `HVideoComponent` will always be transparent through to the video in the component, rather than through to the background or to any background video.

Mattes and Alpha Compositing

Sometimes, using transparency at the level of individual colors may be too much work. If we want to display our entire UI as partially transparent, for instance, this approach is not practical.

In these cases, we can make use of two different approaches to transparency. MHP and OCAP both add support for the `DVBGraphics` class. As you can probably guess, this extends the `java.awt.Graphics` class, and from our point of view the most interesting thing it adds is support for alpha compositing (following a pattern similar to that used in Java 2). This lets us define a rule that says how transparency is applied to that graphics context, and how it is blended with any components beneath it in the AWT hierarchy.

Like the `DVBColor` object, instances of the `DVBGraphics` class will be used in place of normal AWT `Graphics` objects in MHP and OCAP implementations, both in the APIs and internally. Every `Graphics` object you will encounter is actually a `DVBGraphics` object.

The `setDVBComposite()` method lets us tell the middleware what alpha-blending rules should be applied to that particular graphic context. This takes an instance of the `org.dvb.ui.DVBAlphaComposite` class as an argument, which represents one of the Porter-Duff compositing rules for digital images and whose interface is based on the `java.awt.AlphaComposite` class from Java 2. Each `DVBAlphaComposite` object consists of two main parts: an alpha value (which says how much transparency should be applied) and a rule, which specifies how that alpha value should be applied. This rule will be one of a subset of the Porter-Duff rules. As for the `DVBColor` object, an alpha value is specified as a floating-point value in the range 0.0 (fully transparent) to 1.0 (fully opaque), with a default value of 1.0. Table 7.15 outlines which Porter-Duff rules are supported, and the impact they have when drawing objects.

For each of the rules of Table 7.15, the alpha value of what is drawn will be calculated from a combination of the following three values.

- The alpha value of the source pixel
- The alpha value of the destination pixel
- The alpha value of the `DVBAlphaComposite`

Table 7.15. Supported Porter-Duff rules for image compositing and blending.

Rule	Description	Example
SRC	Source is copied to the destination. Any overlapping areas are replaced by the source.	
CLEAR	Color and alpha of the overlapping area are cleared. Nonoverlapping areas of the source are cleared.	
SRC_IN	Color of the overlapping area is taken from the source, whereas both alpha values are used to determine its alpha value. Nonoverlapping areas are not drawn.	
SRC_OUT	Color of the overlapping area is taken from the source, whereas both alpha values are used to determine its alpha value.	

Table 7.15. *Continued*

Rule	Description	Example
DST_IN	Color of the overlapping area is taken from the destination, whereas the product of both alpha values is used. Nonoverlapping areas of the source are not drawn.	
DST_OUT	Similar to DST_IN, but with a different calculation for the final alpha value.	
SRC_OVER	The source is composited over the destination.	
DST_OVER	The destination is composited over the source.	

The results of these are outlined in Table 7.15. The right-hand column shows the effect of alpha values (in this case, the source image has an alpha value of 0.89, the destination image has an alpha value of 0.61, and the DVBAlphaComposite has an alpha value of 1.0). Each rule defines a specific mathematical formula for calculating the color of each pixel in the final image and its transparency value. These are too complex to reproduce here, but the MHP specification gives the full formula for each rule.

Each of these operations requires a different number of calculations to perform compositing, and thus different rules will have different performance penalties. Table 7.15 outlines these operations in order of performance, from fastest to slowest. Every receiver must support the SRC, CLEAR, and SRC_OVER rules, although exactly how these are supported may vary from platform to platform. Those receivers that use a CLUT for drawing graphics may approximate the final result, whereas those that support 24-bit color will produce a final result that follows the specified rules. Of course, carrying out these types of operations on a 24-bit display will take more CPU power than would be needed for a CLUT-based display.

These alpha-compositing rules may be suitable for some cases, but depending on our needs they may not: applying these to existing UI elements such as list groups or text boxes can be pretty difficult, even if you define your own HLook. In that case, we may want to use the HAVi approach to this problem, which involves mattes. The HMatte class defines a matte that can be applied to any HComponent or HContainer object. A matte tells the middleware what alpha value should be applied to that component when compositing it with the rest of the AWT hierarchy.

We have several types of mattes available to us, depending on the effect we want. The simplest of these is the flat matte (HFlatMatte). This specifies one floating-point value that is used as a scaling factor for the alpha value of every pixel in the component. Therefore, a matte with a value of 0.5 applied to a pixel with an alpha value of 0.5 will result in a final alpha value of 0.25 for that pixel.

Sometimes, we do not want to use a simple matte like this. An image matte (HImageMatte), an example of which is shown in Figure 7.11, uses a grayscale image to apply a scaling factor to pixels within the component. The lightness of a particular pixel in the image defines the scaling factor for the corresponding pixel in the component. This gives us a scaling factor that varies over the area of the component, depending on the image we use for the matte.

If we want a special effect, we can apply a matte that varies over time. This can either be a flat matte whereby the alpha value varies over time (implemented by the HFlatEffect-Matte) or a matte whereby the alpha values vary in time and over the image (using an HImageEffectMatte). Both of these have similar interfaces. The matte values are passed in as an array (either of floating-point values or Image objects) and the application can (as well as

Figure 7.11. Using an HImageMatte to control transparency.

performing more obvious tasks such as starting and stopping the animation) set parameters such as the speed of the animation, the position in the animation, and whether the animation repeats. We can use these more advanced mattes to add effects such as fades, wipes, and dissolves to our UI, and we can apply them to UI widgets as well as to purely static elements.

To set the matte for a component, we use the `HComponent.setMatte()` method, as shown in the following example. We can apply a matte to any type of HAVi component.

```
// create our matte
HMatte myMatte;
myMatte = new HFlatMatte(0.3);

// now apply it to our component
myComponent.setMatte(myMatte);
```

For image mattes, only the part of the component that is covered by the image will be matted. Other parts will be given a matte value of 0.0 (i.e., transparent).

We have already seen that HAVi components can overlap, and there is nothing stopping us from setting mattes on two overlapping components. In this case, the middleware should calculate the final alpha values of all overlapping components, and then draw them in the order AWT normally specifies.

Depending on what we want to do, this may not give us the effect we desire. If we have many components, it may also be a very complex approach. To get round this, the `HContainer` class lets us change how mattes are applied. Normally, mattes are applied as follows.

1. The middleware draws the background of the `HContainer`, using the container's matte to composite the container and any object beneath it.
2. Each component contained in the `HContainer` is rendered in back-to-front order, using its matte to composite it with what has already been drawn.
3. The `HContainer.group()` method groups the objects in the container and tells the middleware to draw things differently.
4. The receiver draws the container's background into an off-screen buffer. No matte is applied at this point.
5. Each component contained in the `HContainer` is rendered in back-to-front order, using the matte associated with that component to composite it with any components that have already been drawn.
6. The middleware renders the content of the off-screen buffer into the screen, using the `HContainer`'s matte to composite it with the existing screen content.

In this case, the container's matte is applied to every object in the container, rather than just to the container's background. This lets us set a single matte on the container, which is then applied to every component within it. If we want to apply the same matte to every object in the container (if components do not overlap), this makes things far simpler for the application. It also makes things quicker, in that we are applying just one matte rather than applying separate mattes to every component. Figure 7.12 depicts the use of mattes in the display hierarchy.

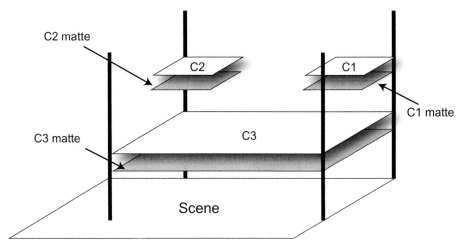

Figure 7.12. Using mattes in the display hierarchy. *Source:* HAVi 1.1 API specification, courtesy of HAVi Organization, November 2004.

If we have grouped a set of components in this way, we can ungroup them just as easily by calling the HContainer.ungroup() method. This leads us to something else we should think about when we are developing applications. Heavy use of mattes and transparency has two problems. First, it is extremely CPU intensive and thus you should only use complex mattes on platforms with enough CPU power to support them. Second, bad use of mattes can make your UI much more difficult to understand. Mattes have a great deal of power, but as with so many powerful techniques it is possible to overuse them and have a worse result than not using them at all.

Generally, applications should stick to the simpler mattes and compositing rules both for simplicity and performance. Mattes are an optional part of MHP and OCAP, and thus some or all of the mattes we have seen here may not be available on a given receiver. The only way to find out is to have the application try setting a matte. If a particular matte type is not supported, calls to HComponent.setMatte() will throw an HMatteException.

Middleware implementations should generally support a range of mattes most suitable for the hardware platform. Although more advanced mattes may be too slow for low-end receivers, they can be used to great effect in platforms powerful enough to support them. Given the importance of a good UI, middleware developers should attempt to provide application developers with as many tools as possible to produce a good-looking UI.

Both MHP and OCAP accept that not every hardware platform will be able to implement the compositing rules exactly as they are shown here. In particular, different hardware platforms will provide slightly different support for alpha blending between graphics and video. This means that some platforms will come closer to achieving correct compositing than others. Obviously, it is better for an implementation's behavior to be as close to the specification as possible, but some slight differences are acceptable if the hardware does not allow for the ideal.

Images

An MHP receiver can display images in a number of formats, such as GIF, JPEG, and PNG. Alternatively, we can use MPEG I-frames to display a full-screen image (such as a background image) or video "drips" to display several full-screen images that only have small differences. In both cases, we need to be aware of a few limitations. Some receivers may use the hardware MPEG decoder for decoding I-frames and video drips, and for broadcasting video, and thus displaying images that use these formats may disrupt any video that is playing. This may be a short glitch while a frame is decoded, or it may last for the entire time the image is displayed, depending on the receiver.

The second thing we need to remember is that decoder format conversion will not be applied to any I-frames or video drips. These will be treated like full-screen images, no matter what aspect ratio or other format information is included in the images.

Text Presentation

Text presentation in DTV systems is complicated by the different display aspect ratios and resolutions a receiver can use. This means that text may be a different width depending on the resolution and aspect ratio of the display, and thus developers need to take this into account when designing UIs.

We can use the low-level AWT operations for drawing text, just as we can on other AWT platforms. The `drawstring()`, `drawChars()`, and `drawBytes()` methods from `java.awt.Graphics` are all available to us, but there are a number of things we need to know before we can draw strings reliably. The first of these is the way font metrics are mapped onto display pixels. For the vertical resolution, this is easy: each pixel corresponds to one point in the font size, and thus text with a font size of 12 points will be 12 pixels high. For the horizontal resolution, things get a little more complicated. The size of each pixel varies with the aspect ratio of a display. For a 4:3 display with a resolution of 720×576, each pixel is 48/45 points wide. For a 14:9 display, each pixel is 56/45 points wide, and for a 16:9 display each pixel is 64/45 points wide.

OCAP does not define this mapping, although it does use square pixels for its default resolution (at a 4:3 aspect ratio), and thus each pixel is 1 point. Because this resolution is also used for 16:9, but with non-square pixels, we can assume that the point-to-pixel mapping in this case is distorted so that each pixel is 4/3 points wide.

As you can imagine, this soon gets complicated when calculating text wrapping, and thus applications may want to make sure that text is always the same number of pixels wide, no matter what the aspect ratio of the display. By setting the configuration of the graphics device to an `HEmulatedGraphicsConfiguration` with an aspect ratio of 14:9, applications can make sure that text will always be the same width in pixels. Of course, this has a couple of problems. First, text will be distorted slightly, but consistently, in every case. Second, other display elements may also be distorted slightly. The second set of errata to the MHP 1.0.3 specification clarified the behavior in this case, so that only text presentation will be affected by this: all other UI elements are considered to have a 1:1 mapping between pixels in the display device and pixels in the graphics device with no change in the aspect ratio. As of MHP 1.0.3, all graphics devices must be capable of emulating a 14:9 aspect ratio, although this is not required for earlier versions, or for GEM- or OCAP-based receivers.

Luckily, we have a tool to help us with text rendering that lets us ignore these problems: the HAVi text layout manager. Layout managers will be familiar to anyone who has done some AWT development on other platforms, and the basic principles are the same in this case. A layout manager normally operates on AWT components rather than text, laying them out within a container according to a predefined set of rules. A text layout manager does basically the same thing, but lays out text within an `HVisible` object according to a set of rules defined by the font and the layout manager. This is a useful tool for MHP and OCAP developers because it lets us display text in an area of the screen without worrying about text wrapping and other factors for which we would have to think about using the low-level text drawing functionality. Furthermore, it even lets us apply simple markup to the text using escape sequences, so that we can change the color and transparency of the text easily.

The basic functionality of a text layout manager is defined by the `org.havi.ui.HText-LayoutManager` interface. This is implemented by the `org.havi.ui.HDefaultText-LayoutManager` class, which provides the default text layout rules used by HAVi classes (such as `HText` and `HTextButton`). Each `HTextLayoutManager` defines the `render()` method that lets us draw text into a component. The interface to this method follows.

```
public void render(java.lang.String markedUpString,
                   java.awt.Graphics g,
                   HVisible v,
                   java.awt.Insets insets);
```

Apart from the string to be drawn and the `HVisible` object it should be drawn into, this method takes a `Graphics` object that determines the graphics context into which the text should be drawn. Together with the `HVisible` object, this forms the boundaries into which text will be drawn. The string will be laid out within the boundaries of the `HVisible` object (subject to any insets defined by the `insets` parameter, and following the text alignment and justification set for that `HVisible` object), but only that text that falls within the clipping rectangle of the graphics context will actually get drawn on the screen. If the text is too long to fit in the `HVisible` object, the layout manager will clip it and insert an ellipsis (. . .) to show that it has been clipped.

MHP extends the text layout manager by defining the `org.dvb.ui.DVBTextLayoutMan-ager` class. This extends the HAVi text layout manager, and adds support for setting layout attributes such as indentation, line and letter spacing, and text wrapping, as well as more complex features such as orientation. Another important feature added by the `DVBText-LayoutManager` is the ability to register a `TextOverflowListener` that will be notified when a text string is too long to fit in the specified `HVisible` object. This lets applications take account of this fact, either by notifying the user that they may be missing some text or by redrawing the UI so that the text fits or so that the user can scroll the text. The full interface to this class follows.

```
public class DVBTextLayoutManager
   implements org.havi.ui.HTextLayoutManager {

   public static final int HORIZONTAL_START_ALIGN;
   public static final int HORIZONTAL_END_ALIGN
   public static final int HORIZONTAL_CENTER;

   public static final int VERTICAL_START_ALIGN;
   public static final int VERTICAL_END_ALIGN;
   public static final int VERTICAL_CENTER;

   public static final int LINE_ORIENTATION_HORIZONTAL;
   public static final int LINE_ORIENTATION_VERTICAL;
   public static final int START_CORNER_UPPER_LEFT;
   public static final int START_CORNER_UPPER_RIGHT;
   public static final int START_CORNER_LOWER_LEFT;
   public static final int START_CORNER_LOWER_RIGHT;
   public DVBTextLayoutManager();

   public DVBTextLayoutManager(
      int horizontalAlign,
      int verticalAlign,
      int lineOrientation,
      int startCorner,
      boolean wrap,
      int linespace,
      int letterspace,
      int horizontalTabSpace);

   public void setHorizontalAlign(int horizontalAlign);
   public void setVerticalAlign(int verticalAlign);
   public void setLineOrientation(int lineOrientation);
   public void setStartCorner(int startCorner);
   public void setTextWrapping(boolean wrap);
   public void setLineSpace(int lineSpace);
   public void setLetterSpace(int letterSpace);
   public void setHorizontalTabSpacing(
      int horizontalTabSpace);
   public void setInsets(java.awt.Insets insets);

   public int getHorizontalAlign();
   public int getVerticalAlign();
```

```
public int getLineOrientation();
public int getStartCorner();
public boolean getTextWrapping();
public int getLineSpace();
public int getLetterSpace();
public int getHorizontalTabSpacing();
public java.awt.Insets getInsets();

public void render(
   String markedUpString,
   java.awt.Graphics g,
   HVisible v,
   java.awt.Insets insets)

public void addTextOverflowListener(
   TextOverflowListener l);
public void removeTextOverflowListener(
   TextOverflowListener l);

}
```

Using a DVBTextLayoutManager is very simple. Before creating the HText object we wish to manage we create a DVBTextLayoutManager and then apply the settings that should be used for the text we will be managing. After we have done this, we can simply create an HText object representing the text in question and pass the layout manager to the HText object as an argument to the constructor. Alternatively, we can attach a text layout manager to any HVisible object using the setTextLayoutManager() method. In the following example, we attach it to an HText object.

```
// Create our layout manager
DVBTextLayoutManager layoutManager
layoutManager = new DVBTextLayoutManager();

// Now set the options we want for laying out the text.
// In this case, we will display centered text that wraps
// if it's too big for one line
layoutManager.setHorizontalAlign(
   DVBTextLayoutManager.HORIZONTAL_CENTER);

layoutManager.setTextWrapping(true);

// This constructor for HText takes the text used for
// this component, the position of the component, and its
// dimensions.
HText text = new HText(
   "This is a test for our layout manager",
   10, 10,
   300, 100);
```

```
// set our text layout manager to manage the layout of
// this object
text.setTextLayoutmanager(layoutManager);
```

Multilingual Support

MHP started out as a purely European standard, but because it has been adopted worldwide many people have asked whether an MHP receiver can support non-European character sets such as Chinese, Korean, or Hebraic characters. Although MHP uses the Western European character set by default, it can support other character sets very easily. A receiver may optionally include support for other locales and character sets, and all strings in MHP (including those found in application signaling) use the UTF8 format for handling multi-byte characters. This means that MHP implementations will quite happily support Asian or other multi-byte character sets, as long as the receiver has the appropriate locale information and a suitable font. This is one area in which downloadable fonts become most useful, although fonts for a particular market are often built into the receiver. Downloading a Chinese font to receivers deployed in China would be a waste of both time and bandwidth, no matter how easily the middleware supported it. All of the text layout managers and other text presentation tools will support multi-byte characters, and thus application developers should have few problems in customizing their applications for various markets.

Using Fonts

As well as being able to change the look and feel of the widget set, we can use various fonts to give our UIs distinct looks. MHP and OCAP receivers will include only the Tiresias font by default (Java applications can use the name `SansSerif` to create it), but others can be downloaded as part of the application. Tiresias is designed to be legible on TV displays, and thus it offers a good basic font for many applications that use the Western European character set.

However, if you are not happy with Tiresias or are developing a product for a market that uses a different character set, fonts in Bitstream's PFR (Portable Font Resource) file format can be downloaded to the receiver and used by an MHP or OCAP application. Because the middleware does not know about downloaded fonts, we cannot use the normal mechanism to instantiate these fonts. Instead, we use the `org.dvb.ui.FontFactory` class, as follows.

```
public class FontFactory {
  public FontFactory()
    throws FontFormatException, IOException;

  public FontFactory(java.net.URL u)
    throws FontFormatException, IOException;

  public java.awt.Font createFont(
    String name, int style, int size)
    throws FontNotAvailableException,
```

```
      FontFormatException,
      IOException;
}
```

Each font factory is associated with one source of fonts. This can be the default source or a source specified by the URL in the constructor. Both constructors are synchronous, and thus they will block until the font tile has been loaded or until the constructor throws an exception. To set the default source for fonts, we use the font index file. This XML file, named dvb.fontindex, is stored in the application's root directory. It contains a list of fonts available to the application, their attributes, and the file they can be loaded from. We will not describe this file in any detail here, but the following example should give you an idea of what it looks like.

```xml
<?xml version="1.0"?>
<!DOCTYPE fontdirectory
   PUBLIC "-//DVB//DTD Font Directory 1.0//EN"
   "http://www.dvb.org/mhp/dtd/fontdirectory-1-0.dtd">

<fontdirectory>
   <font>
      <name>Verdana</name>
      <fontformat>PFR</fontformat>
      <filename>fonts/Verdana.pfr</filename>
      <style>PLAIN</style>
   </font>

   <font>
      <name>Small Screen Font</name>
      <fontformat>PFR</fontformat>
      <size min="14" max="20"></size>
      <filename>fonts/screenfontsmall.pfr</filename>
   </font>

   <font>
      <name>Large Screen Font</name>
      <fontformat>PFR</fontformat>
      <size min="22" max="40"></size>
      <filename>fonts/screenfontlarge.pfr</filename>
   </font>

</fontdirectory>
```

Once we have created a font factory, we use it to create an instance of a Font object. The createFont() method does exactly what you would expect, creating a new font in the size and style specified.

MHP and OCAP support only a subset of the complete PFR format, and thus any fonts that cannot be interpreted by the receiver will cause a FontFormatException to be thrown.

ABCDEFGHIJKLMN

Figure 7.13. An example of the Tiresias font.

Depending on the problem, this may be thrown when the constructor opens the file (e.g., if the entire file cannot be read) or when the application tries to create a font (if that particular font description does not include data in the correct format). If the font file is not available for any reason, the constructors can also throw a `java.io.IOException` to indicate that there is a problem reading the file.

Using downloaded fonts can give your application a distinctive look, but developers need to be aware of several issues when using fonts on a TV display. The most important of these is font sizes. Televisions typically have much lower resolutions than a PC, and are viewed from much farther away. This means that small font sizes are often difficult to read, and generally text should be at least 18 points. Even this is very small on a TV display, and MHP and OCAP implementations do not have to support font sizes smaller than 24 points. The default font size is 26 points, and we recommend that this size be used for most text on screen.

The other major issue is font design. Thin lines may flicker and appear faint on a TV, and thus fonts that have a lot of thin lines (such as those used in bookmaking) are not really suitable. Sans serif fonts typically function best for a TV or computer display intended to convey a limited amount of text and do so such that the eye grasps the characters immediately, and fonts with thick straight strokes are more legible and viewable at distance than serif fonts. As you will see by looking at the example of Tiresias shown in Figure 7.13, it is a sans serif font with thick strokes.

Fonts have to be used carefully, and as with other types of graphic design less is often more. Using one or two well-chosen fonts, even the default font, is better than using a very distinctive font that is difficult to read (or using a number of fonts to which the eye must adjust often in moving from one to another).

Handling User Input

As you would expect for a device connected to a TV, a typical MHP or OCAP receiver will not use a keyboard and mouse for input. Instead, the most common input method will be a remote control unit. If middleware implementers and application developers are not careful, this can cause a number of problems.

Earlier in the chapter we saw that it is important to design a UI so that a user can easily navigate around it using just the arrow keys on a remote. This means setting sensible navigation targets for each component in our UI, but it also means using shortcuts where we can. Sometimes we need to present information on several different screens, and we have to give the user a way of moving among those screens easily.

Table 7.16. Standard keys in an MHP receiver.

Key	Constant Name
Up arrow	VK_UP
Down arrow	VK_DOWN
Left arrow	VK_LEFT
Right arrow	VK_RIGHT
Enter (also known as Select or OK)	VK_ENTER
Number keys	VK_0 to VK_9
Teletext key	VK_TELETEXT
First colored key	VK_COLORED_KEY_0
Second colored key	VK_COLORED_KEY_1
Third colored key	VK_COLORED_KEY_2
Fourth colored key	VK_COLORED_KEY_3

Luckily for us, HAVi defines a number of classes we can use to handle user input and help solve some of these problems. The org.havi.ui.event package extends AWT's event mechanism to add support for a number of the common features we find on a DTV receiver. The most commonly used class is the HRcEvent class. This extends the AWT KeyEvent class to add support for these keys in a way that is compatible with the AWT event mechanism. It also defines a number of constants for keys that will typically be found on a remote control unit, and MHP and OCAP define common subsets of these that will be present on every remote. Table 7.16 outlines the standard keys, along with the constants used to represent them.

OCAP uses a different set of keys, to reflect the differences between receivers in the U.S. market and in the European and Asian markets. It also defines some additional keys that are not part of the HAVi specification, as outlined in Table 7.17.

Many more events may be available, but every receiver must support this basic set. If your application uses a key that is not listed here, do not forget to provide some other method for carrying out any function implemented using that key. For instance, if your application uses the VK_COLORED_KEY_4 button to navigate to a certain screen, you should also provide a button that lets the user get to that screen as well.

Table 7.18 outlines all of the key codes defined by HAVi. In addition to these, all of the key codes defined in the java.awt.event.KeyEvent class can be used.

Many of these keys are used for normal navigation when watching a TV show, such as adjusting the volume, changing channels, and other functions. Because of this, MHP states that applications using any keys other than VK_TELETEXT and VK_COLORED_KEY_0 to VK_COLORED_KEY_3 must make it clear to the user that they are not in TV-viewing mode any more and that the normal navigation keys may not do what the user expects. To do this, any application that uses any other keys must have a visible component that covers at least 3% of the viewable area of the screen.

Table 7.17. Standard keys in an OCAP receiver.

Key	Constant Name
Power on/off	VK_POWER
Channel down	VK_CHANNEL_DOWN
Channel up	VK_CHANNEL_UP
Number keys	VK_0 to VK_9
Up arrow	VK_UP
Down arrow	VK_DOWN
Left arrow	VK_LEFT
Right arrow	VK_RIGHT
Enter (also known as Select or OK)	VK_ENTER
Volume down	VK_VOLUME_DOWN
Volume up	VK_VOLUME_UP
Volume mute	VK_MUTE
Pause	VK_PAUSE
Play	VK_PLAY
Stop	VK_STOP
Record	VK_RECORD
Fast forward	VK_FAST_FWD
Rewind	VK_REWIND
EPG	VK_GUIDE
RF bypass (if the receiver has channel 3/4 RF output)	VK_RF_BYPASS
Menu	VK_MENU
Info	VK_INFO
Exit	VK_EXIT
Last	VK_LAST
Function key 0	VK_COLORED_KEY_0
Function key 1	VK_COLORED_KEY_1
Function key 2	VK_COLORED_KEY_2
Function key 3	VK_COLORED_KEY_3
Page up	VK_PAGE_UP
Page down	VK_PAGE_DOWN
On-demand (OCAP version I10 and later only)	VK_ON_DEMAND
Next favorite channel	VK_NEXT_FAVORITE_CHANNEL

In regard to this, the navigator should not map any important functions to these keys, in that other applications may use them. This also means that applications should use these keys to activate the application, if the user has to press a key to start it. In the United Kingdom, for instance, many interactive ads use the red Teletext button to launch the application associated with the ad. To get notification of these keys even when they do not have input focus, applications can use the DVB user event API (discussed in material to follow).

It is unlikely that a remote will actually support all of the keys defined by HAVi, MHP, or OCAP, simply because any remote that did would be too complicated to use. We can use the

Table 7.18. Key codes defined by HAVi for remote control buttons.

VK_BALANCE_LEFT	VK_BALANCE_RIGHT	VK_BASS_BOOST_DOWN
VK_BASS_BOOST_UP	VK_CHANNEL_DOWN	VK_CHANNEL_UP
VK_CLEAR_FAVORITE_0	VK_CLEAR_FAVORITE_1	VK_CLEAR_FAVORITE_2
VK_CLEAR_FAVORITE_3	VK_COLORED_KEY_0	VK_COLORED_KEY_1
VK_COLORED_KEY_2	VK_COLORED_KEY_3	VK_COLORED_KEY_4
VK_COLORED_KEY_5	VK_DIMMER	VK_DISPLAY_SWAP
VK_EJECT_TOGGLE	VK_FADER_FRONT	VK_FADER_REAR
VK_FAST_FWD	VK_GO_TO_END	VK_GO_TO_START
VK_GUIDE	VK_INFO	VK_MUTE
VK_PINP_TOGGLE	VK_PLAY	VK_PLAY_SPEED_DOWN
VK_PLAY_SPEED_RESET	VK_PLAY_SPEED_UP	VK_POWER
VK_RANDOM_TOGGLE	VK_RECALL_FAVORITE_0	VK_RECALL_FAVORITE_1
VK_RECALL_FAVORITE_2	VK_RECALL_FAVORITE_3	VK_RECORD
VK_RECORD_SPEED_NEXT	VK_REWIND	VK_SCAN_CHANNELS_TOGGLE
VK_SCREEN_MODE_NEXT	VK_SPLIT_SCREEN_TOGGLE	VK_STOP
VK_STORE_FAVORITE_0	VK_STORE_FAVORITE_1	VK_STORE_FAVORITE_2
VK_STORE_FAVORITE_3	VK_SUBTITLE	VK_SURROUND_MODE_NEXT
VK_TELETEXT	VK_TRACK_NEXT	VK_TRACK_PREV
VK_VIDEO_MODE_NEXT	VK_VOLUME_DOWN	VK_VOLUME_UP
VK_WINK		

`HRcCapabilities` class to find out which of these keys are supported by a particular remote. This class has the following interface.

```
public class HRcCapabilities extends HKeyCapabilities {

    public static boolean getInputDeviceSupported();

    public static boolean isSupported(int keycode);
    public static HEventRepresentation
      getRepresentation(int keycode);

}
```

The `getInputDeviceSupported()` method tells us whether the receiver supports a keyboard (either a real keyboard, an on-screen keyboard, or a cellphone touchpad-like keyboard). An application can use this to decide what functionality to offer. For instance, if an application supports instant messaging functionality it may choose to provide this only on those receivers that support some type of text input. Alternatively, it may choose to implement its own virtual keyboard.

Applications can also use the `isSupported()` method to find out whether a particular key is supported. This offers a useful way of tailoring the navigation model to fit the capabilities of the receiver: an application could query whether the two optional colored keys are available, and draw a different UI based on whether or not they are present. If they are, they could be used to offer additional navigation shortcuts.

Because much of our UI is graphical, we may want to draw a representation of a key on the screen as an aid to navigation (after all, what one manufacturer may name OK may be named SELECT by another, and the graphical representation of a key as it is drawn on the remote may also differ between receivers). Using the `getRepresentation()` method, we can find how to draw a given key. This returns an `HEventRepresentation` object for the key in question. The `HEventRepresentation` class lets an application find out how a key should be represented on the screen. Depending on the type of key, it may be represented by a color, a text string, an image up to 32 × 32 pixels square, or a combination of these. By getting the representation for that key, we can draw a UI that matches the symbols and text on the user's remote control unit.

Keyboard Events and Input Focus

The AWT event model offers a pretty good basis for user input in MHP, not least because of its familiarity. In the conventional AWT event model, a component may only receive keyboard events when it has user input focus. In an environment such as MHP, this restriction can cause some problems.

To prevent newly loaded applications from interfering with any applications that are already running, any `HScenes` created are not visible by default, and do not automatically gain input focus. This lets the user keep interacting with any application that currently has focus.

Unfortunately this is not enough, in that DTV applications may want access to key events without having a graphical component. Imagine the case of an EPG that only pops up when the GUIDE button is pressed. If this application wanted to maintain an invisible AWT component on the screen to detect any AWT events, it would have to be contained in an `HScene`. As we saw in the section on graphics, `HScenes` may not be allowed to overlap. This means that the EPG would be restricting the space available to other applications, even though it had no real visible component on screen.

Because this is not very polite, having no components in the AWT hierarchy is a far better idea. Even if we had the invisible component, that component would need focus to get the event anyway, and thus this approach is less than useful. Therefore, we need an API to get user input events without having to use AWT.

The `org.dvb.event` package defines an API for allowing applications to access events before they enter the AWT event mechanism, and both MHP and OCAP applications can use this API to provide extended user interaction. A class that implements the `org.dvb.event.UserEventListener` interface can receive input events even though that application does not have user input focus.

The events generated by this API are not AWT events, but are instead instances of the `org.dvb.event.UserEvent` class. Although other event types may be added in the future, at present only keyboard events are supported. To applications, these are very similar to AWT

KeyEvents, and applications can find out which key was pressed and what modifiers (Shift, Ctrl, Alt, or META keys, for example) are active (just as they can with a KeyEvent).

Before an application can receive user events, it must define a UserEventRepository that defines the group of events the application wishes to receive. As we can see from the following API for this method, it contains methods to allow an application to add and remove various combinations of keys.

```
public class UserEventRepository {

    public UserEventRepository (String name);

    public void addUserEvent (UserEvent event);
    public UserEvent [] getUserEvent ();
    public void removeUserEvent (UserEvent event);

    public void addKey (int keycode);
    public void removeKey (int keycode);

    public void addAllNumericKeys ();
    public void addAllColourKeys ();
    public void addAllArrowKeys ();

    public void removeAllNumericKeys ();
    public void removeAllColourKeys ();
    public void removeAllArrowKeys ();
}
```

If an application wants to be notified of all key events (which is pretty rare in practice), it can use the OverallRepository class instead. This subclass of UserEventRepository automatically contains all user events the receiver can generate.

Once the application has defined the set of keys it wishes to receive events for, it can use the EventManager class to request access to these events. This class is a singleton object, which can be obtained by calling the EventManager.getInstance() method. The interface for this class follows.

```
public class EventManager implements
    org.davic.resources.ResourceServer {

    public static EventManager getInstance ();

    public boolean addUserEventListener (
        UserEventListener listener,
        org.davic.resources.ResourceClient client,
        UserEventRepository userEvents);

    public void addUserEventListener (
        UserEventListener listener,
        UserEventRepository userEvents);
```

```
    public void removeUserEventListener(
       UserEventListener listener);

    public boolean addExclusiveAccessToAWTEvent(
       org.davic.resources.ResourceClient client,
       UserEventRepository userEvents);

    public void removeExclusiveAccessToAWTEvent(
       org.davic.resources.ResourceClient client);

    public void addResourceStatusEventListener(
       org.davic.resources.ResourceStatusListener listener);

    public void removeResourceStatusEventListener(
       org.davic.resources.ResourceStatusListener listener);
}
```

After it has obtained a reference to the `EventManager` class, an application can use the `addUserEventListener()` and `removeUserEventListener()` methods to add and remove listeners for the set of events defined by a given `UserEventRepository` object. Access to these events is a scarce resource because applications can also request exclusive access to a set of key events, and thus an application may lose access to some or all key events specified in a repository.

Exclusive Access to Keyboard Events

When entering sensitive information (e.g., credit card numbers or passwords), allowing other applications to receive keyboard events can be a big security hole, especially if those applications do not need an AWT component. For this reason, applications can request exclusive access to specific keyboard events.

Applications can get exclusive access to events in two ways. By calling the following variation of the `addUserEventListener()` method (shown previously), applications can get exclusive access to a set of events defined in a `UserEventRepository`.

```
    public boolean addUserEventListener(
       UserEventListener listener,
       org.davic.resources.ResourceClient client,
       UserEventRepository userEvents);
```

This method takes a `ResourceClient` instance as an additional parameter, and this resource client will be notified whenever the middleware wants to remove exclusive access to one or more keys (or give exclusive access to another application). In this case, those events will be delivered via the DVB event mechanism to our application. To give up exclusive access, we call the `removeUserEventListener()` method.

Sometimes, though, we do not want to use this. We want to use the normal AWT event mechanism, but guarantee that only our application gets that particular set of events. In this case,

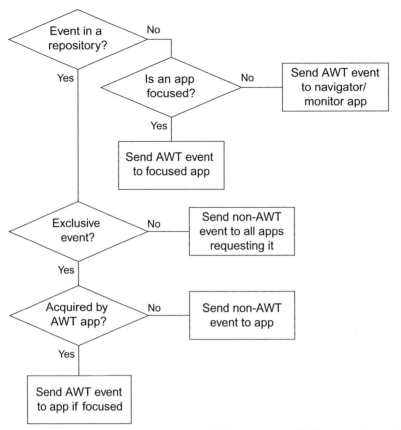

Figure 7.14. Handling user input events in MHP and OCAP. *Source:* ETSI TS 101 812:2003 (MHP 1.0.3 specification).

we can call the `addExclusiveAccessToAWTEvent()` method. This has the same behavior as the `addUserEventListener()` method, but any events are delivered via the AWT event mechanism instead.

When an MHP receiver gets a user input event, it has to decide which application (or applications) should get that event. The flowchart shown in Figure 7.14 illustrates just how incoming events are routed to the appropriate applications.

Because the application can request exclusive access to user input events, these are treated as if they were a scarce resource (by definition, only one application can have exclusive access to a given user input event). The manner in which these resources are managed is platform dependent, although it will probably follow the basic rules we saw in Chapter 6. Any application that requests exclusive access to user input events must take into account the fact that it could lose that exclusive access at any time.

One thing you may have realized is that this approach assumes that the middleware can determine which application is making the request. This is not necessarily easy, but it is something many different APIs in MHP need. One approach to doing this is described in Chapter 14.

Practical Issues for DTV Graphics

As we have seen, working with graphics on a DTV receiver can get very complicated due to the hardware limitations of the receiver and the differences from a PC platform. Application developers and graphic designers need to consider a number of other things.

As we have already discussed, the different shapes of video and graphics pixels may mean that the video and graphics are not perfectly aligned. The video and graphics layers may not even be the same resolution. Setting the VIDEO_GRAPHICS_PIXEL_ALIGNED preference in the configuration template will give us the platform's best effort at aligning the two layers of the display, although there is no guarantee a receiver can actually support this.

As if this were not enough, other factors can cause display problems. Overscan in the TV can mean that only the central 90% of the display is visible, and that only the central 80% of the screen is free from distortion. These numbers apply to 4:3 television sets only. 16:9 displays have different areas that may be hidden or distorted. To avoid problems, developers should stick to the "graphics-safe" area, also known as the "title safe" area. This is the area in the center of the screen that will always be visible and free of distortions on any TV. It has the dimensions outlined in Table 7.19.

If you really want to put graphics over the entire display, be aware that some of them may not appear on the screen. Many graphics tools will provide guides to show the areas of the screen that are within the safe areas.

Another problem is that what the receiver outputs may not be what is actually displayed on the screen. Given that many modern TV sets allow the viewer to control the aspect ratio, the receiver may be producing a 4:3 signal that is being displayed by the TV in 16:9 mode. Even then, there are various options for how the TV will actually adjust the display to the new aspect ratio. Although there is nothing your application can do about this, because it will not even be aware of it, it is something that you as a developer may need to be aware of because even if the application is not broadcast as part of a 16:9 signal now it may be in the future.

Table 7.19. The title safe area for PAL and NTSC displays with an aspect ratio of 4:3.

Standard	Safe Area Width	Safe Area Height
PAL	562	460
NTSC	384	512

Australia has taken some steps to avoid this problem, because broadcasters and receiver manufacturers have agreed to broadcast all digital content in 16:9 format, with black bars at the sides of 4:3 content to avoid stretching the image. This means that designers and application developers do not have to contend with changing aspect ratios in the broadcast signal. This meant resolving some issues related to wide-screen signaling that resulted from the use of the same signal to generate both analog and digital content, but this was not difficult with agreement from all parties.

Finally, as we have seen earlier, the graphics, video, and background configurations are not always independent, and changing one may have an effect on the others. Applications that care about the graphics or video configuration should monitor them, to make sure the configuration stays compatible with what the application is expecting, and to be able to adapt when the configuration does change.

We must also remember that a TV is not a computer display. Interlacing means that thin horizontal lines will flicker, and small text will have the same problem. Designers should also avoid using colors that are too saturated, because on a TV display these can bleed and result in blurred edges and ugly artifacts. Similarly, we can use only a subset of the available colors even if a set-top box can display 24-bit color: not all colors are legal in the NTSC color space. Adobe Photoshop and other high-end graphics tools include filters that will use only NTSC-compatible colors in an image, and it is worth using this to make sure you have no problems.

Another color-related problem is the use of very saturated colors in a UI. This can cause colors to bleed on TV displays, and thus the use of colors with a high level of saturation (typically, above 85%) should be discouraged. It is also important to remember that some MHP receivers will be CLUT based, and will thus not be able to reproduce all possible colors. MHP defines a standard color palette that every receiver must support, similar to the Web-safe palette used by web designers. Where possible, designers should choose colors from this palette to make sure the UI looks as it should. This stops the receiver from carrying out any dithering, thus reducing the quality of the graphics on the screen. On high-end receivers this may not make a difference, but it means that your application will display nicely on all receivers.

As well as these issues, it is important to consider the audience that will be using your application. They will often be less technically minded than a PC user, and may be more interested in a passive experience than an active one requiring a lot of interaction. Clear instructions and consistent UI are important to make the applications as easy to use as possible. It is also important to remember that watching TV is a social activity for many people, and thus more than one person may want to interact with an application. An application associated with a quiz show may support more than one player, for instance. TV is a very different medium from the PC, and the nature of the user's interaction is also very different.

8 Basic MPEG Concepts in MHP and OCAP

The ability to access broadcast content is a feature of many applications. This chapter introduces the basic concepts associated with the various types of media content in MHP and OCAP systems, and discusses the core classes that let us refer to elements of a broadcast stream.

Given that a number of the APIs in MHP and OCAP deal with concepts related to MPEG, and particularly to MPEG in a broadcast environment, the classes that describe these concepts are important to both application and middleware developers. They are not very glamorous, but you will often see them being used by other APIs in the two middleware stacks. These classes describe an MPEG-2 transport stream and the various elements that make up that transport stream. Although they use the MPEG service information (SI) to discover the relationships between these elements, these classes are not part of an SI API. They provide only the most basic information about the MPEG streams being received, and are more useful as a means of uniquely identifying a stream.

Many of these classes can be found in the `org.davic.mpeg` package, and there are three classes that especially interest us: the `TransportStream` class, the `Service` class, and the `ElementaryStream` class. As you have probably guessed, these represent, respectively, an MPEG-2 transport stream, a service, and an MPEG-2 elementary stream.

Each of these classes has a relationship with the other two, and thus an application calls the `TransportStream.getServices()` method to retrieve an array of `Service` objects describing the services in that transport stream. Similarly, by calling the `Service.get-TransportStream()` method, an application can get the `TransportStream` object that carries that service. Similar methods allow you to get the `ElementaryStream` objects representing the components of a given service.

These classes are not designed to give us access to SI, and thus we can get only very limited information from them. The only parts of the SI you can access from these classes are the transport stream ID (from a `TransportStream` object), the service ID (from a `Service` object) and the PID and DSM-CC association tag for an elementary stream (from an `ElementaryStream` object).

One of the reasons these classes expose so little information is that different SI standards contain different information. The classes we have seen so far provide the basic information that can be found from any transport stream, but other information may be available. For DVB systems such as MHP, each of the classes we have just seen has a subclass in the `org.davic.mpeg.dvb` package. These have the same names as the classes we have already seen, with the prefix Dvb, as in the following: `DvbTransportStream`, `DvbService`, and `DvbElementaryStream`. These classes give us access to extra information available in DVB networks. The `DvbTransportStream` class, for instance, adds methods to get the network ID and original network ID for a transport stream. The `DvbElementaryStream` and `DvbService` classes add methods that allow applications to get the component tag for an elementary stream, or to get the elementary stream with that component tag from a given service.

OCAP adds one class of its own in the `org.ocap.mpeg` package. The `PODExtendedChannel` class is a subclass of `org.davic.mpeg.TransportStream` that represents an MPEG stream received over the POD extended channel interface (an out-of-band channel between the head-end and the receiver). Because this is an out-of-band channel, we do not have the SI that would normally be present, and thus the transport stream ID for a stream represented by `PODExtendedChannel` is always −1. All other methods in this class will return a null reference, because this channel is only used for SI and thus will not contain services in the way we would normally expect. `PODExtendedChannel` is an abstract class in later versions of OCAP, and thus the middleware may provide a subclass of this for use by applications.

Some of the JavaTV APIs also contain some of this information, although their focus may be slightly more high level and focus more on SI. The main difference between the JavaTV classes and the other classes is this integration with JavaTV SI, as we will see in the next chapter when we examine the JavaTV classes in more detail. Table 8.1 outlines the classes available on MHP and OCAP receivers.

Content Referencing in the MHP and OCAP APIs

There are many instances in which an MHP or OCAP application will want to refer to a piece of content, such as when it is playing media, loading a file, or displaying a new service. In many cases, APIs will allow applications to use several different types of content, and thus we need a standardized way of describing how the receiver can access that content.

Table 8.1. MPEG infrastructure classes in MHP and OCAP.

Class	DVB	OCAP
`org.davic.mpeg.TransportStream`	✓	✓
`org.davic.mpeg.dvb.DvbTransportStream`	✓	
`org.davic.mpeg.Service`	✓	✓
`org.davic.mpeg.dvb.DvbService`	✓	
`org.davic.mpeg.ElementaryStream`	✓	✓
`org.davic.mpeg.dvb.DvbElementaryStream`	✓	
`org.ocap.mpeg.PodExtendedChannel`		✓
`javax.tv.service.Service`	✓	✓
`javax.tv.service.navigation.ServiceComponent`	✓	✓
`javax.tv.service.transport.TransportStream`	✓	✓
`javax.tv.service.transport.Network`	✓	✓

The classes from `org.davic.mpeg` and `org.davic.mpeg.dvb` are suitable for many purposes, but applications cannot create their own instances of these classes. Similarly, we cannot use these to refer to a file in a broadcast file system, or to a specific event on a service. How can applications create a reference to a broadcast stream that is useful to both the application developer (who has to write the code that creates it) and to the middleware (which actually has to use it)?

We could use a URL, but this entails other problems. We still need some internal representation for a URL, because we cannot keep parsing the URL string every time. Extending the existing `java.net.URL` class would add unnecessary functionality that would cause a lot of pain for middleware developers, mainly because of the design of Sun's standard URL class. This is aimed very much as IP-based connections, and adapting this for digital TV use is not easy. There are other reasons this class is not the best choice, and we will explore some of these later in the chapter.

To overcome these problems with the URL class, MHP, JavaTV, and OCAP all use the concept of a locator. A locator is an object that represents a reference to some content but does not define how this reference is stored. It is up to the middleware to choose the format that suits it best and that is most efficient for a given type of locator. This makes life easier for middleware implementers, because they can store any information they need to about where to find a given piece of content.

Of course, this does not help an application developer who needs to work with locators, and thus locators also have a standardized string representation known as an external form. In every case so far, that external form is a string representing a URL. This does not have to be the case, however, and thus applications should not make the assumption that every locator will have an external form that is a URL.

We will take a look at examples of some of these URLs in material to follow, but for now we should identify the types of content that can be referred to using a locator. In MHP and OCAP systems, locators are used in MPEG-related APIs for referring to the following types of content.

- Transport streams
- Services
- Elementary streams within a service
- Events
- Files in a broadcast file system
- Video drips

Most of these will be familiar to you already. Video drips are a special type of content used by the media control API to display images in an efficient way. These are discussed in Chapter 10.

References to broadcast streaming content (transport streams, services, elementary streams, and events) all use the same basic URL type. Unfortunately, that is as simple as it gets, because DVB systems use a format different from those of OCAP and ACAP systems.

Locators for DVB Streaming Content

Locators for streaming content in a DVB system use the dvb:// URL format. The syntax of a dvb:// URL is as follows.

```
dvb://<onID>.<tsID>.<sID>[.<ctag>[&<ctag>]][;<evID>][<path>]
```

The various elements of this URL format are outlined in Table 8.2.

All of these elements (except for the path parameter) are represented as hexadecimal values without the leading 0x. The path component uses the standard URL path format as defined

Table 8.2. Elements of a URL for a DVB locator.

URL Element	Description
onID	The original network ID, which identifies the broadcaster or network that produced the content (not the network currently broadcasting it, if they are different)
tsID	The transport stream ID, which identifies a specific transport stream the network is broadcasting
sID	The service ID, which refers to a service within that transport stream
ctag	The component tag, which refers to a specific elementary stream that has been tagged in the SI
evID	The event ID, which identifies a specific event that is part of the service
path	The path to a file in a broadcast file system being transmitted on that elementary stream or service

Table 8.3. Examples of DVB URLs.

DVB URL	Purpose
dvb://123.456.789	Identifies a DVB service
dvb://123.456.789.66	Refers to an elementary stream within a service using its component tag
dvb://123.456.789;42	Refers to a specific event on a service
dvb://123.456.789/images/logo.gif	Identifies a specific file in the default broadcast file system for a given service

in RFC 2396. All of the numeric identifiers used for the various elements of the URL match the identifiers used in the DVB SI that is part of the transport stream. Table 8.3 outlines a few example DVB URLs.

Only the first three elements (the original network ID, transport stream ID, and service ID) are required. This means that a DVB URL will always refer to a service. Those APIs that deal with transport streams use locators in any of the formats we have just seen, but they ignore any parts of the locator that are more specific than they need. Using the same URL format means that we can use the same URL for referring to a transport stream and to something contained within that transport stream, and in general this behaves the same as a URL in the Internet world. A URL for a web site may identify a specific file on that web site but it will also identify the server that contains it.

The last locator in Table 8.3 refers to a file in broadcast file system, but the file system in question is not specified. In this case, the receiver will assume that the locator refers to the default file system. It is not specified anywhere which file system is the "default," however, and thus it is generally best to avoid using this type of locator. There are not many cases in which this locator is necessary, and thus it is not commonly used. A safer approach is to include the component tag that identifies the object carousel you are referring to. This way, even if another file system is added to the service your locator will still be correct.

You may have noticed that there is no way to refer to a transport stream using this locator syntax. Usually an application or middleware is not dealing with transport streams. They are dealing with a service or component within that transport stream.

Locators for Streaming Content in OCAP Systems

The DVB locator format is linked fairly tightly with the SI available in a DVB signal, and thus it is not really suitable for other environments such as OCAP. For this reason, OCAP uses its own protocol called (not surprisingly) ocap://. The full format for the external form of an OCAP locator is as follows.

Table 8.4. Elements in an OCAP URL.

OCAP URL	Purpose
service	Identifies an OCAP service
service_component	Identifies an OCAP component (an MPEG-2 elementary stream)
event_id	Identifies a single event in the service
path	Gives a path to a file in a broadcast file system

Table 8.5. Referring to a service with an OCAP locator.

Syntax	Service Identification Method
ocap://0x37F2	Source ID
ocap://n=News	Source name or short name
ocap://f=0x00A3FE2778.0x03	Frequency and program number

```
ocap://<service>[.<service_component>][;<event_id>][/<path>]
```

The elements of this locator format are outlined in Table 8.4.

Unlike the DVB URL format, we can refer to OCAP services in several ways. The simplest of these is a non-zero 16-bit hexadecimal value that corresponds to the source ID field carried in the Virtual Channel Table of ATSC SI. This uniquely identifies that service across the entire network.

This is not very friendly, however, and thus there is also a more human-readable way of identifying a service. Applications can also use the name of the source, as signaled in the source_name or short_name fields carried in the Virtual Channel Table entry for that source. Exactly how the name is resolved depends on the profile of SI being used on the network. The OCAP specification describes in detail how this mapping is carried out.

If these approaches get too friendly for you, an OCAP locator can also refer to a service by the frequency of the transport stream that carries it and the program number within that transport stream. In this case, the frequency is listed in the URL as a 32-bit hexadecimal number and the program number as a 16-bit hexadecimal number. Table 8.5 outlines the syntax that should be used for each of these methods.

There are also several methods of identifying a component within a service. The simplest of these is to use a hexadecimal string that identifies the type of component. The values for the various component types are listed in ANSI/SCTE specification 57.

If there are several components of the same type in the service, the component identifier refers to only the first one in that service. To refer to another component of that type, we can employ two possible methods.

Table 8.6. Syntax for methods of identifying an elementary stream in an OCAP URL.

Syntax	Component Identification Method
`.<component_type>`	Component type only
`.<component_type>,<ISO_language_code>`	Component type and language code
`.<component_type>,<index>`	Component type and index
`@<component_tag>`	Component tag (versions I10 and later only)
`+<PID>`	PID

Some streams (such as audio tracks) are associated with a specific language. For these streams, the locator can include a three-letter language code after the stream type in order to identify a stream of the given type in that language.

Obviously, this does not work for all cases. For example, what if you have two audio tracks in the same language (e.g., audio tracks for a movie with and without commentary from the director), or what about two video tracks that have no language setting? In this case, the locator can append a hexadecimal value to the stream type. This tells the receiver the index of the stream it should use (e.g., the second stream of the specified type in that service, or the third, or the fifth).

As with services, there is also a user-unfriendly way of referring to components if we need it. If we want, we can specify the PIDs of the service components we are referring to. Just like any other number in the locator, the PID is written as a hexadecimal value. If this is too unfriendly, we can use component tags to identify a specific stream component in receivers that support versions I10 and later of OCAP. This still lets us identify a specific PID, but avoids any problems caused by remapping PIDs when a stream is remultiplexed. Table 8.6 outlines methods of identifying a service component, and the syntax that should be used.

We can identify a set of components from the same service by separating each component with an ampersand. For instance, the following URL identifies the video and audio components of the service.

```
ocap://0x37F2.0x02&0x04
```

We can even mix different ways of identifying components in the same URL, but this tends to get confusing. Generally, it is better to stick to the same notation unless there is a good reason for doing otherwise. The OCAP locator format is confusing enough without making things more difficult for yourself.

After the complexity of referring to services and service components, you will be glad to hear that referring to events is easy. The event ID is simply a hexadecimal number giving the event ID as listed in the SI. The following URL identifies the event with event ID 0x12 in the specified stream.

```
ocap://0x37F2;0x12
```

Table 8.7. Example locators for OCAP content.

Locator	Description
`ocap://n=News`	Service identified by name
`ocap://0x213.0x81,eng`	Audio stream identified by service ID and language
`ocap://n=News.0x02&0x81,eng`	Video stream and audio stream identified by language
`ocap://n=News.0x02,0x01`	Video stream identified by index
`ocap://0x213+0x110`	Service component identified by PID
`ocap://n=News;0x12`	Event on a specific service
`ocap://0x213&0x81,eng;0x12`	Audio track for a specific event, identified by language
`ocap://n=Movies.0x0D/myApplication/background.jpg`	A file in a broadcast file system

The path component of a URL is also straightforward, following the standard defined in RFC 2396 (the same format MHP uses for path components). The complexity of OCAP locators may look scary to a developer used to DVB systems, but it is not significantly more complex in practice. It actually makes things slightly more flexible, in that developers can refer to an elementary stream without knowing exactly where it is in the service. It also means that component tags or PIDs can be remapped without causing any problems for applications, while still allowing developers to refer to streams in a very low-level way if that makes more sense for that application.

Generally, though, these low-level approaches cause more trouble than they are worth. If a transport stream is moved to a different frequency, or if the PIDs are remapped for any reason, the locator is no longer valid. This is only a problem when there is no valid SI for a particular stream, and there are not many cases in which you would want to refer to those streams anyway. Overall, it is much better practice for network operators to include SI for those streams than for applications to use low-level references to them. Table 8.7 outlines some full examples of OCAP locators that refer to different pieces of content.

Locators for Files

In some cases, we need to use a locator that refers to a file. Although the `dvb://` or `ocap://` URL formats can refer to a file, sometimes we need to tell the middleware that we are specifically referring to a file on a local file system, or on a broadcast file system we are currently connected to. Because the `dvb://` or `ocap://` locator formats can refer to any file on any broadcast file system on any service, they are a bit too broad for our needs.

Table 8.8. Typical file locators in an OCAP system.

Syntax	File Location
`file://carousels/012/034/background.jpg`	A file in the root directory of a broadcast carousel (this path is implementation dependent)
`file://local/monitorApp/background.jpg`	A file in a subdirectory of a local file system (this path is implementation dependent)

OCAP and MHP both use the same format for referring to files in a mounted file system. To refer to a file on a broadcast file system we are connected to (where we have already mounted the file system) we use a `file://` URL. The format for this URL is the standard format specified in RFC 2396, and thus we will not look at it in too much detail. Some typical file URLs are outlined in Table 8.8.

As with any `file://` URL, we need to use the absolute path (we cannot use relative file names in a URL). This gets interesting in our case because we do not know where the OCAP or MHP middleware will mount a given object carousel in its file system. To successfully (and interoperably) find the correct URL for a file in a DSM-CC object carousel, it is necessary to get this from the file itself. Do not panic if some parts of the code in this example are not clear to you. We will look more closely at how we manipulate broadcast file systems in Chapter 12.

```
// Create a DSM-CC object that refers to the file in the
// carousel. Note that the file name is usually
// relative to the application's root directory.
org.dvb.dsmcc.DSMCCObject myDsmccObject;
myDsmccObject = new org.dvb.dsmcc.DSMCCObject(
  "my/file/path/filename.mp2");

// We can call getURL() on that object to get a file://
// URL that points to it
java.net.URL url = myDsmccObject.getURL();

// Now we can create the Locator
LocatorFactory locatorFactory;
locatorFactory = LocatorFactory.getInstance();

Locator myLocator;
try {
  myLocator = locatorFactory.createLocator(
    url.toString());
}
catch (javax.tv.locator.MalformedLocatorException e) {
  // Catch the exception in whatever way we see fit.
```

```
    // Since we should always get a valid URL from getURL(),
    // this exception should not get thrown in this
    // situation but a little paranoia is always good.
  }
```

The locator we get back from this will use a `file://` URL of the correct format to refer to that file.

Locators for Video Drips

Video drips are a new content format unique to the DTV world. The main aim of this format is to provide a memory-efficient way of displaying several similar images. Each drip is a very short piece of MPEG-2 consisting of a small number of frames. The first thing in the file is an MPEG-2 I frame that can be decoded and presented to the user. One or more P frames follow this, and these are decoded based on the preceding I frame. This allows the decoder to update a static image in a very memory-efficient way.

In this format, the data is passed to the media control API via an array of bytes, and thus the content must already be in memory. Because of this, the format does not really have a locator that identifies the place the data is loaded from. We still need a URL, though, to create locators and the appropriate objects to decode this content format. For this reason, the URL for every piece of video drip content is as follows.

```
dripfeed://
```

Unbelievable as it may seem, this is a valid URL, and this tells the locator what the content format is without needing to refer to a specific piece of data. We will explore how this works in Chapter 11.

Locator Classes

One of the biggest problems with locators in JavaTV, OCAP, and MHP is the sheer variety of classes that implement them. These classes follow.

- `javax.media.MediaLocator`
- `org.davic.media.MediaLocator`
- `javax.tv.locator.Locator`
- `org.davic.net.Locator`
- `org.davic.net.dvb.DvbLocator`
- `org.davic.net.dvb.DvbNetworkBoundLocator`
- `org.davic.net.TransportDependentLocator`
- `org.ocap.net.OcapLocator`

Each of these is used to refer to a piece of content, but they are all used in subtly different ways. For instance, the media control API uses a `javax.media.MediaLocator` to refer to a piece of media that will be presented, and this may refer to a service or service component, an audio clip, or a video drip.

`org.davic.net.Locator` and its subclasses (including `org.ocap.net.OcapLocator` and `org.davic.net.dvb.DvbLocator`) are designed to refer to transport streams or services. `javax.tv.locator.Locator` is a more general locator defined by JavaTV that can refer to any piece of DTV content (a service, transport stream, or file in a broadcast file system).

Given all of these different locator types, they have to be related somehow. The good news is that there are only two main hierarchies of locators, and these are related. The first hierarchy is used by everything except the media control API. This has `javax.tv.locator.Locator` as its parent class, and can refer to any type of broadcast content (although it is usually used for referring to services or service components).

The second hierarchy is used only for media control, and it consists of `javax.media.Media Locator` and `org.davic.media.MediaLocator`. The DAVIC class was designed to bridge the gap between JMF and the rest of the system, and thus it defines an additional constructor that takes an `org.davic.net.Locator` as an argument. This allows applications to easily take a locator from the SI API (for instance) and construct a JMF player for it. Figure 8.1 shows the relationship between the various locator classes.

As you can see, not all of the locator classes are present in each platform. Some classes are specific to DVB systems, whereas others are specific to ATSC systems.

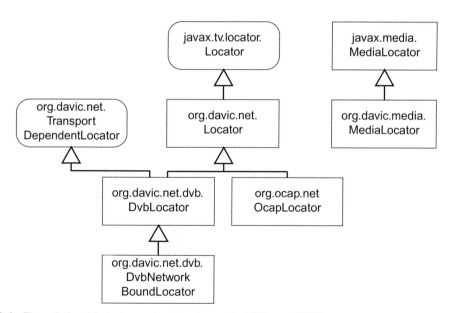

Figure 8.1. The relationship between locator classes in MHP and OCAP.

Creating a Locator

Some of these locator types (javax.media.MediaLocator and org.davic.net. Locator and their subclasses) can be created directly from a URL. The org.davic.net. dvb.DvbLocator class also has a number of other constructors that allow the various components of the locator to be specified directly, as follows.

```
public class DvbLocator extends org.davic.net.Locator {

    public DvbLocator(String url)

    public DvbLocator(int onid, int tsid)

    public DvbLocator(
        int onid, int tsid, int serviceid);

    public DvbLocator(
        int onid, int tsid, int serviceid, int eventid);

    public DvbLocator(
        int onid, int tsid, int serviceid, int eventid,
        int componenttag);

    public DvbLocator(
        int onid, int tsid, int serviceid, int eventid,
        int[] componenttags);

    public DvbLocator(
        int onid, int tsid, int serviceid, int eventid,
        int[] componenttags, String filePath);

}
```

For clarity, we are showing only the constructors here. Out of the constructors shown here, only the first one is valid for an org.davic.net.Locator. Any of these constructors (as well as the constructor for org.davic.net.Locator) can throw an org.davic.net. InvalidLocatorException if the locator does not refer to a valid stream or file. This means that the middleware should do some checking of the locator before it creates it, but practical issues mean that there is only so much checking we can do. This is especially true if the locator refers to a different transport stream we currently know nothing about. At the very least, the middleware should check that the URL is valid for that particular middleware stack (a DVB locator is not valid for an OCAP middleware implementation, and vice versa).

Instances of javax.tv.locator.Locator, unlike DAVIC or DVB locators, can be created using the javax.tv.locator.LocatorFactory class only. This factory class provides the createLocator() method, which takes a URL of the same format as the previous locator classes and returns a JavaTV locator, as follows.

```
public abstract class LocatorFactory
    extends java.lang.Object {
```

```
    public static LocatorFactory getInstance();

    public javax.tv.locator.Locator createLocator(
        String locatorString)
        throws javax.tv.locator.MalformedLocatorException;

    public javax.tv.locator.Locator[] transformLocator(
        javax.tv.locator.Locator source)
        throws javax.tv.locator.InvalidLocatorException;
}
```

On OCAP systems, any locators returned by this class will actually be an instance of `org.ocap.net.OcapLocator` (and therefore also a DAVIC locator). In an MHP receiver, any locators returned by calls to the JavaTV `LocatorFactory` will be subclasses of `org.davic.net.dvb.DvbLocator`.

Network-bound Locators

By default, locators do not tell the receiver how to access a piece of content. If the content is available from two networks, the locator does not identify which network should be used. For instance, let's take a receiver that has both a cable network interface and a terrestrial network interface. If a locator refers to a service that is available over both network interfaces, the middleware can get that service from either source.

This has a number of advantages, because it lets the receiver decide where to get content. By doing this, we can potentially reduce the number of tuning operations we need. In the previous example, if the terrestrial tuner is already tuned to the transport stream containing the content we want, the receiver does not need to use the cable tuner. For DVB systems, this can even refer to content available from different networks over the same interface (e.g., when a satellite or terrestrial interface can see more than one network). The locator only identifies the content, and says nothing about where the receiver should get it.

In some cases, however, we do care about how we get at the content. If our receiver has a cable tuner and a terrestrial tuner, we may want to get the content over the cable network if the same content is available on both. To tell the middleware to get the content using a specific network interface, we need to use a network-bound locator. This locator refers to both a piece of content and to the network interface we use to access that content.

MHP systems use the `org.davic.net.dvb.DvbNetworkBoundLocator` class to represent network-bound locators. OCAP systems, on the other hand, do not have a special class to represent a network-bound locator. In both cases, we use the `javax.tv.locator.LocatorFactory.transformLocator()` method we have just seen in the class definition above to convert a normal locator to a network-bound locator. This takes a normal locator and returns an array of network-bound locators that refer to that content on different networks. For instance, if the locator for a service that is passed to `transformLocator()` is available through two different interfaces in the receiver, `transformLocator()` will return an array of two locators, each representing that service as it would be accessed over a specific network interface.

9 Reading Service Information

Service information (SI) is the mechanism that lets a receiver know about the channels and shows available on a DTV network. Here, we will discuss the APIs that MHP and OCAP applications can use to read this information, and look at the differences in SI between the two platforms. Because these APIs are such a central part of the middleware, we will also look at SI from the middleware perspective and discuss how we can build the most efficient and reliable implementation of these APIs.

For a DTV system, SI serves three purposes. The first of these is to tell the receiver what services are contained in that transport stream and how they are organized. The second purpose is to tell the receiver how the rest of the network is organized: the other transport streams that make up the same network, their frequencies, and any other parameters necessary for the receiver to tune to them correctly.

These first two purposes are both focused on the needs of the receiver, whereas the third purpose is to help the user. SI can carry human-readable information such as the names of the services in the transport stream, the audio languages that may be available, and information about shows currently playing.

Some of this information is optional, but the basic principle is the same in every case. Some systems define several levels of SI, ranging from the most basic information needed by the receiver to complete information about services and future shows. In every case, however, there will always be some SI present.

SI and Other System Components

Not just applications will use SI. SI is important to many of the other components in an MHP or OCAP receiver. The most obvious of these is the application manager. In fact, this typically uses an extension to the usual SI specifications to carry out most of its work, as we saw in Chapter 4, but it also relies on conventional SI. Without the SI that identifies the broadcast file systems used in MHP and OCAP, or the other data broadcasting standards (such as IP multicast) that are supported, the receiver would not be able to download an application.

Another important component is the section filtering component. We will discuss this in much more detail in the next chapter, because that component actually reads the SI tables from the transport stream and passes them to the SI component for parsing. In this case, there is a mutual dependency between the two components, because the SI component relies on the section-filtering component to get the SI tables, but the section-filtering component relies on SI to map a reference to an elementary stream into a PID it should filter.

We have already mentioned that access to broadcast file systems relies on SI in order to identify how and where the file system is carried, but there is more to it than this. Without SI to tell us what types of streams are part of a service, we would have no way of knowing which stream actually contained a broadcast file system, or what type of broadcast file system it contained. SI also gives us the mechanism to identify relationships between different broadcast file systems, and we will see more of this in a later chapter.

This affects much more than just broadcast file systems. The same thing applies to other forms of data broadcasting, including parental ratings for services, subtitles and closed captions, and teletext information services. All of these functions in a DTV receiver rely on SI to perform their functions.

Another important component that relies on SI is the Emergency Alert System (EAS) component found in OCAP receivers. This uses a standardized SI table to describe the EAS messages, and the receiver must be able to detect and process these messages. In an OCAP system, the middleware itself must be able to handle these messages, but it must also be able to pass on the messages to the monitor application or any other application that has assumed this functionality.

Why Do We Need Two SI APIs?

Okay, so SI is useful stuff. As always, life is not as simple as we have made it seem here, because there are several flavors of SI. The most basic level of SI is called program-specific information (PSI), defined by ISO as part of the MPEG-2 standard. This tells the receiver how the services in a transport stream are organized, and how to demultiplex them.

On top of this, we have more SI that tells the receiver about life outside that particular transport stream. This SI is not part of the MPEG standard, and thus we have two separate and largely incompatible versions of it. DVB developed one version, which is in use in Europe and parts of Asia, and ATSC developed another version that is used for the most part in North America. We have simplified this slightly, and some systems in the United States (most notably some satellite networks) will follow the DVB standards, whereas some Asian operators in countries such as South Korea use the ATSC standard for cable and terrestrial broadcasting.

The two standards do the same thing, however: they give the receiver a description of the entire network so that it can tune between different transport streams, and they give the receiver some human-readable information about the services that make up the network and the shows that are on those services. For the rest of this chapter, we will assume that you are familiar with the SI standard that is in use in your part of the world. If you are not, appendices A and B provide an overview of DVB and ATSC service information, respectively.

To access most of this SI, an application needs to use a special API. In MHP receivers, there are actually two APIs for reading SI. The first of these is defined by MHP, and this gives access to DVB SI in a way that will be familiar to anyone with some experience of DVB SI. The second API is defined by JavaTV, and is therefore part of the OCAP standard as well. This is a more general API that can be used for either ATSC or DVB SI. We will look at both APIs later in the chapter.

The only part of the SI we cannot access via the SI APIs is the AIT (and the XAIT in OCAP receivers). This may seem like an oversight, but there are good reasons for doing this. The AIT and XAIT are not really part of the standard SI you would normally find on a DTV network, and thus both SI APIs stick to the job of reading the SI people expect to find there. Because application signaling is a little bit special, applications need to use the application listing and launching API to gain access to this. We will look at this API in detail in Chapter 17.

Caching Strategies

It may take a while to transmit a complete SI table, and thus the receiver will usually cache the important tables in its memory to avoid too many delays. These tables could be quite large, though, and the memory available for caching may be quite small. Each receiver will have a different strategy for caching data, and thus we cannot be sure when data will actually be in the cache. When we need some information that is not in the cache, our only option is to wait for it to be transmitted again.

Because this can take a little time, the caching strategy for SI can be an important factor in the response time of the receiver. The network operator will update different SI tables at different rates, and this can play an important part in what is cached. The streams that make up a service, for instance, may be updated more often (different shows may include several audio languages, or data streams for applications) than the services in a transport stream. Event information may be updated once every few hours, as time passes and the schedule is updated. Other tables, such as those describing the basic organization of the network, change very infrequently and thus the receiver may cache this information even if it is reset.

There are also a number of options regarding how much data you cache. Some receivers will cache entire tables, others will cache specific rows in the table, and still others will cache specific descriptors (the range of different strategies available is pretty broad). We will look at some of the options open to us later in the chapter.

There are some tables we would be foolish not to cache because of their importance to the operation of the receiver. These include the Network Information Table (NIT), the PAT, and the PMT for the current service. Another table we should cache is the Service Description Table (SDT) in DVB systems or the Virtual Channel Table (VCT) in ATSC systems for the current transport stream. Beyond this, it is really up to the receiver how much it caches.

In a DVB network, it is not always necessary to cache all of the SI describing services. Caching the SDT for the current transport stream (SDT-actual) is usually a good idea, but caching the SDT for other transport streams (SDT-other) may not be worth the memory it takes. In an OCAP network, on the other hand, this is all contained in the VCT and is thus difficult to separate.

Caching some elements of the event information table (EIT) is often useful, although in most cases this can be limited to either the current and next events, or at most the current EIT in an ATSC system. Caching full event information is usually not necessary, and it can take up too much space for it to be worthwhile.

Luckily, OCAP receivers can save some space by not caching the Extended Text Table (ETT), which contains extended text descriptions of services, events, and transport streams. Given that this is needed infrequently, the memory is best used for other things. An alternative is to only cache those elements that have been requested (or specific subtables), rather than caching the entire table.

Some parts of the SI can even be cached when the receiver is rebooted, if we are careful. For instance, the NIT probably will not change very often and thus we can assume that caching that particular table across reboots will not cause problems. Similarly, the Rating Region Table (RRT) from ATSC PSIP probably has a small enough set of changes to be cached across reboots.

Other tables change so fast that it is not worth caching them at all. The ATSC System Time Table (STT) is one of these, as are its DVB counterparts the Time and Date Table (TDT) and the Time Offset Table (TOT). In both cases, these are only used to set the system time, and so caching them would be pretty pointless.

One special case is the ATSC Master Guide Table (MGT). We will not cache this as an SI table, but the information it contains will be used by the SI component to make sure that enough memory is allocated for each table that will be cached, and to make sure that any cached data is up to date.

At the most basic level, the repetition rates of the various tables gives us an idea of which SI tables are most important. Table 9.1 gives you an idea of the importance of the various tables relative to one another, based on the repetition rates of the various tables. Tables closer to the top are repeated more often than tables below them.

We can use this to predict the most important and commonly used data in advance, and thus middleware developers should use this knowledge where they can. Given the nature of the challenges involved in building a reliable middleware stack we need to employ every advantage available.

Table 9.1. Relative importance of service information tables, based on repetition rate.

MPEG (Common)	ATSC/OCAP	DVB
←——————→	PAT	←——————→
←——————→	PMT	←——————→
←——————→	TSDT	←——————→
←——————→	CAT	←——————→
	MGT	NIT
	VCT	BAT
	DCCT	SDT-actual
	EIT-0	EIT-actual (present/following)
	STT	SDT-other
	EIT-1	EIT-actual (scheduled events)
	ETT	EIT-other
	EIT-2	TDT
	EIT-3	TOT
	EIT 4 and higher	RST
	DCCST	

In-band Versus Out-of-band SI

What we have discussed so far is in-band SI: information delivered as part of the transport stream and associated with that transport stream or others on the same network. This is the only type of SI MHP developers need to worry about, but OCAP developers do not have it as easy.

In an OCAP system, the network operator can also deliver SI out of band via a channel other than the transport stream. Out-of-band SI will normally be transmitted via the extended channel interface of a CableCARD device or POD module.

The relationship between in-band and out-of-band SI is a complex one: if out-of-band SI is present, the receiver should use that instead of any in-band SI. When a CableCARD device is not installed, or when no out-of-band SI is being transmitted, the receiver should use the in-band SI instead.

Receivers will choose one of these two sources of SI and will ignore any tables they receive on the other channel. When a table is present in the in-band SI but not in the out-of-band SI, the receiver will ignore it.

To complicate things further, OCAP developers have to worry about the various profiles of SI that may be used on a network. ANSI/SCTE document 65 defines six different profiles for SI carried out of band, and OCAP uses this specification for its out-of-band SI. Each of these profiles has different requirements in terms of which tables are mandatory and which are optional, as well different requirements for the content of those tables. We will not discuss this here (Appendix B covers this topic in slightly more detail), but OCAP developers need to be aware that they may not always have as much SI as they may hope for.

The DVB SI API

For MHP developers, the DVB SI API may be the most familiar of the two SI APIs. This gives an application access to all of the SI tables that are part of the standard DVB SI. In other words, it provides access to everything but the application signaling information.

The DVB SI API is contained in the `org.dvb.si` package, and it maps more or less directly onto the underlying SI tables. This is very useful if you are familiar with these tables, but it may not be the most efficient way of carrying out a given task. As we will see later, the JavaTV SI API takes a task-based approach to this that may be easier to use in some cases.

The other big limitation of the DVB SI API is that it is obviously tied pretty tightly to the DVB SI format. For this reason, it is not a part of the OCAP standard. If you want your applications to be portable between OCAP and MHP systems, it is better to use the JavaTV SI API.

The SI Database

Both of the SI APIs revolve around the same core concept: a database that contains all of the SI tables. In reality, this entire database may not exist in the receiver's memory at any one time. Given memory limitations, the receiver will probably cache a subset of the SI data and fetch the rest from the transport stream as necessary.

We have slightly simplified things here, and the two APIs handle the complexities in slightly different ways. We will look at how this is handled in JavaTV later, but for now we will take a closer look at the DVB approach. In the DVB SI API, the database of SI is represented by the `org.dvb.si.SIDatabase` class. This represents the entire database of SI the receiver has access to on a given tuner. The `getSIDatabase()` method will return references to all of the available databases, one for each tuner in the receiver. In most cases, there will only be one SI database, in that most receivers will only have one tuner. The relationship of instances to tuners is illustrated in Figure 9.1.

MHP uses the `org.davc.net.tuning.NetworkInterface` class (which we shall see in more detail in Chapter 17) to represent any tuners in the receiver. The `org.dvb.net.tuning.DvbNetworkInterfaceSIUtil` class gives us a way of finding which tuner is associated with which `SIDatabase`, as we can see from the following interface.

```
public class DvbNetworkInterfaceSIUtil {

  public static org.dvb.si.SIDatabase getSIDatabase(
    org.davic.net.tuning.NetworkInterface ni;
```

```
    public static org.davic.net.tuning.NetworkInterface
      getNetworkInterface(org.dvb.si.SIDatabase sd);
}
```

Once we have a reference to an `SIDatabase`, an application can then get information from the various SI tables that are being received by the tuner associated with that database. To get this information, we have a number of methods at our disposal. A simplified version of the interface to the `SIDatabase` class follows. For the sake of brevity, we have not shown method arguments, exceptions, and some of the methods that do not interest us right now.

```
public class SIDatabase {

  public static SIDatabase[] getSIDatabase();

  public SIRequest retrieveSIBouquets();

  public SIRequest retrieveActualSINetwork();
  public SIRequest retrieveSINetworks();

  public SIRequest
    retrieveActualSITransportStream();
  public SIRequest
    retrieveSITransportStreamDescription();

  public SIRequest retrieveActualSIServices();
  public SIRequest retrieveSIServices();
  public SIRequest retrieveSIService();

  public SIRequest retrievePMTServices();
  public SIRequest retrievePMTService();

  public SIRequest
    retrievePMTElementaryStreams();

  public SIRequest retrieveSITimeFromTDT();
  public SIRequest retrieveSITimeFromTOT();
}
```

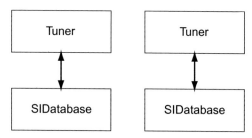

Figure 9.1. The relationship between `SIDatabase` instances and tuners.

All of the methods we show here have some common parameters, and we will take a closer look at some of these methods in more detail later. For the full details of this class, though, you should check the API specification.

Making an SI Request

Given that the receiver can cache SI data to improve performance, applications can choose whether they want the cached data, the most up-to-date data from the transport stream, or simply whichever is available (i.e., the receiver will check the cache first, and then wait for the data from the stream if it is not in the cache). This allows the application to trade speed for accuracy, and given that SI may be broadcast at a low bit rate this feature is extremely useful.

One implication of this is that retrieving data from the SI database may take some time, because the receiver may have to wait for the information to be broadcast before it can return it. For this reason, nearly every method in the SIDatabase class is asynchronous. Each retrieval method listed previously has a similar interface, and thus the following simply represents a typical example.

```
public SIRequest retrieveSIService(
    short retrieveMode,
    Object appData,
    SIRetrievalListener listener,
    org.davic.net.dvb.DvbLocator dvbLocator,
    short[] someDescriptorTags)
    throws SIIllegalArgumentException;
```

All of the methods associated with the interface to the SIDatabase class throw an SIIllegalArgumentException. This will be thrown if any of the arguments is not valid.

The first three parameters are also common to all of the methods that retrieve SI from the SIDatabase. The retrieveMode parameter indicates where the middleware should get the data. This can take one of three values, which are all constants from the org.dvb.si.SIInformation class, as follows.

- FROM_CACHE_ONLY (The receiver will retrieve the data from the cache, or fail if it is not cached.)
- FROM_CACHE_OR_STREAM (The receiver will retrieve it from the cache, or wait for it to be broadcast in the stream if it is not cached.)
- FROM_STREAM_ONLY (The receiver will ignore any cached data and wait for the data to be broadcast in the stream.)

The appData parameter is an application-specific object that identifies the request to the application. The content of this parameter is passed back in the event that signals completion of the request, thus allowing an application to use the same listener for multiple requests while being able to identify which result is associated with which request.

The `listener` parameter indicates which object should be notified of the completion of this request. By passing this at the time of the request, rather than registering a specific listener separately, the application has much more control over how it processes the results. It adds an extra parameter to every method call, but this overhead is generally worth the trade-off.

The `dvbLocator` parameter tells the API which service we are interested in. This locator should represent the service we want to get information about. Remember that the `SIDatabase` only contains information from one tuner, and thus if the locator you pass in is a network-bound locator it must refer to a service on the network that is received by that tuner.

Finally, we pass in an array of descriptor tags. This tells the receiver which descriptors we are interested in, and it should contain the descriptor tags we want. If we are interested in all of the descriptors, we can use an array with one item in it, whose value is `-1`. Of course, we may not be interested in any descriptors, and in this case we can use a null reference instead of passing in an array.

Out of these parameters, the `retrieveMode`, `appData`, `listener`, `dvbLocator`, and `someDescriptorTags` are common to all of the retrieval methods seen previously. These parameters will be in the same order we have just seen, with any other method-specific parameters going between the `dvbLocator` and `someComponentTags` parameters.

Given the similarity in the arguments of all of these methods, it is probably no surprise that they all work in the same way. When an application issues a request to the SI database using one of the methods we have seen, the method will return an `SIRequest` object, as follows.

```
public class SIRequest {

   public boolean isAvailableInCache();

   public boolean cancelRequest();
}
```

This has two purposes. First, the `SIRequest` object uniquely identifies the request that has just been made, and if the application wants to cancel that request it can do so using the `cancelRequest()` method. Second, the `isAvailableInCache()` method lets the application know if the data is in the receiver's cache, or whether the data has to be fetched from the stream.

Getting the Results of a Query

When the information is available, the SI component in the middleware notifies the application by sending an `org.dvb.si.SISuccessfulRetrieveEvent` to the listener that we specified when we made the request. This event contains references both to the `SIRequest` object that was returned when we made the request and to the object passed in the `appData` argument to the request. As we have already mentioned, the `SIRequest` gives the application a way of identifying exactly which request this result is for, whereas the `appData` object can be used for any purpose the application developer chooses. Often, the `appData` object is used to provide a mechanism for filtering different types of requests we might make. By using instances of different classes (or even an `Integer` with a different value) for each type of request, we can easily filter the different types of results so that a single listener object can handle many different SI requests.

Once the application receives a `SuccessfulRetrieveEvent` (which indicates that the data has been fetched), it calls the `getResult()` method on the event to actually get the results of the query. This returns an `SIIterator` object that contains all values that have been returned by the request. This is very similar to other Java iterators, and thus we will not describe it here in too much detail.

The `SIIterator` will contain zero or more objects that implement the `SIInformation` interface. These represent elements of the SI database. Each object will contain the content of one row from a particular SI table. This interface has a number of subclasses, corresponding to the various tables in the SI database. Therefore, the process for actually using the results from a query of the SI database looks like that shown in Figure 9.2.

Sometimes, we will need to make another request based on the results of our first query. Because all SI requests are asynchronous, we can do this directly from the event handler of our first request. We can even use the same listener, provided we use a different value for the query's `appData` argument so that we can distinguish between the two requests. Later in the chapter we will see an example of how this can be achieved.

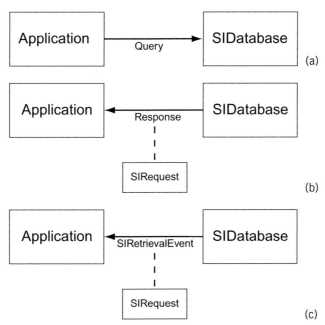

Figure 9.2. Making a query for SI information from the DVB SI API: (a) making the initial query, (b) the `SIManager` returns an `SIRequest` object for the query, and (c) an event signals the completion of the request.

Table 9.2. SI classes and the methods that return them.

`SIInformation` *Subclass*	*Returned By*

`SIEvent`	`SIService.retrievePresentSIEvent()`
	`SIService.retrieveFollowingSIEvent()`
	`SIService.retrieveScheduledSIEvents()`

`SINetwork`	`SIDatabase.retrieveSINetworks()` `SIDatabase.retrieveActualSINetwork()`
`SIBouquet`	`SIDatabase.retrieveSIBouquets()`
`SITransportStream`	`SINetwork.retrieveSITransportStreams()`
`SITransportStreamBAT`	`SIBouquet.retrieveSIBouquetTransportStreams()`
`SITransportStreamDescription`	`SIDatabase.retrieveSITransportStream` `Description()`
`SITransportStreamNIT`	`SIDatabase.retrieveActualSITransportStream()`

`SIService`	`SIDatabase.retrieveActualSIServices()`
	`SIDatabase.retrieveSIServices()`
	`SIDatabase.retrieveSIService()`
	`SIEvent.retrieveSIService()`
	`SITransportStream.retrieveSIServices()`

`SITime`	`SIDatabase.retrieveSITimeFromTOT()` `SIDatabase.retrieveSITimeFromTDT()`

`PMTService`	`SIDatabase.retrievePMTService()`
	`SIDatabase.retrievePMTServices()`
	`SIService.retrievePMTService()`

`PMTElementaryStream`	`SIDatabase.retrievePMTElementaryStreams()` `PMTService.retrievePMTElementaryStreams()`

Table 9.2 outlines the possible subclasses of `SIInformation` that could be returned in the content of the `SIIterator`. Each iterator will contain objects of only one of these classes. No request will return objects of more than one type.

Because we are relying on the broadcaster inserting the information the application wants, we cannot be certain a request will succeed. An optional SI table may not be broadcast, or a table that is being broadcast may contain only a subset of the information it can carry.

Although a number of the DVB SI tables are mandatory and must be broadcast in a DVB service, unfortunately the SI specification does not say that these tables actually have to contain anything. This can lead to a situation in which the table is present but empty, which can be extremely frustrating to application developers and viewers in some circumstances. Although carrying SI data may seem like it wastes bandwidth that could be used for other things, for ITV systems it is often wiser to carry enough of the optional SI data that a system can still give some mostly sensible responses to SI queries.

Events

We have already seen that the `SISuccessfulRetrieveEvent` is generated when an SI database query succeeds. If the application cancels the request, the SI component will notify any listeners using an `SIRequestCancelledEvent`. Similarly, if the request cannot be completed because of a resource problem in the receiver the application will be told via an `SILackOfResourcesEvent`. No specific resources must be reserved in order to make a query, but the SI component itself may need to use some scarce resources to make the query under some circumstances.

The other events that may be generated (`SITableNotFoundEvent`, `SIObjectNot-InTableEvent`, and `SINotInCacheEvent`) tell us that the information we requested could not be found. The first two can be generated by any SI query, if the information we requested is not being broadcast. The last one, however, is only generated when we look at the cached SI data but not the stream (by using `FROM_CACHE_ONLY` as the retrieval mode for the query). In this case, we may get the data we want by looking in the stream as well as the cache.

An Example of Retrieving SI Data

The process of getting SI data may look complicated, but it is actually pretty straightforward in practice. If we look at how this API is actually used, things may become a bit clearer. First, we will look how we make the request for the SI data, as follows.

```
// get a reference to an SIDatabase. Since the
// receiver probably only has one network interface,
// we will take a shortcut and simply use the first
// one in the array, rather than querying every
// database
SIDatabase[] databases = SIDatabase.getSIDatabase();
SIDatabase database = databases[0];

// Now we issue a request for the data we want. In
// this example, we want to retrieve the Program Map
// Table for the current service in this case.
//
// The first three arguments are the standard ones
// described above (retrieval mode, application-specific
// request data and the listener to be notified when
// the request completes).
//
// The next argument gives the locator of the service
// that we're requesting the information for (
// as an org.davic.net.dvb.DvbLocator), while the final
// argument is an array of 'short' values that lists
// the descriptors that we're specifically interested
```

```
// in. We leave this null to indicate that we don't
// care about retrieving any descriptors (an array of
// one element with the value -1 would indicate that
// we want information about all descriptors, or we
// could pass the descriptor tags of specific
// descriptors we are interested in)
SIRetrievalListener listener;
listener = new MySIListener();

try {
  database.retrievePMTService(
    SIInformation.FROM_CACHE_OR_STREAM,
    new Integer(1),
    listener,
    myServiceLocator,
    null);
}
catch (SIIllegalArgumentException siiae) {
  // do nothing
}
```

Once we make the request, we obviously need a listener to receive the results of the query. In this case we implement that in a separate class, as follows, but there is no reason this could not be implemented by the class that makes the query.

```
public class MyListener
  implements org.dvb.si.SIRetrievalListener {

  public void postRetrievalEvent(
    SIRetrievalEvent event) {

    // first, check that the requests actually
    // succeeded. This request would probably
    // only fail due to resource limitations,
    // since a stream that didn't contain this
    // information would be hopelessly messed up
    if (event instanceof SISuccessfulRetrieveEvent) {

      // now we check the application-specific data to
      // see what kind of request this is. The reason
      // why we do this becomes apparent further down in
      // the code, when we issue another request (for a
      // different type of data) that also uses this
      // object as a listener
      Integer appData = (Integer) event.getAppData();
      if (appData.intValue() == 1) {
```

```
// since its value is 1, we've got a result
// from the original request

// get the iterator that contains the results of
// the request
   SISuccessfulRetrieveEvent ev;
ev = (SISuccessfulRetrieveEvent) event;
SIIterator myIterator = ev.getResult();

// now loop through the iterator, checking every
// element and using the results as we want to
while (myIterator.hasMoreElements()) {

   // the information that we care about is
   // returned in a set of PMTService objects
   PMTService service =
      (PMTService) myIterator.nextElement();

   // get the information about the service that
   // is contained in the PMTService object.

   // Now get the information for the elementary
   // streams. This involves issuing another
   // request to the database. This time, we
   // don't pass in any application-specific data,
   // and we use the current object as the
   // listener for the request notification.
   //
   // In this case, we don't make the request
   // directly to the SIDatabase. The PMTService
   // class allows us to issue the request via
   // that class as well, and the method used for
   // that request is simpler and takes fewer
   // arguments.
   //
   // The first three arguments are the same as
   // for all other requests, while the final
   // argument is again an array of 'short'
   // integers describing the descriptors that
   // we're interested in.
   service.retrievePMTElementaryStreams (
   SIInformation.FROM_CACHE_OR_STREAM,
   new Integer(2),
   this,
   null);
   }
```

```
                else if (appData.intValue() == 2) {
                // here is where we would handle the
                // results from the second request that
                // we issued (for the elementary streams
                // from the service).
                }
            }
        }
    }
    //other methods of the class go here
}
```

In this case, the listener makes another SI request once it has received the first set of results. To make things easier for ourselves, we reuse the listener to handle this second request.

As you can see, we use the `appData` parameter of the request (which is passed back to use as part of the notification that the request has completed) to determine which request we are handling and what we should do with the results. This approach means that our application can use a single class to handle as many types of SI queries as it needs to, and thus we do not need to write several different listener classes to handle our SI queries.

Monitoring SI

In addition to making queries for specific SI, it is often useful to monitor a specific SI element for changes. Monitoring an SI table can tell us when the network operator has added a new service, or when the set of streams that make up a service has changed. Whereas this is obviously useful to EPGs, other applications may also need to use this to track available audio tracks or camera angles, for instance.

To monitor these changes, we implement an `SIMonitoringListener` in our application or middleware component. This interface has a single method that receives events when the SI element we are interested in is updated, as follows.

```
public void postMonitoringEvent (
    SIMonitoringEvent anEvent);
```

To start monitoring an SI element, we must register our listener with the `SIDatabase`. Depending on the type of SI we want to monitor, we should call one of a number of methods that tell the SI database which SI elements we are interested in. Table 9.3 outlines which methods we should call to monitor various types of SI.

Of course, each of the methods in Table 9.3 has a corresponding method that removes the listener. Applications should take care to register a listener on the correct `SIDatabase` instance. Do not forget that there may be more than one of these in the system, and thus you need to make sure that the SI element you want to monitor is actually available via the SI database you are registering the listener with.

Table 9.3. Methods for monitoring different types of SI elements.

Type of Information	Method

Bouquet information	`addBouquetMonitoringListener(`
	`  SIMonitoringListener listener,`
	`  int bouquetId)`

Present/Following event information	`addEventPresentFollowingMonitoringListener(`
	`  SIMonitoringListener listener,`
	`  int originalNetworkId,`
	`  int transportStreamId, int serviceId)`

All event information	`addEventScheduleMonitoringListener(`
	`  SIMonitoringListener`
	`  listener, int originalNetworkId,`
	`  int transportStreamId,`
	`  int serviceId,`
	`  java.util.Date startTime,`
	`  java.util.Date endTime)`

Network information	`addNetworkMonitoringListener(`
	`  SIMonitoringListener listener,`
	`  int networkId)`

Components of a service	`addPMTServiceMonitoringListener(`
	`  SIMonitoringListener listener,`
	`  int originalNetworkId,`
	`  int transportStreamId,`
	`  int serviceId)`

Services	`addServiceMonitoringListener(`
	`  SIMonitoringListener listener,`
	`  int originalNetworkId,`
	`  int transportStreamId)`

One thing to remember when requesting or monitoring information about events is that the start and end times of the event must be specified in UTC time. This avoids any potential confusion regarding time zones and daylight savings time, but it does mean that the application needs to calculate the current offset from UTC time before it makes a request.

Whenever a change is made to an SI element we are monitoring, we will be notified through an `SIMonitoringEvent`, as follows. These events give us some details about the element(s) for which SI has been updated.

```
public class SIMonitoringEvent
   extends java.util.EventObject {

   public SIMonitoringEvent(
     SIDatabase source,
```

```
        byte objectType,
        int networkId,
        int bouquetId,
        int originalNetworkId,
        int transportStreamId,
        int serviceId,
        java.util.Date startTime,
        java.util.Date endTime);

    public Object getSource();

    public byte getSIInformationType();

    public int getNetworkID();
    public int getBouquetID();
    public int getOriginalNetworkID();
    public int getTransportStreamID();
    public int getServiceID();
    public java.util.Date getStartTime();
    public java.util.Date getEndTime();
}
```

`SIMonitoringEvents` come in several types. These tell the application what type of SI has changed, as indicated by one of the constants in the `SIMonitoringType` interface outlined in Table 9.4.

Depending on the type of SI that has changed, some of these methods may not return useful results. Getting the start time for a change to a transport stream or a service is meaningless, for instance. The application should use the SI type to make sure it will get valid results from any methods it calls on the event.

At first, it may seem confusing to have different event listeners for handling SI requests and for handling SI monitoring. In practice, however, this is a sensible thing to do. `SIRetrievalListener` objects are just callbacks used by the middleware. Thinking of them as events tends to hide their true purpose a bit.

Table 9.4. `SIMonitoringType` constants and the SI they represent.

Constant	SI Type
BOUQUET	Bouquet association table
NETWORK	Network information table
PMT_SERVICE	Information about a particular service in the PMT
PRESENT_FOLLOWING_EVENT	EIT present/following (either actual or other)
SCHEDULED_EVENT	EIT schedule (either actual or other)
SERVICE	Service description table

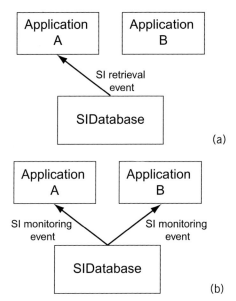

Figure 9.3. Dispatching `SIRetrievalEvents` and dispatching `SIMonitoringEvents`: (a) SI events are sent to one listener only, and (b) SI monitoring events are sent to all registered listeners.

An `SIRetrievalEvent` will be sent to one listener only, when the request completes. On the other hand, a change in an SI table may result in an `SIMonitoringEvent` being sent to many listeners, depending on the number of them monitoring that SI.

This means that we need two different event-dispatching strategies to handle these situations. Depending on the frequency of updates, `SIMonitoringEvents` may be dispatched much more frequently, to a larger number of listeners, than `SIRetrievalEvents`. This means that events may need to be dispatched more quickly, and thus we may need either a larger thread pool for event dispatching or a quicker response time from applications before threads are killed and replaced.

Low-level Access to SI Data

From time to time, we may need lower-level access to SI than we have seen so far. The `SIInformation` interface defines the method `retrieveDescriptors()` to allow retrieval of some or all of the descriptors associated with an SI element. Depending on the subclass of `SIInformation` we call this method on, we will get the descriptors for virtually any piece of SI.

Each descriptor is represented by a `Descriptor` object as shown. Using these objects, we can get at the content of the descriptors themselves. Unfortunately, we have to do this the hard way. The `Descriptor` class does not do anything except the most basic parsing of the descriptor content, and thus we must carry out any detailed parsing ourselves.

```
public class Descriptor {

   public byte getByteAt(int index)
      throws IndexOutOfBoundsException;

   public byte[] getContent();
   public short getContentLength();
   public short getTag();

}
```

The values for the descriptor tag (which are returned by the getTag() method) are defined as constants in the org.dvb.si.DescriptorTag interface. The sheer number of possible descriptors tags makes it impractical to give a full list here, and thus you should refer to the org.dvb.si API documentation for details of these.

The reasons we do not have more detailed parsing of the descriptor's content should be pretty clear. There are a lot of descriptors defined by DVB SI and PSI, and other standards can add new descriptors as long as they do not conflict with those already defined. Having a separate class for every descriptor would increase the size of the API tremendously, and because other parts of the API will parse the standardized descriptors it is not necessary to define dedicated classes for them. Applications that need to know detailed information about descriptors will know how to get it using the existing SI queries, or will understand the format of those descriptors so that they can parse them themselves.

By letting us get at the descriptor data, however, this API does give applications (or the middleware) a chance to handle descriptor types that are not included in the SI specifications. We can use this to read any private descriptors that are specific to a given network, and thus an application can extend the functionality of a receiver to handle new features and services simply by adding support for the appropriate descriptors. Applications will have to do the descriptor parsing themselves, but that is not a big problem in many cases.

This reflects the entire design philosophy of SI and of the descriptor-based approach. By allowing applications to extend the SI handling capabilities of the receiver (for a few tables, at least) we are giving the broadcasters the ability to extend the standardized SI tables. This lets them use these tables for carrying information about types of services and applications that have not been thought of yet.

Sometimes when we request a descriptor, the SI data we are interested in may have been updated before we make the request and the content of the descriptor (or even its presence) may have changed. In this case, attempting to retrieve the descriptors will generate an SITableUpdatedEvent. Should we receive this in response to a request for some descriptors, the application should get an updated version of the SIInformation object (by reissuing the SI request for that object) and then retrieve the descriptors from that.

Using the JavaTV SI API

The DVB SI API is easy enough to use if you are familiar with the structure of DVB SI, and if you are working in a pure MHP environment. It is nice to have an alternative, however, and MHP and OCAP give us the JavaTV SI API. This takes a different approach to accessing SI that can be useful for MHP developers as well. Instead of following the structure of the SI tables in the system, the JavaTV SI API has a structure based more around the tasks you want to accomplish with SI. This has two advantages. First, it makes the API far more flexible in the SI formats it can handle, and thus supports ATSC SI as well as DVB SI. Second, it makes some tasks easier simply by not forcing you to make all of the requests to individual SI tables yourself.

In reality, this is not much of an alternative for OCAP developers because it is the only SI API you get. Having said that, it is useful enough that MHP developers can use it just as easily as the DVB SI API, providing a portable approach to SI access across the two platforms.

The `javax.tv.service.*` package and some of its subpackages contain the JavaTV APIs that deal with services. Generally, the classes in this hierarchy are organized into packages based on specific functions, and one of those functions is accessing SI. This is not just contained in one package: the `javax.tv.service` package itself contains the high-level concepts common across all of the APIs that deal with services, whereas some of the subpackages concentrate on giving us access to SI for specific tasks. Other packages are completely unrelated to SI. For now, we will concentrate on the following packages.

- `javax.tv.service`
- `javax.tv.service.guide`
- `javax.v.service.navigation`
- `javax.tv.service.transport`

The other package in this hierarchy is `javax.tv.service.selection`, which we saw in Chapter 5. We will not look at that in any more detail here.

Because the JavaTV SI API does not think about the individual tables that make up the SI, some developers may have to start thinking a little differently in order to make the most of it. In practice, this is not difficult. It is mainly a question of thinking about the task you want to perform, rather than the SI you need to perform that task. Let the API worry about that, and you have an application that will work equally well on OCAP and MHP systems.

To use these packages effectively, we first need to look at how they were designed to be used. As we have already said, applications typically use SI to carry out several specific tasks, including the following.

- Finding out what services are currently available and how those services are organized into transport streams and bouquets
- Accessing information about a service, such as the service name, type of content, and other information

- Accessing information about events on a service or about the schedule of the service (e.g., for display in an EPG)
- Changing to a new service (service selection)

Apart from the last item in the list, these tasks correspond roughly to the various packages just mentioned. This mapping is not exact, but it is pretty close.

Basic Concepts

The main class you will use for getting any SI is the `javax.tv.service.SIManager` class. This represents all of the SI available to the receiver. Unlike the `SIDatabase` in the DVB SI API, this contains all of the SI received from any tuner in the system (or from out-of-band channels such as the POD extended channel interface). We may still have more than one instance of the `SIManager`, however, because each instance will contain SI in one language only. If we want to access SI in two languages, we need to create two instances of the `SIManager`.

To get an instance of the `SIManager`, applications call the `SIManager.createInstance()` method. This is slightly different from the `getInstance()` design pattern other APIs use, because every call will create a new instance. Each instance of this class may use a lot of memory, and thus applications should not create more than one instance unless they have to. The costs of doing so probably far outweigh the benefits.

Although this class is supposed to represent the entire database of SI, middleware developers can make life easier for themselves by having a single underlying class that actually implements this database. We do not need a separate instance of the entire database for every application, and thus it is usually much more memory efficient simply to have the `SIManager` class redirect any queries to the main database.

There may be cases in which this mapping gets slightly more complicated; for instance, where different instances of the `SIManager` class have different preferred languages. Applications should generally not change the preferred language unless there is a very good reason to do so. If no preferred language is set, the receiver will probably cache SI in the language the user has chosen for the rest of the system. Although none of the standards actually says that this is what you should do, it is pretty sensible behavior.

We will take a look at the interface to the `SIManager` in detail in material to follow, but before we go any further let's look at how JavaTV handles the process of retrieving SI. Quite a few methods in the `SIManager` class (and other classes) will retrieve some type of SI, but they all take the same basic approach.

This approach is very similar to that used in the DVB SI API: most requests for SI are asynchronous, so that the receiver can try to retrieve uncached SI from the broadcast stream without blocking the application. This means that we often need some way of telling the application when an SI query has finished, and thus each asynchronous request takes an instance of a `javax.tv.service.SIRequestor` as a parameter. This serves the same purpose as the `org.dvb.si.SIRetrievalListener` we saw earlier in the DVB SI API, and has a similar interface.

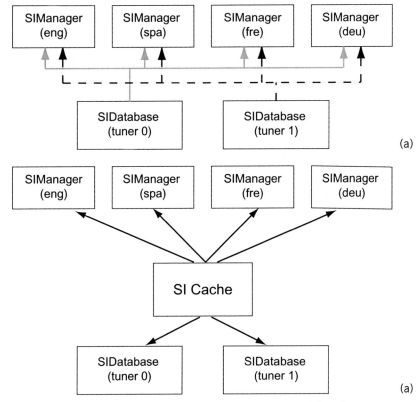

Figure 9.4. Mapping `SIDatabase` instances to `SIManager` instances: (a) each `SIManager` can have a view onto several `SIDatabase` objects, and (b) alternatively, `SIManager` objects and `SIDatabase` objects can all have different views onto the same SI cache. (We will look at this in more detail later in the chapter, when we look at different ways of implementing the SI APIs.)

Any method that makes a request for SI data returns an `SIRequest` object. This provides one method, `cancel()`, which gives us a way to cancel a pending SI request. This is similar to the `SIRequest` object in the DVB SI API.

Different classes in the API will provide different methods to request SI, and we will take a look at these later in the chapter. Before we do that, let's take a closer look at how we handle the results.

Handling the Results from an SI Query

The `SIRequestor` is an event listener that will be notified when an SI query is finished. Because there may be several SI queries happening simultaneously, there may be several instances of the `SIRequestor` class in use by the same application. Unfortunately, there is no way of uniquely identifying an SI request, and thus application developers must take care to make sure that an application can differentiate between the different SI requests that may be pending at any time by using different `SIRequestors` for different requests. This is a weakness when compared to the DVB SI API, and it can mean that applications need more event handlers than would otherwise be necessary.

Once the middleware has the SI data that was requested (or knows that the SI is unavailable), it notifies the `SIRequestor` object associated with that request. The interface for this class provides two methods, to tell the application whether the request succeeded or failed and to give it the SI data it requested or to tell it why the request failed. The interface to `SIRequestor` looks as follows.

```
public interface SIRequestor {

  public void notifyFailure(SIRequestFailureType reason);
  public void notifySuccess(SIRetrievable[] result);
}
```

As you can see from this, notification of a successful SI request includes an array of `javax.tv.service.SIRetrievable` objects. This interface is the base class for a hierarchy of classes that describe the different types of SI data we can retrieve from the broadcast stream. We will look at some of these subclasses later as we explore the various parts of the JavaTV SI API.

Unlike the DVB SI API, we have a different method to tell us that the SI request failed. In some ways, this does make the application code simpler, just because it lets us separate error handling from result processing.

The Core SI API

Now that we have seen how we can handle the results from a request we have made, let's look at how we can use the API to make a request in the first place. The core of the API is the `javax.tv.service.SIManager` class. This is the starting point for an application to get SI, be it SI about services, available transport streams, or (broadcast) network interfaces. The interface for this class follows.

```
public abstract class SIManager {

  protected SIManager();

  public static SIManager createInstance();

  public abstract java.lang.String getPreferredLanguage();

  public abstract void setPreferredLanguage(
     java.lang.String language);

  public abstract RatingDimension getRatingDimension(
     String name);
```

```
    public abstract java.lang.String[]
        getSupportedDimensions();

    public abstract ServiceList filterServices(
        ServiceFilter filter);

    public abstract Service getService(
        javax.tv.locator.Locator locator);

    public abstract javax.tv.service.transport.Transport[]
        getTransports();

    public abstract void registerInterest(
        javax.tv.locator.Locator locator,
        boolean active);

    public abstract SIRequest retrieveProgramEvent(
        javax.tv.locator.Locator locator,
        SIRequestor requestor);

    public abstract SIRequest retrieveServiceDetails(
        javax.tv.locator.Locator locator,
        SIRequestor requestor);

    public abstract SIRequest retrieveSIElement(
        javax.tv.locator.Locator locator,
        SIRequestor requestor);
}
```

One thing you will notice about the JavaTV SI API as we look at it in more detail is that it does not just give us SI. It also gives us a way of navigating around the transport streams and services that make up the network, giving us access to a set of objects that represent the various components of the network. We can get an object that represents a transport stream, and from that we can get a set of objects representing the services within that transport stream, and so on. These objects are similar to those we saw in Chapter 8, and they share many of the same functions. Having said that, there are also a number of important differences, and the DAVIC classes do not have the same level of integration with JavaTV. In JavaTV, these classes represent SI only, and do not represent a way of referring to a particular transport stream (except via the locators returned by the getLocator() method).

If we were thinking about this just as a way of getting access to SI, this may not be the cleanest design we could use. A more common approach (as used by the DVB SI API) would be to have a single API that retrieves only SI. That is not what this API is about, though. It is about giving us a way of using that SI to carry out the tasks we will usually want to perform, and we normally have to use more than SI in the process. It is also about platform independence, and thus the API does not make any assumptions about the type of SI used within the system.

Given that this set of packages follows this philosophy, we will not take a class-by-class walk-through of the API. Instead, we will look at how we use it to carry out some common tasks.

Access to Transport Information

Applications can use the `SIManager` class to get information about the various network interfaces in the receiver and the transport streams available on them. From this, an application can find out how the transport streams are organized, both physically (how we tune to them) and logically (which bouquet and which network they are part of). Many of the classes that provide this functionality are to be found in the `javax.tv.service.` `transport` package, including all of the classes we describe in the following.

Using the `SIManager.getTransports()` method, we can get an array of `Transport` objects that describe how services are grouped. Each `Transport` object corresponds to one content delivery mechanism (typically, a broadcast network interface), and we can use this to get information about the type of network interface (cable, satellite, terrestrial, or something else) or to register a listener to be notified when the content available on that network interface changes.

We can also get some other information about the organization of the network, depending on the type of `Transport` object we have. The three subclasses of the `Transport` object give us several ways of showing the structure of the network, and some or all of these may be available to us depending on the receiver and on the amount of SI being broadcast. The `TransportStreamCollection` interface lets us find out about the set of transport streams available on a given network interface, thus showing the physical organization of the network.

The `BouquetCollection` interface gives us a list of the bouquets available for a given network interface. For those of you who are not familiar with bouquets, they provide a logical way of grouping services (all of the movie channels may be collected in a single bouquet, for instance, whereas all of the sports channels may be collected in another). Network operators can also use bouquets to group the services they offer in a specific subscription package.

The final subclass of `Transport`, the `NetworkCollection` interface, represents the set of networks available for a given transport. Although this is meaningless in some cases (a cable network, for instance), it can be very useful in satellite or terrestrial broadcasting in which several networks may be available, either because a terrestrial receiver is within the transmission area of more than one network or because a satellite receiver is using a satellite shared by more than one network operator. Networks are a logical way of grouping transport streams, just like a `BouquetCollection`.

`Transport` and its subclasses are all defined as interfaces in the JavaTV specification, and thus a `Transport` object can implement all of these interfaces if all of the necessary SI is available. Some of these may never be available in a network, however. Bouquet information is only a part of DVB SI, for instance, and thus an OCAP receiver will never implement the `BouquetCollection` interface.

Now that we have seen the classes that make up this hierarchy, let's take a more detailed look at one of them. These classes all work in a similar way, and thus we will take the `Transport StreamCollection` class, as follows, as an example of how they work.

```
public interface Transport {

    public void addServiceDetailsChangeListener(
        ServiceDetailsChangeListener listener);

    public void removeServiceDetailsChangeListener(
        ServiceDetailsChangeListener listener);

    public javax.tv.service.navigation.DeliverySystemType
        getDeliverySystemType();
}
```

The `TransportStreamCollection` interface extends this, as follows, to add some methods related specifically to the transport streams available on a transport.

```
public interface TransportStreamCollection
    extends Transport {

    public void addTransportStreamChangeListener(
        TransportStreamChangeListener listener);

    public void removeTransportStreamChangeListener(
        TransportStreamChangeListener listener);

    public SIRequest retrieveTransportStream(
        javax.tv.locator.Locator locator,
        javax.tv.service.SIRequestor requestor);

    public SIRequest retrieveTransportStreams(
        javax.tv.service.SIRequestor requestor);
}
```

We can use the `TransportStreamCollection` interface to retrieve information about either a single transport stream (using a locator to identify the transport stream of interest) or about all of the transport streams available on the network interface associated with that `TransportStreamCollection`. When we get the results of these queries, each transport stream is represented by a `TransportStream` object. This encapsulates some of the information about a transport stream, such as a brief description and the transport stream ID. A `TransportStream` object has the following interface.

```
public interface TransportStream extends SIElement {
    public int getTransportStreamID();
    public java.lang.String getDescription();

}
```

The class that implements this interface may include more information about the transport stream, but this is not available to the application. In this sense, it is similar to the `org.davic.mpeg.TransportStream` class we saw in the last chapter. Both of these classes may carry extra information that is private to the middleware on top of information visible to applications. In fact, because the JavaTV `TransportStream` class is actually an interface it could even be implemented by the same underlying class.

Although doing this may seem like a good idea, beware of the additional complexity this may impose on your API implementations. Doing this will make it more difficult to separate the two components later, should you need to, and you should think about other options such as having implementation-specific information contained in another class that is referenced by both the JavaTV and DAVIC `TransportStream` classes.

This goes for other classes, which we will examine in this chapter as well. Middleware implementers should examine the trade-offs involved in combining elements of the SI APIs and the core MPEG API before doing this, to make sure that it really is the best architectural solution for their middleware stack.

Access to Information About Services

Although it can be useful to get information about transport streams, the most common use for SI is to find out about a specific service or group of services. Individual services are represented in JavaTV by the `javax.tv.service.Service` class. This provides the most basic information about the service, such as its name, the type of the service (DTV, digital radio, analog TV, or another type of service entirely), and its locator. This is similar to the `SIService` class from the DVB SI API, as follows.

```
public interface Service {

    public javax.tv.locator.Locator getLocator();
    public java.lang.String getName();
    public ServiceType getServiceType();
    public boolean hasMultipleInstances();
    public SIRequest retrieveDetails(SIRequestor requestor);
}
```

We can get a `Service` object that represents the service we are interested in by calling the `SIManager.getService()` method. This takes a single parameter: a locator that refers to the service we are interested in. Like the `getTransports()` method, this is a synchronous call and thus returns a `Service` object representing the service we are interested in. As you can see from the previous definition, this lets us get only the most basic information about a service.

To get any other information, we need to use the `javax.tv.service.navigation` package. This contains a number of classes that give us detailed information about a service. If we want to see any more information about the service itself, we need to get the details of this service contained in the `ServiceDetails` class. We can get a `ServiceDetails` object for a particular service using either the `Service.retrieveDetails()` method or the `SIManager.retrieveServiceDetails()` method. Both of these are asynchronous calls, and thus take an `SIRequestor` object that will get notified when the request is complete. `SIManager.retrieveServiceDetails()` also takes a locator that refers to the service we want the details for.

When our `SIRequestor` object gets notified that the request has succeeded, the objects containing the result will implement the `ServiceDetails` interface. This gives us some detailed information about the service, such as the long name of the service and the CA (conditional access) system used to encrypt it (via the `javax.tv.service.navigation.CAIdentification` interface). We can also use the `ServiceDetails` interface to retrieve further information about that service, such as the program schedule. The `ServiceDetails` interface follows. You can see from this the other type of information we can get about a service.

```
public interface ServiceDetails
   extends javax.tv.service.SIElement,
           javax.tv.service.navigation.CAIdentification {

   public void addServiceComponentChangeListener(
      ServiceComponentChangeListener listener);

   public void removeServiceComponentChangeListener(
      ServiceComponentChangeListenerlistener);

   public javax.tv.service.DeliverySystemType
      getDeliverySystemType();

   public java.lang.String getLongName();

   public javax.tv.service.guide.ProgramSchedule
      getProgramSchedule();

   public javax.tv.service.Service getService();

   public javax.tv.service.ServiceType getServiceType();

   public javax.tv.service.SIRequest retrieveComponents(
      javax.tv.service.SIRequestor requestor);

   public javax.tv.service.SIRequest
      retrieveServiceDescription(
      javax.tv.service.SIRequestor requestor);
}
```

The `retrieveComponents()` method will retrieve a list of the components in the service. In this case, the result of the SI request will be an array of objects implementing the `javax.tv.service.navigation.ServiceComponent` interface, as follows. Each of these objects will represent an elementary stream that is part of the service.

```
public interface ServiceComponent extends SIElement {

  public java.lang.String getAssociatedLanguage();
  public java.lang.String getName();
  public StreamType getStreamType();

  public javax.tv.service.Service getService();
}
```

The `ServiceComponent` class provides the basic information you would expect to get for an elementary stream: the type of stream, a reference to the service with which it is associated, and the language associated with this stream. Of course, not all streams have a language component (video streams or data streams, for instance), and `ServiceComponent` objects that represent those streams will return a null reference when an application calls `getAssociatedLanguage()`.

Readers who know something about how SI works in a DVB network may notice a limitation imposed by this API. There appears to be no way for a `ServiceComponent` to belong to more than one service, in that the `getService()` method can only return one result. In practice, this is not a major limitation because several `ServiceComponent` instances may refer to the same underlying elementary stream. The main implication of this is that applications should not assume that just because two `ServiceComponent` instances are not equivalent they do not refer to the same elementary stream.

Access to Information About Events

For an EPG, knowing about services is not enough. It also needs to know about individual events within a service. The `ServiceDetails` interface (seen earlier) includes the `getProgramSchedule()` method, which returns a `ProgramSchedule` object (as follow). This object provides a number of methods for requesting information about the current and next event, and about all future events on that service.

```
public interface ProgramSchedule {

  public void addListener(
    ProgramScheduleListener listener);

  public void removeListener(
    ProgramScheduleListener listener);

  public javax.tv.service.SIRequest
    retrieveCurrentProgramEvent(
    javax.tv.service.SIRequestor requestor);

  public javax.tv.service.SIRequest
    retrieveNextProgramEvent(
    ProgramEvent event,
    javax.tv.service.SIRequestor requestor);
```

```
public javax.tv.service.SIRequest
  retrieveFutureProgramEvent(
  java.util.Date time,
  javax.tv.service.SIRequestor requestor);

public javax.tv.service.SIRequest
  retrieveFutureProgramEvents(
  java.util.Date begin,
  java.util.Date end,
  javax.tv.service.SIRequestor requestor);

public javax.tv.service.SIRequest
  retrieveProgramEvent(
  Locator locator,
  javax.tv.service.SIRequestor requestor);
}
```

As you can see from the interface, the methods for retrieving event information are all asynchronous. When the `SIRequestor` is notified about the successful completion of a request for event information, the notification will include one or more objects implementing the `ProgramEvent` interface. For those methods that request information about a single event, there will be just one `ProgramEvent` object, but calls to `retrieveFutureProgramEvents()` will result in an array of `ProgramEvent` objects getting returned—one for each event that matches the query.

The `ProgramEvent` interface following provides access to information about that specific event: its name, rating, start and end times, and duration, among other things. Like the `ServiceDetails` class, it also includes the `retrieveComponents()` method. In this case, though, the method returns a list of components for the event rather than for the service. Depending on when this method is called in relation to the start of the event in question, this information may or may not be available. It is pretty unlikely that the receiver will have information about the components of an event scheduled to start six hours from the time of the request.

```
public interface ProgramEvent extends SIElement {

  public java.util.Date getStartTime();
  public java.util.Date getEndTime();
  public long getDuration();

  public java.lang.String getName();
  public ContentRatingAdvisory getRating();

  public javax.tv.service.Service getService();

  public javax.tv.service.SIRequest
    retrieveComponents(
    javax.tv.service.SIRequestor requestor);
```

```
    public javax.tv.service.SIRequest
      retrieveDescription(
      javax.tv.service.SIRequestor requestor);
  }
```

When `retrieveComponents()` is called for the current event on a service, we will always get a result, though. This will be the same as if `retrieveComponents()` were called on the `ServiceDetails` object for the service the event is part of.

Program schedules can change over time, and thus the `ProgramSchedule` interface allows an application to register a `ProgramScheduleListener` with the middleware. This lets the application receive notification of any changes in the program schedule. This could let an EPG know when an updated schedule has been transmitted, so that it could update the information it displays to the user. This will report both periodic updates (for instance, when more schedule information is available in the SI) or special updates, such as when an event is delayed and its start time is changed in the broadcast SI.

Monitoring SI

As you will have noticed, the JavaTV SI API takes the same basic approach as the DVB SI API to monitoring changes in SI. Applications can register listeners for changes in any piece of SI they are interested in, and these listeners will be notified when a change is detected.

Unlike the DVB SI API, the methods for adding and removing listeners are attached to the various classes representing the streams and events, and the events themselves are specific to the various types of SI. Having said that, the event listeners are all subclasses of `javax.tv.service.SIChangeListener` and the events are all subclasses of `javax.tv.service.SIChangeEvent`. Table 9.5 outlines which listeners receive

Table 9.5. SI monitoring listeners and the events they receive.

Listener	Event
javax.tv.service.transport. BouquetChangeListener	BouquetChangeEvent
javax.tv.service.transport. NetworkChangeListener	NetworkChangeEvent
javax.tv.service.guide. ProgramScheduleListener	ProgramScheduleEvent
javax.tv.service.navigation. ServiceComponentChangeListener	ServiceComponentChangeEvent
javax.tv.service.transport. ServiceDetailsChangeListener	ServiceDetailsChangeEvent
javax.tv.service.transport. TransportStreamChangeListener	TransportStreamChangeEvent

which events. In each case, the event object is in the same package as the corresponding listener.

As with the DVB SI API, more than one application can register a listener for changes in the same SI element, and thus the underlying event-dispatching mechanism is also very similar. Events notifying an application about the success or failure of an SI request are callbacks that are always sent to a single listener only, whereas monitoring events are conventional Java events. We will discuss the similarities between these parts of the two APIs later in this chapter, when we look at how middleware developers can use a common approach to handling the different sets of events we need to dispatch.

The OCAP SI Extensions

As if this were not complicated enough, OCAP developers have something more to worry about. OCAP adds a few SI-related classes of its own, from the `org.ocap.si` package. These extend the functionality of the JavaTV SI API, and provide descriptor-level access to parts of the PSI, similar to that provided by parts of the DVB SI API. The OCAP SI API is a bit more limited in this respect, however, and thus we can only get access to the PAT and PMT tables. Within these tables, applications can access any field or descriptor they choose. As we have seen in our discussion of the DVB SI API, this can be used to handle nonstandard descriptors used by a particular application or by the monitor application. From a user's perspective, the two main classes are `ProgramAssociationTableManager` and `ProgramMapTableManager`. These let an application retrieve the available PATs and PMTs, respectively, and listen for changes in them.

Apart from descriptor-level access to SI, the main addition of the OCAP SI API is the ability to retrieve specifically the in-band or out-of-band versions of a table. This is something that is not an issue for an MHP receiver, or for some other JavaTV-based systems. An OCAP receiver with a CableCARD device installed can receive either in-band or out-of band SI. In an OCAP receiver, out-of-band SI data will always take precedence over in-band SI if it is available, and thus the OCAP SI API provides the only way for an application to get at the in-band data in preference to any out-of-band data being received by the CableCARD interface. ACAP receivers in cable systems will also follow this approach.

The basic process for retrieving out-of-band SI data is the same as in the JavaTV SI API. The methods that make an SI request take an object implementing the `javax.tv.service. SIRequestor` interface as a parameter, and the middleware will notify this object when the data is available. The data returned will be either a `ProgramAssociationTable` or a `ProgramMapTable` object, both of which implement the `javax.tv.service.SIElement` interface. The basic structure used by these two objects is similar, as follows, and thus we will only take a detailed look at one of them here.

```
public interface ProgramMapTable
   extends org.ocap.si.Table {

   public org.ocap.si.Descriptor[]
      getOuterDescriptorLoop();
```

```
    public int getPcrPID();

    public org.ocap.si.PMTElementaryStreamInfo[]
       getPMTElementaryStreamInfoLoop();

    public int getProgramNumber();
}
```

The `org.ocap.si.Table` interface is a subinterface of `javax.tv.service.SIElement`. As you can see from the interface definition, we can get a variety of low-level information about the table. Possibly the most useful of these is the ability to get information about the elementary streams that make up the service. Calling `getPMTElementaryStreamIn-foLoop()` will return an array of `PMTElementaryStreamInfo` objects, each of which describes one elementary stream in the service, as follows.

```
  public interface PMTElementaryStreamInfo {

     public org.ocap.si.Descriptor[] getDescriptorLoop();

     public short getElementaryPID();

     public java.lang.String getLocatorString();

     public short getStreamType();
  }
```

Each `PMTElementaryStreamInfo` instance gives some basic information about a particular elementary stream and corresponds roughly to the `javax.tv.service.locator.ServiceComponent` interface. Most of these methods are fairly obvious, although it is worth pointing out that the `getStreamType()` method will return one of the constants from the `org.ocap.si.StreamType` class, outlined in Table 9.6. These constants are similar to the stream types we saw in our discussion of OCAP locators in Chapter 8.

When working with the PMT at this low level, we must remember that an elementary stream can appear in more than one PMT (as we saw earlier). Luckily, we can compare the PIDs of two elementary streams to find out whether they are in fact the same stream.

To get more information from the table or to get the elementary stream descriptions, we need to look at the descriptors. These are stored as two groups, reflecting the internal structure of the table. The `ProgramMapTable.getOuterDescriptorLoop()` method returns an array of descriptor objects representing the descriptors in the common loop of the PMT. To get the descriptors for a particular elementary stream, we need to call the `PMTElemen-taryStreamInfo.getDescriptorLoop()` method. This also returns an array of `org.ocap.si.Descriptor` objects, as follows.

```
  public abstract class Descriptor {

     public abstract byte getByteAt(int index)
        throws IndexOutOfBoundsException;
```

```
    public abstract byte[] getContent();
    public abstract short getContentLength();
    public abstract short getTag();
}
```

If this looks familiar to you, it should. It is an almost exact copy of the `org.dvb.si.DescriptorTags` and `org.dvb.si.DescriptorTag` classes from the DVB SI API. The only differences are that the `org.ocap.si.Descriptor` class is abstract and the set of constants defined in the `org.ocap.si.DescriptorTags` interface is different. As with the `org.dvb.si.DescriptorTags` class, the list of constants is too long to enumerate here, and thus you should check Annex T of the OCAP specification for details.

System Integration

One of the problems middleware implementers face when building the SI component is the number of different APIs it supports. All GEM implementations will support the JavaTV SI API, whereas OCAP adds the `org.ocap.si` package and MHP adds the DVB SI API. Other GEM-based standards may add their own SI APIs.

Table 9.6. Stream types in OCAP SI.

Constant	Stream Type
ASYNCHRONOUS_DATA	Asynchronous data following the general instrument standard *Data Service Extensions for MPEG-2 Transport* (STD-096-011)
ATSC_AUDIO	ATSC A/53 audio
AUXILIARY	MPEG-2 auxiliary
DSM_CC	DSM-CC
DSM_CC_MPE	DSM-CC multiprotocol encapsulation (DSM-CC type A)
DSM_CC_SECTIONS	DSM-CC sections of any type (DSM-CC type D)
DSM_CC_STREAM_DESCRIPTORS	DSM-CC stream descriptors (DSM-CC type C)
DSM_CC_UN	DSM-CC user-network messages (DSM-CC type B)
H_222	H.222.1
ISOCHRONOUS_DATA	Isochronous data following the general instrument standard *Data Service Extensions for MPEG-2 Transport* (STD-096-011)
MHEG	MHEG applications
MPEG_1_AUDIO	MPEG-1 audio
MPEG_1_VIDEO	MPEG-1 video
MPEG_2_AUDIO	MPEG-2 audio
MPEG_2_VIDEO	MPEG-2 video
MPEG_PRIVATE_DATA	MPEG-2 PES packets containing private data
MPEG_PRIVATE_SECTION	MPEG-2 private sections
STD_SUBTITLE	Standard subtitles
VIDEO_DCII	DigiCypher II video

We can divide the various APIs into two categories: low-level APIs (such as the OCAP and DVB APIs) and high-level APIs (such as the JavaTV SI API). Unfortunately for middleware developers, the higher-level API is the one that is common across all platforms. This causes a few problems when we are trying to make our components as general as possible. In the rest of this chapter we will mainly discuss the DVB SI API and the JavaTV SI API. The challenges of implementing the `org.ocap.si` package are similar to those involved in implementing the DVB SI API, and thus any lessons from one can be applied to the other.

Ideally, we would like to reuse as many elements as possible between our implementations of the different APIs and between the different SI standards we may have to support. All of the SI standards use the same private section format for carrying SI, and this gives us an important starting point in our API design.

Each standard also shares the same basic concepts of tables, subtables, and descriptors, and thus we can build one component for parsing SI tables and descriptors and use it across all of the standards we wish to support. The only challenge here is to remain as general as possible by using concepts such as generic SI tables rather than implementing specific data structures for the SDT, VCT, and other tables being broadcast.

Should we want to use a higher-level representation of the tables, we can keep this separate from the cache implementation and share that representation between the various API implementations in the component. In many implementations, it may be easier to use the implementations of existing classes (e.g., the subclasses of `org.dvb.si.SIInformation`) for this purpose.

Caching SI

Because each standard uses the same basic data structures for carrying SI, we can reuse our implementation of the SI cache and the functionality that monitors specific tables for version changes. This is general across all SI implementations even though the actual SI tables may be different, and thus using a common implementation can save a lot of work.

The SI cache is likely to be one of the most complex pieces of the SI component. Several different applications and middleware components may request overlapping elements from SI tables, and thus it is important to make sure we do not flush data from the cache too early. We must also take care to use an appropriate caching strategy. Some implementations may choose to cache entire tables, whereas other implementations may choose to cache specific entries. Similarly, some implementations may choose to update cached content when they detect a new version, whereas others may simply flush out-of-date information from the cache until it is requested again. These different factors mean that building the SI cache is not a simple task.

The good news is that we can reuse our cache implementation very easily across different SI implementations, provided we design it properly. The basic functionality for monitoring SI tables and reading specific SI data is identical across all DTV standards. By building a cache implementation that only knows about the most basic SI concepts—tables and descriptors

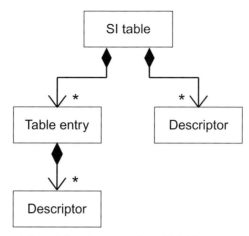

Figure 9.5. The simplest description of the structure of an SI table.

—we can keep any knowledge about specific tables in the API implementations and thus ensure that our cache implementation is as general as possible. The cache does not need to know that tables with table ID 0x42 are instances of the SDT in a DVB system, or that instances of an ATSC RRT will have a table ID of 0xCA. We can describe every SI table using the basic structure shown in Figure 9.5, and any knowledge about the format of a specific table can be kept outside the SI cache.

We may still want to use some standard-specific information such as table-ID-to-PID mappings within the SI cache, but we can encapsulate this in a single class that may be shared between the API implementations and the cache.

Building the API Implementations

The SI cache and a simple parser for descriptors and tables gives us a reusable set of components for reading SI data from the transport stream. We do not have any APIs for accessing this data yet, and thus the next stage is to consider how we could implement the appropriate SI APIs on top of this core functionality.

One disadvantage of generalizing the core concepts in the way we have described is that each API implementation needs to be a little more intelligent. The different SI implementations need to have some understanding of the specific SI format being used in order to request the correct data from the SI cache, but this does not add much complexity. We have already seen that the OCAP and DVB SI APIs both operate at quite a low level and are very SI specific. Because of this, the API implementations form a fairly thin layer over the basic caching functionality because they do not need to perform any real processing on the SI data. Most of the methods in the two APIs will retrieve specific data from a specific table, and the only complexity lies in whether the data should be fetched from the cache or from the stream. The following code shows how this could be implemented for the DVB SI API. We could optimize this code a little further (e.g., by setting the parameters on the request as part of the constructor), but we have not done so for clarity. Similarly, we have not shown any error handling that is necessary.

```
public SIRequest retrieveActualSINetwork(
    short retrieveMode,
    Object appData,
    org.dvb.si.SIRetrievalListener listener,
    short[] someDescriptorTags)
    throws org.dvb.si.SIIllegalArgumentException {

    com.steve.si.SITable cachedNIT;
    com.steve.si.SIRequestImpl request;
    com.steve.si.SIIterator result;
    com.steve.si.SICache cache;
    bool isCached;

    cache = com.steve.si.SICache.getInstance();

    if (retrieveMode != FROM_STREAM_ONLY) {
        cachedNIT = cache.getTableFromCache(
        SITableTypes.NIT, someDescriptorTags);

      if (cachedNIT != null) {
        // we can dispatch the event immediately because
        // we have the result
          isCached = true;
      request = new SIRequestImpl (
        isCached, appData, listener);

      // first prepare the result, then dispatch the
      // event
          com.steve.si.SINetworkImpl nit;
      nit = new SINetworkImpl(cachedNIT);
      result = new org.dvb.si.SIIterator(nit);

      org.dvb.si.SISuccessfulRetrieveEvent ev;
      ev = new SISuccessfulRetrieveEvent (
        appData, request, result);

      com.steve.si.SIEventDispatcher.dispatchEvent(ev);
          return request;
      }
    }

    if (retrieveMode != FROM_CACHE_ONLY) {
      isCached = false;
```

```
    // we need to request the data from the stream. In
    // this case, result handling and event dispatching
    // will be done elsewhere
    request = new SIRequestImpl(
      isCached, appData, listener);
    SICache.getTableFromStream(
      SITableTypes.NIT, someDescriptorTags, request)
    return request;
  }

  // we generate an SINotInCacheEvent
}
```

The temptation when dealing with the JavaTV SI API is to implement it in terms of the lower-level APIs (such as the DVB SI API) in the component. This is generally a mistake, because it makes one API implementation dependent on another and thus causes problems when we want to implement a different middleware platform that uses a different SI standard. It is generally better to keep the JavaTV API implementation separate and implement both APIs using a set of common components at the lower level, as shown in Figure 9.6.

The JavaTV SI API has a very similar structure to the other SI APIs that may be present, and thus building this set of common components is relatively straightforward. All of the APIs are asynchronous, they all use intermediate objects to identify a particular request, and they all return their results via events. The main difference is that the JavaTV SI API may use information from more than one table to satisfy a request, or it may use only a subset of the information from a given table. This adds to the complexity of our API implementation slightly, but it will not affect the core components such as the SI cache and the table parser. In general, a JavaTV API implementation will take a very similar approach to that shown previously for the DVB SI API.

The MHP specification has devoted an entire annex to the rules governing what information should be returned by methods in the JavaTV SI API and how some of the API classes relate

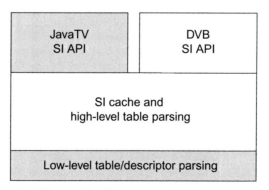

Figure 9.6. A possible way of splitting up the SI component to improve reusability.

Table 9.7. DVB SI classes and the JavaTV interfaces they implement.

DVB SI Class	JavaTV SI API Class
`org.dvb.si.SINetwork`	`javax.tv.service.transport.Network`
`org.dvb.si.SIBouquet`	`javax.tv.service.transport.Bouquet`
`org.dvb.si.SITransportStreamNIT`	`javax.tv.service.transport.TransportStream`
`org.dvb.si.SIService`	`javax.tv.service.navigation.ServiceDetails`
`org.dvb.si.SIEvent`	`javax.tv.service.guide.ProgramEvent`

to one another. Most of these rules are obvious and we will not discuss them here except to say that they do make it easier to integrate the two APIs. One notable change is that the `retrieveProgramEvent()` methods from `javax.tv.service.SIManager` and `javax.tv.service.guide.ProgramSchedule` will always fail when called in an MHP receiver, as will `javax.tv.service.SIManager.retrieveSIElement()` when referring to an event rather than a service.

The main area of interest for us is the description of how objects in the DVB SI API relate to objects in the JavaTV SI API. DVB has decided that for the JavaTV SI API to be truly useful in MHP it must have some connection to the DVB SI API so that applications can get access to DVB-specific information. Several classes in the `org.dvb.si` package will implement their counterparts from the JavaTV SI API, in order to provide a unified approach to reading SI. Table 9.7 outlines the DVB SI classes and which JavaTV interfaces they will implement.

Handling Event Handlers

These structural similarities between APIs mean that it is easy to use a common implementation for some parts of the SI component and then use adaptor classes to provide that functionality in the correct form for the different APIs. At the low level we have seen how SI caching and table parsing can be generalized, but we can also generalize event dispatching. In the previous example, we use a centralized event dispatcher to handle the process of sending events. This has a number of advantages in terms of security and simplicity, and it makes our code easier to reuse. We can split the events and their corresponding listeners into the following two groups.

- *Callbacks:* These are used to notify an application or middleware component that a request for SI data has completed. These are one-to-one relationships (each event is dispatched to one and only one listener).
- *Monitoring events:* These are used to notify applications or middleware components that an SI element has been updated. These are one-to-many relationships (the middleware will send the same event to one or more listeners).

Managing the events and handlers for callbacks is fairly simple. All of the APIs have an object that represents an outstanding request, and by adding private fields to this object we can easily associate the listener with the request. If we use a common implementation to repre-

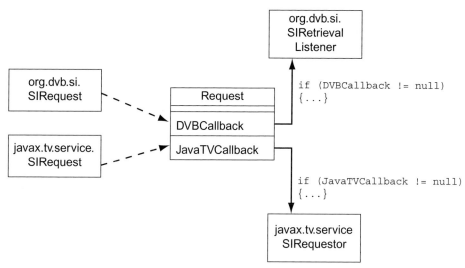

Figure 9.7. Integrating requests from the two SI APIs.

sent all of the different types of requests, we can combine the event dispatching for both APIs by dispatching the events based on the parent class of the listener. Figure 9.7 shows one possible approach to this.

It is a little more difficult to tame the event handling related to SI monitoring, partly because of the number of different listeners we could have. MHP makes this slightly more manageable by using the `SIDatabase` as the central point at which all listeners are registered, no matter what type of SI element we wish to monitor. This concentrates the entire event-dispatching functionality in one class, which generally makes it slightly more manageable.

JavaTV spreads this functionality across many classes, but this is not a significant difference in practice. Each instance of an `SIElement` that can be monitored contains enough information to identify that element uniquely. These `SIElements` can forward the registration of listeners to a central class using the information they contain to identify the SI that should be monitored. Instead of making a complex request directly to a central database, the application makes a simple query to an intermediate object (the `SIElement` it wants to monitor), which then adds extra identifying information and passes on the request.

This adds an extra step to the complexity of registering an `SIChangeListener`, but it has the big advantage of harmonizing the approaches used by the different API implementations. Table 9.8 outlines roughly which methods in the DVB and JavaTV SI APIs monitor the same SI elements.

There is not an exact match between the two APIs, but there are some fairly close similarities. For example, when monitoring networks the DVB SI API registers a monitor for a single

Table 9.8. SI monitoring methods in the JavaTV and DVB SI APIs.

SI Element	JavaTV SI API	DVB SI API
Bouquet	`BouquetCollection.` `addListener()`	`SIDatabase.` `addBouquetMonitoringListener()`
Network	`NetworkCollection.` `addNetworkChangeListener()`	`SIDatabase.` `addNetworkMonitoringListener()`

Transport stream	`TransportStreamCollection.`
	`addTransportStream`
	`ChangeListener()`
		No equivalent

Service (SDT)	`Transport.`
	`addServiceDetails`
	`ChangeListener()`
		`SIDatabase.`
		`addServiceMonitoringListener()`

Event	`ProgramSchedule.addListener()`	`SIDatabase.addEventPresent`
		`FollowingMonitoringListener()`
		`SIDatabase.`
		`addEventSchedule`
		`MonitoringListener()`

Elementary streams	`ServiceDetails.`
	`addServiceComponent`
	`ChangeListener()`
		`SIDatabase.`
		`addPMTServiceMonitoringListener()`

network, whereas the JavaTV API registers a monitor for all networks in a `NetworkCollection`. In this case (as indicated in Figure 9.8), the JavaTV SI API could be implemented using the DVB SI API simply by registering separate listeners for all networks in the `NetworkCollection`.

For all types of events, our biggest problem is event dispatching. Implementations should maintain a pool of threads for notifying applications and other middleware components about events generated by the SI APIs. This requires a delicate balance between the number of threads used and the amount of memory these threads take up. If we had just one thread for event dispatching, a misbehaving application could delay event delivery to every other application and middleware component in the system.

Ideally, every application and client would have its own thread for dispatching events. In this way, applications that do not return quickly from an event notification will only delay themselves. This may require too many threads, however, and thus we may need to share them between applications. In this case, a middleware implementation must make sure that applications return from event notification within a reasonable time and do not delay other applications too badly. If necessary, the only way to achieve this may be to kill the offending thread and create a new one.

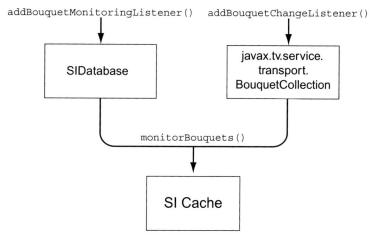

addBouquetMonitoringListener()   addBouquetChangeListener()

SIDatabase

javax.tv.service.
transport.
BouquetCollection

monitorBouquets()

SI Cache

Figure 9.8. Integrating IS monitoring functionality.

Performance Issues

By thinking carefully about how other middleware components use SI, you may be able to improve the performance of your middleware. After all, this is a central component of the middleware and it is only logical to look here first to make any performance improvements.

One of the things that need to be considered when you are thinking about the design of the SI component is the level at which you store your SI data. Because MHP and OCAP are both Java-based middleware stacks, that may be a reason to lean toward a Java implementation of the SI database. At the same time, if the SI database is getting updated or queried frequently it may be better to use native code because the underlying section filtering and basic table parsing will probably also be carried out in native code. To complicate things further, we have to consider the penalty of accessing native data structures from Java code using the Java native interface. All of these things have to be considered, and there is no right answer. Too much will depend on the choice of virtual machine, the performance of the host platform, and the design of other middleware components for us to give you any specific advice.

One approach is to use a Java implementation of the SI database, but to duplicate some of the data needed by lower-level elements of the implementation in native data structures. For instance, by storing the version numbers of cached SI tables in native code the native code that fetches those tables from the transport stream can avoid making too many references to Java data structures. Those readers familiar with the Java native interface will know that one of the most important optimizations is simply to make the interface between Java and native code as small and simple as possible. Avoiding too many calls here—in either direction— can be an important step in improving performance.

Implementing most of the API in Java may have its own problems, of course. Even in these days of just-in-time compilers and highly optimized virtual machines there is still a slight performance penalty for using Java. Depending on your platform, this may or may not be a big problem for you. In any case, it is something else that can affect the decision about how to implement the SI database. The best advice we can give is to implement the SI database in the same language you are using for the middleware components that refer to it. If your section-filtering component and your media control component are implemented using native code, it is best to take the same approach for at least some of your SI database.

10 Section Filtering

To read SI or other data from a broadcast stream, a receiver must use one or more section filters to find the appropriate parts of an MPEG stream. This chapter looks at how a receiver can efficiently use section filters to get this information. Furthermore, we will look at how applications can use the section-filtering API to get access to data streams that are not available in other ways. We will also examine how applications can do the most efficient job of filtering the information they need.

MPEG streams can carry a large number of different types of data in addition to audio and video. In the last chapter we discussed SI, but this is just one type of data we might see. MPEG elementary streams can also carry subtitles, broadcast file systems, control information for the conditional access system, and time code information for synchronizing applications and broadcast media. Applications can also use MPEG elementary streams to deliver data in a proprietary format, if carrying the data in files is not suitable. For instance, private data streams could carry EPG data, or data for a home shopping application, or even applications for another middleware platform.

All MPEG elementary streams must be split into packets so that they can be multiplexed into a transport stream. For audio and video, this is pretty straightforward, with each packet containing a certain number of frames or audio samples. Data streams normally have very different requirements and characteristics, such as the following, that make splitting them into packets much more difficult.

- Each packet in an audio or video stream will be broadcast only once, and will be followed by the next packet in the stream (although packets from other streams may be inserted between them). Compared with audio or video streams, data streams typically consist of a smaller number of packets that are repeated, and thus the receiver may see the last packet of data before it sees the first.

- There may be more than one item of data in a given stream, such as different SI tables or different files in a broadcast file system. We need some way of knowing how the data in the stream is organized.
- There may be more than one version of the data. SI tables will be updated, or files in a broadcast file system may be replaced.
- There is a need to carry many different types of data, and thus the format used must be flexible enough to support this.

To help solve these problems, MPEG defines a concept called "private sections," or sometimes just "sections." These are used to packetize the data for multiplexing, and are equivalent to the PES packets used for carrying audio or video data.

Sections follow a standard format (discussed in material to follow). This is optimized for carrying SI tables, but sections can be used to carry other types of data. Each section will contain up to 4,096 bytes of data, which may be part of an SI table or some other type of data stream. It will also contain a standardized section header that tells the receiver how many sections are being used to carry the data and how they should be assembled.

Because private sections are used to carry SI as well as application-specific data and broadcast file systems, any DTV system will need to read data from them. To do this, DTV receivers provide a number of section filters. These give us a way to search an MPEG stream for sections that meet a set of criteria we specify, such as a specific PID or SI table ID.

A number of different components will rely on these section filters. They will be used by the SI component for filtering SI tables, and by the conditional access system for filtering any conditional access messages targeted at that receiver. The middleware will use them to access broadcast file systems (standardized file systems in the case of the open middleware standards, or proprietary systems that may be used by other middleware stacks). Electronic Program Guides (EPGs) or pay-per-view systems may use them to read private data streams, and a number of other applications may use them for their own purposes. Given this, you can see that section filtering is something that is critically important to the quality of the receiver. A receiver that is not very good at section filtering often has other performance or reliability problems.

Hardware Versus Software Section Filters

Section filtering can be carried out either in hardware or in software. Because section filtering is such an important task in many receivers, most MPEG decoder chips designed for DTV use will include hardware support for a number of section filters. Older chips may only include eight section filters, whereas most modern chips will include up to 32.

Although eight or more are probably enough for most needs, those receivers with fewer hardware section filters (or with no hardware MPEG support at all) may need to use software section filtering instead. The results of the two methods are identical in principle, although in practice there is obviously a performance penalty involved in filtering high-bit-rate streams in software, especially when there will usually be several section filters active at once.

This performance penalty may cause more problems than just slowing down our applications. Given the speed at which sections will be arriving, they must be filtered in a reasonable time or the receiver has to discard sections when its buffers become full. Discarding sections is always bad, because it means that we may not get all of the data we need and in a broadcast system we cannot ask the server to resend it. The result of this is that software section filtering is something that is best not used unless you have a fast CPU to handle it and really do not have enough hardware section filters to meet your needs.

MHP and OCAP specify that a receiver should have at least two section filters available for applications for every service being decoded. In other words, if your receiver can decode one service at a time there should be two section filters available for applications on that service to use. If your receiver can decode two services, there should be four section filters available: two for applications on each service. If only one application is using section filters, it may be able to use all four. Two filters per service is the minimum, not the maximum, and receivers may choose to share the available section filters intelligently between services. Applications on the same service must share the available section filters between them, however, and thus there is no guarantee that an application can get access to a section filter if it needs it. Note that this number does not include any section filters used for parsing SI or for any other tasks carried out by the middleware. These section filters are reserved for applications. This means that middleware implementers may have to be slightly careful about how other middleware components use section filters. Given the number of filters typically available in the hardware, this is not likely to be a big problem, but it is something we should keep in mind.

Using Section Filters

MHP and OCAP applications can make use of these section filters through the DAVIC section-filtering API. This is contained in the `org.davic.mpeg.sections` package, and it provides several different ways for applications to access any private sections that interest them. Each of these ways is designed to cope with a specific type of situation in which an application may want to access private sections.

Before we look at these mechanisms in detail, we need to know a bit more about the format of an MPEG-2 private section. Each section has a header that will for the most part follow the standard format outlined in Table 10.1. This provides a framework for identifying a private section and for telling us something about the type of data contained in the section. The second part of the section—the section body—contains the payload. This may or may not follow a standard format, depending on the type of data it contains.

Sections belonging to SI tables will usually have the section syntax indicator set to 1. When we filter a section, we usually want to filter on values from the header (anything before the private data bytes) if we are filtering something like an SI table. In particular, we will often filter sections based on the table ID and table ID extension, in that these are used to distinguish different SI tables and subtables.

For private data streams, we may not be able to get the accuracy we want by filtering on just the header fields and thus we can also filter based on the first few bytes of the section data.

Table 10.1. The MPEG-2 private section format.

Syntax	No. of Bits	Identifier
`private_section() {`		
`  table_id`	8	uimsbf
`  section_syntax_indicator`	1	bslbf
`  private_indicator`	1	bslbf
`  Reserved`	2	bslbf
`  private_section_length`	12	uimsbf
`  if(section_syntax_indicator == '0') {`		
`    for(i=0; i<N; i++) {`		
`      private_data_byte`	8	bslbf
`    }`		
`  }`		
`  else{`		
`    table_id_extension`	16	uimsbf
`    Reserved`	2	bslbf
`    version_number`	5	uimsbf
`    current_next_indicator`	1	bslbf
`    section_number`	8	uimsbf
`    last_section_number`	8	uimsbf
`    for(i=0; i<private_section_length-9; i++) {`		
`      private_data_byte`	8	bslbf
`    }`		
`    CRC_32`	32	rpchof
`  }`		
`}`		

Source: ISO 13818-1:2000 (MPEG-2 systems specification).

This gives us a way of filtering private data streams. If those streams have some unique identifying pattern in the first few bytes of the section data, we can use this to make our filtering more accurate and save work for the application performing the filtering.

The section-filtering API allows applications to use both positive and negative filters, and thus an application can filter for sections that match a specific pattern or for those that do not meet a specific pattern. Sometimes, using a negative filter can give better results than a positive filter, and thus application developers should not rely solely on positive filters.

Section filtering can be a time-consuming task, just because we do not know when we will get some sections that match our filter. Given this, it is no great surprise that section filtering is an asynchronous operation. The application sets up some filters, sets a listener for the events generated when a matching section is found, and then waits for events to start arriving.

In practice, life is not quite that simple, of course, but you will be pleased to hear that it is not much more complicated in this case. The main complication is the fact that section filters are scarce resources, and thus applications have to take care to manage section filters correctly and not use the underlying resources when they do not need to.

The Section-filtering API

As with many of the APIs that deal with MPEG, the section-filtering API uses the resource notification API to help share resources among the applications and middleware components that need to use them. In this case, the API uses it in a slightly different way, as we will see later. Resource management is very important for the section-filtering API, simply because so many components within the system use section filters. We have already seen that many middleware components will use them, in addition to any application that may wish to carry out any section filtering.

The API consists of four major classes. The first of these is the Section class. This represents an MPEG-2 section, and lets us read the various fields and the payload in a private section. There is nothing really complicated about this class, and because we have just taken a look at the format of private sections we will not describe this class here in any detail. The interface to the section class follows (in a slightly simplified form), and every method that gets data from the section may throw a NoDataAvailableException to indicate that the section contains no data, or that the requested data is not available due to the format of that section.

```
public class Section implements java.lang.Cloneable
{
  public byte[] getData();

  public byte[] getData(int index, int length)
    throws java.lang.IndexOutOfBoundsException;

  public byte getByteAt(int index)
    throws java.lang.IndexOutOfBoundsException;

  public int table_id();
  public boolean section_syntax_indicator();
  public boolean private_indicator();
  public int section_length();
  public int table_id_extension();
  public short version_number();
  public boolean current_next_indicator();
  public int section_number();
  public int last_section_number();

  public boolean getFullStatus();
  public void setEmpty();
  public Object clone();

}
```

Section Filters

The `SectionFilter` class represents a real section filter, which the receiver can implement either in hardware or in software. To an application, both of these approaches are the same, although software section filtering may be slower and may not be able to filter all sections if there is too much else going on in the receiver. The decision to implement section filters in hardware or software is really up to the receiver manufacturer, and depends mainly on the performance of the platform and the number of hardware section filters it has. The interface to the `SectionFilter` class is as follows.

```
public abstract class SectionFilter {

    public void startFiltering(
        java.lang.Object appData,
        int pid,
        int table_id,
        int offset,
        byte[] posFilterDef,
        byte[] posFilterMask,
        byte[] negFilterDef,
        byte[] negFilterMask);

    public void setTimeOut(long milliseconds);
    public void stopFiltering();

    public void addSectionFilterListener(
        SectionFilterListener listener);
    public void removeSectionFilterListener(
        SectionFilterListener listener);
}
```

For brevity, we only show the most complex version of the `startFiltering()` method here. There are six versions in total, each having a subset of the arguments listed in the version shown previously. The MHP or OCAP specification gives a full description of the variations of the `startFiltering()` method that are available.

As you will have noticed, the `Section` class has no public constructor. Any instances of this class will be created by the middleware rather than the application.

The `startFiltering()` and `stopFiltering()` methods start and stop the processing of a transport stream, although they may not do quite what you expect. Before a section filter can begin filtering, we must attach it to a transport stream. If a section filter is already attached to a transport stream when you start filtering, filtering will begin immediately. Otherwise, it will simply store these settings and then begin filtering when the filter is successfully attached to a transport stream. Attaching to a transport stream simply means that we associate a section filter with a specific transport stream, and reserve any resources we need in order to start filtering on that transport stream.

The startFiltering() method lets us set a range of parameters we can use to filter sections, and we will take a closer look at what these parameters actually mean. The arguments to the startFiltering() method are outlined in Table 10.2.

As we can see, the section-filtering API supports both positive and negative filtering, and actually allows both types of filtering to be carried out at the same time. This is a very useful feature that can help to make sure only the appropriate sections get filtered. We can use this to say (for instance) "Filter those sections with a table ID of 16, but ignore any with a table ID extension of 42." When we are parsing a DSM-CC broadcast file system, we can use this to search for a specific subset of DSM-CC messages while ignoring those that do not interest us at that time.

The setTimeOut() method allows the application to specify that filtering should end after a certain period. For some types of section filters, filtering will stop if no sections are matched within the timeout period, or if the timeout period has passed since the last section was matched. In other cases (table section filters), filtering will stop if the filter has not managed to filter all of the necessary data before the end of the timeout period.

We have already mentioned that the section-filtering API is an asynchronous API, and the final two methods of the SectionFilter class allow an application to register and unregister listeners for events from this particular filter. These can indicate that a new section matching the filter pattern has arrived, or they can be used to indicate a problem with the section filter or a change in its state. Before we start a filter, we should register at least one listener if we want to receive notifications about the progress of section filtering (most, if not all, applications will want to do this). Later in the chapter we will take a closer look at the events a section filter can generate.

Types of Section Filters

There are three types of section filters available to an application, each designed for a different purpose and with different strengths and weaknesses. These are implemented by subclasses of the SectionFilter class we have just seen, and the interfaces to them are fairly similar.

The first type of filter we can use is the simple section filter. This is implemented by the SimpleSectionFilter class, and is just a one-shot filter that filters a single section before stopping. When a section matches the filtering criteria, filtering stops and a SectionAvailableEvent is generated. An application may then get a reference to the filtered section by calling the SimpleSectionFilter.getSection() method.

If you want to filter more than one section, it is a little inconvenient to use a filter that stops after every section filtered (because the application will need to reset the section every time). If the filter is not reset quickly enough, some sections may have passed before filtering begins again. For this reason, it is best to use the simple section filter in those cases in which only a small number of sections will match the filter that has been set, and in which the gap

Table 10.2. Parameters of the `SectionFilter.startFiltering()` method.

Parameter	Purpose
appData	An object that uniquely identifies this filter to the application. This is used to identify section filter events in a way similar to that used in the SI API, so that a single listener can accept events from more than one section filter and still be able to distinguish between them. This is not a parameter that will be used for filtering sections, and the API only uses it to identify which filtering operation a particular section filter event belongs to. Because of this, the `appData` parameter can contain anything the application wants to use.
pid	The PID of the elementary stream that should be filtered.
table_id	The value of the table ID for sections that should be filtered.
offset	An offset into the section that indicates the part of the section the filter should operate on. The offset can be a value between 3 and 31, indicating that the filter should start matching *n* bytes from the start of the section, where *n* is the value of the offset. This is indexed from zero, and thus an offset of 3 will actually start filtering on the fourth byte of the section. This range of values means that the filter can cover the interesting parts of the section header and the start of the section data. Out of the first three bytes, only the table ID is interesting to us and we can filter that using the parameter we have just seen. The section header is 8 bytes long, and thus we can filter on the first 24 bytes of the section data as well as on the headers. By supporting filtering on part of the section, applications can use the first few bytes of the data for their own private identification while still being able to filter the data easily.
posFilterDef	The data that should be matched by the section filter. This array of 8 bytes defines the values that should be matched. We may not care about every part of those 8 bytes, however, and thus we use the filter mask (the next parameter) to identify which parts of the filter definition we care about. A positive filter is triggered when the parts of the filter definition that are enabled by the filter mask completely match the content of the section, starting at the appropriate offset into the section.
posFilterMask	A bit mask indicating which bits in the section should be matched to the filter definition. Like the filter definition, this is an array of 8 bytes. Any bits with a value of 1 will be matched, whereas any bits with a value of 0 will be ignored.
negFilterDef	A filter definition for negative filtering. Negative filtering causes the filter to be triggered when any part of the filter definition that is enabled by the filter mask fails to match the content of the section, starting at the appropriate offset into the section. Like the positive filter definition, this is an array of 8 bytes.
negFilterMask	A bit mask for negative section filtering. This serves the same purpose as the positive filter mask, but it is applied to the negative filter definition instead. Again, this is an array of 8 bytes.

between sections in the stream is long enough to allow the application to receive the event and reset the filter.

For those cases in which many sections have to be filtered from the stream, the ring section filter is usually a better choice. This is implemented by the `RingSectionFilter` class. This provides a circular (ring) buffer of sections whereby more than one section can be stored before the filter will stop. We tell the filter how big its buffer should be when we create the filter, and thus it can be as large or as small as we need.

Whenever the filter matches a new section, it places that section in the next free slot in the buffer and the section filter then sends a `SectionAvailableEvent` to any registered listeners. An application can use the `RingSectionFilter.getSections()` method to read the buffer and get access to the newly filtered section.

Before a slot in the buffer can be reused, the application has to tell the section filter that it has finished with the section in that slot. This is done by calling the `Section.setEmpty()` method on the section that occupies that slot. When no more slots in the buffer are available, the section filter will stop. Thus, it is important that the application call `Section.setEmpty()` as soon as possible after the section is received if it wants filtering to continue. The best approach is for the application to copy the section data using the `Section.clone()` method as soon as it receives a `SectionAvailableEvent` and then

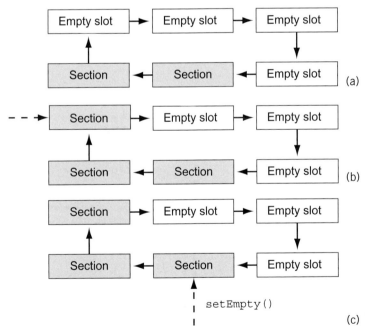

Figure 10.1. The ring section filter allows filtering of more than one section at a time without needing to be reset: (a) the ring section filter can store several sections at once, (b) new sections are added to the next empty slot, and (c) slots must be freed before the buffer fills, or the section filter will stop.

free the section in the buffer. That way, if it takes a little bit longer to process the section there is less chance of accidentally causing the filter to stop.

The final type of section filter available to applications is the table section filter, which allows applications to filter entire SI tables easily. This is implemented by the `TableSectionFil-ter` class, and this type of section filter will automatically define a buffer big enough to hold the entire SI table it is filtering.

As with the `RingSectionFilter`, a `TableSectionFilter` will generate a `Section-AvailableEvent` after it receives a new section that matches the filter. Each section is placed into the filter's buffer, and then the `TableSectionFilter` will continue filtering until it has found every section that makes up that table.

This has a couple of advantages over the `RingSectionFilter` in some cases. In particular, the application does not need to worry about the size of the table. The section filter will make sure it allocates buffers that are big enough, and it will make sure it gets all sections for the table. Because the buffer is guaranteed to be big enough to hold the entire table, the application does not need to worry about processing the sections as soon as possible, or freeing any sections in the buffer.

Filtering tables is not as easy as it may seem, however, because the version of a table may change while we are filtering it. To simplify things, the `TableSectionFilter` only filters those sections belonging to the version to the table that is current when filtering starts. In other words, if version 10 of the table is current when we start filtering any sections belonging to later versions of the table will be ignored. If the filter detects that the version number of the table has changed it will generate a `VersionChangeDetectedEvent`, which is covered in more detail later in the chapter.

Any processing of sections received by an application or middleware component should be done in a separate thread, rather than in the `SectionFilterListener`. The reason for this is pretty simple once we think about how methods in the listener get called. Event listeners will be called by a thread that belongs to the middleware, and this thread may be used for delivering events to other applications as well. Doing too much work in the event listener will mean that the delivery of other events may be delayed, and this could cause problems for your own applications or for other applications.

This is similar to the problem we discussed in the last chapter, but it is even more important here. A filter that matches many sections means that a large number of events could be generated very quickly, and this means that we could fill our section filter's buffer. Even if we have not received the events, any sections filtered will still be placed in the section filter's buffer and the filter will stop if no more slots are available in its buffer.

Section Filter Groups

Section filters are a scarce resource, and thus we need a way of managing them. To do this, the section-filtering API uses the `SectionFilterGroup` class, as in the following. In addition to managing the use of section-filtering resources, section filter groups allow several section filters to be grouped (hence the name) to improve performance and make life easier for applications and middleware components that want to use them.

```
public class SectionFilterGroup
    implements org.davic.resources.ResourceProxy,
               org.davic.resources.ResourceServer {

    public SectionFilterGroup(int numberOfFilters);

    public SectionFilterGroup(
       int numberOfFilters,
       bool priority);

    public void attach(
       org.davic.mpeg.TransportStream stream,
       org.davic.resources.ResourceClient client,
       Object requestData);

    public void detach();

    public org.davic.mpeg.TransportStream getSource();

    public SimpleSectionFilter newSimpleSectionFilter();

    public SimpleSectionFilter newSimpleSectionFilter(
       int sectionSize);

    public RingSectionFilter newRingSectionFilter(
       int ringSize);
    public RingSectionFilter newRingSectionFilter(
       int ringSize,
       int sectionSize);

    public TableSectionFilter newTableSectionFilter(
       int sectionSize);

    public void removeResourceStatusEventListener(
       org.davic.resources.ResourceStatusListener listener);

    public void addResourceStatusEventListener(
       org.davic.resources.ResourceStatusListener listener);
}
```

As you can see, `SectionFilterGroup` implements both the `ResourceProxy` and `ResourceServer` interfaces from the resource notification API. This makes the section-filtering API slightly unusual in the way it uses the resource notification API, although it makes little difference in practice.

Applications can create section filter groups using the public constructor, which takes an argument that specifies how many filters are in that group. By allowing applications to group filters like this, it makes it easy to reserve more than one filter at a time and thus reduce the potential for deadlocks among applications that use section filters. If an application knows it will need to have three section filters active on the same stream at the same time, it can reserve all three filters in one operation.

For instance, suppose we have two section filters available in our middleware. If we have two applications, both of which need two section filters, there is the potential for deadlocks. Without the ability to group section filters, we could have the case in which each application has reserved one section filter, and thus neither can reserve the second it needs and both applications may not be able to carry out the filtering they want. If one application uses a section filter group, it will try to reserve both filters in a single atomic operation. If one filter is already reserved, the operation will fail and that application will not claim any filters. This leaves the second filter free for the other application, and thus at least one of the two applications can run properly.

The downside to this approach is that it can work against the application and is not very friendly to other applications. By possibly requesting more resources than it currently needs, the application runs the risk of not getting any of those resources. It also means that any resources the application is not currently using are not available to other applications.

For these reasons, it is generally better to have section filter groups that contain the exact number of filters you need at that time. If you need more filters for another task, but not immediately, it is best to use a separate section filter group for them. This gives more flexibility to your application and to the middleware in the way resources are allocated. Remember that the more filters you request the more likely it is your request will be denied.

Optionally, we can give a section filter group a priority when we create it. By default, every section filter group will have high priority, but it is possible to define some filter groups as low priority. To do this, applications can use the second version of the constructor shown previously, using the value `false` for the `priority` argument. Low-priority section filter groups will be reclaimed before high-priority groups, although application priority will probably play a much more important part in determining which application loses access to resources. The priority of the section filter group will only be used to resolve resource conflicts between applications with the same application priority or within the same application. As with so many other resource management issues, setting the right priorities for your applications is the most important thing you can do to make applications work well together.

Creating a section filter group does not reserve any filters. This only happens when we attach the section filter group to a transport stream. We will take a look at this in material to follow.

Once we have a `SectionFilterGroup` that contains the number of section filters we need (see Figure 10.2), we can create individual section filters within that group. Simply reserving the section filters does not tell the middleware what types of section filters we want and we have to do this separately. To do this, we use the `newSimpleSectionFilter()`, `newRingSectionFilter()`, or `newTableSectionFilter()` method, depending on the type of filter we want to create. For ring section filters, we also need to tell it the size of the ring buffer it should create. We can also make sure that the buffer does not use more memory

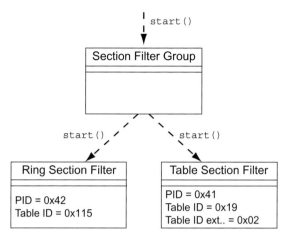

Figure 10.2. Section filter groups allow more than one filter to be manipulated at the same time.

than necessary by telling it the maximum size of the sections it will hold, but to be honest this is usually best left at the default.

Before a section filter can be used, we have to attach the group that contains it to the transport stream we wish to filter. The `attach()` method attaches a section filter group to a particular transport stream. At this point, the API reserves all section filters needed (if it can) and associates those filters with that transport stream. Section filters attached to a transport stream will only filter sections from that transport stream, and if we want to filter another transport stream we must detach the section filter and attach it to the new transport stream.

When the section filter group is attached to a transport stream, any section filters in that group that have been started will start filtering sections. Other section filters will only begin to filter sections when they are explicitly started.

Once an application has finished with a section filter group, it can release the filters and stop any active section filters in the group by calling the `detach()` method. Simply stopping a filter by calling `SectionFilter.stopFiltering()` does not detach it from the transport stream, and does not necessarily release the resources it is using. Only detaching the section filter will release any scarce resources.

Section Filter Events

The final piece of the section-filtering puzzle is the `SectionFilterListener` interface. Applications should implement this interface so that they can receive events from any section filters they create. Because the section-filtering API uses events to tell the application when a new section has been filtered, this is a vital part of the API.

Table 10.3. Events generated by a section filter.

Event	Meaning
IncompleteFilteringEvent	The section filter cannot perform any filtering because of an incorrectly defined set of filter parameters.
TimeOutEvent	No sections have been filtered within the specified timeout period, and thus the section filter has stopped itself.
SectionAvailableEvent	A new section has been filtered and is available from the filter. The application should call the getSection() method (for a simple section filter) or getSections() method (for a ring or table section filter) to get the new section data.
VersionChangeDetectedEvent	The version number of the sections has changed. The filter will ignore sections with a different version number to previous sections that have been filtered, to make sure that there are no inconsistencies in the filtered data. This event is only generated by a TableSectionFilter.

The possible events an application can receive are outlined in Table 10.3. We will not look at all of these events in detail, but we will take a closer look at the more interesting ones in material to follow.

The most interesting events from an application developer's perspective are the Section-AvailableEvent (which we have already seen), the TimeOutEvent, and the Version-ChangeDetectedEvent. Each private section that follows the standard syntax has a version number used to identify the version of the data in that section. For instance, the version number of sections making up an SI table will get incremented when the data in the SI table changes. Any sections with a data version number that is different from that which was current when filtering started will be ignored. This may not be what we want to happen, and thus the VersionChangeDetectedEvent tells us that an updated version of the data is available. This gives the application a chance to stop and restart the section filter if it is interested in getting the newer version of the data.

For a simple section filter, this is not important and can be ignored. Because a simple section filter only filters one section at a time, it will simply detect the first section that matches its pattern. If it is restarted and detects a newer version of the section, this is a problem the application needs to handle. Applications using ring section filters must track the version number of the sections themselves.

The TimeOutEvent tells the application that no packets have been filtered in the time limit specified by the application. A timeout can be set for each section filter using the Section-Filter.setTimeout() method we saw earlier, and thus the application can decide to only wait for a certain time for any data. For example, this could be used by the SI component to find an optional SI table. The SI component can set a timeout that is longer than the repetition rate of the table in question, and if the section filter generates a `TimeOutEvent` the middleware knows that the table is not present.

An Example

Now that we have taken a look at the various parts of the section-filtering API, let's look at a practical example of how they fit together. To filter sections successfully, we need two pieces of code: one to set up the filters and start filtering, and one to handle any events we get back from the section-filtering API. To begin, let's see how we set up the filter, as follows.

```
// Create a SectionFilterGroup that contains two
// section filters
SectionFilterGroup filterGroup;
filterGroup = new SectionFilterGroup(2);

// Create a couple of section filters within the filter
// group. First, we will create a simple section filter
SimpleSectionFilter simpleFilter;
simpleFilter = filterGroup.newSimpleSectionFilter();

// Now we create a ring section filter that can hold
// up to five sections
RingSectionFilter ringFilter;
ringFilter = filterGroup.newRingSectionFilter(5);

// Since the section filter isn't much use to us if we
// don't know when we've got a new section, we register
// an event listener with the filter. In this case, we
// will only register a listener with the ring section
// filter. We will see the listener in more detail later.
SectionFilterListener myListener;
myListener = new MyListener();
ringFilter.addSectionFilterListener(myListener);

// We've now created all the section filters that we'll
// use, so we can set some parameters. Since we haven't
// attached to a transport stream yet, calling the
// startFiltering() method will simply set the parameters
// on the filter. In this case, we set the ring section
// filter to match table ID 116 (the application
// information table) on PID 100. This method can throw
// a lot of exceptions that we need to catch.
try {
  ringFilter.startFiltering(null, 100, 116);
}
```

```
catch (FilterResourceException fre) {
  return;
}
catch (org.davic.mpeg.NotAuthorizedException nae) {
  return;
}
catch (ConnectionLostException cle) {
  return;
}
catch (IllegalFilterDefinitionException ifde) {
  return;
}

// Having set the parameters, we can attach the section
// filter group to a transport stream. This will
// automatically start the filters once the group is
// attached.
//
// The second parameter to attach() is an
// org.davic.resources.ResourceClient that will be
// notified if the section filter is needed elsewhere.
// The third parameter is optional application-specific
// data that can be passed to the resource client.
//
// This can also throw a lot of exceptions.
try {
  filterGroup.attach(ourTransportStream, myResourceClient, null);
}
catch (FilterResourceException fre) {
  return;
}
catch (InvalidSourceException ise) {
  return;
}
catch (org.davic.mpeg.TuningException te) {
  return;
}
catch (org.davic.mpeg.NotAuthorizedException nae) {
  return;
}

// At this point, our ring section filtering is active
// and we just have to wait for some events that tell us
// what is happening. Our simple section filter is not
// filtering anything, because we have not started it.
```

After we have set up our section filter, we need something that listens to section-filtering events. The following class interface shows how we can do this, and although we have implemented it as a separate class in this example it does not have to be. The same class could be responsible for setting up the section filters and for listening to section-filtering events.

```
public class MyListener
   implements SectionFilterListener {

   // tells us where in the buffer we can find the next
   // section to read
   public int currentSectionIndex = 0;

   public void sectionFilterUpdate (

      SectionFilterEvent event) {

      // get the filter that the event was received from.
      // In this case, we assume it's a ring section
      // filter.
      RingSectionFilter filter;
      filter = (RingSectionFilter) event.getSource();

      // Check that we have filtered a section. We will
      // ignore other events for now.
      if (event instanceof SectionAvailableEvent) {

        // get the sections from the filter
        Section[] sections = filter.getSections();

        // now get the current section;
        Section currentSection;

        currentSection = sections[currentSectionIndex];

        // get the section data. We could also choose to
        // get some of the section headers if we were more
        // interested in those.
        try {
           byte[] sectionData = currentSection.getData();
      }
      catch (NoDataAvailableException ndae) {
        // do nothing;
      }

        // if we've got all the data we need from this
        // section, we free it so that this slot in the
        // buffer can be re-used and increment the index
        // for the current slot in the buffer
        currentSection.setEmpty();
```

```
        currentSectionIndex++;
        currentSectionIndex = currentSectionIndex % 5;

        // now we can do what we want to with the data from
        // the section without having to worry so much
        // about the buffer filling.
      }
    }
  }
```

If your application will be filtering many sections, it is best for the event handler to do nothing except clone the section and pass it to another thread for processing. This allows the event handler to return quickly, while still allowing the application to perform a reasonable amount of processing on the data. In the cases in which only simple processing is being done, such as extracting a single field from the section, this is not necessary.

The Middleware Perspective: Event Dispatching

We have seen the event model from the application's perspective, but the middleware also needs to know a few things about the event model. The middleware faces the following two main problems when dispatching events related to section filtering.

- A large number of events may be generated in a very short time, depending on the filter that is set.
- The application may not process these events as quickly as they are generated.

This gives middleware developers a number of potential headaches. Even polite applications can cause problems if they are just a little too slow processing the events that are generated. This delay can mean that the middleware has a backlog of events waiting to be dispatched or that it needs to use a large number of threads to dispatch them on time.

We saw in the last chapter the types of problems this can cause, and they only get worse in the section-filtering API. The speed with which events can be generated is potentially much higher, and thus it is more likely that events will be waiting before they can be dispatched. Similarly, a ring section filter with a large buffer may have a large number of events waiting to be dispatched if its listener is slow to respond.

This is another place where the middleware needs to make a trade-off about how it delivers events. It is essential that the middleware use more than one thread to dispatch section-filtering events, but exactly how many threads should be used will depend on the architecture of other parts of the middleware. If other middleware components use the standard section-filtering API to handle their section-filtering needs, a large number of threads will be needed. A large number of sections will be filtered, and events will need to be dispatched to many middleware components in a short time. If the middleware uses a private API for section filtering, fewer threads will be needed because fewer applications will be using this API.

The other part of the trade-off is how aggressive the middleware should be in killing event-handling threads that take a long time to return from dispatching events. Although there is always the chance that a malicious application will not return from an event notification, this is going to be uncommon. At the same time, if the middleware is delivering three separate events to that application, using three separate threads, a large part of the API's resources may be caught up dealing with that application and other applications may suffer.

Finding a balance between these factors is tricky, but the simplest answer is to use one event-dispatching thread per application. This means that if an application is too slow in processing events only that application is affected. We could use more than one thread per application but this introduces another problem. With several threads delivering events to the same application, there is a danger that events can be processed in a different order than they should be, even if they are dispatched in the same order. Although there are ways to avoid this type of race condition in the event listener it is generally easier to use a single thread.

An alternative solution is to have two separate thread pools. One thread pool contains a small number of threads used for dispatching events to applications. The second pool contains more threads, and it is used purely for dispatching events to other middleware components. This approach lets the middleware use two different strategies for the two different cases. Middleware components we know to be trustworthy can be given a little more latitude, whereas applications that may not be quite as trustworthy can be dealt with in a more secure way. One version of this approach, which we apply to individual applications in this case, is shown in Figure 10.3.

Of course, this is not just a problem for the section-filtering API. Any other API that dispatches events to an application has the same problem, as we saw with the SI component in Chapter 9. We could use a single thread for dispatching events from all APIs to one application, but this adds a great deal to the complexity of our middleware. Although using a separate thread for each API means that more threads may be active, it is generally easier for middleware designers and implementers to take this approach. Keeping the modularity between different APIs is usually more helpful than the minor efficiency savings we would get otherwise.

Managing Section Filter Resources

Given the number of applications and middleware components that will use section filters, they need to be managed carefully both by applications and by the middleware. In the following material we will see how application developers and middleware implementers can do their part to make things go as smoothly as possible.

Managing Section Filters in Your Application

The application's view of section filter management is simple: Use as many as you need to get the results you need, but not too many. Exactly how you do this is slightly more complex.

The first rule is a basic one: Make your section filters as specific as possible to reduce the number of sections that match them. By filtering only those sections you really need, you can

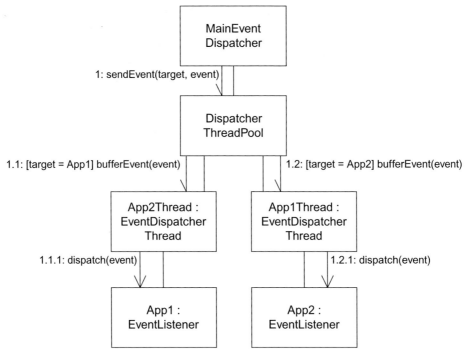

Figure 10.3. A possible strategy for event dispatching.

keep the number of events the middleware generates as small as possible. This helps your application, because it is not bogged down processing events it does not care about, and it helps the middleware because it means that less memory will be wasted filtering unnecessary sections.

The second rule is also simple: Use the smallest number of filters you need to. Using two filters with the same filtering rules for the same transport stream will not help you, and in fact it will probably reduce the performance of the middleware and your application because you will get two events for each section that matches the filters. Filters always work independently, and so there is no point using more than one filter with the same rules.

This also means that you should not reserve filters unless you are about to use them. We saw earlier how section filter groups let applications reserve more than one filter at a time. The downside is that if you do not need all of the filters in a group immediately some of them are wasted. If another application wants just one filter, you may be depriving it of a resource you are not actually using. Given that only a small number of filters may be available for a given service, if your application reserves a filter it is not using another application may not be able to run.

Generally, there is no need to have more than one filter in a section filter group. The only exception to this could be those used by the SI component (assuming it uses the section-filtering API for access to SI tables). In this case, the component usually needs more than one section filter, and it usually needs several filters at the same time. For applications, this type of situation is rare.

This also introduces us to the complex part of managing section filters. What happens if we want to filter two sets of sections but we have just one section filter available? In this case, it may be sensible to relax the first rule. For those cases in which it is easy to find a set of parameters for a section filter that will filter out all of the sections you need and only a few extra, it is probably worth the overhead of dealing with the false matches. Again, the best way is to make the filter as specific as possible, but you may have to accept that you will get some sections you do not really want.

Managing Section Filters in the Middleware

Section filters are a critically important resource for the middleware, which is why there are so many of them in a hardware implementation. This means that a lot of the time the middleware will be using many of the available section filters, and designing a middleware stack that uses fewer filters is not easy.

Items such as SI are so important they should always get the section filters they need, and the same goes for the section filters needed by the CA system and for broadcast file system access. Once the middleware has all of the section filters it needs, it can start giving them to applications. Generally, it is much more important for the middleware to get enough section filters than it is for applications to have three section filters available instead of two.

The main issue here is not to be afraid of giving the middleware what it needs. We only need to reserve two section filters for every service we can decode (and if we can decode two services, we will usually have twice the number of section filters to start with because we will have two hardware MPEG decoders). Giving applications more is a nice gesture, but it is more important that the middleware can do its job efficiently.

The next question is only slightly less fundamental. Should the middleware use the DAVIC section-filtering API at all, or should it use a lower-level API to get what it needs? To be honest, there is no right answer. The advantage of using the DAVIC section-filtering API is that it already exists. The disadvantage is that it is a Java API, and in particular it is an asynchronous Java API. This means it is not as efficient as it could be, and there may be better ways to do this given the number of sections we may have to filter.

A lower-level API written in C or C++ may be a lot faster, even if it is doing the same thing. Many of the arguments from Chapter 9 that apply to the design of the SI component also apply to the section-filtering module, and we will not cover those again here. Figure 10.4 shows some of the possible approaches, and we suggest that you pick the approach that leads to the least communication between Java and native code in order to maximize performance.

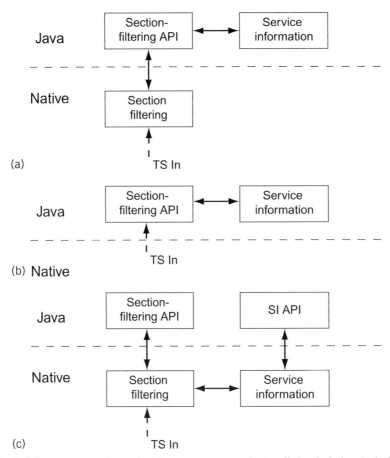

Figure 10.4. Building components so that they can communicate efficiently is key to building an efficient middleware stack: (a) section filtering in native code, and service information in Java, (b) doing everything in Java, and (c) doing everything in native code with a Java API only.

The only thing to add is that the section-filtering API may deal with a lot more data than just SI, and thus you need to think very carefully about efficiency. This does not mean just the efficiency of the actual filtering operations but the efficiency in getting data from native code to Java and from the section-filtering module to other modules in the middleware.

For example, if your section-filtering component is written largely in C++, a Java-based data broadcasting component may not be the most efficient solution to that particular problem given the overheads of the Java native interface. For issues such as these, you really cannot think in terms of the individual components as we are doing here. It may make more sense for all of these APIs to be simple Java wrappers that sit on a lower layer of C or C++ code, instead of being separate Java components that communicate only through Java objects. Like so many other cases, the limitations of the platform will have an influence on your middleware architecture.

11 Media Control

MHP and OCAP applications will often want to control the video the receiver is showing. In this chapter, we introduce the Java Media Framework (JMF) and look at how we can control what the user sees and hears. We will see how we can show various video streams, and how we can resize, clip, and position that video on the screen.

One of the most important functions of a DTV receiver is decoding and controlling broadcast video and other multimedia content. Both MHP and OCAP adopted Sun's JMF for this task. JMF adds support for decoding and displaying different types of media from a number of sources. These sources include both streaming media and media from a local file, and thus it is easy to see why JMF was adopted. As usual, MHP took the approach of not developing a new API when one already existed to do the job.

The version of JMF used in MHP and OCAP is based on JMF 1.0. Unlike the most recent version available for PCs and workstations, version 1.0 concentrates purely on receiving content from a server, decoding it, and displaying it. The JMF API lives in the `javax.media` package, and although the API has not changed much from Sun's original JMF specification there are some new restrictions on what does and does not work, and on the return values from some of the methods. These were needed because JMF was designed for systems in which it has complete control of the media source, and this is not the case in a broadcast environment.

For instance, a normal JMF-based application playing a media clip from a hard disk can pause the playback and start it later, choose which point in the stream to start the playback from, and fast-forward and rewind through the clip. With a broadcast stream, this is not possible because the receiver has access only to the data currently being broadcast (ignoring PVR devices for now, which are not pure broadcast systems). In a broadcast environment, if you pause the video stream and then start playing it again after a few minutes the video will jump to what is currently being broadcast instead of playing from where it stopped.

This is only one of the areas involving a major change in philosophy between the PC and broadcast worlds. JMF was arguably not the ideal choice for a media control API given its original strong focus on media playback for desktop systems and Internet-connected devices, but it was a lot easier to use JMF than to define something new, particularly because many people have prior experience with JMF.

Content Referencing in JMF

As we saw in Chapter 8, MHP and OCAP both use locators as a way of referring to broadcast content, which could be a video stream, a file in a broadcast file system, or another type of content entirely. Locators are platform-dependent references to local or broadcast content, and the internal representation of a locator may vary between implementations. Every locator will have a standardized form (called the external form) represented by a string containing a URL. This may not be compatible with the `java.net.URL` class in the middleware (i.e., you may not be able to construct a valid `java.net.URL` object from this URL string), but it will follow the format laid down in RFC 1738.

JMF uses its own type of locator to represent content it will play. This is represented by the `javax.media.MediaLocator` class. Like some other types of locators, `MediaLocators` can be created directly from a string containing a URL that refers to the content. In MHP and OCAP systems, these can refer to three different types of content, but all of them use the same basic method for referring to that content. Later in the chapter we will take a closer look at the content formats supported.

Basic JMF Concepts

JMF involves three major concepts you need to understand before you can really use it effectively. The most important element of JMF is a player. This class is actually responsible for decoding and playing a piece of media. Any players in a JMF system are subclasses of the `javax.media.Player` class.

Every player has a set of zero or more controls associated with it. A JMF control is an extension to the player that allows optional functionality to be added to a player without having to create a subclass. For instance, controls are typically used to provide things like freeze-frame functionality or language choice on top of the built-in functions of the `Player` class. Each control will be a subclass of the `javax.media.Control` class.

The final element is the data source. This entity actually gets the media data the player will decode. By separating this from the player, we get a design that enables a player to receive data from more than one source without having to handle all of the various sources itself. Each data source is implemented by a subclass of the `javax.media.protocol.Data-Source` class, and thus the operations performed on the data source are always the same.

Data sources come in two flavors: push data sources and pull data sources. The pull model means that the data source has to request data from a source such as a server. A data source that used HTTP to get its data would be a pull data source, for instance. Push data sources do not need to make an explicit request. As soon as they connect to the server or source of the data that data is transferred automatically. Broadcast streams are a good example of the push model: once the data source connects to the right stream, the data just keeps coming. Whereas desktop JMF implementations make an explicit distinction between the two (the `javax.media.protocol.PushDataSource` class and `javax.media.protocol.PullDataSource` classes, as well as classes representing push and pull data streams), MHP and OCAP do not. Both specifications are silent about how the receiver treats broadcast data, and an application should never make any assumptions about the type of data source in use. The two classes we have just mentioned may be present, but they may not actually do anything.

The `DataSource` class can be considered a little like the unified I/O model in a UNIX platform, wherein everything appears like a file and can be manipulated with file operations. The application (and the player) can manipulate the `DataSource` object and get the media data through a single interface, without having to care if that data is coming from memory, a local hard disk, a broadcast stream, or some other source.

Every player is always associated with a `DataSource` object. It is not possible to create a player that does not have a data source attached to it. We have already seen that data sources for broadcast streams will follow the push model. Data sources that get their data from a file will always be pull data sources, even if that file is stored on a broadcast file system. We will see what impact the push and pull models have on our applications and middleware a little later in the chapter.

Figure 11.1 shows the relationships among the three main components of JMF, using some of the data sources and players you might find in a JMF implementation for DTV. On the

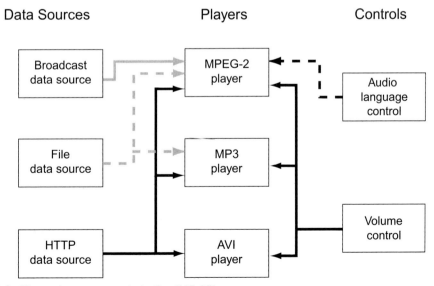

Figure 11.1. The major components in the JMF API.

left, we have three possible data sources from which we can choose. These load data from, respectively, an MPEG stream, a local file, memory, and an HTTP connection. The byte stream we will get out of these data sources will be identical. An application could use any data source and get data in the same way using the same interface. Only the means used by the middleware to load the data would change.

The data source then sends its data to a player. In this example, we have three players from which we can choose. One handles MPEG-2 data, the second handles AVI movie clips, and the final one handles MP3 audio data.

Finally, we have two controls available. The audio language control is only applicable to those players that handle multiple audio tracks, which in this case is the MPEG-2 player. The volume control, on the other hand, is applicable to all three players.

Now that we have seen the overall architecture, let's take a look at how this works in practice. The `javax.media.Manager` class is the main entry point for any application wishing to use the JMF API. For reasons we will explore later, we cannot directly create instances of a player or a data source, and thus we use this class to create them instead. This class has the following interface.

```
public class Manager {

  public static Player createPlayer(
    java.net.URL sourceURL)
    throws java.io.IOException, NoPlayerException;

  public static Player createPlayer(
    MediaLocator sourceLocator)
    throws java.io.IOException, NoPlayerException;

  public static Player createPlayer(DataSource source)
    throws java.io.IOException, NoPlayerException;

  public static javax.media.protocol.DataSource
    createDataSource(java.net.URL sourceURL)
    throws java.io.IOException, NoDataSourceException;

  public static javax.media.protocol.DataSource
    createDataSource(MediaLocator sourceLocator)
    throws java.io.IOException, NoDataSourceException;

  public static TimeBase getSystemTimeBase();

  public static java.util.Vector getDataSourceList(
    String protocolName);

  public static java.util.Vector getHandlerClassList(
    String contentName);
}
```

The most interesting methods are the `createPlayer()` and `createDataSource()` methods. `createDataSource()` creates a new instance of a `DataSource` that will fetch data from the location referred to by the URL or the `MediaLocator` that is passed in. The API implementation may convert URLs to `MediaLocator` instances in order to make life easier, but the result is the same. Exactly which subclass of `DataSource` gets returned depends on the location you are fetching the content from. Content from a broadcast MPEG stream will use a different data source than content fetched from a file, even one that is contained in a broadcast file system.

Similarly, the `createPlayer()` method creates a new `Player` object that can be used to play a media clip. Players can be created either directly from a `MediaLocator` (the easiest way) or from a `DataSource` object. If a player is created directly from a `MediaLocator`, the `Manager` will create a `DataSource` object of the appropriate type first.

The Player Creation Process

Given that a JMF implementation can support multiple data sources, and may have players to handle multiple media types, how do we get from a URL or a locator (or a data source, for that matter) to a player that is presenting a specific type of media from a specific source? Let's consider the process of creating a player from a URL. We will take a URL as our example, because creating one from a `MediaLocator` follows the same basic principles. First, the application calls the static `Manager.createPlayer()` method. In this case, let's assume it calls it with the following URL.

```
http://www.example.org/media/SomeMPEG2Content.mpg
```

The first thing that happens is that the `Manager` examines the protocol part of the URL. The protocol indicates the data source needed to access this content, and the `Manager` uses this to construct the class name of the `DataSource` object it needs to create. This is constructed using the following pattern.

```
<class_prefix>.<protocol_name>.DataSource
```

In this case, the protocol is `http`. The class prefix is simply a fixed string that points to the top of the class hierarchy containing the classes for the data source implementation. Therefore, if our class prefix were `com.stevem.media.protocol` the resulting fully qualified class name would be as follows.

```
com.stevem.media.protocol.http.DataSource
```

Classes representing data sources always have the name `DataSource`. This makes it easy for the middleware to construct the right class name and instantiate the class using the Java reflection API.

Once the class name has been created, the class is loaded and instantiated. The `DataSource` object will be used to get the media data that will be presented. After we have created an instance of the `DataSource` class we need, the `Manager` uses this data source to connect to the location specified in the URL (this process is explored in further detail in material to follow). Once a connection has been established, the `Manager` calls the `getContentType()` method on the data source to find the MIME content type of the data the data source is fetching. The MIME content type is then used to construct the class name for the `Player` class that will be loaded. This is very similar to the process used to create the class name for the data source. The class name takes the following form.

```
<class_prefix>.<MIME_type>.<MIME_subtype>.Player
```

Thus, if the class prefix were `com.stevem.media.players` and the MIME content type `video/mpeg` (because we had connected to a URL for an MPEG-2 file) the resulting class name for the player would be as follows.

```
com.stevem.media.players.video.mpeg.Player
```

Classes implementing JMF players always have the name `Player` to make it easy for the middleware to construct the right class name and load the class. Now that the `Manager` has constructed the class name for the `Player` object, it loads the class and instantiates it. Once the player is instantiated, the manager calls the `Player.setSource()` method to associate the data source with the player. Once it has done this, the `Manager` has a player that is ready for use by the application. Figure 11.2 depicts the player creation process.

If you are an application developer, by now you may be sitting reading this in a cold sweat wondering what this all means for you. Relax; you are safe. From an application developer's perspective, this is not important. Calls to `Manager.createPlayer()` or `Manager.createDataSource()` will return a `Player` or `DataSource` instance you can use, and that is all you have to know about the process.

Middleware developers should not get too worried either. This is not actually as complex as it seems, although you may need to think through a few more concrete examples to understand it properly. The most difficult part of the entire process is getting the MIME type for the content, and in many cases MHP and OCAP make life easier for us by restricting the content formats we have to support.

In many cases, the content type can be determined from the URL, without needing to query the remote server or look at the data itself. Sometimes, looking at the protocol is enough. A `dvb://` or `ocap://` URL will usually represent a service or an elementary stream when we are using JMF (unless it includes a path to a file). For files, we can sometimes determine the MIME type by looking at the file extension. MHP standardizes a number of MIME content types for specific types of content (and file extensions that go with them), and OCAP has adopted some of these.

In these cases, the MIME type can be hard-coded in the `DataSource.getContentType()` method. This makes life easier for the middleware, in that it can avoid having to dig too deeply into the file data to work out what the content type is. Table 11.1 outlines which file extensions have been standardized by MHP and OCAP. Files with certain extensions, however, may still need some analysis of the file before the content type can be identified.

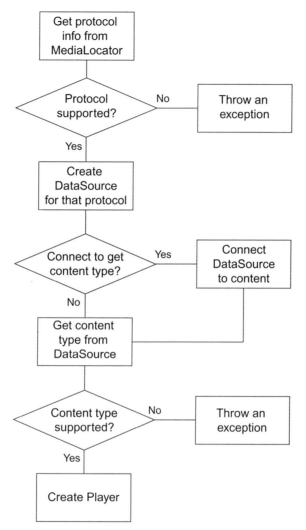

Figure 11.2. Flowchart of the player creation process.

Note that not all of the formats listed in Table 11.1 will be supported by JMF. Some of these are general-purpose file types wherein the MIME type has been standardized for other purposes. Even if the MIME type is used in other APIs, it may not be possible to use JMF read data from files with that extension.

The most obvious example of this is the `multipart/dvb.service` MIME type used by MHP as the MIME type for broadcast DVB services. These will usually not be available as a file in a transport stream, but the MIME type will still be used for every DVB service in the receiver.

Table 11.1. MIME content types defined by MHP and OCAP.

Content Format	MIME Type	File Extension
JPEG image	`image/jpeg`	*.jpg*
PNG image	`image/png`	*.png*
GIF image	`image/gif`	*.gif*
MPEG I-frame	`image/mpeg`	*.mpg*
MPEG-2 video	`video/mpeg`	*.mpg*
MPEG-2 video drips	`video/dvb.mpeg.drip`	*.drip*
MPEG-2 audio data	`audio/mpeg`	*.mp2*
Plain text	`text/plain`	*.txt*
UTF8 text	`text/dvb.utf8`	*.txt*
DVB subtitle data (MHP only)	`image/dvb.subtitle`	*.sub*
	`text/dvb.subtitle`	
DVB teletext data (MHP only)	`text/dvb.teletext`	*.tlx*
Font data file	`application/dvb.pfr`	*.pfr*
DVB-HTML file	`text/xml`	*.xml*
	`application/xml`	
CSS style sheet	`text/css`	*.css*
ECMAScript script file	`text/ecmascript`	*.js*
Xlet class file	`application/dvbj`	*.class*
DVB service (MHP only)	`multipart/dvb.service`	*.svc*

The MIME content type for OCAP services is not standardized. Out of the content types listed in Table 11.1, only MPEG audio, MPEG-2 drips, and DVB or OCAP services need to be supported by JMF. All other content formats will be accessible through other APIs only.

Given the growth of the Internet and the delivery of streaming content over IP connections, some receivers may want to support streaming media from IP-based protocols such as RTP or HTTP. Although this is possible (but not in an interoperable way for now), it adds some to the burden of the JMF implementers. In the case of HTTP, the `DataSource` implementation must read the MIME content type from the HTTP response headers. RTSP data sources should also get this from the RTSP response headers.

A Closer Look at Data Sources

As you will have noticed from the previous discussion, we glossed over a few steps in the player creation process. Probably one of the most important of these is the actual mechanics of the data source. The other is the mechanics of the player, which we will examine in the next section. First, let's take a closer look at the interface to the `DataSource` class, which follows.

```
public class DataSource {

    public DataSource();
    public DataSource(javax.media.MediaLocator source);
```

```
    public void setLocator(javax.media.MediaLocator source);

    public javax.media.MediaLocator getLocator();

    public abstract String getContentType();

    public abstract void connect()
       throws java.io.IOException;

    public abstract void disconnect();

    public abstract void start() throws java.io.IOException;
    public abstract void stop() throws java.io.IOException;
}
```

This is not very complex, but it is important to understand it. The first thing to notice is the two constructors. Although a `DataSource` object can be constructed directly using a `MediaLocator`, this is not the typical process by which it is created. The middleware does not usually know which class it will need to load until it parses the URL, and thus we need to take an alternative approach. The class name is constructed as a string, and thus it is usually loaded and instantiated using the `Class.newInstance()` method from Java's reflection API.

For this reason we have separate methods to set and get the `MediaLocator` associated with the `DataSource`, as well as being able to set it using the constructor. In those cases in which we call the default constructor (e.g., when `Class.newInstance()` is used to create the object), we can set the `MediaLocator` after creation. Each `DataSource` object must have an associated locator before we can do anything useful with it. Without this, it is the JMF equivalent of a null reference and is not terribly useful. Applications should never set the locator themselves, however. This is the responsibility of the middleware, and the middleware should never create a `DataSource` object without setting a compatible `MediaLocator`.

Once the middleware has set a locator, neither the middleware nor the application can change it. The way `DataSource` objects are created means that this would add a great deal of overhead to the use of data sources. The class that handles the data source is very dependent on the protocol of the associated `MediaLocator`, as we have seen. Although we can trust the middleware to create a `DataSource` object of the correct type and set a compatible `MediaLocator`, we cannot trust an application to do this. We would have to check that any `MediaLocator` is compatible with that `DataSource` object.

Although that does not seem so bad, things get worse when we take the next step. The `DataSource` is associated with a `Player` that decodes a specific media type, and the middleware learns the MIME type of that media from the data source, possibly based on (you guessed it) the media locator. We can verify that a `MediaLocator` uses the same protocol as the one used by an existing `DataSource` simply by parsing the URL. Working out whether it is compatible with the associated `Player` may take a lot more work, and it could take a long time and use scarce resources that are not available at that time.

Given these problems, it is easier to impose a total ban on applications changing the `Me-diaLocator` of an existing `DataSource`. We can trust the middleware to get it right under specific conditions (e.g., when there is no `Player` associated with that `DataSource` yet, and when it can be certain that it has created a `DataSource` instance of the right class), but there is no way we can guarantee that an application will get it right. In that creating a new `Data-Source` object is a trivial operation, it is easier to force applications to do this instead and make sure that all of the objects used by JMF are of the correct type and in the correct state.

After we set the locator, the middleware needs to connect the data source to the location specified by the locator. Until it has done this, we have no way of accessing the data. This method may do different things depending on the data source. For instance, in a file data source the `connect()` method will open the file, whereas in an HTTP data source the `connect()` method will set up an HTTP connection to the URL specified in the locator. By the time the `connect()` method returns, the data source will be ready to start fetching data.

Both middleware implementers and application developers should take care to remember that the `connect()` method may reserve any scarce resources the data source needs to make the connection. If a data source is connected when it does not need to be, it may be depriving another application or middleware component of a resource it needs. Ideally, applications and middleware implementations should connect a data source as late as possible before it is needed for media playback, and release it as soon as possible after media playback is stopped. This is not always practical, but it is a good guideline to follow if you can.

Having made a connection, we can actually find the type of data the locator refers to by using the `getContentType()` method we have already seen. In a PC-based JMF implementation, it is almost impossible for the JMF implementation to know the content type of the data until a connection exists. For example, it may need to get the information from the HTTP headers or even from header information contained in the data itself.

In a JavaTV, OCAP, or MHP implementation, things are a little different and are generally simpler. We have already seen that there is a limited number of standard data sources and content types supported by the receiver, and that the content type can sometimes be known just from the type of data source used to access it. The data source may need to make other checks, however, such as checking that this is indeed a valid URL (e.g., that it points to a service or an audio or video elementary stream, rather than to a data stream), but this is not related to finding the content type. A valid `dvb://` or `ocap://` URL that does not include a path component will always refer to the same content type.

Middleware stacks based on other standards may use a different protocol and different content types for accessing broadcast services (ACAP uses the `acap://` scheme, for instance). Because the JavaTV specification says nothing about the format of the locators to be used, a project using a pure JavaTV middleware stack may need some further implementation guidelines to define the MIME content types unless there is a way of getting this information from the protocol being used.

The `start()` and `stop()` methods on the data source start and stop, respectively, data transfer. It is only when the `start()` method is called that the data source actually has data it can pass to a player. Obviously, the data source must be connected to the actual source of the media data using the connect() method before start() can be called.

Once the data source is no longer in use, the disconnect() method can be called to disconnect the data source from the actual source. The advantage of doing this is that if some scarce resources are needed to keep the connection alive (e.g., a modem connection to a server) the resources can be explicitly released when they are not in use. By separating the connection/disconnection process from the process of actually getting data, the time-consuming parts (such as connection setup) can be carried out before the data is actually needed, thus saving time when the data is required.

JMF Players

We have seen how to create a player, and thus now it is time to see how we can actually use it. First, let's take a look at the complete interface to the Player class, as follows.

```
public interface Player
    extends MediaHandler, Controller, Duration {

    public abstract void setStopTime(Time stopTime);
    public abstract Time getStopTime();
    public abstract void setMediaTime(Time now);
    public abstract Time getMediaTime();
    public abstract long getMediaNanoseconds();
    public abstract Time getSyncTime();

    public abstract float getRate();
    public abstract float setRate(float factor);

    public abstract int getState();
    public abstract int getTargetState();

    public abstract void realize();
    public abstract void prefetch();
    public abstract void start();
    public abstract void syncStart(Time at);
    public abstract void stop();
    public abstract void deallocate();
    public abstract void close();

    public abstract Control[] getControls();
    public abstract Control getControl(String forName);
    public abstract GainControl getGainControl();

    public abstract void setSource(
        javax.media.protocol.DataSource source)
        throws java.io.IOException,
            IncompatibleSourceException;
```

```
    public abstract void addControllerListener(
        ControllerListener listener);
    public abstract void removeControllerListener(
        ControllerListener listener);
}
```

As you can see, this is a big interface, and we have only shown part of it here. Before you start panicking too much, do not worry. You do not need to understand or even know about most of the things a player can do. If you are really interested, there are other books out there that cover the topic of JMF in much more detail.

There are some differences between players in DTV middleware implementations of JMF and players in the desktop PC version. The first of these differences is that some of the other standard features of players in a desktop implementation are not available. These include such items as the `setSource()` method (inherited from the `MediaHandler` class). We have already seen why setting the data source in this way is not practical.

Another difference is that the selection of media may be driven more by user preferences or platform settings than would be the case in a desktop implementation. This includes items such as the choice of audio track or subtitles based on user preferences and the language settings in the receiver firmware.

The largest difference, however, is that players in a desktop JMF implementation will usually have a user interface or control panel attached, whereas in a DTV receiver or other consumer device they probably will not. The `Player.getVisualComponent()` and `Player.getControlPanelComponent()` methods will usually return a null reference in MHP and OCAP, because the media is typically played in the video layer. Chapter 7 discusses how the various layers of the display relate to one another and to any video that is presented.

It is possible for `getVisualComponent()` to return an AWT `Component` object under certain circumstances, however. Some high-end receivers may allow the true integration of graphics and video, and in these cases `getVisualComponent()` may return a component that can be used by the application to handle scaling and positioning of the video. We will look at this in more detail later in the chapter.

The Player State Machine

A player can be in one of several states, depending on what it is doing at the time. A player can be in one of the following five major states.

- Unrealized
- Realized
- Prefetched
- Started
- Closed

In addition to these, a player may temporarily be in one of several minor states. We will take a closer look at these minor states later.

When a player is first created, it will be in the Unrealized state. Calling the `realize()` method moves the player into the Realized state, whereby the player has all the information it needs in order to get the resources needed for playing a media clip. This means that the player knows what type of data is in the clip, and finding this may require some scarce resources.

The `prefetch()` method moves a player in to the Prefetched state. At this point, the player has got all the resources it needs (including scarce resources) to play the media, and it has also fetched enough of the media data to be able to start playing immediately. In PC implementations of JMF, this state lets the player download enough data from a remote connection to start playing a clip as soon as the `start()` method is called. Thus, an application can know that network latency issues or other sources of delays will not become an issue when the player is started.

This is not a problem for broadcast streams, but it probably will be for audio clips loaded from a file. The file may have to be loaded from a broadcast file system (for which latency can be very high), and thus using the `prefetch()` method allows applications to handle some of these potential delays effectively.

The last useful state is the Started state, which is entered when the `start()` method is called. When a player is in the Started state, it is actually playing media and is doing everything you would expect.

The final state for any player is the Closed state. When an application calls the `close()` method on the player, it will stop the player, release any resources it was using, and permanently dispose of any data structures created by the player. Once a player enters the Closed state, it cannot be used again and calling any methods on that player may generate an error.

In most cases, an application will not care about these states very much. It will just be interested in playing some video or audio, possibly prefetching the media first to avoid any delays. Even though JMF gives us this control over the state of a `Player`, we do not have to use it. Each of the methods we have seen here (`realize()`, `prefetch()`, and `start()`) can be called from any preceding state, and they will automatically call the other methods as necessary to get the player into the appropriate state.

These methods may take some time to execute due to latency in the receiver or in the broadcast media. For instance, we have no idea how long it may take to prefetch an audio file from a broadcast file system. For this reason, each of these state changes is asynchronous. Whenever one of the three methods that causes a forward state transition (`realize()`, `prefetch()`, or `start()`) is called, the player goes into an intermediate state (Realizing, Prefetching, and Starting states, respectively) and the method returns immediately. These are the minor states we mentioned earlier, and the player will automatically transition out of these states when it can move to the next major state. As we will see in material to follow, the player uses events to tell any interested applications when it moves between states.

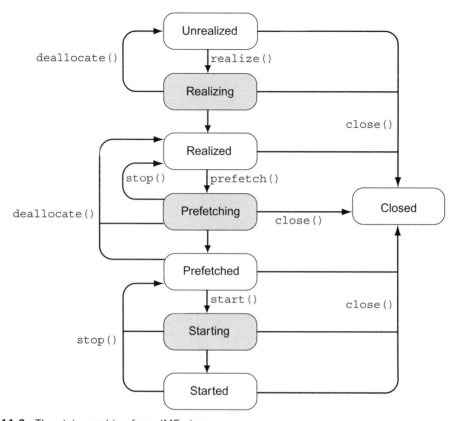

Figure 11.3. The state machine for a JMF player.

State transitions the other way, however, are all synchronous. The `stop()` method takes a started player to the Prefetched or Realized state, depending on the type of media being presented. The `deallocate()` method is used to free any scarce resources the player may have acquired for prefetching or playing the content. Calling `deallocate()` when the player is in the Realized state (or higher) causes the player to return to the Realized state. If a player is not yet in the Realized state, it will return to the Unrealized state. The `close()` method can be called from any other state, and will always take the player to the Closed state. This is the final state for any player, and once a player is in the Closed state it cannot go back to any of the other states. Figure 11.3 shows the full state model for a `Player`.

The shaded states will automatically transition to the next state when all necessary steps have been carried out.

Player Events

As with other asynchronous APIs, JMF uses events to notify interested applications about changes in the state of a player. The `addControllerListener()` method lets an application register to receive events from the player. These will all be subclasses of the `Con-trollerEvent` class.

Many of the events generated correspond to state transitions within the JMF player. The `javax.media.TransitionEvent` and its subclasses (`PrefetchCompleteEvent`, `RealizeCompleteEvent`, `StartEvent`, `StopEvent`, `DeallocateEvent`, and `ControllerClosedEvent`) correspond directly to state transitions within the player, and the `StopEvent` class has subclasses that indicate why the player stopped. These subclasses are outlined in Table 11.2.

MHP and OCAP add the `org.davic.media.MediaPresentedEvent` to the list of JMF events, which is generated when the player actually starts presenting media to the user. This may be subtly different from the time the player starts due to the buffering model of the MPEG decoder, and thus the separate event allows applications to synchronize exactly with the start of media presentation.

JMF defines other events to notify the application about problems with the connection to the data source, or about changes to the player such as a change of rate or a change in the media time the player should stop at. These are too complicated to discuss here in any detail, and thus you should look at the JMF specification to learn the gory details. Only certain subclasses of `Player` will generate these events, however. Players for broadcast services will typically generate fewer events because the operations that generate them will have no meaning for broadcast players.

The `org.davic.media` package also defines the `ResourceWithdrawnEvent` and `ResourceReturnedEvent`, which allow an application to know about resource issues related to the player that cause changes to the content being presented. These events do not affect the state of the player (unlike the events listed previously) because even though a player has lost some resources it may still be rendering some content. We will take a look at these events in more detail later in the chapter, when we take a closer look at resource management issues.

Table 11.2. Subclasses of `javax.media.StopEvent`.

Event	Reason
`DataStarvedEvent`	The player stopped because data was not delivered quickly enough from the data source (or because there was no more data to deliver).
`EndOfMediaEvent`	The end of the media clip was reached.
`RestartingEvent`	The player has stopped so that it can process a change of rate or change in media time and needs to prefetch more data before it starts playing again.
`StopAtTimeEvent`	The player has reached the stop time set for it and has stopped automatically.
`StopByRequestEvent`	The player's `stop()` method has been called.

The `org.dvb.media` package included in MHP and OCAP adds the `Presentation-ChangedEvent`. This deals with the situation (almost unique to DTV) in which the content being presented may change due to events outside the control of the application. This could be due to the end of a free preview period, for instance, or where the set of streams changes at the end of an event because an audio language is no longer available. Applications have no other way of knowing that the presented content has been changed, and thus this event gives the application a chance to find out about these changes and react to them.

Time Bases, Clocks, and Media Time

It is generally useful to have some concept that tells you how far through the media clip you are, and thus most APIs for playing streamed media implement the concept of current media time. JMF is no exception in this, although the concept of media time is often different in the broadcast world. Like most other media control APIs, JMF has a set of methods in the `Player` interface that let an application get the current media time, and then jump to a new point in the clip and set how fast a clip should be played.

This is fine for a PC application in which we can download the entire video clip before we start watching it, or where the video clip at least has a definite start and end, because we can calculate the elapsed time since the beginning of the clip. Unfortunately, this model does not work very well in a broadcast environment. If the media is always being streamed, what do we mean by media time? Is it the time since the start of the event? The time since the user selected the current service? The time since the receiver was last switched on? Potentially, it can be anything the broadcaster or box manufacturer wants!

For broadcast streams wherein the stream is running 24 hours a day, seven days a week, the concept of a media time that reflects the time since the start of the clip is meaningless. The only thing that may be useful is the time since the start of the current show, and even that has complications. Should that time include advertising breaks? What happens if the show is edited, so that some action that is 10 minutes into the show on one network may only be 9 minutes into the show on a different network?

There is no nice way around these problems, and thus MHP and OCAP take the simplest approach and simply ignore the issue. In both standards, the media time for broadcast streams is defined as a time value that always increases in a way that is platform dependent, and that may or may not resemble any values from the real world. Any time value will be meaningless to the application, and thus we might as well use something completely arbitrary from the very beginning.

Despite this, media time in MHP and OCAP is not completely useless. After all, we have got media clips from other sources that do have a definite start and end time, such as audio clips. For these clips, the media time behaves as you would expect it to, and an application can use it to tell how much of the clip has been played or to tell the receiver to jump to a different part of the clip. The `Player.setMediaTime()` method tells the player it should jump to that point in the media, and depending on the state of the player at the time it will either continue playing the media from that point or start playing from that point when it is finally started. Applications can also use the `setStopTime()` method to choose a media time at which a clip should stop playing.

JMF uses two elements to work out what the current media time actually is for a clip and how that time changes. The first of these is a time base. This is simply a time reference (a free-running clock) that increases at a constant, unchanging rate. When we say constant, we mean constant: the time base will continue to run even if the player it is associated with has been stopped. An MHP receiver may have several different time bases existing at the same time, depending on what media types it is handling. The actual value of the time base does not have any direct relationship to the media time value. It simply provides a steady set of ticks that tell the player when the media time should change.

The other part of the calculation is a clock. This is used together with the time base to tell the player how much media time has passed for each increment of the time base. The JMF `Player` class is a subclass of the `javax.media.Clock` class, and thus every player acts as its own clock. Several clocks may share the same time base, however.

Each clock has a rate, which is a floating-point value that defines the mapping between the time base and the media time. If the rate is 1.0, the media is being played at normal speed. A rate of 2.0 indicates double-speed playback, and a rate of 0.0 means that the clock (and the player) is stopped. The rate does not have to be positive. A rate of –1.0, for example, means that the stream is being played backward at normal speed. Similarly, rates between 0.0 and 1.0 tell the player to play a clip in slow motion.

We can set the rate for a given player using the `Player.setRate()` command. If the player is started, this will cause the rate of playback to change according to the new rate, whereas in other cases the new rate will be used when the player is started. Players do not have to support every rate, and thus the middleware will set the actual rate to the one that is closest to the desired rate. `setRate()` returns a floating-point value that gives the actual rate that was set, so that the application can see if the receiver has not been able to exactly meet the request. Figure 11.4 depicts the mapping function of clock rate.

This is one of the main areas of difference between JMF in DTV systems and other JMF implementations. Unlike conventional PC media types, you cannot stop a broadcast stream. Similarly, you are very limited in how you can change the rate of the player for a broadcast stream. In some cases, like a hard-disk-based PVR, you may be able to rewind the stream or play a previously recorded stream slower or faster than the normal rate. Despite this, once you reach the part of the stream currently being broadcast you cannot advance any further.

Given the nature of broadcast streams, MHP and OCAP both only require a JMF player for broadcast content to support a rate of 1.0 (normal playback) and 0.0 (stopped). Other rates may be supported, but applications should not count on it.

At the same time, built-in applications may be able to use different rates to control PVR functionality, and may even be able to use the media time value of recorded clips to calculate how to jump to a specific piece of content. Just because the media time is arbitrary does not

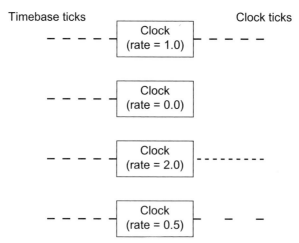

Figure 11.4. The rate of a clock controls the mapping between time base ticks and clock ticks.

mean it cannot be useful in some cases. If the application knows how the media time value changes in a particular implementation, it may be able to make use of it.

There is much more to clocks and time bases than we are discussing here, in particular the role of clocks and time bases when synchronizing several players so that they can be controlled as one. We will take a short look at this later in the chapter.

DSM-CC Normal Play Time

If we actually need to use the media time, what options do we have? JavaTV, MHP, and OCAP are silent about what media time actually means, which solves the problems we saw earlier but is not very helpful.

Luckily for us, MHP and OCAP do give us a possible solution. Even though the JMF media time is almost useless, broadcasters can insert their own media time value into a service and applications can use this to work out what point they are currently at in the broadcast.

To do this, we need to use DSM-CC normal play time (NPT) values. These are time codes embedded in an elementary stream as descriptors. They tell the receiver the value of the current media time. We will discuss this more in the next chapter, and see how we can use them for synchronizing applications with broadcast content.

Controls

We saw earlier in the chapter that JMF uses the concept of a control to add extra functionality to a player. Each control is a subclass of the `javax.media.Control` class, and each control typically adds one specific piece of functionality. This approach has two immediate advantages from the point of view of object-oriented development. First, it allows us to generalize the implementation of a common control, and thus several different players may use a single control class that implements a common function such as a volume control. Second, there is a great deal of separation between the player and its controls. There are no standard interfaces between a player and a control, and thus the player does not have to implement a big API that may not be used. This interface is completely private and thus implementers can define the interface that best suits their needs for individual controls.

This approach separates the functionality into two main areas. The player is responsible for controlling the process of getting media, decoding it, and presenting it at the right time. Controls are responsible for manipulating how that media is presented, and for making changes that do not affect the basic life cycle of the player. This allows us to add some important functionality without causing problems with scarce resources, as we will see later in the chapter.

By keeping controls separate from the player, we can keep a simple interface for basic players and use controls to extend the functionality on more advanced platforms. It also means that we can take an incremental approach to building an implementation of JMF. Middleware implementers can concentrate on developing the basic player functionality first, and then add extra controls to perform advanced manipulation of the content when they are sure that basic functions are working, or when they are sure the platform is powerful enough to support them.

This modular approach also makes it very easy for application developers to see which functionality is present in a given receiver and change the behavior of their application accordingly. Later in this section we will explore how controls can add features such as video scaling and positioning, language selection, and finer control over media playback.

The use of controls to add functionality means that the interface to a particular player is very extensible. The functionality of that player can even change while the application is running, as some controls become available and others are no longer useful. This rarely happens in practice, but there are a few cases in an MHP or OCAP implementation in which you may encounter this.

To get a list of available controls, applications can call the `Player.getControls()` method. This returns an array containing all controls available for that player. JMF itself defines a couple of controls that may be available, depending on the platform and the type of player, but a lot more are added by DAVIC, MHP, and OCAP (explored in material to follow). Alternatively, we can use the `getControl()` method to get a specific control. By passing the fully qualified name of a control as an argument, we can find out whether the player supports the control in question. If it does, we get a reference to that control; if not, we get a null reference.

Applications should not try to do things except via the predefined controls, because any other interface is nonstandard by definition. This is especially true for consumer systems, for which there is no de facto standard implementation.

JMF Extensions for DTV

The biggest changes from the standard desktop JMF implementation come when you start looking at the extra controls the various DTV standards have defined. JavaTV defines a couple of extra JMF controls, whereas MHP and OCAP define several more that applications may use. Some of these are new to MHP and OCAP, and some are borrowed from the DAVIC 1.4 specification.

These controls are largely designed to fix the problems caused by the differences between the PC and broadcast environments. Let's meet some of the more useful ones.

javax.tv.media.MediaSelectControl

This control gives applications more detailed control over the presentation of the elementary streams that make up a DTV service. The `MediaSelectControl` allows an application to add or remove streams from the set of streams being presented by a player, and allows one elementary stream to be replaced with another (e.g., replacing the current video stream with one showing a different camera angle). This can be extended to replace every part of the stream if the application developer so desires.

Because controls do not affect the state of the player, using a control to change the set of streams being presented can have a number of advantages. There is no need to stop the player (or worse, create a completely new player) just to change the service components being presented, and thus there is no chance that resources will be taken away from the player during this operation.

The biggest limitation of this control is that it can select new components only from within the current service. Being able to select components from a different service would introduce a number of possible conflicts with the basic functionality of the player and would go against the philosophy of JMF (namely, create a new player if you want to play a new clip).

The `MediaSelectControl` lets you switch between any elements of the service, but limitations elsewhere in the system mean that you cannot always be as flexible as you would like to be. Although you could use this control to switch between different video streams, this can be very difficult in practice. Synchronizing multiple video streams to a single audio stream is extremely difficult, and thus the best way of supporting multiple camera angles is to broadcast the various video streams as separate services. This uses slightly more bandwidth, but it is a much easier way of achieving the desired effect. Other solutions usually entail technical problems for both content production chain and receiver that outweigh any bandwidth requirements.

javax.tv.media.AWTVideoSizeControl

This control allows an application to control how video is scaled and positioned in the video layer of the graphics hierarchy (see Chapter 7 for details of the graphics model and the various layers). This does not include any video being displayed in the graphics layer in an

AWT component. In these cases, applications should use the normal AWT methods for changing the size and position of a component.

This has probably left you wondering why it is called an AWTVideoSizeControl when it has nothing at all to do with AWT, and to be honest we cannot give you a good answer. It made sense when the API was being developed, but during the development of the JavaTV specification things changed enough that the original name is not very appropriate.

All of the scaling and resizing operations carried out with this control take place in the screen coordinate space. As we saw in Chapter 7, this is the full-screen coordinate space used by the video device. For an MHP receiver, this will have a resolution of 720×576 pixels. OCAP devices will support a resolution of at least 640×480 for standard-definition services, and possibly higher resolutions for high-definition services.

To use this control, an application creates a javax.tv.media.AWTVideoSize object to represent the operations it wants to carry out. This class takes two rectangles as arguments to the constructor. The first of these is the source rectangle, which represents the area in the source video stream that should be displayed. The second is the destination rectangle, which tells the receiver where on the screen this content should be displayed, and at what size. The AWTVideoSize object will automatically calculate the transformations needed, and thus applications do not have to worry about the messy details of this. The interface to the AWTVideoSize class is as follows.

```
public class AWTVideoSize {

  public AWTVideoSize(java.awt.Rectangle source,
    java.awt.Rectangle dest);

  public java.awt.Rectangle getSource();
  public java.awt.Rectangle getDestination();
  public float getXScale();
  public float getYScale();
  public int hashCode();
}
```

As you can see, we can also query the receiver to find out what scaling factors are applied to the video because of this transformation. The AWTVideoSizeControl.setSize() method lets us apply these transformations to the video. Alternatively, we can get the current transformation being applied to the video by calling the getSize() method. We can also find the default transformation the receiver will apply using the getDefaultSize() method. This is a handy way of removing any transformations we have applied, such as when our application exits.

Using AWTVideoSizeControl is not as simple as just setting a size for the video, however. The receiver hardware may impose limitations on the positioning and scaling of the video, and the setSize() method will fail if it cannot support the transformations needed. To get around this, application developers should always check that a transformation is supported by calling the checkSize() method. This takes an AWTVideoSize object as an argument, and returns the closest match the receiver supports. By doing this, applications can be sure that any transformations will work, even though they may not be exactly what the application wants.

org.dvb.media.VideoPresentationControl

This control is used by players that are presenting the video in an `HVideoComponent` to control how the video is clipped. Although we can use AWT methods on the component to tell the middleware where to display the video and at what size, we cannot use them to tell the middleware how to clip the video. This is where the `VideoPresentationControl` comes in. This lets us set a clipping region on the decoded video, and only that part of the video will then be visible in the `HVideoComponent` (after having been scaled to fit the component as closely as possible). The `setClipRegion()` method takes a `java.awt.Rectangle` instance that represents the region that should be visible.

The receiver hardware may not support arbitrary clipping for video, and thus `setClipRegion()` will return the clipping region that was actually set so that an application knows how closely its request has been met. This can be important for positioning other components relative to the decoded video. For the same reason, we can query the `VideoPresentationControl` to find out the positioning and scaling capabilities of the receiver, so that our application can make sure it manipulates the `HVideoComponent` only in ways the receiver can support.

It is important to remember that setting a clipping region on the source video does not mean that the clipping region will always stay in that location. When the receiver is in pan-and-scan mode (converting 16:9 video to 4:3 by cropping parts of the video), any pan-and-scan information in the MPEG stream will cause the clipping rectangle to move. It will not be resized, but it may change position with respect to the incoming video in order to follow the pan-and-scan vectors.

Another useful feature of this control is the ability to convert between coordinate spaces. Component-based players are positioned using the coordinate space of the application's `HScene`, and sometimes we need to know where that is in other coordinate systems. The `VideoPresentationControl` gives us the following four methods for finding out the normalized coordinates of the video being displayed.

- `getTotalVideoArea()` returns an `HScreenRectangle` representing the entire video area (including any lines used for nonvisible data such as VBI data), whether it is in the viewable area of the screen or not.
- `getTotalVideoAreaOncreen()` returns an `HScreenRectangle` representing the entire video area positioned in the visible area of the screen and visible to the user.
- `getActiveVideoArea()` returns an `HScreenRectangle` representing the active video area (only those pixels containing visible data), whether it is in the viewable area of the screen or not.
- `getActiveVideoAreaOnScreen()` returns an `HScreenRectangle` representing the part of the active video area visible to the viewer.

Because this operates only on component-based players, most applications will use the `BackgroundVideoPresentationControl` instead.

org.dvb.media.BackgroundVideoPresentationControl

This control extends the `VideoPresentationControl`, but is only used with video presented in the video plane (i.e., the player associated with this control must be a background player). Rather than setting the size and position of the video using AWT, this control lets us set the size, position, and clipping region using a single method call.

Like the `AWTVideoSizeControl`, this control allows the user to control the location on screen of decoded video. Instead of operating in the screen coordinate space, this control operates in the HAVi normalized coordinate space, which may be more consistent for the application than using the screen coordinate space. The `setVideoTransformation()` method lets us apply a set of transformations to the decoded video, whereas `getVideo-Transformation()` tells our application what transformations are currently being applied.

As you can see, this approach is conceptually similar to that used by the `AWTVideoSize-Control`. Instead of using an `AWTVideoSize` object to set the transformation, this control uses an `org.dvb.media.VideoTransformation` object, as follows. This is very similar to the `AWTVideoSize` class, but the parameters are slightly different and the transformations must be partially calculated by the application.

```
public class VideoTransformation {

    public VideoTransformation();

    public VideoTransformation(
        java.awt.Rectangle clipRect,
        float horizontalScalingFactor,
        float verticalScalingFactor,
        org.havi.ui.HScreenPoint location);

    public void setClipRegion(java.awt.Rectangle clipRect);

    public java.awt.Rectangle getClipRegion();

    public float[] getScalingFactors();

    public void setScalingFactors(
        float horizontalScalingFactor,
        float verticalScalingFactor);

    public void setVideoPosition(
        org.havi.ui.HScreenPoint location);

    public org.havi.ui.HScreenPoint getVideoPosition();

    public boolean isPanAndScan();
}
```

Instead of taking a source and destination rectangle as arguments to the constructor, the `VideoTransformation` takes as arguments a pair of scaling factors, a clipping rectangle in the source video, and a position for the clipped and scaled video. The clipping rectangle in the source video operates in the screen coordinate space, just like the `AWTVideoSize` class, but the destination location is given by an `org.havi.ui.HScreenPoint` object. Applying a transformation has the same effect as making three separate method calls on the control to set the position, size, and clipping region.

Like the `AWTVideoSizeControl`, limitations in the receiver mean that we may not be able to apply the exact transformation we want to. The `getBestTransformation()` method takes a `VideoTransformation` object as an argument, and returns a new `VideoTransformation` object that represents the closest transformation the middleware can support.

Like the `AWTVideoSizeControl`, we cannot use this control to adjust the size or position of video displayed in an AWT component. In this case, applications must use the normal positioning and resizing operations on the `Component` object.

org.dvb.media.VideoFormatControl

As we mentioned when we looked at the graphics APIs in Chapter 7, one of the problems of a TV environment is the sheer number of different picture formats and aspect ratios receivers can handle. The display may have an aspect ratio of 4:3, 14:9, or 16:9. The incoming video may also use any one of these aspect ratios. We also have to deal with all of the conversions between these: letterboxed, anamorphic, pan and scan, pillar-boxed, and many others.

To display all of these possible combinations, the receiver uses a technique called decoder format conversion (DFC). This resizes the incoming video based on what the receiver can support and what the user prefers. 16:9 broadcasts may either be cropped or letterboxed, for instance, depending on how the user wants to display them on a 4:3 screen. Similarly, 4:3 images may be stretched for display on a 16:9 screen, or they may be displayed with black bars on each side. Even how they are cropped or stretched may depend on the decoder format conversion.

If you have seen a widescreen TV that lets you choose the picture format, you have seen something similar to decoder format conversion in action. The only difference is that decoder format conversion is carried out by the receiver rather than by the TV.

The `VideoFormatControl` provides a way for the application to find out some information about the DFC being applied to the video. This process takes the incoming signal and transforms it into whatever format the receiver should display. In addition to providing a way for the application to get the format of the original video from the active format descriptor in the MPEG stream, it allows the application to monitor for changes in the active format description, the aspect ratio of the output, or the decoder format conversion that is applied.

Decoder format conversion will be set in one of two ways. Most of the time, it will be set automatically by the receiver, based on the format of the incoming video and any preferences the user has set for the output format. This is known as the TV behavior control.

When the application sets a transformation on the video (a scaling, positioning, or clipping operation, or any combination of these), this takes over from any decoder format conversion that has been automatically set. This is known as application behavior control. In this case, only the transformations set by the application will be applied to the video.

An application can query which DFC is currently being applied by calling the `getDecoderFormatConversion()` method. This returns a constant that indicates the type of conversion being applied. Similarly, the `getActiveFormatDefinition()` method returns a constant describing the active format description currently being transmitted for the incoming video. Application developers should be aware that OCAP and MHP define different sets of constants for the possible DFCs.

Sometimes, an application may want to apply a specific DFC to the video. By calling the `getVideoTransformation()` method with the constant representing the desired conversion, it can get a `VideoTransformation` object that can then be used with the `BackgroundVideoPresentationControl` to apply that conversion.

The process of decoder format conversion in an MHP receiver is based on the EACEM E-Book, and is not very clear. This is a known problem with the MHP specification, and DVB hopes to fix this in a future revision of the standard when enough people with the right knowledge can be gathered. OCAP refers to ANSI/SCTE specification 54 instead.

org.davic.media.LanguageControl

Subclasses of the `LanguageControl` class control allow the application to choose which language the receiver uses for audio tracks and subtitles. By default, the receiver will make those choices based on a number of user preferences, and thus in most cases the application will not have to make any changes to ensure that viewers get what they want. In some cases, however, the application may need to change some of these settings itself.

The `org.davic.media.AudioLanguageControl` allows the receiver to choose which language is used for the audio track, if more than one language is available. OCAP provides another way to do the same thing using the `MediaSelectControl`, in that locators in OCAP can specify the type of stream and the language that should be used. Using this approach is simpler and clearer, however, and is also more compatible with MHP systems for which other approaches are not possible.

MHP also includes the `org.davic.media.SubtitlingLanguageControl`, which lets the application choose which language should be used for subtitles using a similar technique. Because OCAP uses a different subtitle format, OCAP receivers use the `org.ocap.media.ClosedCaptioningControl` for any subtitle manipulation.

org.ocap.media.ClosedCaptioningControl

As we have just mentioned, this OCAP-specific control lets the application control the presentation of closed-caption subtitles. An application can set whether subtitles should be displayed or not via the `setClosedCaptioningState()` method, and can register for events regarding the presentation of closed-caption services. Applications can also use the `setClosedCaptioningServiceNumber()` method to select the closed captioning service that is presented. The `ClosedCaptionAttributes` class can be used in conjunction with this control to change how closed-caption services are presented. Applications should be very careful when doing this, however. In U.S. markets the FCC has set very strict rules about how closed captions should be displayed, and application developers should be extremely careful to follow these rules if they change the way closed captions are presented.

The `ClosedCaptioningControl` has changed significantly in recent versions of OCAP, and thus the functionality we describe here may not be available in older versions of the OCAP middleware. As we have already mentioned, these are only some of the controls available. The full list of controls that may be present for MHP and OCAP receivers is outlined in Table 11.3.

Not all of these controls are mandatory for all content formats, however. In particular, `org.davic.media.MediaTimePositionControl` is only required for nonbroadcast media clips. Similarly, players for video drips only need to support `BackgroundVideo-PresentationControl` and `AWTVideoSizeControl`.

Similarly, some of these controls will only have an effect when the player is in the correct state. Using the `FreezeControl` on a player in the Realized state will probably have no effect, for instance. `VideoPresentationControl`, `BackgroundVideoPresentation-Control`, and `AWTVideoSizeControl` will all control a player no matter what its state, and any scaling, clipping, or positioning operations will apply even when the state of the player changes (until the player is deallocated or closed). Depending on the implementation,

Table 11.3. JMF controls supported by MHP and OCAP receivers.

Control	Mandatory in MHP	Mandatory in OCAP
org.davic.media.LanguageControl	✓	✓
org.davic.media.AudioLanguageControl	✓	✓
org.davic.media.SubtitlingLanguageControl	✓	
org.davic.media.FreezeControl	✓	✓
org.davic.media.MediaTimePositionControl	✓	✓
org.dvb.media.VideoPresentationControl	✓	✓
org.dvb.media.BackgroundVideoPresentationControl	✓	✓
org.dvb.media.VideoFormatControl	✓	✓
org.dvb.media.DVBMediaSelectControl	✓	✓
org.dvb.media.SubtitlingEventControl	✓	
javax.tv.media.AWTVideoSizeControl	✓	✓
javax.tv.media.MediaSelectControl	✓	✓
org.ocap.media.ClosedCaptioningControl		✓

the effect of other controls may or may not apply after a state change. If the `FreezeCon-trol` has been used to stop the video before it is scaled or positioned, the middleware should scale, position, and clip the frozen video as appropriate. Not all receivers will be able to do this, however, and in this case the video will resume playing long enough for the middleware to get the next frame and scale, position, or clip it.

Using Players to Control Players

One thing we can do using JMF is to use one player to control other players. This means we can synchronize different players to each other, and start or stop them at the same time.

The JMF `Player` class is a subinterface of the `javax.media.Controller` interface, which is in turn a subclass of `javax.media.Clock`. We have already seen what a clock does, and the `Controller` interface extends this with the basic state model and resource management model used for players. From our perspective (both as application developers and middleware implementers), we will probably never directly use an object that only implements the `Controller` interface. Instead, we will use a `Player` instance to do everything we want.

The main piece of functionality a player has and that a controller does not is the ability to control other controllers. What this means is that one player can take responsibility for controlling the state and media time of several other players. The `Player.addController()` method takes a `Controller` object as an argument, and tells the player that it should manage the life cycle of that controller as well as its own. One player can have several other controllers added to it, and they can be removed using the `Player.removeController()` method. In practice, these controllers will all be other `Player` objects, and thus this gives applications or the middleware a useful way of controlling several players at once.

Why is this useful? Imagine that your middleware implementation uses separate `Player` objects for presenting the video and audio associated with a service. Every time you want to start or stop the player, you have to carry out the same operation on both players. By giving one player control over the other, you only have to call these methods once to get the same effect. Any methods you call on the parent `Player` to change its state or the state of its clock (e.g., the rate or the media time) will be propagated to all of the players the parent player is controlling.

This may seem like a good idea to middleware implementers, but it has a number of drawbacks in practice. The biggest problem is that it only synchronizes the state of the players and their clocks. If you call `getControls()` on the parent player, you only get back a list of the controls for that player. Any controls for players the parent player is not controlling are not included. This means that applications and the middleware still have to know about all of the different players. Even worse than that, they only need to know about it some of the time.

The other disadvantage of this approach is that the controllers only get their state synchronized when the managing player next changes its state. If a player in the Unrealized state is

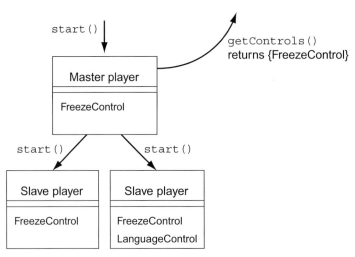

start()

getControls()
returns {FreezeControl}

Master player

FreezeControl

start()   start()

Slave player

FreezeControl

Slave player

FreezeControl

LanguageControl

Figure 11.5. Synchronizing players using a master player has a number of limitations.

added to a player in the Realized state, the first player will not automatically be realized. If the controlling player is started, both players will move to the Started state. Similarly, we can add a started player to a realized player, and even though the first player is a slave of the second it will remain started. The limitations involved in synchronizing players are depicted in Figure 11.5.

Ideally, all of the important parts of a service (video, audio, and subtitles) should be presented by a single player, and this is mandatory when using a `ServiceMediaHandler` to present components, as we saw in Chapter 5. We could do this using several players that are slaved to a single controlling player, but this is usually not worth the effort.

Even if the internals of a middleware implementation handle it differently, the middleware should expose this functionality through JMF as a single player. This will make the lives of middleware implementers easier in the end, and application developers will thank them. Avoiding the use of multiple players for a broadcast service reduces the complexity of the middleware stack (because you only have to manage one player internally), and means that applications are also simpler. Most applications will assume that there is only one player, and the MHP specification implicitly assumes this because of the way controls are used. This does not mean that it cannot be implemented using multiple players slaved to a single controlling player, but this is probably introducing extra complexity with no real benefit.

There are other reasons it is a bad idea, but these are generally less important. This functionality is really only useful in a PC or workstation environment and should be avoided in MHP and OCAP unless there is a very good reason for using it. Anything that complicates applications or middleware for no good reason is best avoided.

A DTV Special Case: The Video Drip Content Format

In our discussion of locators in the previous chapter, we introduced the video drip content format used in MHP and OCAP. This is a slightly special case for JMF because of the way data is loaded and presented.

Usually, broadcast content in DTV systems is treated as if it is coming from a push data source, whereas content loaded from files (even files in a broadcast file system) is regarded as coming from a pull data source. Players for video drips use a push data source in that they do not explicitly request the content, but their data has to be pushed to them by the application. This may seem confusing, but if you look at the big picture it soon starts to make sense.

As we saw earlier in this section, a JMF player gets its media data from a `DataSource` object. This data is usually loaded automatically by the `DataSource` object, but it does not have to be. When presenting video drips, the data source used is an instance of the `org.dvb.media.DripFeedDataSource` class. This class has a special method, `feed()`, that takes an array of bytes as an argument.

What happens is that a player is created and started in the usual way, but will not present any media because as yet it has not been told where to get its media data from. As we saw in Chapter 8, the locator used to create a data source for video drips is `dripfeed://`, which describes the content format but nothing else.

What happens is that an application can create and start a player, and then use the `DripFeedDataSource.feed()` method to provide some data to the player. The player will then decode and present that data. Of course, you can also call the `feed()` method before starting the player, and in this case what happens is dependent on the state of the player. If the player is in the Prefetched state when the `feed()` method is called, the data will be stored and displayed when the player is started. If the player is in any other state except for starting or started, or if the data source is not attached to a player, the content fed to it is thrown away.

Why is this useful? Well, imagine you have a set of large background images you want to display progressively. You could encode all of these as separate images, but this uses a lot of memory in the receiver and may take a long time to load from the broadcast file system. If there are only small differences between the images, we can encode the first one as an MPEG I-frame, and then encode the others as P-frames so that we only have to store the differences from the first frame. This is a much more efficient use of space, and is the type of thing that is easy to do on a DTV receiver.

Table 11.4 indicates how this works. In this example, we only add some extra buttons to the image, and thus the changes we are making are fairly small. Instead of sending the entire image again, we can encode the second image as a P-frame and only send the differences.

In a DTV receiver that has a limited amount of memory, this is a valuable technique that can save an application a lot of time and memory that would otherwise be needed for loading and storing images. Video drips are not a way of letting applications play video from a file

Table 11.4. Using video drips to update a background image.

Description	Image
I-frame, complete image.	
P-frame, update to previous I-frame.	
After displaying both frames.	

because they may be too slow to meet the frame rate needed for the video, and there are no guarantees about how long it will take to decode a frame. Most importantly, MHP places a number of restrictions on the format of an MPEG clip used for video drips. Video clips may not include an audio stream, and must consist only of I-frames and P-frames. Section 7.1.3 of the current MHP specification gives a full description of the restrictions placed on video drips.

JMF in the Broadcast World

We have already seen that JMF in the broadcast world is very different from JMF in a PC, and we have seen how this affects the architecture. Now we will move on to look at other practical issues application developers and middleware developers will need to deal with. As with so many other parts of JMF, these are driven by the difference between the broadcast world and the PC world, in terms of the hardware capabilities of the underlying platform and of the tasks the platform has to accomplish.

Getting a Player for the Current Service

Sometimes, you do not want to create a new player for your media, because it is easier to modify a player that already exists. In many cases, you will already have a reference to the Player object because your application created it, but sometimes you will need to work with a player your application did not create. If your application is associated with a service, how do you control the player that is showing the service your application is associated with?

There is no easy way to do this using JMF, but this is not really a problem for JMF. To do this, we use the relationship among service contexts, services, and the ServiceContentHandlers for various elements of the service (see Chapter 5).

As we saw in Chapter 5, the javax.tv.service.selection.ServiceContext class lets us get a list of the presenters that are displaying components for the service with which this class is associated. One or more of these presenters will implement the javax.tv.service.selection.ServiceMediaHandler interface, which implements javax.media.Player.

JavaTV uses the ServiceMediaHandler interface to differentiate those players that handle parts of a service from other JMF players that may exist in the system. The distinguishing feature of a ServiceMediaHandler as opposed to other JMF players is that a ServiceMediaHandler will present all elements of the service that use the same clock (e.g., audio, video, and subtitles if they are being presented). The following code shows how we get the ServiceMediaHandler for a given service.

```
// Get a reference to the JavaTV ServiceContextFactory
ServiceContextFactory factory;
factory = ServiceContextFactory.getInstance();

// From this, we can get a reference to the parent
// service context of our Xlet. To do this, we need a
// reference to our Xlet context. It's at times like
// this that we see why our application should always
// keep a reference to its Xlet context
ServiceContext myContext;
myContext = factory.getServiceContext(myXletContext);

// ServiceContentHandler objects are responsible for
// presenting all elements of the service that share the
// same clock. This includes all of the media components
ServiceContentHandler[] handlers;
handlers = myContext.getServiceContentHandlers();

for(int i=0; i < handlers.length ; i++) {
  if (handlers[i] instanceof ServiceMediaHandler) {
    // This is a Player for part of the service, since
    // ServiceMediaHandler objects implement the JMF
    // Player interface
```

```
        // we can do whatever we want with the
        // ServiceMediaHandler now that we've found it
    }
}
```

One thing that is obvious from this code is that you may get several `ServiceMediaHandlers` for a given service. This is unlikely for most DTV services, but you should not ignore the possibility.

Players and Service Selection

When an application selects a new service, this will have an effect on the way the receiver presents video and audio. Apart from the obvious change in the content, any changes made to the player using its controls will be reset. Although this is not a big problem for some options such as audio language, it can cause problems if we have set any video scaling, positioning, or clipping operations.

The MHP specification states that applications should reacquire references to any players or controls following a service selection operation, and at the same time players should be reset to their default values. This may seem like a problem with the MHP specification, but there is a good reason for requiring this behavior. By specifying the behavior of a player like this, MHP and OCAP give implementations the freedom to reuse a player when a new service is selected, or to create a new one. Although it is generally better for the middleware to reuse players where possible, MHP allows alternative approaches.

But how do we know when a new service has been selected? We cannot use a `Service-ContextListener`, because the service selection may happen in another service context. Instead, we can use the current player to tell us through a `PresentationChangedEvent`. Although this will get generated under other circumstances as well, it is the most reliable way of telling an application that a new service has been selected. To make sure that video is displayed in the right place, applications that change the size or position of the video should listen for `PresentationChangedEvents`, and when they receive a `PresentationChangedEvent` should reacquire the player and reset the options they need.

Creating a completely new player can lead to some flickering when a new full-screen player is created by the middleware and then set to the old size and position by the application, and thus middleware implementers need to design their middleware to minimize this flickering. Although the middleware has to reset all of the player's options to their default values, one way of avoiding any flickering is to wait a short period before applying these so that an application can reset any options it needs to. This way, an application can avoid carrying out unnecessary scaling or positioning operations. This may seem like a minor issue, but things like this make the difference between a high-quality middleware implementation and a poor one. Using the approach described previously should avoid flickering on many of the current MHP and OCAP implementations, and thus this appears to be widely regarded as the approach application developers should take. Because of this, middleware implementers need to handle this situation well.

Integrating Video with AWT

In Chapter 7, we saw how a receiver will usually display video in a layer separate from the layer in which AWT components are displayed. The layer in which video is displayed is known as the video layer.

Most of the time, we will want video to be displayed behind our AWT components as the background to our application. Even in those cases in which the video may only take up a small piece of the screen as an inset, we can still keep it in the background by scaling and positioning the video properly. Players that present their video in the video layer are known as background players because the video is in the background with respect to any AWT components.

Sometimes, though, applications really need the video to behave as though it were part of the AWT hierarchy. If we want the video to overlap other AWT components, this is the only way we can do it. Some high-end receivers may allow the true integration of graphics and video, depending on how MPEG decoding is implemented, and in these cases we can display the video in an AWT component. This lets us do all the things we can normally do with an AWT component, such as change its Z order with respect to other components, and change the size and position just like we would with any other component. In this case, the video will behave as if it were displayed in the graphics layer of the display even if it is actually still being drawn in the video layer. This type of player is known as a component player.

All players start their lives as background players. To convert a background player into a component player, all we need to do is call the `getVisualComponent()` method on the player we want to convert. If the player can be converted, this method will return an instance of `org.havi.ui.HVideoComponent`. If the player cannot be converted, `getVisualComponent()` will return a null reference. This process does not create any new players. All that has happened is that a background player has been changed into a component player. This is a one-way operation, and there is no way of converting a component player into a background player.

When a player is converted into a component player, some of the controls it previously had (especially those related to scaling and positioning video in the background layer) may not be useful anymore. For this reason, the player will dispose of some of its associated controls and may create some new ones. Applications cannot tell which controls will be kept and which will be recreated, and therefore must get new references to those controls they want to use for the component player.

Instances of `HVideoComponent` follow the usual rules of AWT. Moving or resizing the `HVideoComponent` using AWT operations will move or scale the video associated with that component, and hiding the component will hide the video. The first call to the `paint()` method for that component will move and resize the decoded video from the video layer of the display to the location and size of the player's component.

There is no guarantee that any `HVideoComponent` will be where you expect it to be when it gets created, and thus the first thing an application should do is move and resize the component. By doing this before drawing the component, applications can avoid some potentially ugly glitches when the video is moved and resized. Because some receivers may not support pixel-accurate positioning for MPEG video, the actual video presented may overlap the bounds of the `HVideoComponent` by 1 pixel in each direction. This should not be visible to users, however, and thus the middleware has to handle the cases in which this is true and redraw damaged elements of other components as necessary.

The main difference from other AWT components comes when we start to consider compositing. `HVideoComponents` will always have an alpha value of 1, and thus application developers may need to be careful when they are compositing these with other AWT components. Depending on the state of the player it is associated with, an `HVideoComponent` may have different behavior at different times. When the player is in the Prefetched, Starting, or Started state, the `HVideoComponent.getVideoDevice()` method will return the HAVi video device used by that `HVideoComponent`. When the player is in any other state, `getVideoDevice()` will return a null reference.

Component players are an optional feature of MHP, and thus not every receiver will support them. Because of the hardware requirements for this type of functionality, only a few receivers support component players at the time of writing. As processors become faster and receivers more powerful, though, we will see fewer platforms with restrictions on video scaling and positioning.

Subtitles, Closed Captions, and JMF Players

Subtitles and closed captions are a common feature in many DTV broadcasts, just as they are in analog broadcasts. However, the move to digital gives us some new options. Subtitles may be available in more than one language, and we may even be able to decide where and how they are shown. We have already seen the controls that allow us to do this.

In most cases, the decision to display subtitles, and the language they should be displayed in, will be made by the receiver based on a combination of several user preferences, such as the default language setting and the choice of whether subtitles are enabled by default or not. Applications can still have some control over how subtitles are handled, though, and earlier in the chapter we saw the controls applications can use to control subtitles.

Applications should take care that any UI elements they draw do not interfere with subtitles. Unfortunately, there is no way for an MHP application to know where subtitles are being drawn and thus application developers and network operators need to make sure that applications and subtitles do not draw to the same piece of screen. In cases in which the application cannot be repositioned, it can turn subtitles on or off using the `SubtitlingLanguageControl.setSubtitling()` method. This can cause problems for users who need subtitles, though, and thus it is generally better to avoid this.

OCAP applications have slightly more control over how closed captions are presented, although they cannot control where on the screen they are presented in version I12 and later of OCAP (older versions could do this). We have already seen that OCAP developers should take great care when changing the appearance of closed captions in order to make sure that they do not violate any of the FCC's rules governing this.

The appearance of closed captions can be changed using the `org.ocap.media.Closed-CaptioningAttributes` class. This is a singleton class that provides an interface to the closed-caption settings. After getting a reference to this object using the `ClosedCaptioningAttributes.getInstance()` method, applications can use the `setCCAttribute()` method to set the various attributes for closed captioning. The full signature for this method follows.

```
public void setCCAttribute(int[] attribute,
                           java.lang.Object value,
                           intccType);
```

This takes an array of one or more integers identifying the attributes to be set (which are described by constants in the `ClosedCaptioningAttributes` class), a value to set those attributes to, and a flag identifying the type of closed captioning for which attributes should be set. This can take a value of `ClosedCaptioningAttributes.CC_ANALOG` or `ClosedCaptioningAttributes.CC_DIGITAL` to control analog and digital closed captions, respectively.

In general, it is best not to change the presentation or settings of subtitles and closed captions unless there is a very good reason to. In certain cases, this can even violate broadcasting rules or guidelines governing accessibility.

Managing Resources in JMF

Decoding video and audio uses a lot of hardware resources in the receiver, and thus any developer using or implementing JMF has to be aware of the resource management issues involved. The main resource a JMF player will use is the MPEG decoder. This is normally a hardware component, because doing it in software is so CPU intensive. Even for those receivers that do use a software MPEG decoder it usually takes so much CPU power that a receiver can only decode one video and possibly two audio streams at any time.

Unlike most of the APIs in MHP or OCAP, JMF actually has a standardized algorithm for managing the MPEG decoder resource. If a JMF player is started and needs to use the MPEG decoder, the MPEG decoder resource will be taken away from any other players that may be using it. By operating on this last-come/first-served basis, the middleware can make sure the player that requested the resource last (which is probably part of an application the user is actually interacting with) will get access to it.

Any other players that were using the resource will send an `org.davic.media.ResourceWithdrawnEvent` to any listeners that registered themselves with those players. This tells the applications that a player has lost access to a resource it needs, but it does not mean that it is a fatal error. The player will remain in the state it was in, and it may still present some media if it is in the Started state, depending on which resources were lost. One `ResourceWithdrawnEvent` will be sent for every resource lost.

This event lets applications know that a player may not be doing what they expect, and gives them a chance to recover by stopping the player, or by using a static image instead of any decoded video. When the resources are returned to the player, the player will send an `org.davic.media.ResourceReturnedEvent` to any listeners to tell them that the player now has the resource again. Players in the Started state will also resume playing media (assuming they now have all of the resources they need to do so).

In those cases in which the loss of the resource is permanent and the player cannot continue, the player will generate a `javax.media.ResourceUnavailableEvent`, and the middleware will close the player and then release any resources used by the player. `ResourceUnavailableEvent` means that the player cannot recover from the loss of a resource, and that the application should create a new player. When this event is generated, the middleware automatically stops, deallocates, and closes the player that generated the event.

Restrictions on Another Resource: Tuning

JMF may use scarce resources other than the MPEG decoder. You may think that the tuner would be one of these, but this is not actually the case. The MHP and OCAP specifications place an important restriction on how JMF uses the tuner resource, and this may not be obvious unless you read the specification very carefully.

The MHP specification says that no API will make the receiver tune to a different transport stream unless the specification explicitly says that this can happen. If you read the MHP specification carefully, you will find that JMF is not explicitly allowed to tune. This is not an oversight: the MHP specification really does mean that you cannot use JMF to automatically tune to a new transport stream. This means that a JMF player will never use the tuner resource because it has no need to use it.

OCAP also follows these restrictions, and thus a JMF player in an MHP or OCAP implementation cannot decode a service or elementary stream that is not part of the current transport stream. To do this, the application must explicitly tune to that transport stream first, using the tuning API or JavaTV service selection API.

This is a major implication, and unless you are careful it is easy to miss. If you are wondering why your application will not display that service correctly in your JMF player, this could be the reason.

Although this behavior may seem strange at first, it is actually a natural extension of the restrictions designed to make sure that several applications can coexist. Even if your application does not need any parts of the current transport stream, other applications may. Tuning away from the current transport stream means that applications cannot load more files from broadcast file systems, that service information will not be kept up to date, and that section filters set by other applications may no longer be valid. In short, tuning to a new transport stream can have a big impact on every application that is running. By only allowing tuning to happen through specific APIs, and by limiting which applications can use those APIs,

MHP and OCAP give applications the most flexibility they can while still maintaining some control over what applications can do.

From a resource management perspective, there are even more reasons for doing it this way. We have already seen how JMF allocates the MPEG decoder on a last-come/first-served basis, and we have seen why it is done like this. If an application were allowed to tune to another transport stream using JMF, there would be a much higher chance of applications having other resources taken away from them at the same time (resources such as access to their data files, service information they may have been accessing, and private data streams they may have been filtering). Once we start looking at the consequences of allowing JMF players to tune, this restriction starts to make more sense.

In other standards, it is even less obvious what should happen when JMF players want to present services from another transport stream. Where OCAP and ACAP inherit this part of the specification from MHP, JavaTV is silent about what should happen, and thus JavaTV implementations may or may not tune. For pure JavaTV systems, some other implementation guidelines may be needed to make sure that every receiver (and application) behaves in the same way.

Playing Audio from Sources Other Than Files

JMF is a good interface for controlling broadcast streams, but it has a number of limitations when it comes to audio playback. The biggest of these is that it can only play audio from a file, or at least from something the application can refer to using a `file://` URL. This means that we cannot use it for playing audio from memory, or from any other source.

Luckily, there is another way of doing this. The `org.havi.ui.HSound` class is part of the HAVi UI API, even though it is not a GUI widget. This gives applications an alternative way of playing audio clips that entails fewer restrictions on data sources. The `HSound` class has the following interface.

```
public class HSound {

  public HSound();

  public void load(String location)
    throws java.io.IOException, SecurityException;

  public void load(java.net.URL contents)
    throws java.io.IOException, SecurityException;

  public void set(byte data[]);

  public void play();
  public void stop();
  public void loop();

  public void dispose();

}
```

As you can see, this is much simpler than the JMF API and is thus not suitable for all cases. However, if we want to play audio from a web server or from an array of bytes, this is the only option open to us.

The two versions of the `load()` method let us load audio clips from a file or from a URL, whereas `set()` lets us create the audio clip from an array of bytes. In both cases, these operations are synchronous, and thus data loaded from a broadcast file system may cause the `load()` method to block. Once we have loaded or set a clip, we can play that clip once (using the `play()` method) or repeatedly (using the `loop()` method). Both `play()` and `loop()` are asynchronous methods.

An `HSound` object can be used to play more than one clip. Calling `load()` or `set()` when a clip has been loaded will stop the `HSound` if it is currently playing, and then dispose of any audio data the clip is using. Once it has done this, it will load the new clip. Similarly, the `dispose()` method will remove any previously loaded sample data without loading a new clip.

The file formats supported by `HSound` are the same as those supported by JMF, which for now limits it to MPEG-1 audio. Middleware manufacturers may support other formats as well, of course, but these will not be available to interoperable applications.

12 DSM-CC and Broadcast File Systems

Almost every MHP and OCAP application will be downloaded to the receiver via the broadcast stream. This chapter takes an in-depth look at how broadcast file systems work, and discusses their strengths, their limitations, and how middleware developers can build the best implementation using this technology. We also look at how application developers can make the most of these file systems in ensuring that applications load as quickly as possible.

For any ITV system, we need some way of loading classes and data files that are needed by the application. This usually means a file system that applications can use. Although many receivers will include some type of return channel, we cannot rely on having a two-way connection, and thus any file system we use has to work over a broadcast-only channel.

To do this, MHP chose to use the DSM-CC (Digital Storage Media, Command and Control) standard to provide file system access, and OCAP followed this choice. DSM-CC is an ISO standard, part 6 of the MPEG-2 standard (ISO/IEC 13818-6). JavaTV does not specify any type of technology for broadcast file systems, although it does include the `javax.tv.carousel` package that adds some basic support for carousel-based file systems.

DSM-CC Background

DSM-CC was originally supposed to be a standard for controlling networked video sources, such as video-on-demand servers or networked VTR machines. From this original idea, it has grown into a more general set of protocols for providing multimedia services and applications over a broadband connection. Many more features have been added to support this, as we will see in the rest of this chapter.

Why Choose DSM-CC?

Anyone who has looked at the DSM-CC specification knows that it is extremely complex and covers many areas. Although DSM-CC covers the same ground as some other well-known standards, DSM-CC has a number of advantages that make it useful in DTV systems. The first of these is that DSM-CC is specifically designed to work with MPEG, and was standardized by ISO with this in mind. This ready-made solution meant that DVB did not have to invest any time in making a group of disparate protocols work together correctly.

The second big advantage is that DSM-CC works in a heterogeneous environment, unlike many other protocols. DTV content often travels over many types of networks before it reaches the viewer. These may include Asynchronous Serial Interface (ASI) connections, IP networks, or ATM networks in the head-end; hybrid fiber-coax networks in a cable system; satellite transponders in a satellite system; and almost any other networking technology you can think of, including both one-way (broadcast) and two-way networks. When several broadcasters use a piece of content, it may traverse even more types of physical networks and networking protocols. DSM-CC is a high-level protocol that can work on top of any of these networks and is thus not dependent on any other network technology.

A final advantage of DSM-CC is that it is a complete suite of protocols, covering all of the elements needed by a DTV system. Elements in the DSM-CC toolkit include the following.

- Network configuration and session setup
- Control of media streams
- Synchronization with media content
- Access to files in both one-way and two-way networks
- Lower-level data broadcasting technologies not based on files
- Carriage of data streams (e.g., IP datagrams) on a broadcast connection

Many of these features could be replaced by other protocols such as DHCP, RTSP, and RTP, but a single standard covering all of the needed functionality can make sure that there are no gaps or inconsistencies between the different elements. All of the elements of DSM-CC are designed to work together to provide the features specifically needed by multimedia applications and services, rather than being general-purpose protocols that happen to work together.

Isn't There Better Documentation?

The disadvantage of this approach is that DSM-CC does everything in a single standard, which means that it is very complex and can be difficult to read. Many people complain that the DSM-CC specification is almost impossible to understand, and they are largely correct. Because DSM-CC is such a complex specification, most applications only use a subset of the standard, and this is what DSM-CC was designed for. It is a toolkit, rather than a single standard whereby every element is needed by every platform that uses DSM-CC.

DVB and ATSC both define subsets of DSM-CC for use in data broadcasting. CableLabs follows the DVB data broadcasting standard (with a few minor differences) for use with

OCAP. For DVB, ACAP, and OCAP systems, it is best to start with the following two specifications.

- ETSI EN 301 192 (DVB Specification for Data Broadcasting)
- ETSI TR 101 202 (DVB Implementation Guidelines for Data Broadcasting)

The ACAP specification also defines elements of the broadcast file system used with ACAP systems. For ATSC systems other than OCAP or ACAP, the following standards are most appropriate.

- ATSC A/90 (the ATSC data broadcast standard)
- ATSC A/91 (implementation guidelines for the data broadcast standard)

Only once you have read and understood these documents should you start looking at the DSM-CC standard itself.

An Overview of DSM-CC

The most common use of DSM-CC is to provide access to objects on a remote server, and DSM-CC uses many concepts from CORBA (Common Object Request Broker Architecture) to do this. Files and directories are treated as objects (but these are not the only types of objects we can access), and clients can use a remote procedure call (RPC) mechanism to manipulate these objects and get access to the data they contain. Because the interfaces to these objects are specified in IDL, the implementation language and network technology does not matter. To the DSM-CC implementation the procedure calls will always look the same.

Object Carousels

Unfortunately, DTV networks often use a one-way broadcast network in which RPC does not work. Not every receiver will support a return channel to the network operator, and return channel technologies such as PSTN modems are not practical for this purpose. To solve this problem, DSM-CC uses a concept called an object carousel to transmit all of the objects to the client where any manipulation is carried out locally.

A directory structure that will be transmitted as an object carousel is split up into a number of modules, and these modules are then transmitted in turn to the receiver. A module is simply a set of objects that is treated as a single unit when transmitting the file system. Each module may contain several files with a total size smaller than 64 Kbytes (storing several files in a module larger than 64 K is not allowed). Splitting files across more than one module is not allowed either, and thus files larger than 64 K must go in separate modules that will contain only that file. Files in a module can come from any part of the directory tree and need not come from the same directory.

Once the entire directory tree has been split into modules, each module is then transmitted in turn, one after the other. After we have transmitted every module, we start from the beginning and transmit every module again. The result of this is a stream of MPEG sections that contain our file system, and if a receiver has missed part of a particular module (because of an error, or because it started reading partway through the transmission of that module) it just has to wait for the module to be retransmitted.

This can take a little time, however, and thus the latency of a broadcast file system can be very high. For a carousel with a total size of 256 Kbytes transmitted at a bit rate of 128 Kbits/sec, it will take approximately 15 seconds to transmit the entire carousel and thus this will be the maximum latency for loading a file from this carousel. To reduce this problem, broadcasters will transmit some modules more frequently than others, so that modules containing common files will be available sooner than modules containing files that are used less frequently.

An Example Object Carousel

Let's look at a concrete example of this. Suppose we wish to broadcast the directory structure shown in Figure 12.1.

To create the modules for broadcast, we start adding files to the module. Adding the first two files (index.html and image1.jpg) is no problem. Adding image2.jpg would take the module above 64 Kbytes in size, and thus we cannot do that. However, we can add the next file (clip1.aiff in the audio directory) and the directory entry for the audio directory itself with no problems. We could also add some of the content of the classes direc-

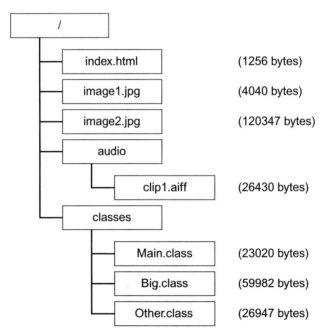

Figure 12.1. An example directory structure.

tory, but we will not for reasons explored later. Therefore, we end this module and start a new one.

The file `image2.jpg` is larger than 64 Kbytes, but we cannot split the file across more than one module and thus it goes in its own module. The resulting module is larger than 64 Kbytes, but this is allowed.

That leaves us with the content of the `classes` directory to be added to the carousel. These files will not all fit in the same module, but we can organize them in a way that makes loading them quicker. The first thing we add to the new module is the directory entry for the `classes` directory. Because we are also likely to need the directory entry to access the files within that directory, we add as many of those classes as possible to this module. In this case, that means the files `Main.class` and `Other.class` get added. As before, the final file (`Big.class`) will go in a module on its own.

Note, however, that this may not be the most efficient way of splitting the files across modules, which depends on when the files are needed and on the relationships between them. Doing this for real would take some careful thought to get the most efficient carousel layout. The carousel layout for our example directory tree is shown in Figure 12.2.

We have already mentioned that we can broadcast some modules more frequently than others in order to reduce latency for commonly used files. Of course, by doing this we increase the total size of the carousel (in that the total size of the carousel is the amount of data we have to transmit in order to transmit every module at least once), and thus the access time for less commonly used files will increase. This trade-off has to be considered carefully when designing a carousel to optimize the download speed, and we will look at how we can improve carousel performance later in the chapter.

In some documents, you will see references to service domains, and in many cases this will be used in a context similar to that of an object carousel. To be precise, a service domain is a group of DSM-CC objects that are used together. In a broadcast network, these are con-

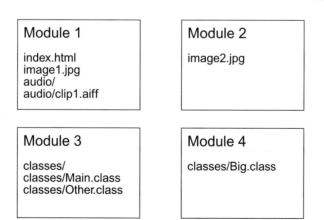

Figure 12.2. The carousel layout for our example directory tree.

tained in an object carousel and thus we can think of them as the same thing. Each service domain will be accessed via a service gateway object that represents the root of the directory structure in that service domain.

More Than Just a File System

Although DSM-CC is often used as a broadcast file system protocol, there is more to it than that. An object carousel can contain objects other than files or directories. Stream objects represent services or elementary streams in the broadcast content, and clients can use these to find more information about those streams. In some DSM-CC applications (such as video-on-demand services), stream objects also let the client control the content to which the stream object refers. MHP and OCAP do not support this, however, and thus we must use the JMF API to do this.

Normal Play Time

One stream-related aspect of DSM-CC that is used in MHP is the concept of Normal Play Time (NPT). NPT is a time code carried in the MPEG stream that can be used by the broadcaster to define a meaningful clock value for a given point in the stream. This may not seem very useful, in that MPEG already carries a PCR (program clock reference) that is used to synchronize the clock in the receiver with the clock used when transmitting the streams. PCR values are useful if you are an MPEG decoder, but in many other cases they are not much help to you because they only let you reconstruct the correct system time clock (STC) value for the receiver. An NPT time code is more flexible than a PCR value, and lets applications synchronize to streams in broadcast content without having to worry how time references are calculated. Although it is based on the PCR value, NPT values are a higher-level time code.

NPT values are carried in NPT reference descriptors within a stream of private sections. They may start at any value, and although they typically increase throughout a single piece of media they may have discontinuities either forward or backward. This means that even if a stream containing NPT values is edited NPT values will not need updating and will remain the same for that piece of media.

This concept is illustrated in Figure 12.3. In this case, the media clip shown at the top is edited for content in two places, which makes the stream shorter (these edits are marked by A and B in the diagram). Advertisements have been inserted, which makes it longer (indicated by C in Figure 12.3). The shaded sections in the diagram show all of these edits, with changes in the NPT values also shown. Despite these changes, the NPT values at each point in the final media correspond to the values at the same point in the original, unedited version. This is a powerful feature of NPT.

One other advantage of NPT is that NPT values can be nested. NPT is carried in descriptors, each containing two main parts: the NPT value itself and a content ID. The receiver uses this content ID to discover the type of edit. These fall into the following three basic categories.

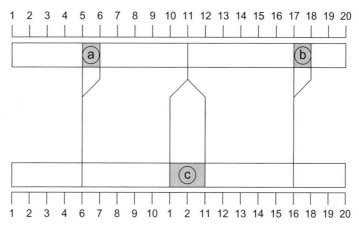

Figure 12.3. Stream editing and NPT.

- Edits in the original stream (the content ID will be the same on both sides of the discontinuity)
- New NPT time codes nested within another NPT time code (e.g., a set of advertisements inserted into a TV show, where both use NPT)
- Boundaries between two different pieces of media (e.g., different shows)

Some NPT descriptors allow us to determine the type of edit in advance. Another descriptor (the NPT endpoint descriptor) defines when a particular content ID is valid, and we can use this to determine whether one set of time codes is nested within another. This lets an MHP or OCAP receiver ensure that NPT values are as correct as they can possibly be, although there may still be a period of uncertainty of up to 1 second in which no NPT descriptors have been received for the current service. Once we have received the first NPT reference descriptor, we can reconstruct the NPT value for any point in the stream.

Stream Events

In addition to NPT time codes, DSM-CC supports other synchronization points called stream events. These are markers embedded in a DSM-CC elementary stream that provide a way for the receiver to synchronize with specific points in the media. This is useful when the NPT value could change, or when our synchronization points may not occur at predictable NPT values.

Understanding stream events properly can take a little time, partly because they consist of several elements and some of these elements have very similar names. Table 12.1 outlines the terminology we use in this book when we discuss stream events.

DSM-CC stream event objects are stored in an object carousel, and are just like any other DSM-CC objects. A stream event object provides a general description of one or more stream

Table 12.1. Definitions of DSM-CC stream event elements.

Term	Meaning in This Text
Stream event	The event triggered within the receiver when a stream event descriptor is received.
Stream event objects	DSM-CC stream event objects as carried in a DSM-CC object carousel. These act as a high-level description of a stream event.
Stream event descriptors	The descriptors carried in a DSM-CC stream, which actually mark synchronization points.

events, each of which includes an event ID (which must be unique within that carousel) and a human-readable name. This allows a receiver to know what events can be generated, and helps it to check that an application is listening for stream events that will actually get used. In effect, this is like a class in object-oriented programming. It is a general description of a type of object, rather than something that represents a specific object.

The stream event descriptor is the second part of the puzzle. This tells the receiver that an event has actually been generated, and can be compared to an instance of a class if we extend our object-oriented programming analogy. The stream event descriptor contains specific values describing a single stream event, and more than one stream event descriptor can refer to the same description in a stream event object (just as a class can have more than one instance).

A stream event descriptor contains three main attributes: the ID of the event, an NPT value at which the event should be generated, and an application-specific payload. The ID allows the receiver to work out which stream event object is associated with this descriptor. Because a broadcaster cannot be sure exactly where a multiplexer will insert the descriptor, each descriptor can carry an NPT value that says when the event should be triggered. This allows the receiver to know in advance that it should generate an event when a specific NPT value is reached. This makes the timing of the events more predictable, and adds a little more reliability into the system because the broadcaster can send several stream event descriptors with the same values in order to make sure at least one of them is decoded properly. The MHP specification says that scheduled stream events should be signaled at least once every second for a minimum of 5 seconds before the time they should trigger.

Stream event descriptors can also tell the system that an event should be triggered immediately. These are known as do-it-now events. This allows broadcasters to insert stream events into a live show much more easily. For instance, a sports application may use stream events to detect when a goal has been scored in a soccer match, and use the payload to identify which team has scored and the current score. We will see exactly how we can use events to carry this type of information later in the chapter, and we also cover this more in Chapter 15. Stream event descriptors containing do-it-now events are only broadcast once per event, and thus some care has to be taken by the receiver to make sure that it receives them properly. Whenever a stream event descriptor is received, the receiver takes the following steps.

1. It checks to see that a stream event object with the same event ID is present in the default object carousel. If an event with that event ID is not present, the descriptor is ignored.
2. If the encoding of the descriptor shows that the event is a do-it-now event, the event is triggered immediately.
3. If the event is not a do-it-now event, the receiver checks the NPT value at which the event should be triggered. If an event with the same event ID is already scheduled to be triggered at the same NPT value, or if the NPT value has already passed, the event descriptor is ignored.
4. When the NPT value reaches the value specified for a scheduled event, the event is triggered.

One advantage offered by this separation of stream event objects and stream event descriptors is that events can be reused. As we have already mentioned, several stream event descriptors can contain the same event ID, even if they are triggered at different times and contain different private data. This allows an application to use the event ID to define classes of events. For instance, in our soccer example the application may use event descriptors with one event ID to indicate that a goal has been scored, whereas a different event ID may signify a penalty kick being awarded and a third event ID may signify the end of the match. Thus, an application can start processing an event just by knowing the event ID. In some cases, no other application-specific data is needed, although the payload can carry extra data if necessary.

The Relationship Between Normal Play Time and Stream Events

As we have discussed, the time reference in a DSM-CC stream event is a reference to an NPT time code value and the event will trigger at that NPT value or the next NPT value past that point. If the trigger time for an event is skipped because of an NPT discontinuity (i.e., that part of the content has been edited out), the event triggers at the first NPT reference past the trigger time.

An NPT signal is only required to be present in those cases in which a stream event needs it. If all of the stream events in a piece of media are of the do-it-now type (wherein the event is triggered immediately rather than waiting for a specific NPT value), we do not need to include NPT time codes.

Do-it-now events are the most common types of stream events in use today for several reasons. They are more flexible, and can be used almost everywhere a scheduled event can be used. They are also simpler, in that they do not rely on the presence of NPT, and application developers and receiver manufacturers have more experience with using them. Finally, there is a danger that NPT time codes can be corrupted when streams are broadcast across networks. In particular, if PCR values in a stream are remapped or recreated there is a danger that the PCR values used in NPT time codes will not be updated to match these. If they are not, the NPT values will be inaccurate with respect to the stream and stream events that rely on these NPT values will be triggered at the wrong time.

For these reasons, it is generally better to use do-it-now events where possible. In most cases, it is possible to replace a scheduled stream event with a do-it-now event at the time the scheduled event should trigger.

Whether using scheduled events or do-it-now events, application developers need to be aware that these events will take some time to reach the application because they must first be processed by the middleware. Because the time this takes will vary between implementations, it is difficult to get frame-accurate synchronization using stream events.

DSM-CC in Detail

DSM-CC is based on a client-network-server model, and it defines two protocols for communication between different parts of the system. Clients and servers use user-to-network (U-N) messages to communicate with the network itself, for configuring a DSM-CC client on the network, and for setting up connections between clients and servers on the network. This can be thought of as a replacement for configuration protocols such as BOOTP and DHCP, and for session management protocols such as RSVP.

User-to-user (U-U) messages allow clients and servers to communicate with each other and manipulate objects or media content. The U-U protocol is a more general version of protocols such as RTP and RTSP (for media control) and FTP or HTTP (for file access).

We will not look at the U-N protocol in too much detail in this chapter, in that MHP and OCAP primarily use the U-U protocol. For more information about this and other elements of DSM-CC we do not discuss here, the DSM-CC specification is probably the best place to look. Be warned, however, that it is not an easy read.

At first, it may not seem easy to use U-U messages in a one-way network such as a TV broadcast. In that this is an object-based protocol, we solve this by broadcasting all of the objects from the server to the client. The client can then manipulate these objects locally, without needing a return path to the server. In a two-way network, this manipulation will be carried out using remote procedure calls and the objects will reside on the server.

MHP and OCAP both use the DSM-CC object carousel format for carrying objects from the server to the client. A number of other transport mechanisms are available to us as part of DSM-CC, however. Although these may not be suitable for all purposes, they can be useful for carrying certain types of data. In the next few sections we will take a closer look at some of these alternative transport mechanisms.

Data Carousels

The DSM-CC data carousel provides a simpler alternative to object carousels for some cases, and in fact object carousels are built on top of data carousels. Data carousels are also used by the ARIB B23 specification, as well as the ATSC A/90 data broadcasting specification.

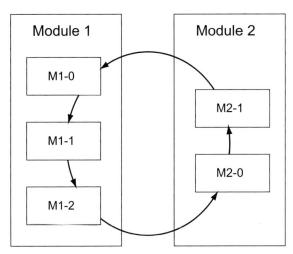

Figure 12.4. Transmitting data via carousel modules as a series of blocks. *Source:* ETSI TR 101 202: 1999 (DVB data broadcasting implementation guidelines).

A data carousel consists of a number of modules of data that are transmitted to the client so that the client can read any data it needs from the carousel. Some modules are transmitted more often than others, in order to improve access time for commonly used data. This is where the concept of modules in an object carousel comes from. Object carousels are a logical layer on top of data carousels, and thus we have to follow the limitations imposed by the data carousel.

To transmit these modules, each module is split into a number of blocks, which are transmitted in turn (as indicated in Figure 12.4). As well as the actual module data, a data carousel contains additional information so that a receiver can find the blocks that make up the module it is looking for. The data and control information are transmitted as a series of data carousel messages, and we will take a closer look at the format of these messages later in the chapter.

Object Carousels

To carry the objects that make up a file system, we need to define a set of higher-level structures that sit on top of the basic data carousel. DSM-CC object carousels use the Object Request Broker (ORB) framework that was defined by CORBA to provide a hierarchical structure of objects.

Several different types of objects can be carried in an object carousel, as we have already seen. Each object within the object carousel is transmitted as a BIOP (broadcast inter-ORB protocol) message as defined by CORBA, and each type of object is represented by a different type of BIOP message. The following message types are supported in an MHP or OCAP object carousel.

- File messages represent real files. They contain the actual data that makes up the file.
- Directory messages represent logical containers for a set of file messages. Each DSM-CC directory contains a number of files (just like directories in a real file system), and a DSM-CC directory message contains a set of references to the files within a directory.
- Stream messages are references to MPEG-2 streams, usually containing video or audio data. Each stream message can refer either to a single MPEG-2 program or to one or more elementary streams.
- Stream event messages describe the stream events contained in a stream. The stream events themselves are described by specific descriptors associated with the media stream, but stream event messages tell the receiver what stream events are present and associate them with a textual name.
- Service gateway messages represent a concept that is similar to a directory (and in fact, service gateway messages are almost identical to directory messages). The main difference is that service gateway messages identify the root directory of the object carousel. This means that every object carousel will have one (and only one) service gateway message.

Several BIOP messages can be contained within every data carousel module. The exact number of messages that can be carried in each module depends on the size of the messages, due to the size limitation on data carousel modules. Figure 12.5 depicts the transmission of an object carousel as a series of MPEG-2 sections.

Multiprotocol Encapsulation

Sometimes, we want to transmit a stream of data to the client rather than a file system or distinct blocks of data. This may happen if we have data that is specifically designed for streaming, or because the data arrives in a format in which it cannot be easily fit into a data carousel or object carousel. Some IP traffic, such as UDP datagrams, is a good example of this. A stream of UDP packets does not fit well into a data carousel or object carousel, and yet transmitting it in an MPEG PES stream containing private sections usually means inventing a new protocol to carry the data.

DSM-CC recognized that there will always be a need to carry streaming data in an MPEG transport stream, and thus DSM-CC supports a data format called multiprotocol encapsulation. In MHP or OCAP receivers, this is often used for carrying IP data in a transport stream, effectively giving the receiver a simple one-way Internet connection. There are limits, of course. This is just a one-way connection for datagram packets, rather than a complete two-way TCP connection.

Although IP data is probably the most common form of data that is carried, other protocols are supported. Non-IP data formats are beyond the scope of this book, however.

Addressing a Receiver

IP uses a combination of IP addresses and hardware MAC addresses to identify a receiver, with MAC addresses being used at the lower levels. This works well and is well understood,

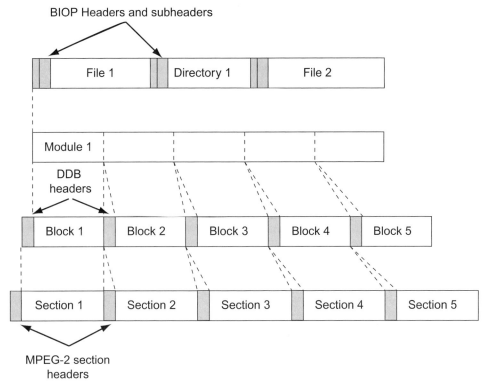

Figure 12.5. Transmitting an object carousel as a series of MPEG-2 sections.

and thus DSM-CC uses the same approach. Each receiver is given a 48-bit MAC address (the same length as an Ethernet MAC address), and this is used to identify which receiver (or receivers) is the destination for a specific packet. DSM-CC says nothing about how a receiver gets its MAC address or how this is assigned, however.

Obviously, a MAC address alone is not very useful. IP datagrams use IP addresses, and there has to be some way of routing an IP address to the right receiver. In fact, there are two ways. Some versions of MHP describe this using a descriptor that is carried in the AIT (called the routing descriptor). Unfortunately, this was not quite enough, and thus MHP 1.0.3 and MHP 1.1.1 use a new mechanism called the IP/MAC Notification Table (INT), which is defined in version 1.3.1 of the DVB data broadcasting specification. This is an additional SI table that carries information used by multiprotocol encapsulation, and we will discuss it in more detail in material to follow.

Transporting IP Packets

To carry IP packets in the MPEG-2 stream, DSM-CC defines a type of section called the datagram section. This is compatible with the normal DSM-CC private section format, but it extends it to make it easier for receivers to filter packets based on the MAC address using hardware section filters.

Datagram sections can support any type of OSI layer 3 networking protocol, but DVB has optimized the section format to make it easier to use with IP datagrams. One example of this is the way the destination MAC address of each packet is carried. It is split into two groups, with the two least significant bytes (bytes 5 and 6) being stored at the start of the section, and the others following later. This allows hardware section filters (which can often only filter on the first part of a section's payload) to filter out datagrams addressed to other receivers.

Each datagram section carries a single datagram for a single MAC address (although this may be a multicast or broadcast address). Although it is possible for an IP datagram to be larger than an MPEG-2 section, this is not supported by DSM-CC. IP datagrams used in multiprotocol encapsulation should have an MTU value of at most 4,080 bytes, but certain values for some header fields may reduce this.

SI and Multiprotocol Encapsulation

SI plays an important part in multiprotocol encapsulation. As well as the descriptors that are defined to handle other types of DSM-CC data (see later parts of this chapter), DVB has defined some SI to optimize the use of DSM-CC multiprotocol encapsulation with IP data.

Routing Descriptors

Versions of the MHP specification up to MHP 1.0.2 (thus including GEM and OCAP) use the IP routing descriptor to define the mapping between multicast IP addresses and elementary streams that carry the traffic for those addresses. These descriptors are carried in the common descriptor loop of the AIT, and thus they apply to every application in that AIT. Of course, this also means that they only apply to MHP systems.

MHP defines routing descriptors for both IPv4 and IPv6 traffic. These are both fairly simple descriptors that map a multicast IP address range (specified by an IP address and subnet mask) and port number onto a given component tag. This is enough to tell the receiver which streams it should monitor, but IP routing descriptors have a number of limitations.

The IP/MAC Notification Table

Between MHP 1.0.2 and MHP 1.0.3, DVB revised the data broadcasting specification and produced version 1.3.1 of the specification. This version of the specification introduced a general-purpose way of defining how IP is carried on DVB transport streams. This is the INT. The INT consists of several subtables that define a set of targets and a set of actions that apply to those targets. Each subtable has the format outlined in Table 12.2.

The data broadcasting specification was defined by people who knew about and cared about IP transport over DVB transport streams, whereas MHP was defined by people for whom

Table 12.2. Format of the INT.

Syntax	No. of Bits	Identifier
IP/MAC_notification_section() {		
table_id	8	uimsbf
section_syntax_indicator	1	bslbf
reserved_for_future_use	1	bslbf
reserved	2	bslbf
section_length	12	uimsbf
action_type	8	uimsbf
platform_id_hash	8	uimsbf
reserved	2	bslbf
version_number	5	uimsbf
current_next_indicator	1	bslbf
section_number	8	uimsbf
last_section_number	8	uimsbf
platform_id	24	uimsbf
processing_order	8	uimsbf
platform_descriptor_loop()		
for (i=0, i<N1, i++) {		
target_descriptor_loop()		
operational_descriptor_loop()		
}		
CRC_32	32	rpchof
}		

Source: ETSI EN 301 192: 2003 (DVB data broadcasting specification).

this was a necessary evil. For this reason, MHP 1.0.3 and 1.1.1 follow the new data broadcasting specification in order to avoid an incompatible solution that is not a major part of MHP. MHP 1.0.2 and OCAP systems must use the older version of the SI, however. The format of the INT is shown in Table 12.2, and sections containing the INT will have the table ID 0x4C.

Because many different types of receivers may be operating on a given network, there has to be some way of choosing which receivers these messages apply to. Each receiver manufacturer has a platform ID that is specified by DVB, and we can use this to identify receivers from a specific manufacturer. Every INT subtable will identify a different platform ID, although there may also be cases in which the same platform ID is used in more than one INT section (for instance, when a subtable is too big for a single section). In this case, sections with the same platform ID belong to the same subtable.

Each INT consists of three types of descriptors. The first of these, known as the platform descriptors, provide some information about the platform. This extends the platform ID to give a little more information about the platform that is the target of this subtable. The platform descriptors are contained in the platform descriptor loop.

The other two types of descriptors are target descriptors (which identify a set of receivers or an IP address range) and the operational descriptors, which define a set of actions that apply to those receivers or IP address ranges. These are grouped in a larger loop, so that every iteration of the outer loop contains a set of target descriptors and a set of operational descriptors that apply to those targets.

Target descriptors can identify either a receiver (by its serial number, smart card number, MAC address range, and IP address range) or some network traffic (by MAC address range, IP address range, or a source/destination address pair). This allows the target descriptors to be used for two purposes. Typically, they identify a set of IP addresses and a port number, and the operational descriptor loop then tells the receiver where to find the data for those addresses. A target descriptor could also be used to force only a small set of receivers to see these packets.

The number and type of target descriptors gives the broadcaster a great deal of flexibility in addressing a receiver, and is the reason the platform ID is so important. A broadcaster could identify one group of receivers by IP address, another by a range of smart card numbers, and a third by their serial numbers.

The operational descriptor loop contains a set of descriptors that specify which actions should be applied to the receivers identified in the target descriptor loop. The most important of these descriptors is the IP/MAC stream location descriptor. This serves basically the same purpose as the routing descriptor defined in early versions of MHP: it tells the receiver how to map a given address range and port number to a given DSM-CC stream. In this case, the descriptor only describes which elementary stream contains the data. IP address and port number mappings are described in the target descriptor loop.

Other descriptors that are part of the operational descriptor loop include the telephone descriptor (which tells the receiver what telephone number should be dialed to access that service) and the ISP access mode descriptor, which tells the receiver how it should access an ISP to use a service. Only dial-up access is defined for now, and thus these descriptors are commonly used together.

The order in which different INT subtables are processed may be important in some cases because of how the target and operational descriptors affect the processing of packets. For this reason, each INT section has a processing order. This must start at 0x00 for the first subtable to be processed, and the receiver will then process subtables in the ascending order of their processing order values. Thus, a subtable with a processing order value of 0x10 will be processed before a subtable with a processing order value of 0x12. There is a lot more to the INT table than we have discussed here, and more information can be found in version 1.3.1 or higher of the DVB Data Broadcasting Specification (ETSI document EN 301 192).

DSM-CC and SI

It is not enough to simply broadcast DSM-CC carousels or streams as part of your service. A receiver needs to know what that data is, and how it can find it. To this end, DSM-CC and the DVB data broadcasting specification define a number of additions to SI the network operator can use to describe the data a stream is carrying. In the previous section, we saw the SI that applies to multiprotocol encapsulation, and in this section we will discuss the more general SI used by DSM-CC.

The current OCAP specification refers to version 1.0.2 of the MHP standard, and thus OCAP receivers must use the SI specified in the old version of the data broadcasting specification for now. This may change in future versions of the OCAP specification in order to align more closely with MHP.

DSM-CC Streams and the PMT

Just like any other stream in a service, DSM-CC streams must be listed in the PMT. Each type of DSM-CC data has its own stream type, outlined in Table 12.3.

DSM-CC streams with the stream type $0 \times 0D$ are not limited to a particular type of DSM-CC data, and thus streams of this type can carry object carousels, data carousels, descriptors, or any other type of DSM-CC data, including private sections. A receiver therefore needs to check streams with a type of $0 \times 0D$ to see what type of DSM-CC data they contain.

DSM-CC Descriptors

Both the DSM-CC specification and DVB define a number of descriptors that can be carried in the PMT or in other tables to describe the data streams that are part of a service. Table 12.4 describes some of the descriptors defined by the DSM-CC specification.

Of these descriptors, the data broadcast ID descriptor is probably the most common in MHP and OCAP systems. The format of this descriptor is outlined in Table 12.5.

In MHP and OCAP systems, if the `data_broadcast_id` field contains the value 0x0006 the selector gives information about a data carousel. If the value of the `data_broadcast_id` field is 0x0007 the selector will contain a description of an object carousel.

In some cases, the same carousel may be contained in more than one service (e.g., if the same application is available in more than one service). The middleware will use the SI from the

Table 12.3. DSM-CC stream types.

Stream Type	DSM-CC Data Type
0x0A	Multiprotocol encapsulation
0x0B	Object or data carousels
0x0C	DSM-CC descriptors (e.g., NPT descriptors or stream event descriptors)
0x0D	Any DSM-CC data

Table 12.4. Some of the SI descriptors defined by the DSM-CC specification.

Descriptor	Description
Data broadcast descriptor	The data broadcast descriptor is a means of describing the format and type of data encoded in an elementary stream that does not contain audio, video, subtitles, or SI data. As well as describing the type of stream, the data broadcast descriptor lets the broadcaster assign a textual description to it. This descriptor can be carried in the SDT (in the descriptor loop for a particular service entry) or in the EIT (in the descriptor loop for a given event entry). Because it relies on optional tables (and on tables not present in ATSC SI), the data broadcast ID descriptor is often used in its place. This descriptor is defined in EN 300 468 (the DVB SI specification), and more detail specific to DSM-CC streams is added in EN 301 192 (the DVB data broadcasting specification).
Data broadcast ID descriptor	The data broadcast ID descriptor is a shorter version of the data broadcast descriptor that does not include a textual description of the stream. Unlike the data broadcast descriptor, this descriptor is carried in the descriptor loop of the stream's entry in the PMT. For this reason, it is slightly more widely used than the data broadcast descriptor. MHP and OCAP can use a slightly different version of this descriptor. Additional data broadcast ID values have been defined for MHP and OCAP object carousels (0xF0) and for multiprotocol encapsulation streams (0xF1). For either of these data broadcast ID values, the data broadcast ID descriptor includes a list of application types that identifies the types of any applications signaled as auto-start. This gives receivers more information so that they can choose which data streams to connect to first in order to improve performance.
Carousel identifier descriptor (also known as the carousel ID descriptor)	The carousel identifier descriptor provides an alternative way for the broadcaster to give some information about the carousel, and may enable the receiver to load the service gateway for an object carousel without having to load the DSI and DII messages for that carousel. We will learn more about DSI and DII messages in material to follow. In ordinary DVB systems this is purely an optimization, but it is mandatory for MHP object carousels and must identify the PID on which the DSI message for the root of the carousel is broadcast.
Association tag descriptor	Association tags are used by DSM-CC to identify elementary streams in a way that is not DVB specific (like component tags) or that will not change during remultiplexing (like PIDs can). An association tag is linked to a stream by including an association tag descriptor in the descriptor loop of the stream's PMT entry.
Deferred association tags descriptor	An object carousel can use (or refer to) elementary streams that are not part of the current service or transport stream. In this case, receivers may have problems mapping an association tag to the correct stream. The deferred association tags descriptor defines which association tags are referred to by a given object carousel, and tells the receiver which streams contain those association tags. This allows the receiver to access those streams much more efficiently. Each descriptor can contain a mapping for more than one association tag, although each of the association tags in the same deferred association tags descriptor must be present in the same MPEG-2 program. To get around the problem of association tags that refer to elementary streams in different programs, a PMT can contain more than one deferred association tags descriptor.

Table 12.5. Format of the data broadcast ID descriptor.

Syntax	No. of Bits	Identifier
data_broadcast_id_descriptor(){		
descriptor_tag	8	uimsbf
descriptor_length	8	uimsbf
data_broadcast_id	16	uimsbf
for(i=0; i< N; i++){		
id_selector_byte	8	uimsbf
}		
}		

Source: ETSI EN 300 468:2000 (DVB SI specification).

PMT to decide whether two services do refer to the same carousel. For the carousels to be considered identical, the following conditions must be met.

- Both services must be part of the same transport stream.
- Both services must list the boot component of the carousel on the same PID.
- All association tags in the service information must map to the same PIDs.
- The carousel ID of both carousels (as carried in the carousel identifier descriptor) must be identical and in the range 0x100 to 0xffffffff.

If the two services are not in the same transport stream, the following conditions must apply.

- The carousel ID of both carousels (as carried in the carousel identifier descriptor) must be identical and in the range 0x100 to 0xffffffff.
- The carousel identifier descriptor for both carousels must be identical.

DSM-CC Messages

Both object carousels and data carousels carry their data as a series of messages. Even though we are not using data carousels directly, we have to understand the format of data carousel messages because they form the foundation on which object carousels are built.

We will cover both object carousel and data carousel messages in some detail in this section, and thus this information will probably be most useful to middleware developers. Application developers may choose to skip this section, although a thorough understanding of how DSM-CC works in practice may be useful in order to get the most from your applications and services.

Data Carousel Messages

The messages that make up a DSM-CC data carousel fall into two basic categories. Download data messages contain the actual data belonging to the modules in the carousel, whereas download control messages tell the receiver how the download data messages are organized into modules.

Download Data Messages

There is only one type of download data message, the `DownloadDataBlock` message. As the name suggests, each `DownloadDataBlock` message corresponds to one block of data that is broadcast as a single unit. Each `DownloadDataBlock` is also the same size (4,096 bytes, or the size of one MPEG-2 section), except for the last block in each module. This makes it easier for the receiver to parse the messages.

Each `DownloadDataBlock` message contains the ID and version number of the module it is part of, the block number within that module, and the data itself. The module ID and version number let the receiver know whether it should pay attention to this message. The receiver will know which modules it should load, and whether the block refers to a current version of the module or an old one. The block number tells the receiver what part of the module is contained in that block.

Download Control Messages

Now that we have seen how data is carried in a carousel, we need to see how the blocks that carry that data are organized in to a sensible structure. The structure of a module is defined by one or more download control messages.

As we have already seen, small modules may fit into a single block, whereas larger modules may need more than one block. In the latter case, and in the case in which there is more than one module in the carousel, some extra control information is needed to describe how the blocks are grouped into modules. This is done using the first type of download control message, known as a `DownloadInfoIndication` message.

DownloadInfoIndication Messages

Each `DownloadInfoIndication` (DII) message contains a description of a number of modules and of the parameters used to transmit them. Any modules listed in the same DII message are said to be members of the same group. This provides a way for broadcasters to identify modules that belong together, for instance, because they all carry different parts of a single block of data.

The DII message gives the size of the `DownloadDataBlock` messages used to transmit the module data, and for each module that it describes it gives the ID of the module, its version number, and its size, as well as a number of descriptors that may give a more detailed description of the module. By knowing the size of a module and the size of each `Download-DataBlock` message, the receiver can calculate how many blocks will be used to transmit each module. This lets the receiver make sure that it has all of the blocks that make up a module. The version number lets the receiver see when the content of a given module has changed, and tells the receiver to replace any cached copies.

DownloadServerInitiate Messages and Two-layer Carousels

One limitation MPEG imposes on DSM-CC concerns the size of the messages. Each DSM-CC message is limited to a total size of 4 K, which can cause problems for DII messages. In particular, it means that a DII message can only describe up to about 150 modules. For small carousels this is not a problem, and carousels that only use a single DII message are called one-layer carousels.

Bigger carousels (and other scenarios we will see later) will need more than one DII message to describe all of the modules. Because we cannot do this using normal DII messages, DSM-CC defines the `DownloadServerInitiate` (DSI) message. This acts as a top-level control message for those carousels that have several DII messages by grouping a number of DII messages, and the modules associated with them, into a single supergroup. Carousels in which a number of DII messages are linked to a supergroup by a DSI message are called two-layer carousels.

Groups and Supergroups

We have already seen that all modules listed in the same DII message are said to be members of the same group (which includes the DII message itself), and that a DSI message and any DII messages the DSI message refers to and their associated modules are members of the same supergroup. How do we tell which group and supergroup contains a given module, though?

Identifying the members of a given group is easy. Each DII message contains a list of the module IDs that identify the modules in that group. Each `DownloadDataBlock` message that is a member of a given module also contains that module ID.

For supergroups, the `download_id` field of every DII and `DownloadDataBlock` message within the supergroup is set to the same value. We also get some more help from the DSI message to identify the DII messages that make up a supergroup. The `group_info_data` field of a DSI contains a set of descriptors that describe how the DSI is linked to the DII messages that are part of its supergroup. The most important of these descriptors from our perspective is the group link descriptor. This identifies a DII message that is part of the supergroup described by the DSI message.

Each group link descriptor identifies one DII message that is linked to this DSI, and thus a DSI message will typically contain several group link descriptors. The `group_id` field contains the transaction ID of the DII, whereas the position field identifies where that descriptor comes in the list. A value of 0x00 indicates that a descriptor is the first one in the list of DIIs, a value of 0x01 indicates that the descriptor is in the middle of the list, and a value of 0x02 indicates that the descriptor is the last one in the list.

Descriptors in Download Control Messages

DII and DSI messages may include descriptors to carry additional information that is not needed for all messages. DVB has defined a number of descriptors that can be used to carry information that is useful to the receiver, and some of these are outlined in Table 12.6. The descriptor tag values for these descriptors are private to DVB, and thus they may have other meanings in other environments. For this reason, network operators should not use these descriptors outside DVB or OCAP networks.

Object Carousel Messages

Objects in an object carousel are carried as DSM-CC BIOP messages that contain the content of the object within the message itself. For instance, a file message will contain the complete content of that file.

Table 12.6. Some of the descriptors for download control messages that DVB defines.

Descriptor	Description
Name descriptor	This descriptor allows the broadcaster to set a human-readable name for a module or group of modules. In complex data carousels, this lets the broadcaster annotate the carousel so that the receiver can use this information to provide some more friendly information to the user. Although users may not care (do you care what a module is called?), it may sometimes be useful. Because this descriptor can be used to name modules or groups of modules, it can be carried in either the DII `ModuleInfo` structure or the DSI `GroupInfo` structure.
Type descriptor	This descriptor identifies the MIME type of data contained in a specific module or group. This may be carried either in the DII `ModuleInfo` structure or in the DSI `GroupInfo` structure. This descriptor is most useful when the module or group contains just one file, or where all files have the same MIME type.
Compressed module descriptor	Data carousel modules may be compressed to save space in the transport stream. This descriptor indicates that the module has been compressed using the `zlib` compression scheme. Because this descriptor only refers to modules, it can only be carried in the DII `ModuleInfo` structure.
Estimated download time descriptor	This descriptor gives the receiver some indication of how long it will take to download the module or group. This is useful for several reasons. First, the receiver can tell the user how long they need to wait before the data is available. Second, a clever application can use this information to know how far in advance it should begin prefetching a module, making it easier for it to fetch data before it actually needs it. Given the latency involved in getting data from a carousel, this can be extremely useful. Like the type descriptor and the name descriptor, the estimated download time may be carried either in the DII `ModuleInfo` structure or in the DSI `GroupInfo` structure.

Table 12.7. Format of the BIOP message header.

Syntax	No. of Bits	Identifier	Value	Comment
BIOP message header {				
magic	4 × 8	uimsbf	0x42494F50	"BIOP"
Biop_version.major	8	uimsbf	0x01	BIOP major version 1
Biop_version.minor	8	uimsbf	0x00	BIOP minor version 0
Byte_order	8	uimsbf	0x00	Big endian byte ordering
message_type	8	uimsbf	0x00	
message_size	32	uimsbf	*	
objectKey_length	8	uimsbf	N1	
for(i=0; i< N1; i++) {				
objectKey_data_byte	8	uimsbf	+	
}				
objectKind_length	32	uimsbf	0x00000004	
objectKind_data	4 × 8	uimsbf		
}				

As well as the payload, each DSM-CC message will have some additional fields that describe the message. The first of these fields is always a 4-byte magic number: the ASCII string BIOP (0x42494F50 hexadecimal). This identifies the start of a BIOP message and allows the receiver to easily check whether a given data carousel module is carrying BIOP messages. A module will always contain a complete BIOP message, and thus any modules containing object carousel data will start with this magic number.

From our perspective, however, the most useful field is the objectKind_data, which identifies the type of data carried in a message. DSM-CC allows a great deal of flexibility in this, but DVB defines this value to be a 4-byte string containing 3 bytes of data and a terminating byte with the value 0x00. We will examine the values that can be given to this field when we examine each message in turn.

The final field we will examine is the objectKey. This is a unique value that identifies the message, and which is used within CORBA interoperable object references (IORs) to identify that particular object. Object carousel generators will use this field to make sure that any references to an object in the carousel refer to the correct object. Each BIOP message starts with a header. This header contains the fields outlined in Table 12.7.

The version numbers refer to the version of the BIOP protocol being used. For MHP and OCAP, this is fixed as version 1.0, and thus the version numbers in the message headers will always be the same.

File Messages

File messages carry the content of any files that are part of the object carousel. These messages have the format outlined in Table 12.8.

Table 12.8. Format of the BIOP file message.

Syntax	No. of Bits	Identifier	Value	Comment
BIOP::FileMessage() {				
magic	4 × 8	uimsbf	0x42494F50	"BIOP"
Biop_version.major	8	uimsbf	0x01	BIOP major version 1
Biop_version.minor	8	uimsbf	0x00	BIOP minor version 0
Byte_order	8	uimsbf	0x00	Big endian byte ordering
message_type	8	uimsbf	0x00	
message_size	32	uimsbf	*	
objectKey_length	8	uimsbf	N1	
for(i=0; i< N1; i++) {				
objectKey_data_byte	8	uimsbf	+	
}				
objectKind_length	32	uimsbf	0x00000004	
objectKind_data	4 × 8	uimsbf	0x66696C00	"fil" type_id alias
objectInfo_length	16	uimsbf	N2	
DSM::File::ContentSize	64	uimsbf	+	objectInfo
for(i=0; i< N2-8; i++) {				
objectInfo_data_byte	8	uimsbf	+	
}				
serviceContextList_count	8	uimsbf	N3	serviceContext List
for(i=0; i< N3; i++) {				
serviceContextList_data_byte	8	uimsbf	+	
}				
messageBody_length	32	uimsbf	*	
content_length	32	uimsbf	N4	
for(i=0; i< N4; i++) {				
content_data_byte	8	uimsbf	+	Actual file content
}				
}				

Source: ETSI TR 101 202: 1999 (DVB data broadcasting implementation guidelines).

The value of the `objectKind_data` field for file messages is 0x66696C00, which corresponds to the ASCII characters fil followed by a terminating zero.

Directory and Service Gateway Messages

BIOP directory and service gateway messages represent directories in the object carousel. Service gateway messages are used to represent the root directory of the carousel. Both types of messages share the same structure, with some minor differences in the values of some of the fields. Table 12.9 outlines the format of these messages.

Directory messages and service gateway messages have exactly the same structure (after all, a service gateway is a directory but just happens to be the root directory of the carousel). The value of the `objectKind_data` field for a directory message is 0x64697200, corresponding to the ASCII characters `dir` with a terminating zero. For a service gateway message, the value of this field is 0x73726700 (`srg`, again followed by a terminating zero).

Directory objects contain a list of references to the files and directories they contain, carried as IORs. We will take a detailed look at the format of IORs later in the chapter.

Stream Messages

The structure for BIOP stream messages can be seen in Table 12.10. Stream messages are references to streams that are being broadcast, either as part of the transport stream containing the object carousel or in another transport stream. In a video-on-demand system, stream objects may refer to streams that are available to be viewed but are not being transmitted at that time.

Stream messages use the value 0x73747200 (`str` with a terminating zero) for the `objectKind_data` field. One element that may not be familiar in this data structure is a list of taps. A tap is a structure used in DSM-CC to refer to a specific MPEG stream (either an elementary stream or a complete program) or to a DSM-CC DII or DSI message. In the case of stream messages, these taps refer to the stream the message describes. There may be one tap that refers to the complete stream, or there may be several taps that refer to a number of elementary streams that make up a single logical stream as described in the object carousel.

To give an example of this, a DVB service may contain two audio tracks for different languages. A DSM-CC object carousel may contain a single stream object that refers to that entire

Table 12.9. Format of the BIOP directory/service gateway message.

Syntax	No. of Bits	Identifier	Value	Comment
BIOP::DirectoryMessage() {				
magic	4 × 8	uimsbf	0x42494F50	"BIOP"
biop_version.major	8	uimsbf	0x01	BIOP major version 1
biop_version.minor	8	uimsbf	0x00	BIOP minor version 0
byte_order	8	uimsbf	0x00	Big endian byte ordering
message_type	8	uimsbf	0x00	
message_size	32	uimsbf	*	
objectKey_length	8	uimsbf	N1	

Table 12.9. *Continued*

Syntax	No. of Bits	Identifier	Value	Comment
for(i=0; i<N1; i++) {				
objectKey_data_byte	8	uimsbf	+	
}				
objectKind_length	32	uimsbf	0x00000004	
objectKind_data	4 × 8	uimsbf	0x64697200	"dir" type_id alias
objectInfo_length	16	uimsbf	N2	Object info
for(i=0; i<N2; i++) {				
objectInfo_data_byte	8	uimsbf	+	
}				
serviceContextList_count	8	uimsbf	N3	serviceContext List
for(i=0; i<N3; i++) {				
serviceContextList_data_byte	8	uimsbf	+	
}				
messageBody_length	32	uimsbf	*	
bindings_count	16	uimsbf	N4	
for(i=0; i<N4; i++) {				Binding
BIOP::Name() {				
nameComponents_count	8	uimsbf	N5	
for(i=0; i<N5; i++) {				
Id_length	8	uimsbf	N6	NameCompo-nent id
for(j=0; j<N6; j++) {				
id_data_byte	8	uimsbf	+	
}				
kind_length	8	uimsbf	N7	NameCompo-nent kind
for(j=0; j<N7; j++) {				
kind_data_byte	8	uimsbf	+	As type_id
}				
}				
}				
bindingType	8	uimsbf	+	0×01 for nobject 0×02 for ncontext
IOP::IOR()+objectRef				
objectInfo_length	16	uimsbf	N8	
for(j=0; j<N8; j++) {				
objectInfo_data_byte	8	uimsbf	+	
}				
}				
}				

Source: ETSI TR 101 202: 1999 (DVB data broadcasting implementation guidelines).

Table 12.10. Format of the BIOP stream message.

Syntax	No. of Bits	Identifier	Value	Comment
BIOP::StreamMessage() {				
magic	4 × 8	uimsbf	0x42494F50	"BIOP"
biop_version.major	8	uimsbf	0x01	BIOP major version 1
biop_version.minor	8	uimsbf	0x00	BIOP minor version 0
byte_order	8	uimsbf	0x00	Big endian byte ordering
message_type	8	uimsbf	0x00	
message_size	32	uimsbf	*	
objectKey_length	8	uimsbf	N1	
for (i=0; i<N1; i++) {				
objectKey_data_byte	8	uimsbf	+	
}				
objectKind_length	32	uimsbf	0x00000004	
objectKind_data	32	uimsbf	0x73747200	"str" type_id alias
objectInfo_length	16	uimsbf	N6	
DSM::Stream::Info_T {		objectInfo		
aDescription_length	8	uimsbf	N2	aDescription
for (i=0; i<N2; i++) {				
aDescription_bytes	8	uimsbf	+	
}				
duration.aSeconds	32	simsbf	+	AppNPT seconds
duration.aMicroSeconds	32	uimsbf	+	AppNPT microseconds
audio	8	uimsbf	+	
video	8	uimsbf	+	
data	8	uimsbf	+	
}				
for (i=0; i=N6-(N2+10); i++) {				
objectInfo_byte	8	uimsbf	+	
}				
serviceContextList_count	8	uimsbf	N3	serviceContext List
for(i=0; i<N3; i++) {				
serviceContextList_data_byte	8	uimsbf	+	
}				
messageBody_length	32	uimsbf	*	
Taps_count	8	uimsbf	N4	
for(i=0; i<N4; i++) {				
Id	16	uimsbf	0x0000	Undefined
use	16	uimsbf	+	
association_tag	16	uimsbf	+	
selector_length	8	uimsbf	0x00	No selector
}				
}				

Source: ETSI TR 101 202: 1999 (DVB data broadcasting implementation guidelines).

Table 12.11. Possible types of taps that may be used in BIOP stream or stream event messages.

TapUse *Field*	*Value*	*Broadcast on PID*
STR_NPT_USE	0x000B	Stream NPT descriptors
STR_STATUS_AND_EVENT_USE	0x000C	Both stream mode and stream event descriptors
STR_EVENT_USE	0x000D	Stream event descriptors
STR_STATUS_USE	0x000E	Stream mode descriptors
BIOP_ES_USE	0x0018	Elementary stream (video/audio)
BIOP_PROGRAM_USE	0x0019	Program (DVB service) reference

Source: ETSI TR 101 202: 1999 (DVB data broadcasting implementation guidelines).

service. However, it may also contain two separate stream objects, each referring to the video portion of the stream and one of the two audio streams. Alternatively, it may contain one stream object that refers to the video stream and one of the audio streams while completely ignoring the other (all of these are possible). Table 12.11 outlines the types of taps that can be used in stream or stream event messages.

Stream Event Messages

The last type of BIOP message is the stream event message, which identifies one or more stream events that may be included in a stream. Stream event messages are logically a subclass of stream messages, and thus they share many common fields. Each stream event message has the structure outlined in Table 12.12.

Stream event messages use the value `ste` for the `objectKind_data` field (0x73746500 hexadecimal, once we have added the terminating zero). As with stream messages, a stream event message contains a list of taps that refer to the stream containing the stream events. More than one tap may be needed, because the stream event object may refer to an elementary stream carrying NPT data as well as to streams carrying stream event descriptors.

Referring to Streams and Objects

Many of the DSM-CC messages we have just seen need to refer to other objects or streams in the transport stream. Although most of the objects we refer to will be in the same object carousel (for instance, directory messages will refer to the objects representing the files and subdirectories contained in that directory), in some cases we will want to refer to objects in other carousels.

We have seen a number of ways of referring to content, and these approaches all have problems from DSM-CC's point of view. Several of the content-referencing schemes depend on a specific type of SI, whereas others are so low level that the references may change if PIDs are remapped when the data crosses a network boundary. In light of these issues, DSM-CC supports several mechanisms that work together so that DSM-CC messages can refer to content no matter what network protocols are used.

Table 12.12. Format of the BIOP stream event message.

Syntax	No. of Bits	Identifier	Value	Comment
BIOP::StreamEventMessage() {				
Magic	4 × 8	uimsbf	0x42494F50	"BIOP"
version.major	8	uimsbf	0x01	BIOP major version 1
version.minor	8	uimsbf	0x00	BIOP minor version 0
byte_order	8	uimsbf	0x00	Big endian byte ordering
message_type	8	uimsbf	*	
message_size	32	uimsbf	*	
objectKey_length	8	uimsbf	N1	
for(i=0; i<N1; i++) {				
objectKey_data_byte	8	uimsbf	+	
}				
objectKind_length	32	uimsbf	0x00000004	
objectKind_data	4 × 8	uimsbf	0x73746500	"ste" type_id alias
objectInfo_length	16	uimsbf	N6	
DSM::Stream::Info_T {				
aDescription_length	8	uimsbf	N2	aDescription
for(i=0; i<N2; i++) {				
aDescription_bytes	8	uimsbf	+	See BIOP::Stream Message()
}				
duration.aSeconds	32	simsbf	+	See BIOP::Stream Message()
duration.aMicroSeconds	16	uimsbf	+	See BIOP::Stream Message()
Audio	8	uimsbf	+	See BIOP::Stream Message()
Video	8	uimsbf	+	See BIOP::Stream Message()
Data	8	uimsbf	+	See BIOP::Stream Message()
}				
DSM::Event::EventList_T {				
eventNames_count	16	uimsbf	N3	
for(i=0; i<N3; i++) {				
eventName_length	8	uimsbf	N4	

Table 12.12. *Continued*

Syntax	No. of Bits	Identifier	Value	Comment
for(j=0; j<N4; j++) { eventName_data_byte	8	uimsbf	+	(Including zero terminator)

}
 }
}
for(i=0; i=N6-(N2+10)-(2+N3+
  sum(N4));i++){objectInfo_byte
}
	8	uimsbf	+	

serviceContextList_count	8	uimsbf	0x00	Empty serviceContext List

for(i=0; i<N3; i++) {
  serviceContextList_data_byte
}
	8	Uimsbf	+	

messageBody_length	32	uimsbf	*	
taps_count	8	uimsbf	N5	
for(i=0; i<N5; i++) { id	16	uimsbf	0x0000	Undefined
use	16	uimsbf	+	See Table 12.11
association_tag	16	uimsbf	+	
selector_length	8	uimsbf	0x00	No selector
}				
eventIds_count	8	uimsbf	N3	(= eventNames_count)
for(i=0; i<N3; i++) { eventId	16	uimsbf	+	
} }				

Source: ETSI TR 101 202: 1999 (DVB data broadcasting implementation guidelines).

Association Tags

As we have just seen, several BIOP messages have a need to refer to MPEG programs or elementary streams. Although the most obvious way to refer to an MPEG stream is by its PID, this has the following two disadvantages.

- PIDs only refer to elementary streams, not entire programs.
- The PID for any stream can be changed when it is remultiplexed. This means we would have to update every reference in an object carousel were we to remultiplex the stream.

DVB introduced the concept of component tags to solve this very problem, but unfortunately component tags are not available in non-DVB systems. To solve this problem, DSM-CC introduces another identifier called an association tag. This serves the same purpose as a component tag in DVB systems: it provides a way to identify a stream that is easy for the broadcaster and receiver to use and that is not affected by remultiplexing. The association tag is attached to a stream by inserting an association tag descriptor in the stream's PMT entry.

Taps

It is not enough to be able to refer to streams alone, however. Some BIOP messages need to refer to specific download control messages within the data carousel that carries the object carousel.

References to streams or download control messages are both coded as taps to avoid defining two separate types of references. A tap uses the association tag that is introduced by DSM-CC to refer to specific streams without relying on PIDs or on data that is specific to one SI system. The format of a tap in a DVB or OCAP system is outlined in Table 12.13.

The id field is a value that identifies a particular tap. This is private to the carousel, and is simply a 16-bit value. The use field indicates exactly how that tap is used. For the sake of this discussion, we can assume that this refers to the type of object the tap points to, such as a DSM-CC data carousel message, an elementary stream, or a complete MPEG-2 program. Table 12.14 outlines the use values that are supported.

Some of the places where these values are used may not be clear now, but we will explain these later in the chapter. The value given in the association_tag field identifies an elementary stream. This identifies the elementary stream carrying the carousel in a way that does not depend on PIDs or other information that may change. Later, we will see how we can link the association tag to a particular elementary stream.

Table 12.13. Format of the DSM-CC tap.

Syntax	No. of Bits	Identifier	Value
BIOP::Tap() {			
id	16	uimsbf	
use	16	uimsbf	
association_tag	16	uimsbf	+
selector_length	8	uimsbf	
selector_type	16	uimsbf	
transactionId	32	uimsbf	*
timeout	32	uimsbf	*
}			

Table 12.14. Possible values for the `use` field in a DSM-CC tap.

Mnemonic	Value	Description
BIOP_DELIVERY_PARA_USE	0x16	Module delivery parameters
BIOP_OBJECT_USE	0x17	BIOP objects in modules
STR_NPT_USE	0x0B	Stream NPT descriptors
STR_STATUS_AND_EVENT_USE	0x0C	Both stream mode and stream event descriptors
STR_EVENT_USE	0x0D	Stream event descriptors
STR_STATUS_USE	0x0E	Stream mode descriptors
BIOP_ES_USE	0x18	Elementary stream (video/audio)
BIOP_PROGRAM_USE	0x19	Program (DVB service) reference

Source: ETSI TR 101 202: 1999 (DVB data broadcasting implementation guidelines).

The `selector_type` field is an optional field that allows a tap to identify objects within a carousel. A `selector_type` of 0x01 identifies a message selector, and this indicates that the selector contains two fields: a transaction ID and a timeout. The `transactionId` field corresponds to the transaction ID of the DSM-CC message to which this tap refers. This enables the tap to identify a specific DII or DSI message in a data carousel, and is used by IORs to identify a specific message within a carousel. In a general DSM-CC implementation, the selector is not as well defined as we have seen here, and is just a sequence of up to 255 bytes.

Interoperable Object References

When one DSM-CC object needs to refer to another object, we need some interoperable way of doing this. Given that DSM-CC is based on CORBA, it uses a CORBA structure called an interoperable object reference (IOR). Put simply, it gives each object a unique identifier that can be used instead of a pointer to that object.

Each IOR contains a profile body for every location from which the file is available. This allows a single IOR to provide multiple sources for the same file. In the case of DSM-CC in a broadcast environment, this feature is not used very often and an IOR normally contains one profile body that points to an object in a DSM-CC object carousel. As broadband IP connections become more common, this may change. In the future, an IOR may provide references to an object either from a DSM-CC object carousel or from a web site, for instance.

For DSM-CC systems, the most common type of profile body is the BIOP. This contains an object location and a `ConnBinder`. The object location locates the object within a given carousel, and contains fields giving the carousel ID, the module ID, and the object key of the specific object in the carousel. These three values taken together give a specific reference to the object in question.

As well as knowing which object we are referring to, we also need to know how to find it. The `ConnBinder` contains one or more taps that identify the data carousel DII messages describing the module containing the object. This allows the receiver to access the object referred to by the IOR as quickly as possible.

The main restriction of the BIOP profile body is that it can only refer to objects in the same carousel. To refer to objects in a different carousel, we must use the Lite Options profile body. This contains a reference to the carousel containing the object and a reference to the full path of the object within the carousel.

The path is carried as a set of path components, with each component having a name and a type (which identifies the type of DSM-CC object referred to by that component). For the carousel reference, we use the NSAP address of the carousel containing the object. This is unique to that carousel, and thus gives us an ideal reference. NSAP addresses are 20-byte structures, which can contain some private data to help locate a DSM-CC carousel. For DVB systems, NSAP addresses will have the format outlined in Table 12.15. For other systems, the dvb_service_location field may be replaced with other private data.

We have already seen NSAP addresses in use, but not actually called them that. For instance, the server ID field in a DSI message is actually an NSAP address (although this is not used in data broadcasting).

If the IOR of an object, or the IOR of one of its parent directories, refers to a carousel that has not been mounted by the receiver (either by the middleware or by an application), any

Table 12.15. Format of an NSAP address for DVB object carousels.

Syntax	No. of Bits	Identifier	Value	Comment
DVBcarouselNSAPaddress()				
AFI	8	uimsbf	0x00	NSAP for private use
type	8	uimsbf	0x00	Object carousel NSAP address
carouselId	32	uimsbf	+	
specifierType	8	uimsbf	0x01	IEEE OUI
specifierData{ IEEE OUI }	24	uimsbf	0x< DVB >	Constant for DVB OUI
dvb_service_location() {				
transport_stream_id	16	uimsbf	+	
original_network_id	16	uimsbf	+	
service_id	16	uimsbf	+	(= MPEG2 program_num ber)
reserved	32	bslbf	0xFFFFFFFF	
}				
}				

Source: ETSI TR 101 202: 1999 (DVB data broadcasting implementation guidelines).

attempts to access the object referred to by that IOR will fail. The middleware will behave as if the file does not exist and will not mount the object carousel automatically. The reason for this is that mounting an object carousel may use a number of resources such as section filters, and may involve tuning to a new transport stream. Because of the effects this will have on other applications, automatically mounting an object carousel in this way is not allowed. The only time a carousel is mounted automatically is when the receiver detects that a new application is being signaled. In this case, it may mount the carousel containing the application.

Transporting Object Carousels in Data Carousels

Object carousels impose a few limitations on the format of data carousel messages. Although these are not significant, there is one we need to be aware of. Data carousel `DownloadInfoIndication` messages used as part of an object carousel must contain a BIOP `ModuleInfo` structure within the `module_info` field. Table 12.16 shows how the `ModuleInfo` structure is organized.

Object carousels also use `DownloadServerInitiate` messages to carry a description of the service gateway. A `DownloadServerInitiate` message used in an object carousel will

Table 12.16. Format of the `module_info` field for data carousel DII messages used in an object carousel.

Syntax	No. of Bits	Identifier	Value	Comment
`BIOP::ModuleInfo() {`				
`  moduleTimeOut`	32	uimsbf	+	
`  blockTimeOut`	32	uimsbf	+	
`  minBlockTime`	32	uimsbf	+	
`  taps_count`	8	uimsbf	N1	
`  for(j=0;j<N1;j++){`				
`    id`	16	uimsbf	0x0000	User private
`    use`	16	uimsbf	0x0017	BIOP_OBJECT_USE
`    association_tag`	16	uimsbf	+	
`    selector_length`	8	uimsbf	0x00	
`  }`				
`  UserInfoLength`	8	uimsbf	N2	
`  for(j=0;j<N2;j++){`				
`    userInfo_data_byte`	8	uimsbf	+	(Including zero terminator)
`  }`				
`}`				

Source: ETSI TR 101 202: 1999 (DVB data broadcasting implementation guidelines).

Table 12.17. Format of the `service_gateway_info` field for data carousel DSI messages used in an object carousel.

Syntax	No. of Bits	Identifier	Value	Comment
`ServiceGatewayInfo() {`				
`  IOP::IOR()`		+		
`  downloadTaps_count`	8	uimsbf	N1	Software download taps
`  for(i=0; i<N1; i++){`				
`    Tap()`	8	uimsbf	+	
`  }`				
`  serviceContextList_count`	8	uimsbf	N2	serviceContext List
`  for(i=0; i<N2; i++){`				
`    serviceContextList_data_byte`	8	uimsbf	+	
`  }`				
`  userInfoLength`	16	uimsbf	N3	User info
`  for(i=0; i<N3; i++){`				
`    userInfo_data_byte`	8	uimsbf	+	
`  }`				
`}`				

Source: ETSI TR 101 202: 1999 (DVB data broadcasting implementation guidelines).

carry a `ServiceGatewayInfo` structure in its `private_data` field. The format of the `ServiceGatewayInfo` structure is shown in Table 12.17.

The `IOR` field contains an IOR to the service gateway BIOP message for that carousel. The use of this structure to describe a service gateway has implications that may not be obvious at first. It means that an object carousel will always be carried by a two-layer carousel. Because every object carousel must have a service gateway (by definition, the root directory of the carousel is the service gateway), it must also have a data carousel DSI message to describe that service gateway.

Parsing DSM-CC Messages

Parsing the two types of DSM-CC messages follows the same basic approach, but there are a few things we need to be aware of. Data carousel messages are designed to be filtered by hardware section filters, and thus this makes life easy for our implementation. We will not discuss in detail how we parse data carousel messages. The ability to use hardware section filters to filter specific message types (or even specific modules) gives a good starting point, and the structure of the data carousel is fairly obvious.

Before we can do this, however, we need to find which data carousel modules contain the data we are interested in. Like data carousel messages, BIOP messages are designed for easy filtering. Unfortunately, we cannot do this just using section filters, because BIOP messages can start at any point in a data carousel module. Despite this, finding BIOP messages in a data carousel module is not very difficult. The header of every BIOP message includes the length of the message, and we can use this to skip past messages that do not meet our search criteria. This gives us a fast way of ignoring messages, but it does mean that we have to filter every section in the module in case the message we are looking for is somewhere other than right at the start.

The biggest problem we have when parsing BIOP messages is name resolution. BIOP file messages do not contain any information about the name of the file they represent because one `File` object can actually be present in more than one place in the directory structure. This allows for structures such as hard links in a UNIX file system, wherein a file can have more than one name.

To resolve the name of a file in DSM-CC, we have to load the directory messages for every directory in the path. This can take some time, and thus it is generally worth caching directory messages (or at least the information they contain). Once we have loaded the directory message for the root directory of the carousel, we can use this to resolve the name of the next subdirectory into an object reference, and so on down the directory hierarchy until we reach the file.

Using the DSM-CC API

As we have already seen, an MHP or OCAP receiver will treat the DSM-CC object carousel containing an application as the default file system for that application, and the JVM will treat the root directory of the application as the current working directory. This means that in many cases an application does not really need to use a different API to access files. So why does MHP provide an API for developers to access the object carousel?

Partly, the DSM-CC API offers new functionality to support the features of an object carousel that are not supported in the standard file API. It also provides access to those parts of DSM-CC that do not relate to files. It is important to remember that a DSM-CC object carousel is used to broadcast objects, not just files. Although files and directories are most common, we may also see streams or stream events. In addition to this, we can also access stream event descriptors and NPT descriptors. In this section, we will only consider the aspects of the DSM-CC API related to file system access.

As with any other type of file system, there are some basic things we will want to do with a DSM-CC file system. The most obvious of these it is to open files and read the data they contain. Because DSM-CC is a broadcast file system, we will have read-only access to any files within the carousel.

Manipulating DSM-CC Objects

The DSM-CC API is contained in the `org.dvb.dsmcc` package. One of the most important classes in this API is the `DSMCCObject` class. Instances of this class (a subclass of `java.io.File`) represent objects in a DSM-CC carousel. As with a standard `File` object, a

`DSMCCObject` is created with either an absolute or relative path name, and can then be used in much the same way as a standard file. The only restrictions are that if an instance of `DSM-CCObject` represents a directory applications should use the methods provided by `java.io.File` to manipulate it. For instances representing files, the `java.io.FileInputStream` or `java.io.RandomAccessFile` classes must be used to read the content of the file.

This approach, familiar to all Java developers, has a number of limitations. In particular, access to a DSM-CC carousel using the standard `java.io` file operations is synchronous, and thus the thread making the request will block until the operation has finished. Given the latency involved in working with DSM-CC, this can cause real problems for a developer, especially if you want to preload content so that you can display it quickly when the user wants it. In a synchronous API, you would either have to load each file on a different thread or wait for one file to finish loading before loading another. Neither of these options is acceptable when you have a large number of files to load.

To help solve this problem, the `DSMCCObject` class supports asynchronous loading of files. The `asynchronousLoad()` method allows an application to load a DSM-CC object into the cache without blocking the current thread. This method takes an `AsynchronousLoadingEventListener` object as a parameter, which will be notified when loading is complete. Thus, an application can get on with other tasks while waiting for content to load and then read the file's content while it is cached.

Similarly, the `synchronousLoad()` method loads an object into the cache synchronously. Why is this useful? First, it will load any DSM-CC object, not just a file (as will the `asynchronousLoad()` method). Second, it allows the application to abort the loading process from another thread using the `abort()` method (which can also be used to abort an asynchronous load). This gives a little more flexibility than simply using the operations provided by the standard Java file operations. By explicitly loading the file first, an application can reduce the latency on other file operations to a known level. If we call `synchronousLoad()` on an object from an object carousel we have not mounted yet, the middleware will throw a `ServiceXFRException`. Using `asynchronousLoad()` on the same object would generate a `ServiceXFRErrorEvent`.

It is possible to load directory information separately from the files in that directory. Hence, we can know the names and sizes of files within a directory but not have any of those files loaded. This directory information is also contained in the object carousel, and will thus have high latency for access. The `loadDirectoryEntry()` class allows an application to load directory information asynchronously so that the information is available to the application when it is needed.

All of these methods only load an object into the cache so that the usual `java.io` file operations can read the object without a delay. The number of objects that can be cached at any time depends on the amount of memory used for the DSM-CC cache, the caching strategy of the receiver, and the size of the objects that have been loaded. If too many objects are loaded, the receiver will flush some objects from the cache. Thus, application developers

331

should take care to load only those objects they will use very soon after loading. This gets more complicated if more than one application is running, in that both applications may be loading objects at the same time.

The DSMCCObject class provides two other methods to control the loading of objects. The prefetch() method allows the application to hint to the DSM-CC subsystem and cache that a file will be used in the near future, allowing the receiver some extra time to load the object before it is needed. On the other hand, the unload() method tells the receiver that an object is no longer needed and that it can be flushed from the cache. Both of these methods only give the receiver a hint, however, and thus the receiver may choose to ignore them in some circumstances.

As we have already mentioned, DSM-CC file systems allow the broadcaster to update the content of a file on the fly. The addObjectChangeEventListener() and removeObjectChangeEventListener() methods allow an application to register and unregister, respectively, for notification when a particular object gets updated. For instance, a news ticker application could use this to receive notification of when the text of a story is updated.

Because an object carousel can contain objects other than files, the DSMCCObject class allows an application to determine what type of object is being referred to using the isObjectKindKnown(), isStream(), and isStreamEvent() messages. We can also get a file:// URL that refers to a particular file using the getURL() method. This can be useful when dealing with APIs (such as JMF) that use a URL to refer to pieces of content.

The full interface for the DSMCCObject class follows. This includes a few methods we have not discussed, but these are much less commonly used.

```
public class DSMCCObject extends java.io.File {

  public DSMCCObject (java.lang.String path);
  public DSMCCObject (
     java.lang.String path, java.lang.String name);
  public DSMCCObject (
     DSMCCObject dir, java.lang.String name);

  public boolean isLoaded();

  public boolean isObjectKindKnown();
  public boolean isStream();
  public boolean isStreamEvent();

  public void synchronousLoad()
     throws InvalidFormatException,
            InterruptedIOException,
            MPEGDeliveryException,
            ServerDeliveryException,
```

```
                  InvalidPathNameException,
                  NotEntitledException,
                  ServiceXFRException;

   public void asynchronousLoad (
      AsynchronousLoadingEventListener l)
      throws InvalidPathNameException;

   public void abort() throws NothingToAbortException;

   public static boolean prefetch(
      String path, byte priority);

   public static boolean prefetch(
      DSMCCObject dir, java.lang.String path, byte priority);

   public void unload() throws NotLoadedException;

   public java.net.URL getURL();

   public void addObjectChangeEventListener(
      ObjectChangeEventListener listener);
   public void removeObjectChangeEventListener(
      ObjectChangeEventListener listener);

   public void loadDirectoryEntry (
      AsynchronousLoadingEventListener l);

   public void setRetrievalMode(int retrieval_mode);
   public static final int FROM_CACHE = 1;
   public static final int FROM_CACHE_OR_STREAM = 2;
   public static final int FROM_STREAM_ONLY = 3;

   public X509Certificate[] [] getSigners();
}
```

Mounting an Object Carousel

There is no requirement for a DVB service to contain just one object carousel. There is not even a requirement for an application to only access object carousels that are part of the same service. Thus, if we can access other object carousels how do we do it?

Just like file systems in UNIX, object carousels can be mounted into a location in the current directory hierarchy. The MHP DSM-CC API represents a service domain using the ServiceDomain class. Before a service domain can be used, it must be attached using the ServiceDomain.attach() method. This mounts the service domain in the file system hierarchy, in the same way that the UNIX mount command does. There are three different versions of this method, as follows, each taking a different set of parameters.

- `public void attach(Locator l);`
- `public void attach(Locator service, int carouselId);`
- `public void attach(byte[] NSAPAddress);`

The first version takes a locator referring to a component of a service that contains an object carousel. The second version takes a locator for a service and a carousel ID, so that the receiver can determine which carousel on that service to use. The final version takes an NSAP address to identify the carousel. Most applications are more likely to use one of the first two versions.

One thing you will notice is that none of these methods allows you to specify where in the directory hierarchy the service domain is attached. This makes sure that the receiver can avoid conflicts and mount the service domain in the location that suits it best. An application can use the `getMountPoint()` method on the `ServiceDomain` to get a DSM-CC object representing the mount point for the service domain.

Once we have attached to a service domain, applications can access files in that service domain using the methods we saw earlier, and when they are finished they can unmount the service domain using the `detach()` method. This will remove the service domain from the directory hierarchy and give the receiver a hint that any files cached from that service domain can be discarded.

The fact that an application has attached to a service domain does not mean that service domain will always be accessible, however. If the receiver tunes away from the transport stream containing the carousel, files in it may not be accessible even though we have previously attached to the service domain. If the receiver decides that it will never be able to connect to the carousel again, it may choose to detach the service domain automatically.

When the service domain is not accessible, any attempts to access that file will fail as if the file never existed. At the level of the DSM-CC API, inability to access an object carousel will cause the API to throw an `MPEGDeliveryException`. This may not be a permanent failure, however, and thus future attempts to access objects in the carousel may succeed.

An Example

The following example shows how an application can use the DSM-CC API to load objects asynchronously from a broadcast file system.

```
// create a new ServiceDomain object to represent the
// carousel we will use
ServiceDomain carousel = new ServiceDomain();

// now create a Locator that refers to the service
// that contains our carousel
org.davic.net.Locator locator;
try {
   // create a DvbLocator because the Locator class is
   // abstract
```

```
    locator = new org.davic.net.dvb.DvbLocator(
      "dvb://123.456.789");
}
catch (org.davic.net.InvalidLocatorException ile) {
  return;
}

// attach the carousel (i.e. mount it) so that we can
// actually access files in the carousel. We only need to
// do this for files that are not in a subdirectory of the
// application's root directory.
try {
  carousel.attach(locator, 1);
}
catch (ServiceXFRException sxe) {
  return;
}
catch (java.io.InterruptedIOException iioe) {
  return;
}
catch (MPEGDeliveryException mde) {
  return;
}

// Now that we've mounted our carousel, we can use it to
// load some objects.

// We can either create an object using a relative path (if
// it's in the same directory structure as the application
// root directory, or we can use absolute paths to load
// files from other carousels. In this case, we need to
// get the mount point of the carousel to create our
// absolute path.
DSMCCObject dsmccObj;
dsmccObj = new DSMCCObject(carousel.getMountPoint(),
                            "graphics/image1.jpg");

// We will load the file asynchronously. The myListener
// variable (which we haven't defined here) is the event
// listener that will receive notification when the object
// is fully loaded.
try {
  dsmccObj.asynchronousLoad(myListener);
}
catch (InvalidPathNameException ipne) {
  return;
}
```

```
// now we can start loading another file in the
// same thread
DSMCCObject dsmccObj2;
dsmccObj2 = new DSMCCObject(carousel.getMountPoint(),
                            "graphics/image2.jpg");

// this one will also use the same event listener
// for notification.
try {
   dsmccObj2.asynchronousLoad(myListener);
}

catch (InvalidPathNameException ipne) {
   return;
}
```

In this case, we load two objects from the same thread without having to wait for the first to finish loading. The myListener object will receive notification that each object has loaded, allowing us to use the data, but in the meantime we can continue doing other useful work in the current thread.

If we wanted to access the data synchronously, we could call synchronousLoad() or use a java.io.FileInputStream to load the data instead, and we could use a java.io.File object for much of the file manipulation (though not for mounting the file system to begin with). Depending on the way data is used, it may not always be necessary to use the DSM-CC API to load it.

Updating Objects

One of the more useful features of DSM-CC is that each object in the carousel has a version number, and objects can be updated as the carousel is being broadcast. This can be useful for updating data files for home shopping applications, questions and answers for a quiz application, or headlines for a news ticker. Applications can detect these updates and then choose to reload the changed file or keep using the old data.

To detect these updates, an application must implement the org.dvb.dsmcc.ObjectChangeEventListener interface. This interface provides the receiveObjectChangeEvent() method, which is called to notify the application of any updates to objects the application has registered listeners with. When the application wishes to monitor an object for changes, it calls the addObjectChangeEventListener() method on a DSMCCObject instance representing the object in question. To stop monitoring for updates, the application can simply remove the listener from the object using the removeObjectChangeEventListener() method.

There is a need to be careful when checking for updates. Depending on the caching strategy the network operator has specified for that carousel, old copies of the object may still be present in the receiver's cache even after the object has been updated. To be completely sure about getting the newest version of the object, the application should set the retrieval mode for the object (using the `DSMCCObject.setRetrievalMode()` method) to `FROM_STREAM_ONLY`, as follows. This forces the receiver to ignore any cached copies of the object and retrieve a fresh copy of the object from the carousel.

```
DSMCCObject myFile = new DSMCCObject(
  "application/some/file/path/file.txt");

try {
  myFile.addObjectChangeEventListener(
    myObjectChangeEventListener);
} catch (InsufficientResourcesException ire) {
  // do nothing
}
myFile.setRetrievalMode(DSMCCObject.FROM_STREAM_ONLY);
```

Some people have reported that it is necessary to load the object before changes can successfully be monitored, but this is not required by the MHP specification and is more likely a feature of some implementations. If your application does not appear to be monitoring updates successfully, it may be worth trying to load the object before calling `addObjectChangeEventListener()`.

The other limitation of `ObjectChangeEvents` is that they may be generated even when the object has not actually been updated. DSM-CC modules have a version number, but the individual objects within a module do not. This means that we can see when a module has changed, but we do not know which object within a module has changed. To be safe, MHP and OCAP will send an `ObjectChangeEvent` for every object in the changed module.

One way around this is to put each file that will be updated in a separate DSM-CC module, but this is a workaround rather than a fix. If the carousel organization is changed and modules are combined, this problem will reappear. Some developers are taking a different approach and using stream events to notify the receiver when a file has changed, but again this is not an ideal solution.

Synchronization: Stream Events and NPT

In Chapter 11, we discussed the use of media time with broadcast content, and concluded that conventional media time values as used by JMF are not very useful for application developers. DSM-CC NPT offers an alternative to JMF media time for application developers, although it does not have the flexibility of JMF media time for many applications.

Applications can get the NPT value for a stream using the `DSMCCStream.getNPT()` method. The `DSMCCStream` object represents a `Stream` object in an object carousel and thus applications must know which object represents the stream they are interested in.

As well as getting the NPT value, we can get the current rate of the NPT. This tells the receiver how fast NPT values should change, and is similar to the clock rate we saw in the last chapter for JMF. The `DSMCCStream.getNPTRate()` method returns an `NPTRate` instance that

Table 12.18. Events that may be received by an `NPTListener`.

Event	Reason for Notification
NPTPresentEvent	An NPT reference has been detected
NPTRemovedEvent	The NPT reference has been removed
NPTRateChangeEvent	The NPT rate has changed
NPTDiscontinuityEvent	An NPT discontinuity has been detected

gives the current rate for that stream. Because NPT values may have discontinuities, or may not be present for all of the stream, we can also register a listener for NPT events using the `addNPTListener()` method. This registers an `NPTListener` object that may receive the events outlined in Table 12.18.

Applications cannot set an event to trigger at a specific NPT value, but they can monitor the current NPT value, either by getting the current NPT value from the `DSMCCStream` or by reconstructing the value themselves. If they choose to do it themselves, applications should listen for NPT events to make sure they are aware of any changes in the status of NPT.

Using NPT to synchronize applications to a media stream is possible, but exact synchronization is very difficult in practice. We have already mentioned that stream events offer a way of synchronizing applications and media, and these are probably a better solution for most applications and one that is easier to implement. For Java applications, there are three classes that do this, which we examine in the following material.

As we saw earlier in the chapter, stream events consist of two parts: stream event objects (which describe which events are available) and stream event descriptors, which actually indicate a synchronization point in a media stream. This distinction also applies in the API. The `org.dvb.dsmcc.DSMCCStreamEvent` class represents a stream event object, and gives us a way of subscribing to and unsubscribing from a particular stream event, as follows. We can also use this class to list the events described by that stream event object. Developers need to remember that more than one stream event object may be in scope at any time, and thus a stream may contain events that are not described in the `DSMCCStreamEvent` object they are currently working with.

```
public class DSMCCStreamEvent extends DSMCCStream {

    public DSMCCStreamEvent(DSMCCObject aDSMCCObject)
        throws NotLoadedException, IllegalObjectTypeException;

    public DSMCCStreamEvent(java.lang.String path)
        throws IOException, IllegalObjectTypeException;

    public DSMCCStreamEvent(
        java.lang.String path, java.lang.String name)
        throws IOException, IllegalObjectTypeException;
```

```
    public synchronized int subscribe(
       java.lang.String eventName, StreamEventListener l)
       throws UnknownEventException,
            InsufficientResourcesException;

    public synchronized void unsubscribe(
       int eventId, StreamEventListener l)
       throws UnknownEventException;

    public synchronized void unsubscribe(
       java.lang.String eventName, StreamEventListener l)
       throws UnknownEventException;

    public java.lang.String[] getEventList();
}
```

Each event has a name and an ID. The event name will be unique only within that stream event object, whereas the ID is unique for the life cycle of that event object. Two or more event objects may simultaneously define events with the same name, but they will never have the same ID. To make sure we are aware of which event we are subscribing to, subscribing to an event using the event name will return the event ID that matches that name. This avoids any confusion caused by two events that have the same name. To avoid problems, applications should always use the ID to refer to a stream event where possible. When a stream event is triggered (because the NPT value listed in a stream event descriptor is reached, or because a do-it-now stream event descriptor is received), that event is represented by a `StreamEvent` object, as follows.

```
  public class StreamEvent extends java.util.EventObject {

    public StreamEvent(
       DSMCCStreamEvent source,
       long npt,
       java.lang.String name,
       int eventId,
       byte[] eventData);

    public Object getSource();
    public java.lang.String getEventName();
    public int getEventId();
    public long getEventNPT();
    public byte[] getEventData();
}
```

This will be sent to the `receiveStreamEvent()` method of the `StreamEventlistener` that was registered when subscribing to an event. This is the only method in the `StreamEventListener` class, and thus we will not examine that in any more detail here.

The interface to the `StreamEvent` class is also pretty straightforward, and thus we will not cover this in too much detail either. As we would expect, it gives us access to the important fields in the stream event descriptor. These correspond directly to the values in the descriptor itself, and thus calling `getEventNPT()` for do-it-now events will return a value of –1. The only thing that may not be obvious is the source of the event. Stream events are associated with the stream event objects that describe them. As we have already seen, stream event names may not be unique (although event IDs must be unique for the period they are used). By associating a `StreamEvent` with the `DSMCCStreamEvent` that defines that event, it makes it easy for applications to avoid confusion by identifying exactly which event has been triggered.

If a stream event relies on an NPT time code (i.e., it is not a do-it-now event), NPT time codes must be present at the time the event is to fire. If the NPT time code is no longer available, or if it has a discontinuity that takes the time code past the time at which the event should be triggered, the receiver will automatically unsubscribe any applications from that event.

One thing application developers must remember is that the middleware must process stream events before they are delivered to the application. Exactly how long this takes will vary from implementation to implementation, and probably from event to event. Thus, events will not be delivered to applications with the same precision they may be triggered with (although even that may not be completely exact, given the way we must reconstruct NPT values). Frame-accurate synchronization is not possible using stream events or any other method, and thus developers should not have unrealistic expectations. Although events will usually be generated within a frame or two of when they should be, there are no guarantees about the level of accuracy that will be achieved in practice.

Practical Issues

As you can see from this chapter, DSM-CC is a complex piece of technology. Even considering only the subset of DSM-CC used by MHP and OCAP, implementing DSM-CC well is a daunting task.

Using DSM-CC in the real world adds many extra considerations on top of just meeting the specification. It is not just middleware implementers that have to worry about this. Application developers and network operators also have a number of issues to consider. Achieving good performance for DSM-CC is one of the most important elements of a successful application. Failure to do this usually means long loading times or pauses in the application, which will frustrate users and cause them to reject the application or service.

Latency and Caching Strategies

Given that the latency of an object carousel is so high, what can we do to mitigate against this? There are several different solutions we can apply, each targeting a different part of the system. To achieve the best effect, we should use all of these techniques where possible.

For middleware implementers, the most important issue is good caching of DSM-CC modules. A receiver may cache entire modules, or it may cache individual objects instead. In both cases, an appropriately sized cache and an algorithm that caches the right data are important. Often, the best approach will be to cache the most recently used files, as well as commonly used objects such as directory entries and the service gateway object. Because the directory object must be loaded before a file in that directory can be loaded, caching directory objects can have an impact that is disproportionate to the cache space used.

MHP supports three different levels of caching for DSM-CC objects, and the broadcaster can choose which level the receiver should use for a given module. This allows the broadcaster to improve the response time, while increasing the risk of returning outdated versions of objects to the application. The caching priority descriptor can be included in a DII message to indicate the priority of a particular module for caching, but it can also indicate which caching strategy the receiver should use. Caching strategies fall into three main categories: transparent caching, semitransparent caching, and static caching.

Transparent caching is the default, and ensures that all information returned to an application will be up to date. In this case, the middleware must assume that DII messages are only valid for 0.5 seconds, and thus any cached information that is more than 0.5 seconds out of date cannot be returned to the application without first checking that the DII message for that module is still valid.

To do this, a middleware stack can take one of two approaches. The approach that gives the best performance for an individual module is known as active caching. In this case, each module the receiver caches will have a dedicated section filter that is used to ensure the validity of that module. Thus, the information in the cache is guaranteed to be completely up to date. Unfortunately, a typical receiver does not have enough section filters to support active caching for many modules, and thus this can decrease overall performance. An alternative approach, and one that is probably the best choice in most cases, is passive caching.

Passive caching allows a middleware stack to make more efficient use of section filters. Instead of monitoring the DII message for every module in the cache, a receiver that uses a passive caching strategy only reads a new DII message from the stream when it needs a file from that module. The module data itself and old DII messages will still be cached, and the receiver will compare the most recent DII message against its cached copy. If the cached copy is still valid, the receiver will return the cached copy of the object. Otherwise, it will read the new copy from the stream. Although this is a little slower than active caching, because the middleware has to read the DII message from the stream (assuming that its cached copy is more than 0.5 seconds old) the middleware may not need to read the module data itself from the stream. This, coupled with the fact that a receiver is usually able to cache more modules using this approach, means that performance is usually better than active caching for a typical receiver and a typical application.

Sometimes, we do not need the accuracy transparent caching gives us, but we do need high performance. In this case, we can use semitransparent caching or static caching. Semitransparent caching is similar to transparent caching, except that DII messages are considered valid if they are less than 30 seconds old. This allows receivers to cache a large amount of data using only a single section filter. Depending on the carousel layout and the way it is broadcast, more than one section filter may be needed, but by using a passive caching strategy fewer section filters are needed to cache a larger amount of data than would be possible with transparent caching. If that data changes infrequently, and if the application or user can tolerate slightly outdated information, semitransparent caching can be a good choice.

In extreme cases, we can use static caching, in which a receiver can ignore any updated versions of modules and use only the original version. This uses even fewer resources, but at the expense of data that can be very out of date. Although receivers may update the content of their cache when a module is updated, applications must assume that when they open a file no updates to that file will occur during the lifetime of the application.

Each of these caching strategies has its advantages and disadvantages. Apart from the age of the cached data, the most important difference is the resource usage of each strategy. Transparent caching will normally need more section filters, whereas other caching strategies need fewer section filters to cache the same amount of data. Although this could make it possible to cache more data, receivers will typically have a fixed amount of memory for DSM-CC caching and thus this normally limits the amount of data a receiver can cache. From an application perspective, the default caching strategy is usually the best, although the quality of this can vary between receivers.

For middleware manufacturers, a combination of active and passive caching may be best. Passive caching allows more objects to be cached using fewer section filters, but by using a more active strategy for caching DII messages (e.g., by actively caching DII messages for commonly used files or directories) the middleware can improve overall performance and reduce the delays caused by loading DII messages. One problem is that the structure of an object carousel may change while it is being broadcast, and thus in some cases we need to make sure the file or directory we want is still present in the carousel. To do this the receiver can simply check that the object and its parent directories have not been updated since they were cached. If this is the case, we know that the object is still in the same place in the directory structure.

The DSM-CC implementation can also use hints from parts of the middleware such as the MHP DSM-CC API or the application manager to cache modules that may be needed soon (e.g., modules containing the initial classes for applications that are signaled in the AIT or XAIT), or to flush objects form the cache. When a new service is selected, any data for object carousels not used by the new set of applications can be flushed.

Latency Issues and Application Design

Application developers and broadcasters can influence loading times quite significantly by organizing their object carousels efficiently. Grouping classes and data files used together in a single module allows the receiver to cache that module and improve access times for all of those files. This is especially useful for the classes and data files used when the application first loads, especially if that module is repeated often enough. Doing this can reduce loading times significantly, and can give the user an impression of responsiveness even when the application may still take some time to load.

Choosing how often to repeat modules can also produce some performance improvements. By repeating modules containing commonly used files more often than those that contain less important files, the broadcaster can reduce latency on most file system accesses and again improve responsiveness. Using a higher bit rate for the stream containing the object carousel can never hurt either, but this is often a luxury that broadcasters cannot afford.

Another possible technique is to design the application appropriately. There are two elements to this: partitioning the use of classes and data files between different parts of the application and ensuring that the application takes advantage of the strengths of DSM-CC. The first of these is a high-level design issue for application developers. By minimizing the number of files used at any time, an application developer can ensure that as much space as possible is free to cache the files that are used. Although this may or may not help, it also helps produce more fault-tolerant applications.

The second element is a lower-level design issue. Applications can use asynchronous loading to make sure that class files and data files are available before they are actually needed, and the DSM-CC API allows applications to provide hints to the middleware about which files should be cached. By using these features appropriately, many of the delays associated with loading files can be minimized or even avoided completely.

There is no magic formula for doing this, however. All of these techniques are an art rather than a science, and different applications will need to apply these techniques in very different ways to achieve the best results. Everyone involved in producing and broadcasting the application needs to be involved in this process, and it is in the interest of everyone involved to make sure the process is a success. An application or service that does not get used because of slow loading times and frustrating delays is one that is not making any money.

One common mistake when trying to improve loading times is to use JAR files for carrying application classes. This is usually harmful rather than beneficial, for several reasons. MHP and OCAP implementations do not have to support JAR files, and thus there is no guarantee that a receiver will be able to load classes from them. The second major problem with JAR files is the way they group files, which means that the entire JAR file must be loaded in order to access any file within it. This can increase latency because more data must be loaded for every file access. Even caching the file may be more difficult, depending on the size of the receiver's DSM-CC cache and its caching strategy. Similarly, the ability to compress modules within an object carousel negates any benefits that may be gained by compressing the content of a JAR file. In general, it is better to examine which files are needed to start the application and group those in modules that are repeated more often.

Application Management and File System Issues

Whenever the application manager loads an application, it must first make sure that the object carousel containing the application files is accessible. This is not a big issue, but it does mean that the middleware needs to manage mount points carefully. Carousels cannot be unmounted unless no application is using them, and ideally if no other applications in the current service use the same carousel. By not unmounting carousels that may be used later, a receiver can slightly improve the startup time for an application, although this happens at the expense of the memory needed to store the service gateway object for that carousel.

Application developers feel the effects of this more than anyone else. Because carousels are mounted dynamically, they will not always be mounted in the same place, and this can change between receivers of the same brand as well as between receivers from different manufacturers.

In some APIs, references to files must include absolute paths. JMF is a good example of this when creating a `file://` URL for an audio clip. There are two options in these cases, depending on how the carousel was mounted. For carousels the application has explicitly mounted, the `getMountPoint()` method on the `ServiceDomain` object representing the carousel will return the appropriate directory (as we have already seen). This can then be used to create an absolute file name for files within that carousel, as follows.

```
ServiceDomain carousel;

// This assumes the carousel variable has already been
// attached to a service domain
DSMCCObject dsmccObj;
dsmccObj = new DSMCCObject(carousel.getMountPoint(),
                           "graphics/image1.jpg");
```

In some cases, the receiver may have mounted the carousel automatically and we cannot get a `ServiceDomain` object for that carousel. This can happen when we need an absolute path to a file in the application's default carousel, for instance. In this case, we can create a `java.io.File` or `DSMCCObject` object that refers to the file in question, and then use the `getAbsolutePath()` method on that object to get the correct absolute path name.

System Integration Issues

As with so many other components in the MHP middleware, the DSM-CC component makes heavy use of other elements of the middleware. The two most important of these are the SI and section-filtering components. We have already seen how DSM-CC uses SI to locate the streams that contain object carousels and to locate objects within those carousels, but once it is located the stream has to use the section-filtering component to read enough of the carousel to do that. Although the format of DSM-CC BIOP messages helps a little with this, it is still not easy. Because one data carousel module can contain more than one BIOP message we may need to search the entire module to find a particular message.

It is not just object carousels that need section filters, of course. Handling NPT, stream events, and IP datagrams transmitted via MPE all take section-filtering resources, and this can impose a heavy load on the section-filtering component. An efficient interface between the DSM-CC and section-filtering components is vital.

Chapters 9 and 10 discussed how this applies to the interface between section filtering and SI, and the same approaches can be taken here. This is one of the areas in which a good solution can have an effect on the entire receiver. Through low-level approaches such as filtering sections efficiently, and high-level approaches such as caching the right DSM-CC objects at the right time, we can make noticeable improvements in the performance of our receivers.

13 Security in MHP and OCAP

Reliability is a key consideration in consumer applications, and MHP and OCAP have a number of security measures to help ensure that applications do not cause problems for the middleware. In this chapter, we look at the security model in MHP and OCAP and see how this affects applications. For application developers and broadcasters, we will discuss how the receiver can authenticate downloaded applications, and how broadcasters can set limits on what an application can do.

It is probably no understatement to say that DTV broadcasters are paranoid about security, which means that sometimes receiver manufacturers have to be as well. There are two major parts to this, as follows.

- Making sure that unauthorized people do not get access to content they have not paid for.
- Making sure no one can tamper with the content before it is broadcast.

The first of these issues is a fairly well-known problem that the pay-TV industry (analog as well as digital) has faced for many years. Encryption and scrambling techniques are used in all pay-TV systems, and this does not significantly change with interactivity. The second issue is related to making sure that you broadcast the right thing. This does change with the introduction of interactive applications. It is fairly easy to tell if a videotape has been tampered with by simply playing it, but it is much more difficult to do this with a Java application.

MHP and OCAP add a third element to the previous list: making sure applications can do only the things they are allowed to do. Interactive applications should not be able to use (either accidentally or deliberately) functionality they are not allowed to, such as using a modem to call a premium-rate phone number. This is a refinement of the Java "sandbox" that many Java developers will already know about.

As we have already seen when looking at the various APIs, security considerations are a factor in many design decisions. In consumer products, it is important that the platform be as reliable as possible, and that gets much more difficult when third parties can download and run code on that platform. Although people will tolerate crashes or compatibility problems with a PC application, they are much less likely to tolerate problems with a consumer device. The architecture of MHP and OCAP receivers is designed to reduce these problems, and the security model is an important part of this.

How Much Security Is Too Much?

If you are a PC application developer, you may think that the security restrictions imposed by MHP and OCAP are very onerous. It is true that security is much tighter in an MHP receiver than in many other platforms, but there are good reasons for this.

Malicious applications are a well-known security problem in the PC world. Viruses and worms are common, and Trojan horse applications that use the modem to call premium-rate numbers, or that connect to fake banking sites to steal credit card details, are just some of the problems facing users. Network operators may broadcast an ITV application to tens or even hundreds of thousands of households, where viewers have little or no technical background. Because of this, reliability and security are extremely important. Television has been extremely reliable for many years now, and thus applications that crash or cause security problems can only harm the perception of DTV and the success of DTV deployments.

The big problem with ITV applications in a horizontal market is that applications from several different suppliers may have to run together on several different types of receivers, in a very large user base. Even in networks running proprietary middleware on only a few types of receivers, a huge amount of testing goes on to ensure applications will run correctly. One of the disadvantages of a horizontal market is that this suddenly gets a lot tougher.

The MHP and OCAP security model makes applications easier to test and more difficult to subvert. Network operators can set the permissions that are needed by an application, and can place limits on those permissions so that they cannot be abused. For instance, an application may only be allowed to call certain telephone numbers with the modem, or only connect to specific servers over the return channel. This means that the receiver can trap some security issues the network operator would otherwise have to test for and find.

The MHP and OCAP Security Model

From the application point of view, security in MHP and OCAP follows the model defined by PersonalJava and J2SE, with a few restrictions. Many of the MHP and OCAP APIs define a number of security-related exceptions that may be thrown when an application does not have sufficient rights to perform certain operations. For instance, applications do not by default have permission to use the application-listing and application-launching API to control other applications. They only have the rights to get information about those applications.

Although the security model in OCAP is based on the MHP security model, there are a few differences between the two. For most of this chapter, we will concentrate on the MHP security model and highlight the OCAP-specific details at the end of the chapter.

Permissions

MHP and OCAP define a number of new permissions that are needed for applications to use many features of the receiver. Almost every API that uses scarce resources in the receiver defines permissions that must be granted before an application can use those resources. By defining these new permissions, MHP and OCAP give very fine-grained control over the functionality applications can use. Table 13.1 outlines the new permission classes MHP and OCAP define.

The receiver assigns permissions to an application based on information from two sources. The user could choose to grant or deny access to specific features via the user settings in the receiver. Similarly, the network operator and application developer can choose to ask for access to specific features via the permissions request file. If the user settings allow the application to be granted those permissions, the application will be given the permissions it has requested (but no more).

This approach ensures a balance of power between the network operator and the user. Both parties must grant permission before applications can access a given feature (although this may be implicit, some receivers do not allow users to set preferences for permissions in this way). By doing this, the potential for an application to carry out unauthorized actions is much smaller. OCAP changes this model slightly to give the network operator some extra control over this, but the basic principle remains the same (as we will see later in the chapter).

In addition to the permissions we have seen here, a number of standard Java permissions are used. Given the need to isolate applications from one another, and from sensitive parts of the middleware, applications will not be granted all of these permissions. Table 13.2 below outlines the restrictions imposed on applications by default.

The receiver will not grant any other permissions to unsigned applications, or to signed applications unless they are explicitly requested.

Permission Request Files

If we want to give an application any permission other than the default set, we need to attach a permission request file to the application. This is an XML document that lists the permissions (and any restrictions on them) the network operator wants to grant to the application. A permission request file is stored in the same directory as the main file for the application (either the initial class file for a DVB-J application or the first HTML file for a DVB-HTML application), and has the name dvb.<*initial_file_name*>.perm, where <*initial_file_name*> is the name of the main file of the application. For OCAP systems, the dvb part of the file name is replaced with ocap, in this and every other case.

Table 13.1. Permission classes added in MHP and OCAP.

Permission Name	Permission Description
`java.util.PropertyPermission`	Access system properties beyond those available to unsigned applications
`java.io.FilePermission`	Access object carousels and/or persistent storage
`org.dvb.net.ca.CAPermission`	Access the CA subsystem and get information about entitlements assigned to the receiver and/or user
`org.dvb.application.AppsControlPermission`	Control other applications using the application-listing and application-launching API
`org.dvb.net.rc.RCPermission`	Use the return channel
`org.dvb.net.tuning.TunerPermission`	Tune to a new transport stream using the tuning API
`javax.tv.service.selection.SelectPermission`	Select a new service
`org.dvb.user.UserPreferencePermission`	Read and/or write user preferences
`java.net.SocketPermission`	Connect to other hosts using an IP connection
`org.dvb.media.DripFeedPermission`	Use video drips for displaying images
`javax.tv.media.MediaSelectPermission`	Select new media clips from the current transport stream
`javax.tv.service.ReadPermission`	Access SI using the JavaTV SI API
`javax.tv.service.selection.ServiceContextPermission`	Control a service context
`org.ocap.system.MonitorAppPermission`	Access APIs that are restricted to the monitor application (OCAP only)
`org.dvb.application.storage.ApplicationStoragePermission`	Store applications and create new abstract services
`org.dvb.smartcard.SmartCardPermission`	Communicate with non-CA smart cards
`org.dvb.internet.HomePagePermission`	Set the home page of the web browser in the Internet access API

We will not look at the format of the permission request file in too much detail in this book. Readers who need to know more about this should examine the DTD for permission request files that is available in Part 12.6.2 of the MHP specification (for MHP applications) or Part 14.2.2 of the OCAP specification (for OCAP systems). To get a flavor of what this file looks like, the following example covers all of the elements that will be found in a typical permission request file for an MHP application.

Table 13.2. Default permissions granted to an MHP or OCAP.

Permission	Default Restriction
`java.awt.AWTPermission`	Denied to all applications. Sensitive parts of the AWT classes may not be accessed by MHP or OCAP applications.
`java.io.FilePermission`	Access to the home carousel only.
`java.net.SocketPermission`	No access to the network.
`java.util.PropertyPermission`	Read-only access to some properties.
`java.lang.RuntimePermission`	Denied to all applications.
`java.io.SerializablePermission`	Denied to all applications.
`javax.tv.media.MediaSelectPermission`	Full access.
`javax.tv.service.ReadPermission`	Full access.
`org.dvb.application.AppsControlPermission`	Applications can launch other applications in the same service, but cannot control applications they did not launch.
`javax.tv.service.selection.ServiceContextPermission`	Access to the home service context only.

```xml
<?xml version="1.0"?>
<!DOCTYPE permissionrequestfile
    PUBLIC "-//DVB//DTD Permission Request File 1.1//EN"
    "http://www.dvb.org/mhp/dtd/permissionrequestfile-1-1.dtd">

<permissionrequestfile orgid="0x000042" appid="0x0001">

  <capermission>
    <casystemid
      id="0x0002"
      messagepassing="true"
      entitlementquery="true"
      mmi="true">
    </casystemid>
  </capermission>

  <applifecyclecontrol value="false"></applifecyclecontrol>

  <returnchannel>
    <defaultisp></defaultisp>
    <phonenumber>+19135550191</phonenumber>
  </returnchannel>
```

```
<network>
  <host action="connect">myserver.mycompany.com</host>
</network>

<tuning value="false"></tuning>
<servicesel value="true"></servicesel>

<userpreferences read="true" write="true"></userpreferences>

<file value="true"></file>

<persistentfilecredential>
  <grantoridentifier id="0x00042"></grantoridentifier>
  <expirationdate date="01/01/2005"></expirationdate>
  <filename read="true" write="true">
    42/2/data/preferences.dat
  </filename>
  <filename read="true" write="false">
    42/15/userData
  </filename>
  <signature>
    023273142977529599703629645104839439432943840
  </signature>
  <certchainfileid>1</certchainfileid>
</persistentfilecredential>

<!- elements below this line are valid in MHP 1.1 only ->

<applicationstorage
  orgid="0x000042"
  store="true"
  remove="true"
  create="true"
  removeservice="false">

<smartcardaccess></smartcardaccess>

<priviligedrce value="internal"></ priviligedrce>

</permissionrequestfile>
```

This is a permission request file for MHP 1.1, and not all of these elements are valid for MHP 1.0.x implementations (as shown in the comments in the code). In particular, MHP 1.0.x applications must use the following doctype declaration.

```
<!DOCTYPE permissionrequestfile
    PUBLIC "-//DVB//DTD Permission Request File 1.0//EN"
    "http://www.dvb.org/mhp/dtd/permissionrequestfile-1-0.dtd">
```

Simply listing the permissions in the permission request file is not enough to give them to the application, however. As its name suggests, the permissions request file only identifies permissions the network operator is *requesting* for that application. The receiver may choose not to grant some or all of these permissions, based on user preferences or other settings. Exactly how this is done will vary from receiver to receiver, but applications should always be aware that they may get only some of the permissions they request.

Table 13.3 outlines the elements that can be used to grant specific permissions to an application. Any permissions not listed in the permission request file will be given only the default rights.

The `persistentfilecredential` element is slightly different from the others. This allows a network operator to give permission for an application to access a specific file or set of files in the persistent file system. Because this file system may contain sensitive information, these credentials must be signed by the organization that provides the application. Before granting access to a persistent file, the middleware will check the validity of this digital signature using the X.509 certificate file that is specified in the credential. We will take a closer look at certificate files in the next section.

Table 13.3. Elements in the permissions request file and the corresponding permission classes.

Element	Permission
file	No equivalent permission.
applifecyclecontrol	org.dvb.application.AppsControlPermission
returnchannel	org.dvb.net.rc.RCPermission
tuning	org.dvb.net.tuning.TunerPermission
servicesel	javax.tv.service.selection.SelectPermission
userpreferences	org.dvb.user.UserPreferencePermission
network	java.net.SocketPermission
dripfeed	org.dvb.media.DripFeedPermission
persistentfilecredential	java.io.FilePermission
applicationstorage	org.dvb.application.storage.ApplicationStoragePermission
smartcardaccess	org.dvb.smartcard.SmartCardPermission
privilegedrce	Not directly available to DVB-J applications (this functionality is only available to DVB-J applications through org.dvb.lang.DVBClassloader), and so no equivalent DVB-J permission. This is aimed at DVB-HTML applications using ECMAScript.

The `org.dvb.internet.HomePagePermission` from MHP 1.1 has no corresponding tag in the permission request file, and thus it is currently not clear how this permission is granted to applications. This may be set via user preferences, or via some other implementation-dependent means, but there is no way for applications to request this permission in MHP 1.1.

In some cases, the user can also choose whether to grant permission to an application for specific behavior. In the case of PSTN return channels, or others where the receiver must dial a specific number to connect, the middleware must ask the user at least once before dialing any number. Numbers that have been approved by the user may be stored in a "white list" so that they are only asked once, but the user must grant permission before an application can connect to that number even if the permission request has been granted by the middleware. To avoid any unnecessary checks, the middleware will only ask the user when the application actually tries to dial that number.

In addition to defining a new DTD (and thus needing a different `DOCTYPE` definition), OCAP adds one extra entity that can be used to request certain permissions. The `ocap:monitorapplication` element can be used to request permissions for privileged APIs that are only available to monitor applications. In the spirit of the rest of the security model, access to these APIs can be granted on a fine-grained level, so that a monitor application may be allowed to use some privileged APIs but not others. To grant or deny access to a specific API, the permissions request file can include an `ocap:monitorapplication` element of the appropriate class that denies or grants permission. Each permission request file can include more than one `ocap:monitorapplication` element to control access to more than one privileged API. Table 13.4 outlines which classes are supported, and the APIs they control access to.

These rights are all represented by the `org.ocap.system.MonitorAppPermission` class. More information about these permissions is available in the documentation for this class that is included in the OCAP specification.

In both MHP and OCAP systems, if the permission request file cannot be parsed the receiver will ignore it and assign the default permissions to the application. Similarly, if the permission request file does not contain a `permissionrequestfile` element the receiver will ignore the entire file. Errors in other elements mean that the element containing the error will be ignored.

Signed and Unsigned Applications

Using a permissions request file is a good way for a network operator to tell the receiver what permissions an application should have, but it does not prevent anyone from tampering with the permission request file or the application. To avoid this, applications must be signed before they will be granted any permission except default permissions.

Network operators can digitally sign the files used by an application in order to help the receiver guarantee that no one has tampered with the application's files. This lets the receiver know that it is safe to give an application access to the resources it claims it is allowed to

Table 13.4. Classes of the `ocap:monitorapplication` element in an OCAP permissions request file.

Class	Permission
`registrar`	Register and unregister unbound applications
`service`	Create abstract services
`servicemanager`	Manage and control all services
`security`	Register a security policy handler with the application manager (see later in this chapter)
`reboot`	Reboot the device
`handler.reboot`	Register a reboot handler
`handler.appfilter`	Register an `AppFilter` with the application manager (see Chapter 4)
`handler.resource`	Register a resource contention handler (see Chapter 6)
`handler.closedcaptioning`	Set a HAVi component to display closed caption subtitles (see Chapter 7)
`filterUserEvents`	Set a `UserEventFilter` (see Chapter 7)
`handler.podResource`	Interact with the Extended Channel and Man Machine Interface resources in the CableCARD device and send or receive application information to/from the device
`handler.eas`	Set a module to handle Emergency Alert System messages
`SetCCIBits`	Override the copy control information on the CableCARD device (not available in version I11 and later of OCAP)
`setDownRes`	Enable or disable downscaling of the output resolution for high-definition services (not available in version I11 and later of OCAP)
`setVideoPort`	Enable or disable video outputs
`podApplication`	Communicate with an application on the CableCARD device
`signal.configured`	Tell the receiver that the monitor application is configured and that the receiver can continue the boot process

have. Unsigned applications have only the default set of rights, which prevents them from using many of the features of the receiver. By default, signed applications also get the default set of permissions. If an application has been signed, however, the receiver will check for the presence of a permission request file to see which additional permissions have been requested.

Network operators can use this feature to broadcast applications without having to check all aspects of the application to find out whether it carries out any forbidden operations. By not signing the application they automatically limit it to the most restrictive set of permissions.

Signed applications may also use more features than unsigned applications, even when no explicit permissions are defined. To avoid any security risks this may pose, unsigned applications are not able to interact with signed applications, and vice versa. Signed applications can communicate with each other via the inter-Xlet communication API (as we will see in the next chapter).

Applications can find out whether a file has been signed (and by whom) by using the `getSigners()` method on the `org.dvb.dsmcc.DSMCCObject` instance representing the file in question. This validates and returns the X.509 certificate chains that have been used to sign that object, or a zero-length array if the object is not signed. Any file that has been signed but cannot be authenticated (e.g., because the file is corrupt, or because a necessary certificate is missing) will appear like an empty file to any applications that try to load it.

Signing Applications

Signing an application is a three-part process, as follows.

1. Verifying the integrity of the application files. This is carried out using hash values to checksum file and directories.
2. Signing the application to verify that these hash values were calculated by the network operator and/or application provider.
3. Allowing the receiver to verify that the signature is genuine using X.509 certificates.

Each of these parts uses a different set of files that must be included in the object carousel for the application. Together, hash files, signature files, and certificate files enable a receiver to verify that an application should be given the permissions it is requesting.

Although MHP and OCAP take the same approach, files have different names on the two platforms. As for other files used by the receiver rather than by the application, MHP uses the `dvb` prefix to identify these files. For OCAP systems, this prefix is replaced with `ocap`, to bring the file names in line with other OCAP system files. Other elements of the file names remain the same between the two platforms.

Hash Files

A broadcaster can use a hash file to provide a checksum for files or directories, thus ensuring that the content has not been corrupted or tampered with. Each hash file contains a set of digest values, wherein each digest value is associated with an object (a file or directory) or a group of objects. These digest values can be computed using the MD5 or SHA-1 algorithm, and the receiver uses these values to tell when a file has been corrupted. Hash files are all binary files, following the format defined in Section 12.4.1 of the MHP specification.

A hash file is only useful if a receiver knows where to look for it, and thus hash files in MHP always have the name `dvb.hashfile` (as discussed earlier, OCAP systems use the name `ocap.hashfile`). One hash file will be present in every directory that contains files authenticated in this way. Subdirectories listed in the hash file will also be hashed, and thus they will include hash files of their own.

Using hash files is straightforward in theory, but may be complicated in practice. Whereas generating the digest values is a quick process at the head-end, checking them at the receiver is normally much more difficult because of the limited CPU power available in the receiver and because of the time needed to load the hash files from the object carousel.

Checking a hash value of a file involves loading the files associated with that hash value, and thus we need to be careful how we calculate these hashes. The hash file format allows us to group several objects (files or directories) and use one digest value for all of the objects in the group. By grouping objects in this way, we have to load all of the objects in the group in order to check the hash value for any object in that group. Because DSM-CC file systems can have a very high latency, this can add a large overhead to the process. In most cases, using a separate digest value for each object is no more difficult, and it can greatly improve performance.

Signature Files

Just creating a hash file is not much use to us, however. Someone who tampers with application files can easily compute a new set of hashes for the changed files. To guarantee that the files are correct, the broadcaster signs them using a digital signature algorithm.

Unlike hash values, we only need one signature file to sign a complete directory. To sign an application, the broadcaster calculates the digest value for the hash file representing the root of the tree of signed directories, and then encrypts this digest value using a public-key algorithm. The network operator uses their private key to encrypt the digest value, and the corresponding public key is stored in a certificate file that is transmitted along with the application. We will look at certificate files in more detail in material to follow.

For MHP applications, each signature is stored in a file called `dvb.signature.`<*id_number*> in the root of the directory structure that it signs. <*id_number*> is a decimal integer with no leading zeroes, which is added to the file name in order to allow the same directory to be signed using more than one key (e.g., from different providers).

To check the signature of a file, we need to check that the entire directory hierarchy up to the signature file is authenticated correctly. This can be a time-consuming process, and we do not want to do that unless we know that the file is actually signed. For this reason, the address space for application IDs in the AIT or XAIT is split into two ranges. Any application ID in the range 0x000 to 0x3FFF is considered an unsigned application, and thus the middleware will not authenticate any files belonging to that application. Applications with an application ID in the range 0x4000 to 0x7FFF will be treated as if they were signed, and files loaded by these applications will be authenticated. Any application with an ID in the signed ID range without a signature file will not be authenticated and will not be allowed to run.

Certificates

As we have already seen, the broadcaster must include their public key in the file system so that the digital signature can be decrypted. This is contained in a certificate file, which has the name `dvb.certificate.`*<id_number>* in MHP systems. MHP and OCAP both use the X.509 certificate format, and in particular a variation of the Internet profile of X.509 that is defined in RFC 2459. The details of X.509 certificates are beyond the scope of this book, and thus we will not look at the format of these certificates in any detail.

If a directory contains a signature file, it must also contain a certificate file with the same ID number so that the signature can be authenticated. Certificate files in other locations will be ignored, even if they have the correct ID number. Each certificate must itself be certified, in a hierarchy leading back to a root certification authority. MHP and OCAP receivers need to check every link in this chain before an application can be authenticated.

X.509 certificates form a chain, wherein a certificate may in turn be authenticated by another certificate, through several iterations to a certificate issued by a certification authority that is known to be trustworthy. This is known as the root certificate. Root certificates will be stored in the receiver during the manufacturing process, although it is possible to update them when the receiver has been deployed. To do this, broadcasters can transmit a set of root certificate management messages (RCMMs).

Each RCMM is contained in a file in the root directory of the object carousel. The most recent message will have the name `dvb.rcmm`, whereas previous messages will have names of the form `dvb.rcmm.`*<sequence_number>*, where the sequence number is a decimal number indicating the order in which the RCMMs should be applied. Older RCMMs will have lower sequence numbers.

Managing root certificates is a sensitive operation, and thus the authentication of an RCMM is even stricter than the authentication process for other files. For an RCMM to be considered valid by the receiver, it must be signed using at least two valid certificates.

The format of RCMMs, and how they are processed, is beyond the scope of this book. More information can be found in the X.509 specification and in the MHP specification.

CableLabs define their own specification for handling root certificate management and other secure updates of the receiver firmware. This makes RCMMs unnecessary in OCAP receivers, and certificate management is carried out using the common download facility. This process is defined fully in the *OpenCable Common Download Specification* and the *OpenCable Security Specification*.

In an OCAP network, stored applications may also be signed, and in this case much of the validation process can be carried out when the application is stored. The only part of the process that has to be carried out when the application is run is the revalidation of any certificates. The receiver must check that the certificates used to sign the application are still valid at that time, but there is no need to recheck that the application's files have been signed correctly.

The root certificate authority for MHP is run by DVB Services Sàrl and WISeKey. This certificate authority is responsible for issuing root certificates to organizations that have entered into an agreement with them. Different agreements will apply, depending on whether a company wants to sign applications, include the root certificates in an MHP receiver, or manage the use of certificates on a DTV network. More information about these agreements and about the process itself is available from the DVB Services web site at *www.dvbservices.org*. The costs associated with these agreements are not publicly available, but more information can be requested via the DVB Services web site.

An Example of the Signing Process

To understand how this works in practice, let's consider the directory tree shown in Figure 13.1. In this case, we wish to sign all of the class files but none of the data files. Not all of the files we are showing here are present in the file system at first (shaded files are added as part of the signing process).

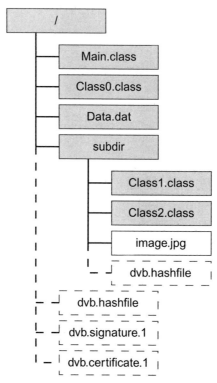

Figure 13.1. A directory structure showing the files related to security.

The following is the process for signing this directory tree.

1. Certificates from each of the authorities that will sign the directory tree are added to the root directory. Because in this case we are the only authority who will sign these files, we only have one certificate. This is stored in the file `dvb.certificate.1`.
2. The lowest-level directories are checksummed first. In this case, this means the `subdir` directory. Only the two class files are checksummed, in that we are not authenticating any data files. The digest values for these files are stored in the `dvb.hashfile` file for this subdirectory.
3. The parent directory is checksummed. To do this, we calculate the digest values for the files in that directory, and the digest value for the hash file of any subdirectories. These are then stored in a hash file (`dvb.hashfile`) in the parent directory.
4. In our case, the top of the directory tree has now been reached and thus we can now sign the application. To do this, we calculate the digest value of the hash file for the parent directory, and then encrypt this using the RSA algorithm. The key used to encrypt the data is the private key that matches the public key given in the certificate file created in step 1. This encrypted digest value is then stored in the file `dvb.signature.1`.
5. Processing is complete and the files can be broadcast.

Revoking Certificates: The Certificate Revocation List

Even though X.509 certificates have an expiration date, the issuer of a certificate may want to revoke a certificate before it expires. For instance, a certificate authority may want to revoke the certificate of a specific broadcaster because it has been compromised, or because the broadcaster does not use that certificate authority any more.

To do this, certificate authorities publish a certificate revocation list (CRL), which is transmitted to the receivers and which identifies certificates that are no longer valid. Each CRL contains a list of certificate serial numbers that are to be revoked, and the receiver will check this to see whether its stored certificates are valid before using them to authenticate an application.

Every certificate (except root certificates) can have an associated CRL that is generated by its issuing certificate authority. MHP receivers must support at least eight CRLs per certificate, wherein each CRL may contain up to 10 entries. When a receiver authenticates an application, it will check the CRLs associated with each certificate in its certificate chain to make sure the certificate is still valid.

Distributing Certificate Revocation Lists

CRLs can be distributed either in the broadcast stream or over the return channel. Each certificate may include an optional field containing a URL that gives the location of a CRL. The disadvantage of using this approach to distribute CRLs is that not every receiver will support a return channel. When coupled with the inconvenience of possibly making a return channel connection and downloading a CRL to authenticate an application, distributing CRLs over the return channel is not feasible in most cases. OCAP receivers or MHP receivers with a permanently connected return channel have a slight advantage here, but it is still not the best solution.

CRLs can also be transmitted in the object carousel. These are contained in a directory called `dvb.crl` that is a subdirectory of the carousel root directory. The name of the file containing the CRL will take one of two forms, depending on how the CRL is authenticated. CRLs authenticated by a broadcast certificate are stored in a file with the name `dvb.crl.<id_number>`, where *<id_number>* corresponds to the ID number used in the name of the certificate file it is associated with. CRLs authenticated by a root certificate use the file name `dvb.crl.root.<id_number>`.

The certificates that authenticate these CRLs must be contained in the `dvb.crl` directory or in the root directory of the carousel. The management of CRLs is closely tied to the management of certificates within the receivers. The receiver must cache CRL information in order to improve performance when authenticating applications and to make sure that CRLs will not be filtered out of the broadcast stream. When a certificate is removed by an RCMM, any CRLs associated with that will also be removed.

Differences Between MHP and OCAP

So far, our description of the security model has focused on MHP systems. OCAP receivers follow the same basic approach, although there are a few differences in file names and other elements.

As we have already mentioned, OCAP receivers do not follow the `dvb.*` naming scheme used by MHP for security files. Instead, they use the prefix `ocap` for these files. Table 13.5 outlines how the file names correspond between the two systems.

We have already seen that the permissions request file has a slightly different format, and this leads into one of the main differences between the two systems. An OCAP monitor application that wants to use an API that is not available to normal applications must have a per-

Table 13.5. Names of security files in MHP and OCAP.

MHP File Name	OCAP File Name
`dvb.hashfile`	`ocap.hashfile`
`dvb.rcmm`	No equivalent (RCMMs not used)
`dvb.rcmm.<sequence_number>`	No equivalent (RCMMs not used)
`dvb.signature.<id_number>`	`ocap.signature.<id_number>`
`dvb.certificate.<id_number>`	`ocap.certificate.<id_number>`
`dvb.crl.<id_number>`	`ocap.crl.<id_number>`
`dvb.crl.root.<id_number>`	`ocap.crl.root.<id_number>`
`dvb.<initial_file_name>.perm`	`ocap.<initial_file_name>.perm`

mission request file that includes one or more ocap:monitorapplication elements that grant permission to use those APIs. To make sure these permissions are not granted to the wrong applications, this permission request file and the monitor application itself must be signed by two parties: the application author and either the network operator or CableLabs.

A final difference is the way permissions are granted to applications. In an MHP receiver, the set of permissions granted to an application is formed by the intersection of the following three sets of permissions.

- The set of default permissions available to the application
- The set of permissions requested by the network operator via the permission request file
- The set of permissions granted by the user, via user preferences or other mechanisms

OCAP receivers change this to give some control to network operators. A monitor application can register a security policy handler with the application manager, which allows the monitor application to control which permissions are actually given to an application before it runs. Security policy handlers are implemented using the org.ocap. application.SecurityPolicyHandler interface, as follows.

```
public interface SecurityPolicyHandler {

   public java.security.PermissionCollection
      getAppPermissions(
         org.ocap.application.PermissionInformation
            permissionInfo)

}
```

The PermissionInformation class provides the security policy handler with information about the certificates used to sign the application and about the permissions it has requested, and the security policy handler can use this information to decide which permissions will be granted to the application. These permissions are returned in a standard Permission-Collection object.

To register a security policy handler, the monitor application must call the setSecurity-PolicyHandler() method on the org.ocap.application.AppManagerProxy class. Of course, a monitor application can only set a security policy handler if it has been granted the appropriate permissions!

14 Communicating with Other Xlets

MHP and OCAP applications cannot communicate with each other using methods that may be available on other platforms. Here, we look at the rationale behind that, and see how applications can communicate with each other. We will also discuss the techniques needed by middleware developers to allow applications to communicate securely and efficiently.

There are times when you may want an Xlet to communicate with other Xlets running on the receiver. This allows a developer to craft a set of Xlets in such a way that some common functions can be implemented in a client-server model, which may be more efficient in some cases. At first sight, this seems quite easy. After all, we can use static methods and static fields to exchange data between different applets in a web site, and thus we can use the same technique for exchanging data between MHP applications, right?

Unfortunately, things are not quite that simple. The OCAP and MHP security model prevents us from doing this, in order to increase the separation between different applications and the security of the platform as a whole. If we could exchange data this simply between applications, we would have a security hole that malicious applications could exploit. To prevent this type of thing from happening, Xlets are completely isolated from each other. MHP and OCAP both specify that classes for every application should be loaded using a different class loader.

This goes beyond simple paranoia over allowing different Xlets to talk to one another. By saying this, MHP and OCAP allow a wider range of design choices when implementers are deciding how Xlets should be run. Explicitly saying that applications will not share a class loader makes it possible for an implementation to run applications in completely different virtual machines, should it choose to do so. Although no implementations to date use this approach as far as we know, it offers a number of advantages in terms of reliability and robustness, although there are a number of disadvantages as well (memory usage and complexity being the biggest). By explicitly saying this, MHP raises a barrier between Xlets while allowing implementers to raise an even taller and thicker barrier should they wish to do so.

Class Loader Physics in MHP

Before we go any further, we need to look at exactly why using multiple class loaders causes us problems. Application developers will not care about this particularly, but middleware implementers need to understand this in order to understand the effects on the overall architecture, and to solve the problems it causes.

A class loader does pretty much what its name suggests: it loads classes. Class loaders allow the system to provide a mechanism for loading classes from several different sources. For example, a PC implementation of Java may have one class loader that can load classes from a local hard disk, and another class loader to load classes via an HTTP connection from a remote server. Every class loader provides the `loadClass()` method, which gives a common entry point for the middleware no matter what source the class is actually loaded from.

The important thing is that class loaders are intimately linked to the identity of a class. The Java language specification states that two classes are only identical if their fully qualified name is the same (the class name prefixed by its package name, such as `java.lang.String`) and if they are loaded through the same class loader. Two classes loaded through different class loaders will be treated by the Java VM (virtual machine) as two different classes, even if they are loaded from the same class file.

Using a separate class loader for each application has a number of security benefits, by making sure that the Java VM keeps classes belonging to one application completely separate from classes belonging to another application. This helps avoid name clashes where two applications have classes with identical fully qualified names, but it also adds an extra security barrier.

Applets on the Web can exchange information using static methods and fields because every applet in a page runs within the same VM, and the classes for every applet are loaded using the same class loader. This means that when two applets refer to a class with the same fully qualified name they both refer to the same underlying class. Thus, both applets see the same values for static fields, and static methods operate on the same object. In an MHP receiver, however, two Xlets will load the class using different class loaders, resulting in two distinct classes. The two Xlets will see different copies of static variables for any application-defined classes and thus there is no way of using these to communicate between Xlets.

The only exception to this is classes that are part of the middleware, known as system classes. In our case, this means the content of the `java.*` packages and any API classes built into the receiver. A Java VM includes a special class loader called a system class loader, and the

VM should always try to load classes using the system class loader before using any other class loaders. Doing this guarantees that system classes such as `java.lang.String` are always loaded from the same source to avoid problems within the VM. If the class cannot be loaded using the system class loader, the VM will try to use any other class loaders that are defined (in the case of an MHP application, this will be the class loader for that application). Only classes loaded by the system class loader are visible to classes loaded through other class loaders. This is shown in Figure 14.1.

As you can probably see, this mechanism allows different MHP applications to use different class paths. Each class loader can use a separate class path, and thus application classes will always be loaded from the correct object carousels. Figure 14.1 shows how different class loaders can be used within the VM to keep applications separate from each other.

If this is starting to give you a headache, that is okay. This can be scary stuff. The example shown in Figure 14.2 may make things slightly easier to understand. Let's assume we have two different MHP applications, each having a dedicated class loader. Each application will try to load two classes: `java.util.Vector` and `appspecific.SomeClass`.

First, let's look at what happens when the two applications load the `java.util.Vector` class. In each case, the runtime system issues a call to the `loadClass()` method on the instance of the DSM-CC class loader used by that application. The DSM-CC class loaders then both try to load the class as a system class by calling their `findSystemClass()` method, which automatically tries to load the class using the system class loader.

In this case, `java.util.Vector` is a system class, and thus the system class loader will successfully load it. The DSM-CC class loaders both get a non-null result from their call to

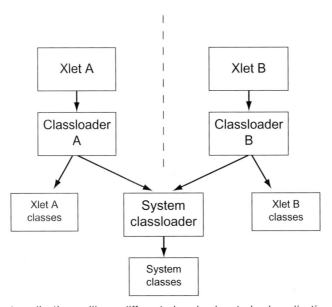

Figure 14.1. Different applications will use different class loaders to load application classes.

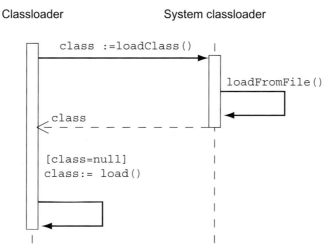

Figure 14.2. Delegating class loading from an application class loader to the system class loader.

findSystemClass(), and thus they simply need to return this result to the part of the runtime system that requested it. Because the class was loaded via the system class loader in both cases, the Java VM will regard it as the same class.

Now let's look at the other situation. When the DSM-CC class loaders call findSystem-Class() to try to load the appspecific.SomeClass class, a null value will be returned because the system class loader cannot find it. In this case, both class loaders will try to load it from the DSM-CC object carousel to which they are currently connected. In the case of class loader A (Figure 14.2), it can find the class on the object carousel and will load it. The class loader then returns an instance of this class to the runtime system.

Class loader B, however, cannot find that class on the object carousel it is connected to. Therefore, it will return a null value to the runtime system, which will in turn throw a java.lang.NoClassDefFoundError.

We have simplified things slightly here. Java platforms also have a bootstrap class loader, which loads classes the VM needs before it can start. This is used for loading classes such as java.lang.Class and java.lang.Classloader, and for the sake of this discussion it can be treated the same as the system class loader.

In a normal Java environment, applications can define their own class loader to load classes from a source that is not supported by the standard Java class loaders. MHP does not allow this because of the security risk it poses. Class loaders written by third parties may not call findSystemClass() during the class loading process, thus providing one way for malicious applications to spoof the middleware into loading harmful classes. Because an application may only load extra classes from an object carousel, MHP defines the org.dvb.lang.DVB-ClassLoader that allows applications to specify an object carousel from which classes should be loaded. Unlike the class path included in the application signaling, this may refer to carousels other than the default one for the application.

The Inter-Xlet Communication Model

So, how do we get around this problem of not being able to access an object from another application? Luckily for us, Java already has something for allowing applications to communicate: remote method invocation (RMI). RMI was originally designed for use in desktop Java implementations so that an application running on one machine could communicate with applications on other machines. This is the Java equivalent of the remote procedure call (RPC) mechanism familiar to many developers.

Using RMI

Every device that supports RMI runs an RMI registry. This is a process that keeps track of what objects are available to be used by remote applications. When an application wants to export an object, it binds the object to a name in the registry, thus allowing other applications to find that object. The only restriction on the type of objects that can be exported is that they must implement the `java.rmi.Remote` interface.

Once an object has been exported, another application can access it. To do this, the application that wishes to use it must perform a lookup on the registry of the machine it wants to connect to. An RMI client cannot simply search all RMI-enabled machines on a network for a given name, and thus it must know which machine it should connect to. Generally, this is not a problem because an application will want to connect to a specific server, and it does this by specifying a host name (or IP address) and port number for the remote registry it wants to search.

This lookup queries the remote registry for a specific object name. If an application on the server has registered an object with that name in the RMI registry, the lookup will return a reference to the remote object to the client. Although this will look like the remote object, it will not actually be a reference to the actual remote object. Instead, the RMI mechanism will create an instance of a stub class that will be used to refer to the remote objects.

Once the stub class has been successfully instantiated, the client application can use it to call methods on the remote object. The stub object acts as a proxy, redirecting all of the method calls to the remote object on the RMI server. As its name suggests, RMI only allows us to call methods on the remote object. The problems involved in synchronizing field values between the remote object and any stub objects outweigh any advantages that can be gained from this, and thus remote objects should define getter and setter methods for any fields that need to be exposed.

Problems with RMI

There are two big disadvantages to using conventional RMI on an MHP receiver. First, RMI implementations in desktop systems typically use TCP/IP for communication between RMI clients and servers. Because TCP/IP is not available in some profiles of MHP, this causes a serious potential problem for us. The overhead of using TCP/IP for communication between two applications on the same device makes this even worse. In a platform such as a DTV receiver, wherein resources are limited, this may not be an acceptable solution.

Second, it requires a set of stub classes to be pre-generated. In a conventional RMI implementation, these are generated using a special compiler included in the JDK tools. To use RMI successfully, these stub classes must either be present in the application's class path or downloadable via HTTP from the RMI server. Because stub classes typically use native code, they are not portable between platforms.

We cannot generate a set of stub classes for every possible receiver, because we do not know which platforms we must support. Even if we did, we would have to include them in the object carousel so that the client application could load them (or make them available on an HTTP server, which causes problems for those receivers without a return channel). The overhead of using stub classes means that this approach is not really practical in an MHP or OCAP implementation.

RMI Extensions

To resolve these problems, DVB added some extensions to the standard RMI API. The most noticeable of these is that RMI no longer needs predefined stub classes (although the MHP specification does not say how middleware developers should work around this). Although this makes things more difficult for middleware implementers, it makes the application developer's life a lot easier. We will take a more detailed look at the effects of this in material to follow.

The second major difference concerns the RMI registry. Conventional RMI implementations use the `java.rmi.Naming` class to provide the interface to the RMI registry, and although there is no fundamental problem with using this class for inter-Xlet communication doing so would interfere with other uses of RMI.

Using a different class also avoids the problem of different naming schemes. The `java.rmi.Naming` class assumes a naming scheme that includes an IP-based component (a host name or IP address, followed by a port number). This is not practical for inter-Xlet communication, and thus using the `Naming` class for this purpose would mean having to support two different naming schemes for object lookup and binding.

To solve this, the inter-Xlet communication API defines the `org.dvb.io.ixc.Ixc Registry` class that provides the RMI registry for inter-Xlet communication. This enables applications to use RMI to communicate with truly remote servers, while exposing a different set of interfaces to other Xlets on the same machine.

The `IxcRegistry` class has a similar interface to `java.rmi.Naming`, but with a few changes to support the unique nature of inter-Xlet communication. In particular, the `Naming` class serves two purposes: as an interface to the local registry on the machine running the application and as the interface to any remote registries. In contrast, the `IxcRegistry` class has nothing to do with remote registries because there is only one registry when we are only dealing with local Xlets. In this sense, it is conceptually closer to the `java.rmi.registry.Registry` interface than to the `java.rmi.Naming` class. The interface to the `IxcRegistry` class follows.

```
public class IxcRegistry {

    public static java.rmi.Remote lookup(
        javax.tv.xlet.XletContext xc,
        java.lang.String path)
        throws java.rmi.NotBoundException,
            java.rmi.RemoteException;

    public static void bind(
        javax.tv.xlet.XletContext xc,
        java.lang.String name,
        java.rmi.Remote obj)
        throws AlreadyBoundException ;

    public static void unbind(
        javax.tv.xlet.XletContext xc,
        java.lang.String name)
        throws java.rmi.NotBoundException ;

    public static void rebind(
        javax.tv.xlet.XletContext xc,
        java.lang.String name,
        java.rmi.Remote obj);

    public static java.lang.String[] list(
        javax.tv.xlet.XletContext xc);
}
```

Methods defined by `IxcRegistry` include the Xlet context of the calling Xlet in order to let the implementation create the stub objects using the correct class loader. We will look at why this is useful later in the chapter, and how we can use it to help improve the efficiency of our middleware implementation.

Unlike conventional RMI, names used in inter-Xlet communication do not include the IP address of the remote host. Instead, we use the organization ID and application ID of the server application to help identify the remote object. This has two advantages. First, by requiring this we force RMI clients to connect to specific servers, just as we do for conventional RMI implementations. Second, the use of the organization ID and application ID provides separate name spaces for each application and avoids problems where more than one application exports an object with the same name. Each name in the inter-Xlet communication API takes the following form.

```
/<organization ID>/<application ID>/name
```

Here, *<organization ID>* and *<application ID>* are encoded as hexadecimal numbers with no leading `0x`. The `list()` method will return the names of every exported object, no matter what application is exporting it. Apart from using the `IxcRegistry` class instead of the `java.rmi.Naming` class, communication between Xlets will use the same mechanisms and the same APIs as other RMI applications.

An Example of Inter-Xlet Communication

The following example should hopefully make inter-Xlet communication appear a little easier. In this case, the server object exposes a method for setting the value of one of its fields, whereas the client uses RMI to get a reference to the server object and then sets the value of that field. This is a simple example, but it is enough to give you a flavor of how inter-Xlet communication works, and how it is different from conventional RMI.

```java
// we need to implement java.rmi.Remote to indicate to
// the middleware that instances of this class may be
// exported over RMI.
public class Exporter implements java.rmi.Remote {

  // some private fields. These could be public,
  // however, and they would still not be directly
  // visible to other applications.
  private int value = 0;
  private Exporter exportedObject;

  // Constructor
  private Exporter() {
  }

  // this method allows us to set the value of one of our
  // private fields. As well as being an example of a
  // remote method, it also shows how we can avoid the
  // restriction of not being able to access fields in
  // remote objects.
  //
  // Any exported method must throw RemoteException.
  public void setValue(int a)
    throws java.rmi.RemoteException {
    value = a;
  }

  // the method that actually sets up and exports an
  // object.
  public void exportMe(XletContext myXletContext) {

    // the Xlet Context is the Xlet context of this
    // xlet, which was passed in to the Xlet.initXlet()
```

```
// method.
XletContext exportingXletContext;
exportingXletContext = myXletContext;

if (exportedObject == null) {
  // create a new Exporter instance. This
  // will be the object that we export.
  exportedObject = new Exporter();
}

// now, we bind the object we will export to a
// name in the registry. We pass the Xlet context of
// the exporting application, the name the object
// should be exported under and finally the exported
// object itself as arguments.
try {
  org.dvb.io.ixc.IxcRegistry.bind(
    exportingXletContext,
    "Exporter",
    exportedObject);
}
catch (AlreadyBoundException abe) {
  // an object has already been exported by this Xlet
  // with this name. In this case, we can use
  // org.dvb.io.ixc.IxcRegistry.rebind()
  // to export the object instead
}
catch (NullPointerException npe) {
  // one of the parameters is null, which is
  // not allowed.
}

// The object is now exported and can be used by
// other applications.
  }
}
```

The class that uses this exported object can do so as follows. It is important to remember that this takes place in a completely separate application, in a completely separate Xlet context.

```
public class Importer {

  public XletContext myXletContext;

  public void useRemoteObject() {
```

```java
// myXletContext is the Xlet context of this Xlet,
// which was passed in to the Xlet.initXlet() method.
XletContext importingXletContext;
importingXletContext = myXletContext;

// first we have to find the remote object. We do
// this by calling the lookup() method on the
// registry, which takes the Xlet context of the
// importing Xlet and the name of the object to be
// imported. We assume that the organization ID and
// application ID of the server are 1 and 2
// respectively.
java.rmi.Remote remoteObject = null;
  try {
    remoteObject = org.dvb.io.ixc.IxcRegistry.lookup(
      importingXletContext,
      "1/2/Exporter");
  }
catch (NotBoundException nbe) {
  return;
}
catch (RemoteException re) {
  return;
}

// Return if we can't find the object
if (remoteObject == null)
    return;

// assuming we get a non-null response, we need to
// cast the imported object into the correct type.
Exporter importedObject = (Exporter) remoteObject;

// now we can use it, although we need to make sure
// we catch any remote exceptions.
try {
  importedObject.setValue(42);
}
catch (java.rmi.RemoteException re) {
  // do nothing in this case
}
  }

}
```

Practical Issues

Although the basic API is the same as for conventional RMI implementations, the inter-Xlet communication mechanism in MHP and OCAP has a number of unique elements that make it rather difficult to implement. Because each application will be loaded through its own class loader, this means that we have to track which class loader is associated with each application. This is a more general issue than it may seem at first. Class loaders are often the only means we have of identifying which application an object is part of, and thus they can be used by other parts of the middleware that need to trace an object back to its parent application (for instance, for checking security).

For any object, we can get a reference to its class loader by calling `<object>.get-Class().getClassLoader()`. Typically, the class loader will be associated with the Xlet context, and hence the use of the `XletContext` objects as parameters in the `IxcRegistry` methods. This lets us immediately identify which application a particular class belongs to, simply by finding the Xlet context that has the same class loader associated with it.

Generating Stub Classes

MHP and OCAP implementations may still use stub classes for communicating between the client and the server, although the MHP specification says that this is not required. Given the practical issues surrounding class loaders, this is often the easiest way. We cannot rely on having externally generated stub classes, however, and thus the middleware may have to generate these classes itself and load them into the correct middleware.

There are two possible approaches to this: generating a new class file for the stub class and then loading it, or loading the class and then modifying its method implementations. Tools such as Javassist from the JBoss project and the Byte Code Engineering Library (BCEL) from Apache's Jakarta project make these approaches easy without too much hard work. Of course, it is still possible for implementations to do it the hard way and generate the byte code themselves, using the Java reflection API to get the correct method signatures, but this runs the risk of introducing bugs into the stub classes. The tools now available make this task far easier and reduce the potential problems caused by badly formed stub classes.

Once a stub class has been created, it can be loaded using the class loader belonging to the client application and then instantiated. This will give us an object the client can use to call methods on the remote parent of that object. Because we only need to worry about method calls, we do not need to copy any fields in the original object. As long as we can map the stub object back to the object it represents, this will meet our needs.

Calling Remote Methods

Once we have created our stub object, we can begin calling methods on it. We cannot directly call methods in the other object, because this would result in threads crossing between applications and violating the security model MHP and OCAP have defined. This means we need some type of intermediary to help us call these methods.

One solution to this is to embed the call within a `Runnable` object that we can then pass from the client to the server. We can do this using the Java reflection API, by embedding code similar to that shown below in an instance of a `Runnable` object.

```
java.lang.reflect.Method targetMethod;
Object remoteObectReference;
Object[] args;

targetMethod.invoke(remoteObjectReference, args);
```

We can then pass this `Runnable` object to the server application and execute it in a thread on the server side (for instance, a worker thread provided by the Xlet context for the server application). We can invoke the correct method very easily, but we must use a dedicated thread to do so because this operation does not map onto any application threads that will be running in the server application.

The important thing to note here is that classes loaded by different class loaders will have different `Method` objects representing their methods. This is understandable, in that classes loaded by different class loaders are never considered to be the same, even if they are loaded from the same class file (after all, that is what got us into this situation to start with). Because of this, the `targetMethod` variable in the previous example must refer to a `Method` object belonging to the class loaded by the class loader for the server application.

This works because system classes (including the `java.lang.reflect.Class` and `java.lang.reflect.Method` classes) are all loaded using the system class loader or the bootstrap class loader, and thus every application will use the same version of these classes. Put simply, applications are separate from one another, but they run on a common layer of system classes. These system classes can talk to one another, even when application classes cannot. Using the reflection API, classes in the middleware stack (such as the `javax.tv.xlet.XletContext` implementation) can "tunnel under the wall" between the two applications and indirectly invoke a method in the server application, as indicated in Figure 14.3.

A more detailed explanation would involve a discussion of class loaders that is beyond the scope of this book. We hope that the description here has given you enough information to make progress in this area.

Arguments and Return Values

If you thought this was complicated, we are not done yet. Any arguments and/or return values will also need to be transferred between the two applications. For the following reasons, this time we cannot simply create a stub class.

- We may need to refer to fields within the object.
- Synchronizing the life cycle of the stub object and the original object would be extremely complicated. We must retain the original object in its original Xlet context until the stub class is garbage collected, even if the application is terminated in the meantime.

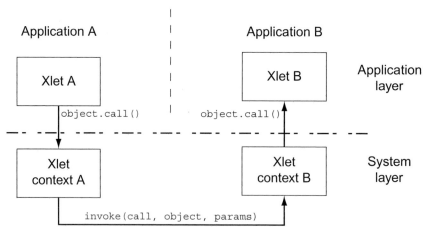

Figure 14.3. Invoking methods in a remote application.

This means that we have to create a deep copy of any objects passed as arguments or as return values. For simple types, this is easy, but it is much more difficult for complex types. Luckily, Java gives us a built-in set of tools for doing this. We can serialize the objects we wish to copy (which will also serialize any objects they refer to), pass the serialized representations of these objects to the appropriate application, and then de-serialize them in the context of that application. Figure 14.4 shows how this works in practice.

This works because any objects serialized by one application and then de-serialized by the other will use classes loaded by the class loader belonging to the second application. Because of this, new copies of the objects will be created with their values and references intact, but using classes loaded by the correct class loader. This assumes that the class files for any exchanged objects are available to both applications, but this restriction will be in force no matter how inter-Xlet communication is implemented.

Of course, this approach has limitations, and copying large structures of objects may use a lot of memory. These are generally issues for the application, however, in that well-designed applications will not be exchanging large amounts of data in this way.

To complicate things further, we need to keep track of the times when a copy of an object that is returned from a remote method call is passed back to the server as an argument in another remote method call. In this case, we need to unwrap the object as we pass it back to the server, so that the original object is used on the remote side instead of any copies of that. We can use a hash table to manage this mapping between stub objects and their remote counterparts, and to make sure that we import each object once only and that both the original objects and their clones are updated correctly.

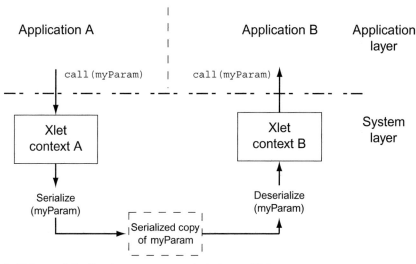

Figure 14.4. Using serialization to pass arguments between Xlets.

Managing Stub Classes

Like other objects in the system, stub classes need to be freed when they are no longer in use. This is typically done when the using application terminates, but there are a few things we must be careful about. Generally, managing instances of the stub classes will be more difficult than managing the classes themselves.

The most important thing to remember is that we will create copies of objects that are passed as arguments or as return values between the two applications. Managing these objects can be a challenge, because we need to make sure we keep just one copy of each object while at the same time allowing them to be garbage collected when they are no longer needed.

As we have already mentioned, one approach is to use a hash table to map imported objects to their stub implementations in the client application, but this means that both objects have an unnecessary reference that may prevent them from being garbage collected. To avoid this, references to these objects within the RMI implementation should always be weak references. A weak reference is a Java reference that is not counted by the garbage collector, and thus any object referred to using only weak references will be garbage collected. These are not officially available in version 1.1 of the Java platform, but some implementations of Java 1.1 may support them as an extension.

15 Building Applications with HTML

HTML is a major part of the new media landscape, and it offers many benefits for some types of ITV applications. This chapter discusses how we can develop HTML applications in MHP and OCAP. As well as looking at the new features HTML applications can use, we will look at how an HTML user agent fits in to the rest of the system and how it interacts with other applications that may be running. We also discuss how broadcasters are using HTML today, and how this differs from the MHP and OCAP vision of how it should be used.

One of the biggest additions to MHP 1.1 is support for DVB-HTML. This defines a standard set of HTML-related technologies for developing MHP applications, instead of relying on Java. Many companies and content developers are interested in using declarative languages such as HTML to develop applications for MHP. For some applications, such as information services, writing them in HTML is far easier than using Java to do the same thing. In other cases, companies are more used to using HTML and want to stick with what they know.

DVB-HTML is a combination of many leading-edge web standards (or at least they were leading edge when MHP 1.1 was published). This means that DVB-HTML supports a selection of XHTML 1.0 modules, CSS level 2, DOM level 2, and ECMAScript, as well as some other features. This is an impressive list, but it is also DVB-HTML's largest weakness. Many of these standards are not yet fully implemented in all of the desktop browsers, and thus the maturity of any embedded solution is likely to be slightly suspect.

Unfortunately for middleware developers who want to include it, DVB-HTML support comes with a lot of baggage. This is an optional part of MHP 1.1, and thus middleware implementers who want to support DVB-HTML need to support many of the other features of MHP 1.1 as well. We have seen a number of non-MHP products that claim to be DVB-HTML compliant, and although this is a nice marketing statement it does not mean very much. DVB-HTML without the rest of MHP is not really meaningful.

To confuse matters slightly, the MHP 1.0.x specification defines the basic application life cycle and signaling for HTML applications, but it does not define the format of DVB-HTML. We will not go into the reasons for this, but it is enough to say that MHP 1.0.x does not support DVB-HTML applications.

Shortly after DVB published MHP 1.1, OpenCable published the OCAP 2.0 profile. The only real addition was support for HTML applications. As you would expect, these are called OCAP-HTML applications, and they follow exactly the same format as DVB-HTML applications. Many features from MHP 1.1 were already included in the OCAP 1.0 profile, and thus there was no need to add these in OCAP 2.0.

Given the similarity between the HTML features both platforms support, we will discuss them simultaneously. Any references to DVB-HTML should be taken as applying to OCAP-HTML as well, unless it is explicitly stated otherwise. One thing to note, however, is that the OCAP 2.0 profile is based on MHP 1.1, rather than on the more recent MHP 1.1.1 specification. The differences between the two are minor, but they do exist.

Application Boundaries

Unlike a web site, a DVB-HTML application has a well-defined boundary the application author specifies in advance. This application boundary takes the form of a regular expression similar to those used in UNIX, which defines the set of files that are part of that application.

The boundary for a particular application is defined by the application boundary descriptor carried in the AIT entry for a DVB-HTML application. Each DVB-HTML application must contain one application boundary descriptor. We will take a closer look at the syntax of the descriptor itself later in the chapter.

This approach has one big advantage over other possible ways of specifying an application boundary. Depending on the regular expressions used, a file may appear within the boundary of more than one application. This gives developers a way of reusing documents across applications without having to include more than one copy of the document in an object carousel.

Application boundaries do not stop an application from referring to a page that is not within its boundary. Like a normal HTML page, a DVB-HTML document can contain a link to any content it likes. The user agent will treat any pages outside the boundary of the current application as if they were separate applications, however, and referencing them may affect the life cycle of the current application. Like service selection in DVB-J, this is something you should use with great care unless you want your application to die. The current application and the new one will not necessarily be active at the same time.

The Core Standards of DVB-HTML

DVB-HTML consists of a number of Internet standards. The most important of these is XHTML, the XML-based version of HTML. MHP supports a modularized version of this, and thus supports some parts of the full XHTML functionality but not others. Table 15.1 outlines which XHTML 1.0 modules are included in the DVB-HTML specification.

More details of these modules are defined in the W3C recommendation "Modularization of XHTML," May 2003. Some people dislike the choice of XHTML 1.0 as the basis for DVB-HTML because of the complexity of the standards. In reality, however, it is probably less complex than HTML 4.0 (thanks to the modularization). It also offers a major advantage in that it is based on XML. This makes it less tolerant of bad practices by developers, and this

Table 15.1. XHTML 1.0 modules included in DVB-HTML.

Modularization	Required
Structure	Yes
Text	Yes
Hypertext	Yes
List	Yes
Applet	No
Presentation	Yes
Edit	No
Bidirectional text	Yes
Basic forms	No
Forms	Yes
Basic tables	No
Tables	Yes
Image	Yes
Client-side image map	Yes
Server-side image map	No
Object	Yes
Frames	Yes
Target	Yes
IFrame	Yes
Intrinsic events	No
DVB intrinsic events	Yes
Meta-information	Yes
Scripting	Yes
Style sheet	Yes
Style attribute	Yes
Link	Yes
Base	Yes
Name identification	No
Legacy	No

Source: ETSI TS 102 812:2003 (MHP 1.1.1 specification).

means we can simplify the HTML parser because many of the ugly shortcuts in HTML are simply not allowed anymore. This improves reliability, and reduces the code size of the DVB-HTML user agent. The formal public identifier for the DVB-HTML `doctype` is as follows.

```
-//DVB//DTD XHTML DVB-HTML x.y//EN
```

Here, x and y are the major and minor version numbers for the version of DVB-HTML. In the case of MHP 1.1, these are 1 and 0, respectively. Each DVB-HTML document must start with the following DOCTYPE declaration.

```
<?xml version="1.0"?>
<!DOCTYPE html PUBLIC
  "-//DVB//DTD XHTML DVB-HTML 1.0//EN"
  "http://www.dvb.org/mhp/dtd/dvbhtml-1-0.dtd"
>
```

It is general good practice to include a DOCTYPE declaration in any HTML document, but it is more important than ever from a reliability point of view with respect to DVB-HTML. Normal HTML documents may include elements that are not indicated in the DOCTYPE definition (or may not have a DOCTYPE at all), but user agents may still choose to display them. For DVB-HTML, this is no longer the case. All applications must validate against the DVB-HTML DTD in order to make sure that they work correctly in an MHP or OCAP receiver.

The MHP specification allows documents that are invalid because they contain elements from a different name space, or because they use a later version of the DTD than is currently specified, but there are strict limits on how the user agent should handle these documents. If the document would validate by removing elements not in the currently defined DVB-HTML DTD, the user agent does not have to display the nonconformant elements. In this case, the elements will still be included in the DOM tree, even though they may not be displayed. Documents that are not valid will be ignored by a DVB-HTML user agent and will not be displayed or have a valid DOM tree.

CSS Support

Content has to look good if it is to be attractive to users, and thus DVB-HTML supports CSS level 2 for formatting and layout of XHTML content. There are a couple of changes from the standard CSS level 2 specification, mainly to handle the differences between a DTV receiver and other web devices.

One of these changes is that a DVB-HTML user agent does not have to support aural style sheets. For many applications, this will not be a big loss because these are not really aimed at the TV anyway.

Probably the biggest change, however, is the addition of a new media type to support displaying content on a TV. Some DTV receivers will be set-top boxes or integrated digital TVs like those we see on the market today. These typically have low resolutions (standard-definition PAL or NTSC resolution is most common) and may only support a limited range of colors. This may either be due to the limitations of the receiver or due to limitations of the NTSC and YUV color spaces. A second class of receiver may have capabilities much closer to those of a PC, supporting higher resolutions with fewer restrictions on colors and on navigation. These may have RGB or digital outputs that can provide high-resolution output to a high-definition TV or a monitor, and which have fewer limitations on color representation.

The `dvb-tv` media type lets a content developer provide different style sheets for receivers that are mainly TV based. This media type means that content developers can provide a look and feel that works well on low-resolution, interlaced displays while using the `screen` media type to provide a look and feel that is more suited to high-resolution PC displays.

Application developers can use this to define items such as different sets of colors or widths of borders. The `dvb-tv` media type also includes support for several new properties and rules that make it very useful for DTV developers. These include features such as support for transparency and compositing of overlapping XHTML elements, as well as the new `@dvb-viewport` rule. We will look at this new rule, and the other new CSS features, later in the chapter.

Scripting Support

Many applications need some type of interactivity, and JavaScript has become the standard way of implementing that in web pages. DVB-HTML uses ECMAScript, the standardized version of JavaScript for this purpose. Using ECMAScript has the advantage that it is a formal standard, rather than a proprietary standard that can change at any time. In particular, DVB-HTML refers to the first edition of ECMAScript.

While ECMAScript and JavaScript are pretty close, they are not identical and both middleware developers and application developers need to be aware of this. Versions of JavaScript after version 1.3 do meet the ECMAScript standard, but they add some extra features not available in ECMAScript. Application developers should take care to use only those elements of JavaScript included in ECMAScript and to avoid any proprietary features of JavaScript.

The ability of ECMAScript applications to extend their code at runtime has been restricted for security reasons. Depending on the permissions assigned to an application, this has implications for the following elements.

- Use of the `eval()` function
- String constructors for `Function` objects
- Setting event handlers
- Modification of `script` elements using DOM
- Use of the `Window.setTimeout()` DOM function
- Use of the `Document.write()` DOM function

To help solve the security problems, MHP divides ECMAScript strings into internal and external strings. Internal strings are those generated by ECMAScript code or string literals in the application, whereas external strings are those produced outside ECMAScript (e.g., by other parts of the middleware). External strings are not considered as safe as internal strings, and thus network operators can decide which classes of strings can be used in runtime code extensions by assigning the appropriate permission to the application (as we saw in Chapter 13). Network operators can choose to enforce no use of runtime code extensions, use of internal strings only for runtime code extensions, or use of both internal and eternal strings.

Dynamic HTML

Scripting languages are often used to modify the content of web pages (so-called "dynamic HTML"). DVB-HTML supports the document object model (DOM) level 2 so that scripts or even other applications can dynamically modify the content of a page.

Only a subset of the modules defined in DOM level 2 have to be supported by a DVB-HTML user agent, but DVB-HTML defines several new modules that must also be supported. These add support to DOM for features that are new in DVB-HTML. Table 15.2 outlines which DOM modules must be supported by a DVB-HTML user agent.

It is not just ECMAScript that can manipulate the structure of a DVB-HTML document. MHP 1.1 defines a set of DOM bindings for Java that allows DVB-J applications to modify DVB-HTML documents. This builds on the standard DOM bindings defined by the W3C, and adds support for the new elements added in DVB-HTML. We will not look at this in much detail here, but if you are interested you can find the Java bindings in the `org.dvb.dom` package.

Table 15.2. DOM modules supported in DVB-HTML.

DOM Module		Required
Package	*Feature String*	
Level 2 core	`Core`	Yes
	`XML`	No
Level 2 views	`Views`	Yes
Level 2 style sheets	`StyleSheets`	No
Level 2 CSS style sheets	`CSS`	No
	`CSS2`	Yes
Level 2 events	`Events`	Yes
	`UIEvents`	Yes
	`MutationEvents`	Yes
DVB-HTML	`DVBHTML`	Yes
DVB events	`DVBEvents`	Yes
DVB key events	`DVBKeyEvents`	Yes
DVB CSS	`DVBCSS`	Yes
DVB environment	`DVBEnvironment`	Yes

Source: ETSI TS 102 812:2003 (MHP 1.1.1 specification).

Developing Applications in DVB-HTML

Developing an application in DVB-HTML is not very different from developing a web site. The changes introduced by DVB-HTML are primarily aimed at the graphics model used by a DTV receiver, and at the differences in navigation a user will encounter.

Navigating Around a DVB-HTML Application

Most MHP or OCAP receivers will not include a free-moving cursor for the user to navigate with, and DVB-HTML user agents may not even be able to smoothly scroll a document within the window it is displayed in. Jump scrolling may be the only way of navigating a document. Generally this should be avoided by designing the content to fit on a single screen and using explicit links to other pages.

To help ease the problem of navigation, DVB-HTML defines a number of additional properties as part of the `dvb-tv` media type. These let the application developer specify how the user can navigate between elements of a document using the four arrow keys typically found on a DTV remote control. The `nav-index` property associates a numeric index with an element. This uniquely identifies that element in that page, and thus applications can use this to help control navigation. Conceptually, this is similar to the `name` attribute, except that it is used purely for navigation purposes.

Applications can use the `nav-first` property to indicate which element should receive input focus when the document is first displayed. If a particular element should always have focus, this can be used to ensure that the behavior is the same on all receivers.

The `nav-up`, `nav-down`, `nav-left`, and `nav-right` properties all identify the element that focus will move to when the corresponding key is pressed. These properties can have an integer value that specifies which `nav-index` should be used as the target, or they can have a string value that identifies a target that is external to the current frame. This string has the following syntax.

```
[frame-id] # [element-id]
```

This identifies the frame and/or element that should gain focus. At least one of these should be present, although the other is optional. The following HTML shows how these can be used.

```
<style type="text/css">
  #myElement nav-right:"frame1#Heading";
  #myImage nav-left:"1"
</style>
```

Alternatively, focus can be passed out of the application by using the string `[outer]` as the target. This lets an HTML inner application that is embedded in a DVB-J application pass focus back out to its parent application. For applications that are not inner applications, this will be ignored. We will discuss inner applications in much more detail in the next chapter.

Using these navigation properties is not as easy as it may appear, however. The user agent may assign its own navigation settings to the document, based on how the document is presented, the size of the document, and the size of the viewport. Changing these could lead to usability problems if the resulting navigation structure causes too much scrolling because the document cannot fit in the current viewport. Unless there is a good reason to change the default navigation order, it is often best to leave this to the user agent. If your XHTML is structured well, this will usually give the user agent enough hints for it to handle navigation in a sensible way.

One common navigation technique is to use the colored buttons on the remote for moving between elements. These keys will generate DOM events, and thus we can use these events to handle navigation for these buttons. This requires a little more work than setting other navigation targets, but this approach will usually be more suitable for the type of navigation we will use these buttons for.

Special URLs

DVB-HTML defines a number of special URLs an application may use to access specific pieces of content. Most of these are variations on the `dvb://` URL format and provide a way to carry out some basic functions that would be extremely difficult otherwise. Table 15.3 outlines these locators and their assignments.

Table 15.3. Special URLs that DVB-HTML applications may use

Locator	Description
`dvb://current`	The current service.
`dvb://current.video`	The video stream being presented on the background video device.
`dvb://current.audio`	The audio stream associated with any video being presented on the background video device.
`dvb://current.av`	Video and audio being presented on the background video device.
`dvb://original`	The parent service for the current application.
`dvb://current.ait/<org_id>.<app_id>` `?param=<val>`	An application that is being signaled on the application's parent service (with parameter passing).
`dvb://current.ait/app_root`	The application root directory as given by application signaling.
`dvb://current.ait/app_icon`	The application's icon, as given by the application signaling.
`exit:`	Exit the application. Any URI elements other than the scheme are ignored.

Displaying an HTML Application

As with Java applications, DVB-HTML applications need to reserve an area of the screen so that they can be presented. With DVB-J and OCAP-J applications, we do this using `HScene` objects, but we cannot use these for an HTML user agent.

Instead, DVB-HTML defines the `@dvb-viewport` rule for the `dvb-tv` media type. This lets DVB-HTML applications request an area on the screen where they will be displayed, and can be thought of as the HTML equivalent of an `HScene`. This is similar to the `@page` rule defined for paged media, in that it tells the user agent what area should be used for rendering the content. If the actual content is larger than the viewable area defined, the user agent should give the user some way of seeing the rest of the content, either by scrolling or by paging the content. Only pages displayed in the parent frame can define a new viewport. Pages or applications in subframes may only use the viewport defined by their parent frame set.

The `@dvb-viewport` rule can include a number of different properties to describe how and where the content should be presented. These are outlined in more detail in Table 15.4.

The initial container for a document may be larger than the viewport being used to display the content (in other words, it may have negative margins). Doing this can cause problems for the user, however, because a DVB-HTML user agent may not support scrolling. It may split the content across multiple pages, with no smooth scrolling between them. For some document layouts, this can seriously harm the usability of an application and thus it is best to avoid this. Figure 15.1 shows how viewports are used to define how a DVB-HTML application is displayed.

One problem with defining a viewport is what happens when the area of the screen an application wants is not available. The DVB-HTML specification is silent about this, but a reasonable assumption is that the user agent will make a best-effort attempt to honor the request. As for Java applications, however, a DVB-HTML application should not rely on being able to get a specific area of the screen.

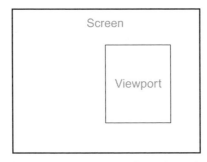

Figure 15.1. DVB-HTML applications use viewports to define where the application will be displayed.

Table 15.4. Properties for the @dvb-viewport rule.

Property	Description
scene	The size and location of the viewport, given in HAVi normalized coordinates. This corresponds to the size and location of the HScene belonging to a Java application. See Chapter 7 for more details of the normalized coordinate space.
horizontal-resolution	The number of pixels the user agent should assume for the width of the viewport. This is used to define the size of the px unit for any content displayed in that viewport.
vertical-resolution	The number of pixels the user agent should assume for the height of the viewport. This is used to define the size of the px unit for any content displayed in that viewport. Neither the horizontal-resolution nor the vertical-resolution properties have to correspond to the actual pixel resolutions of the display device. These are logical units used for CSS positioning only.
initial	An area defining the box that contains the root element of the document. This allows an application to define margins inside the viewport for content.
area	The area of the screen that will contain the viewport (the viewport will be a subrectangle of this area). This allows the application developer to specify that content should be presented only in certain parts of the screen. This can take one of the following values. ● screen (the entire screen, regardless of where video is displayed) ● total-video-area-on-screen (the area of the screen including any decoded video and black bars added by the receiver) ● active-video-area-on-screen (only the area of the screen containing decoded video) ● LV-bar (the vertical bar on the left-hand side of the screen; pillarboxed video only) ● RV-bar (the vertical bar on the right-hand side of the screen; pillarboxed video only) ● TH-bar (the horizontal bar at the top of the screen; letterboxed video only) ● BH-bar (the horizontal bar at the bottom of the screen; letterboxed video only) These values can be combined, as long as the final rectangle is contiguous.
type	Tells the user agent how the viewport should interact with the video layer in terms of preserving quality. This can take one of the following values. ● video (preserve the quality of video presentation, possibly at the expense of graphics quality) ● graphics (give priority to the presentation of the DVB-HTML content over video presentation quality) ● aligned (video and graphics pixels should be aligned) ● none (interaction between the video and graphics players is platform dependent) The aligned value can be used in combination with other values.

Because a TV display can have many different aspect ratios and resolutions, we can set up pseudo-classes for viewports that allow an application to adapt to these differences. The syntax for each pseudo-class name is as follows.

```
display-pseudo-class =
   ":" [ aspect-ratio ] [ resolution ] [ interlacing]
aspect-ratio = integer "x" integer
resolution = "R" integer "x" integer
interlacing = "P" | "I"
```

An application can use this to define viewports that match the aspect ratio and resolution of a specific display (and even distinguish between interlaced and progressive-scan displays). The following rules identify three standard-definition displays, one with an aspect ratio of 16:9 and two with an aspect ratio of 4:3.

- `@dvb-viewport :4x3 {. . .}`
- `@dvb-viewport :4x3R720x756 {. . .}`
- `@dvb-viewport :16x9R720x756 {. . .}`

More than one pseudo-class may match the current display. For example, an application may include pseudo-classes for a 4:3 display and for a standard-definition 4:3 display. In this case, the user agent will follow the normal rules for matching CSS selectors, with one minor extension. CSS selectors will usually be matched according to their specificity, and the number of components included in the `@dvb-viewport` pseudo-class will determine its specificity. In the previous examples, the second pseudo-class is more specific than the first because it specifies the resolution as well as the aspect ratio, and thus it will be used in the case in which a viewport matches both rules.

We can use a viewport for more things than simply displaying the elements of our application. In addition to the properties we have already seen, we can set a number of other properties on a viewport to control what is displayed in it. These properties are outlined in Table 15.5.

A DTV receiver may contain more than one video decoder, and thus there may be more than one video device in the system. Application developers can use any of these video devices by adding the suffix *–n* to any of the background video properties. This suffix is a numeric value that specifies which video device should be used, and video devices are numbered from front to back, with –0 representing the video device closest to the back of the display stack. For instance, the `background-video-1` property will set the URL of the video in the second video device. This suffix is optional, and any properties without this suffix will be applied to the rearmost video device. Table 15.6 outlines constants used to indicate decoder format conversion in the video chain.

Transparent Elements

We saw in Chapter 7 how transparency and opacity can be an important part of a TV-based application. The ability to see the underlying video is often an important feature of the user interface or the application's functionality.

Table 15.5. Display-related properties for the `@dvb-viewport` rule.

Property	Description
`background-image-rectangle`	Defines the area in the viewport where a background image should be placed. The value for this property is a rectangle that specifies (as a percentage) the area of the viewport that should contain the image.
`background`	Defines the background image or color that should be used for the background of the viewport. The behavior of this property is the same as for other CSS background properties.
`background-video-rectangle`	Defines the area in the viewport where background video will be placed. This has the same semantics as the `background-image-rectangle` property.
`background-video-clip`	Defines a clipping rectangle for the background video. This can be defined using `pels` or percentages. The `pel` is a new type introduced by DVB-HTML that corresponds to 1 pixel in the video layer.
`background-video-preserve-aspect`	Tells the user agent whether the aspect ratio of the video should be preserved. A value of `false` will scale the clipped video to the full output rectangle, whereas a value of `true` will scale it as much as possible, and add black bars to fill the rest of the output rectangle if necessary.
`background-video`	Defines the locator or color that should be used for the background video. Applications can specify more than one locator, separated by commas. This lets applications specify alternative pieces of content the user agent will display if the main content is not available.
`background-video-transform`	Defines the decoder format conversion (DFC) the user agent should apply to the video before it is displayed in the output rectangle. This can take one of the values outlined in Table 15.6.

To support this, the `dvb-tv` media type adds some extra CSS properties to support transparency and alpha blending between elements. The `opacity` property can take a floating-point value between 0.0 and 1.0 that defines the level of transparency for a particular element, where a value of 1.0 means that the element is fully opaque and 0.0 means it is fully transparent. As with transparency in DVB-J applications, not all opacity values have to be supported and the alpha value the user agent actually uses may only approximate what is specified.

If an opacity value is set for a container (e.g., a `div` element), the resulting alpha value is applied to all elements in the container. This is similar to the behavior for grouped components in an `HContainer`.

Table 15.6. Constants used to indicate decoder format conversion in the video chain.

Constant	Conversion
DFC_PROCESSING_CCO	A central rectangle is cut out of the source video and fit into the output rectangle.
DFC_PROCESSING_FULL	The entire video frame is displayed, possibly with black bars.
DFC_PROCESSING_LB_14_9	The incoming video is transformed to letterboxed 14:9 video.
DFC_PROCESSING_LB_16_9	The incoming video is transformed to letterboxed 16:9 video.
DFC_PROCESSING_LB_2_21_1	The incoming video is transformed to letterboxed 2.21:1 video.
DFC_PROCESSING_DEFAULT	The default decoder format conversion is used.
DFC_PROCESSING_PAN_SCAN	A part of the video frame is transferred to the output rectangle with the same aspect ratio. Pan-and-scan vectors in the MPEG stream control which part of the video is displayed.
DFC_PROCESSING_UNKNOWN	An implementation-specific decoder format conversion is applied, which does not match any of the other constants listed here. An application cannot set this value, however.

For overlapping elements that are partially transparent, the user agent uses the Porter-Duff rules for compositing. By default, objects are composited using the SRC-OVER rule, although this can be changed using the dvb-composite-rule property. This can take one of the following values, corresponding to the Porter-Duff rule of the same name.

- clear
- dst-in
- dst-out
- dst-over
- src
- src-in
- src-out
- src-over

For details on how these affect the way objects are drawn, see Chapter 7.

Embedding Video in Your Application

Sometimes, application developers will want to embed the video inside an application rather than playing it in the background. Java applications can do this using the org.havi.ui.HVideoComponent class and component players (if they are supported by the middleware), as we saw in Chapter 11. HTML developers have to take a slightly different approach, and thus video can be embedded in an object or img tag by specifying the dvb:// or ocap:// URL for the video that should be presented. This will be treated like any other element, and thus the receiver can scale this element and position it anywhere within the document.

Sometimes, we do not want to display the entire video, and thus we can set a clipping region to display only the part of the video we are interested in. The `dvb-clip-video` property defines the clipping rectangle that will be applied to the video before it is scaled and positioned in the target element. This is specified in `pels` in the source video, and thus a rectangle of (`0pels`, `720pels`, `576pels`, `0pels`) describes a full-screen PAL video.

Depending on the format of the video, this clipping rectangle may not stay the same. For pan-and-scan content, the location of the clipping rectangle will change depending on the pan-and-scan vectors included in the MPEG stream.

There are many reasons a piece of content may not be available, and thus any object that includes video content should also include an `alt` attribute that specifies some alternative text. This does not just apply to video content. The same thing may be true of images or audio clips. Generally, applications should follow the accessibility guidelines the W3C has defined as part of the Web Accessibility Initiative (WAI).

DVB-HTML Application Signaling

Like DVB-J or OCAP-J applications, HTML applications in MHP and OCAP are signaled in the AIT or the XAIT. MHP 1.1 defines some extra descriptors to handle the differences between HTML and Java applications, and these are used by OCAP to ensure compatibility between the two standards.

In addition to the AIT descriptors we have already seen in Chapter 4, MHP defines two descriptors specific to HTML applications. The HTML application location descriptor is similar to the application location descriptor for a Java application. Like the application location descriptor for Java applications, this refers to the base directory of the application (known as the physical root) and to its initial file. This will be an HTML file rather than Java class file, but the semantics are the same.

The main difference in the signaling is the definition of the application boundary. In Java, the class path defines the boundary of an application, but no such mechanism exists for HTML documents and thus we need to use a different approach. The HTML application boundary descriptor is included in the AIT entry for the applications, and includes a regular expression that identifies all of the URLs that can be part of the application. Table 15.7 outlines the syntax of the application boundary descriptor.

Each application boundary descriptor includes a label that can be used by the receiver to prefetch files that are part of the application. Different middleware implementations will use this label in slightly different ways.

Although each application boundary descriptor identifies the files that fall within the boundary of a given application, these boundaries can overlap. Common files can be shared

Table 15.7. Format of the application boundary descriptor.

Syntax	No. of Bits	Identifier	Value
dvb_html_application_boundary_descriptor{			
descriptor_tag	8	uimsbf	
descriptor_length	8	uimsbf	
label_length	8	uimsbf	N1
for(i=0;i<N1;i++){			
label_bytes	8	uimsbf	
}			
for(i=0;i<N;i++) {			
regular_expression_bytes	8	uimsbf	
}			
}			

Source: ETSI TS 102 812:2003 (MHP 1.1.1 specification).

between applications, just like common Java classes can appear on the class path of more than one Java application.

Events and HTML Applications

Other elements of the receiver (or other applications) can interact with a DVB-HTML application using DOM events or using triggers. Triggers are external events such as DSM-CC stream events or user actions such as clicking on a link.

DVB-HTML supports a limited set of DOM level 2 event modules, consisting of the UIEvent and MutationEvent modules, but not MouseEvent or HTMLEvent. Some parts of the HTMLEvent module are useful, however, and thus DVB-HTML defines two new events that inherit directly from the base Event interface. The load and unload events are identical to their counterparts from the HTMLEvent module.

As well as these event classes, DVB-HTML defines a few extra events we can use. The DVB-DOMStable event tells the application that the user agent has finished building the DOM tree for the current document. This lets the application know that no more manipulation of the DOM tree will take place from the user agent, although external applications or scripts within the application may still modify it.

The DVBKeyEvent class describes key presses from the user, and represents many of the standard keys found on a remote control. We will not look at this in too much detail here, but more details are available in the MHP 1.1 specification. As we have already mentioned, this allows an application to use keys such as the colored buttons for navigation or to enhance the application in other ways.

Table 15.8. DOM life cycle events for a DVB-HTML application.

Event	Description
AppStarting	The application has received a `dvb.start` event. We will look at this in more detail later. This event is only generated if the application is signaled as `PREFETCH`.
AppActive	The application has moved from the Loading state to the Active state. The `eventinfo` attribute contains the control code from the application signaling. If this attribute has the value `PREFETCH`, the application has not yet started, and it will only be visible when the application receives an `AppStarting` event. For all other values, the application has already started and is visible to the user.
AppPause	The application has moved from the Active state to the Paused state.
AppResume	The application has moved from the Paused state to the Active state.
AppDestroyed	The application has moved to the Destroyed state.
AppKilled	The application has moved to the Killed state.
AppTerminating	An application in a subframe has terminated. The `eventinfo` attribute will contain the name of the frame that contains the application that has terminated.

Life Cycle Events

The second class of DOM events that are added by MHP are DVB-HTML life cycle events. These events tell the application when its status has changed, and the application can use these to respond to those changes. The base interface for these is the DVBLifecycleEvent, and a DVB-HTML user agent can generate the life cycle events outlined in Table 15.8.

The user agent will always send life cycle events to the root element of the document that is currently displayed in the root frame of the application.

Stream Events and DOM Events

One of the more useful forms of triggers available to DVB-HTML applications is one that is activated by a DSM-CC stream event. This gives the application a way of synchronizing to the broadcast content, just like a DVB-J application that uses stream events.

To use this type of trigger, the MHP middleware lets us map a DSM-CC stream event onto a DOM event. For each application, the mapping from DSM-CC stream events to DOM events is given in an XML file called an event factory file. A typical event factory file might look as follows.

```
<?xml version="1.0"?>
<!DOCTYPE html PUBLIC
   "-//DVB//DTD DVB HTML Event Factory 1.0//EN"
   "http://www.dvb.org/mhp/dtd/htmleventfactoryfile-1-0.dtd">
```

```
<trigger-event
   stream="dvb://105.40.a9/stockPrice"
   event="priceChange"
   type="priceChange"
   cancelable="false">

   <event-attribute
      attribute-name="changedSymbol"
      attribute-data="/[A-Z]{4} new price: \d+/$1:$2" />

</trigger-event>
```

In this case, we set one trigger to monitor changes in the price of stocks (for instance, as part of a stock trading application). The `stream` attribute points to the DSM-CC stream event object that defines this stream event. This should refer to a stream event object in an object carousel. Each stream event object can name more than one stream event, and thus the `event` attribute tells the middleware to which stream event from that object it should subscribe.

The other two attributes tell the DVB-HTML user agent how it should handle those events. The `type` attribute gives the type of DOM event that should be generated when a stream event is received, whereas the `cancelable` attribute tells us whether the event can be cancelled.

Any DOM events generated may include data from the payload of the DSM-CC stream event, and the information that will be included is defined by an `event-attribute` element. This defines how the user agent should map the payload of a DSM-CC stream event onto DOM event attributes. Each `event-attribute` element defines the name of one attribute for the DOM event, and includes a regular expression the user agent will use to map the stream event payload to the value of the specified DOM event attribute. In the previous example, the payload includes the symbol of the stock and its new price. A DSM-CC stream event payload of `ABCD new price: 2463` will construct a DOM event attribute of `ABCD:2463`.

Before we can use the triggers we have defined in an event factory file, we need to associate that file with our application. A DVB-HTML application that uses DSM-CC stream events as triggers can include an event linkage file to tell the user agent what events it should pass to that application. This is an XML file with the same name as the root document of the application it is associated with, but which has the extension `.lnk`. An event linkage file looks as follows.

```
<?xml version="1.0"?>
<!DOCTYPE linkage
   "-//DVB//DTD DVB HTML Event Linkage 1.0//EN"
   "http://www.dvb.org/mhp/dtd/htmleventlinkagefile-1-0.dtd">

<linkage boundary ="*">
   <location URI="./events.evt" />
</linkage>
```

```
<linkage boundary ="news.html">
  <location
    URI="http://www.interactivetvweb.org/newsTicker/events.evt" />
</linkage>

<linkage boundary ="finance.html">
  <location
    URI="http://www.interactivetvweb.org/stockQuotes/events.evt" />
  <location
    URI="http://www.interactivetvweb.org/newsTicker/events.evt" />
</linkage>

<linkage boundary ="sports.html">
  <location
    URI="http://www.interactivetvweb.org/sports/quiz/events.evt" />
  <location
    URI="http://www.interactivetvweb.org/sports/scores/events.evt"
    />
  <location
    URI="http://www.interactivetvweb.org/sports/news/events.evt"
    />
</linkage>
```

In this example, files with the extension `.evt` are the event factory files that are bound to a given document. At first, this approach may seem unnecessarily complex, but it has the advantage of allowing an application author to specify an event mapping once and then reuse it in more than one document.

Not every application needs to define an event linkage file, though. The user agent will automatically subscribe an application to any stream events defined by DSM-CC stream event objects contained in the root directory of that application. These stream events will generate a DOM event of the same name as the stream event, with the payload of the stream event contained in the payload attribute of the DOM event.

Because it may be tricky to define the correct regular expression for converting stream event payloads into DOM event attributes (depending on your knowledge of regular expressions), it is often easier to define an appropriate format for the stream event payload to begin with and avoid processing this in the application.

System Events

So far, we have seen how we can use DOM events to influence the behavior of an application. Sometimes, we do not want the application itself to perform an action. Instead, we want to tell the user agent to do something when it receives a stream event. To do this, DVB-HTML extends the trigger mechanism to add system events. From the point of view of an application developer, system events are just like external triggers except that the user agent handles them internally instead of passing them to the application. An application can use a system event by setting the `type` attribute of a `trigger-event` element to the appropriate type in the event factory file.

The first system event we can use is the `dvb.start` event, which we can use to start the application at a specific time. Applications signaled as PREFETCH can use this event to display the application when the associated stream event is activated. When the user agent receives a `dvb.start` event it will present the application that has associated the `dvb.start` event with that particular stream event. This will only happen when the application has not already started, and applications can only register for `dvb.start` events in the main document (the one signaled in the application signaling). For any other pages, the user agent will ignore this event.

The second event available is the `dvb.page` event. This allows the application author to send the browser to a specific page in response to an event. Unlike the `dvb.start` event, any page can register for the `dvb.page` event.

A `dvb.page` event must include an `href` attribute, which identifies the page the browser should display. Optionally, it can also include a `title` attribute (which gives the title that should be used for the new page) and an `actuate` attribute that tells the user agent how it should go to that page. The `actuate` attribute can take one of two values. If this attribute has the value `onRequest`, the user agent will ask the user before it loads that page. If the user does not agree, the new page will not be loaded. If this attribute has the value `onLoad`, the page will be loaded automatically without asking the user.

By registering this event for certain pages for sets of pages, an application can direct the user to different pages depending on their current page. Alternatively, if the event is bound to the regular expression * in the event linkage file (i.e., to all pages in the application), the user agent will always try to present the new page.

The `dvb.page` event can be a powerful tool for certain types of applications. When combined with external triggers, it becomes possible to synchronize a DVB-HTML application with broadcast content extremely flexibly, and even change the behavior depending on where the user currently is within the application.

Coexistence of HTML and Java Applications

An MHP 1.1 or OCAP 2.0 receiver may run HTML and Java applications simultaneously if the receiver is powerful enough to run a Java VM and a DVB-HTML or OCAP-HTML user agent at the same time. In an MHP system, it is even possible to embed a DVB-J application within a DVB-HTML application, just like using an applet in a web page. It is also possible to use a DVB-HTML application within a DVB-J application. These are known as inner applications (see Figure 15.2), and we will take a closer look at these in the next chapter.

The ability to use both types of applications at the same time makes it easier for application authors to choose the approach that works best in a particular case. For applications such as information services, a declarative approach such as HTML often works better than a Java-

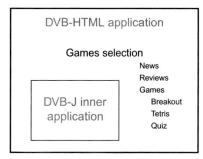

Figure 15.2. Inner applications let us embed DVB-J applications within an HTML application (or vice versa).

based approach. Maintaining a DVB-HTML application may be much less complex than maintaining a Java application that does the same thing. In other cases where a lot of processing is needed, Java applications have an advantage. Inner applications can give us the best of both worlds, allowing us to mix the processing ability of Java with the ease of authoring we get from HTML.

Accessing Java APIs from ECMAScript

Another way to mix Java and HTML is to use the DVB-J APIs directly from ECMAScript. All of the public DVB-J APIs (including all of the standard Java APIs) are exposed to ECMAScript via the `Packages` package. Scripts can use elements from this package to manipulate Java objects, instantiate new objects, and use the DVB-J APIs. An ECMAScript application can create a new Java object using the `new` operator, as follows.

```
new Packages.java.lang.Integer(42);
```

A similar approach can be taken to define subclasses of Java classes. By using the `Subtype` operator, ECMAScript scripts can define a new object representing a Java class. This can then be used to create instances of the newly created class. The following script, for instance, makes a request for information from the JavaTV SI API using a listener defined in the script.

```
<script>
  // first, we define our listener's methods
  function notifyFailure(reason) {
    // do stuff
  }

  function notifySuccess(result) {
    // do stuff
  }
  // the constructor. We need to set which functions in
  // our script correspond to the methods in our Java
  // class
```

```
function listenerConstructor() {
  this.notifyFailure = notifyFailure;
  this.notifySuccess = notifySuccess;
}

// define the class as a subtype of the JavaTV
// SIRequestor interface
SIListenerType = new Subtype(
  "javax.tv.service.SIRequestor", listenerConstructor);

// instantiate an object of our newly-defined class
siListener = new SIListenerType();

// get an SIManager object from the Java APIs.
siManager =
  Packages.javax.tv.service.SIManager.createInstance();

// use that and our listener to make a request
// myLocator is not defined in this example, and should
// be defined elsewhere in the script
siManager.retrieveServiceDetails(
  myLocator, siListener);
</script>
```

Each DVB-HTML application will have its own Java class loader that is used for loading any classes used within a script. We can use a `meta` element with its `name` attribute set to `code-base` to identify a path for loading Java classes, which has the same effect as setting the class path in a Java application or using a `DVBClassLoader` within the application. The `content` attribute of this `meta` element tells the middleware the URL from which it should load any classes it needs. Each `meta` element can contain one URL, and a document can contain more than one of these elements. When dealing with frame sets, however, only those `meta` elements that are part of the root frame are used. The middleware will ignore any that is part of a subframe.

The class loader used by the DVB-HTML application is different from the class loader used by any Xlets embedded within it, and thus we cannot use shared objects to communicate with other applications. If we want to do this, we have to use the inter-Xlet communication API we saw in Chapter 14.

Calling Java methods from ECMAScript scripts has its pitfalls, and the biggest of these is type conversion. ECMAScript and Java use different approaches to types of variables. Java uses strong typing, whereas ECMAScript uses weak typing that allows much easier conversion between types. DVB-HTML defines a number of type conversions that the user agent will apply when converting between values in ECMAScript and Java. These are fairly complex, and thus we will not cover them in much detail here. It is worth noting, however, that DVB-HTML adds three new fundamental types to ECMAScript (`JavaObject`, `Java-Class`, and `JavaArray`) that are used to improve the interoperability between ECMAScript and Java. More detail about these types, and a full list of possible type conversions, is available in Section 8.10 of the MHP 1.1 specification.

This type conversion may have an impact on how we call Java methods. Because the types of the arguments in our ECMAScript function call may not directly match the signature of the Java method we are calling, the middleware needs to convert the ECMAScript arguments into the equivalent Java types. Before it can do this, though, it needs to work out which method we are actually calling.

Each ECMAScript type has a preferred conversion for mapping to a Java type, although other conversions may be possible. For instance, conversion from an ECMAScript `boolean` value to a Java `boolean` value is preferred over conversion to a `java.lang.Boolean` object, although both conversions are allowed. In this case, the middleware will use these preferred type conversions to match the signature of a Java method to the arguments used in an ECMAScript method call. This only happens when more than one method signature matches the argument list used by the ECMAScript code. When only one method signature matches, any conversions that are necessary will take place.

Extending the Document Object Model

In addition to the basic ECMAScript or Java DOM bindings for XHTML objects, DVB-HTML defines some bindings for MHP-specific extensions. These are too complex to cover here, and thus we will refer you to the MHP 1.1 specification for full details of these.

Java bindings for the DVB-HTML DOM extensions are contained in the `org.dvb.dom` and `org.dvb.dom.inner` packages. We will look at a few of the classes from these packages in the next chapter, especially those that relate to embedding DVB-HTML documents within DVB-J applications.

Real-world HTML Support

So far in this chapter we have seen how DVB-HTML and OCAP-HTML should work. It would be very nice if things actually worked this way in practice, but this is often not the case. Before we can develop HTML applications for MHP or OCAP, we need to consider two major issues.

The first of these is support. At the time of writing, no middleware stacks support the full DVB-HTML specification. Several browsers support a subset of DVB-HTML, but they do not all support the same subset. There is also the issue of whether or not this subset is actually useful. It may not include any of the features added by DVB that are useful for TV-based applications, and thus the question arises whether a particular browser is really a DVB-HTML browser or just a normal browser running on MHP.

This is complicated by the fact that a true DVB-HTML implementation should also support all mandatory features of the MHP 1.1 specification. Many people currently ignore this important issue, but it may come back and haunt them later. OCAP does not have this problem, in that the OCAP 2.0 profile only adds support for OCAP-HTML. Of course, no one yet supports OCAP-HTML either.

At the time of writing, network operators cannot rely on having any receivers in the field that support DVB-HTML or OCAP-HTML, let alone having a majority of browsers that support it. There are solutions to this that we will see in material to follow, but these add to the complexity and expense of deploying HTML-based applications.

The second problem that faces DVB-HTML or OCAP-HTML developers is interoperability. Anyone who has read the W3C recommendations will know that they are just that, recommendations. Anyone who has tried implementing a web page using complex CSS for positioning knows that it soon becomes necessary to start using all kinds of nasty workarounds to fix problems in different browsers. The infamous "box model hack" and "Tan hack" for Microsoft's Internet Explorer are probably the most obvious and widespread examples of this, but there are others out there.

Although the W3C have published test suites for CSS level 1, CSS level 2 selectors, and DOM level 2, there are no official test suites for XHTML, ECMAScript, or the DVB-HTML extensions to any of these standards. Consequently, testing interoperability becomes very difficult. Making this even more difficult, some third-party test suites do not themselves validate using the W3C tools, and thus these may themselves cause problems for DVB-HTML user agents. Of course, even though the official test suites exist it is not completely clear whether they validate against the DVB-HTML DTD so that we can use them to test DVB-HTML user agents. There are ways around this, but it adds yet more uncertainty to the process.

Even for a browser that implements the full DVB-HTML specification, there is no easy way to prove that it implements it correctly. Many people are testing their browsers against the Mozilla browser, which is a good starting point but is not enough to be sure. Although Mozilla does a pretty good job of supporting CSS and DOM, it still has a few problems, and testing against Mozilla does not test support for the new features introduced in DVB-HTML.

Testing against another browser like this also introduces the issue of whether interoperability actually means bug compatibility. Just being compatible is not enough to actually meet the specification, and thus this introduces another potential pitfall for browser and application developers. This will become an even bigger issue when more browsers become fully compliant with DVB-HTML, and start to implement features that browsers such as Mozilla do not support.

Given the lack of DVB-HTML browsers out there, it is extremely difficult to guarantee interoperability. Until there are several browsers that are widely deployed, this will continue to be a problem, and it may even be a problem past that point if browser developers do not make sure their browsers conform to the following.

- Meet the specification
- Are interoperable with other DVB-HTML browsers

We mentioned earlier that there is a solution to this. To be honest, it is more of a partial solution, but as long as browser developers try to maintain interoperability it is a workable one.

Many companies offer Java-based microbrowsers that can be downloaded as MHP applications. Network operators can download these browsers to a receiver and use them to display HTML content on an MHP receiver without needing full DVB-HTML support or even an MHP 1.1 receiver. The browser can run as a normal MHP application and can use a command-line argument to go to a specific page.

This approach gives network operators a single platform they can use for developing their HTML applications, and because every receiver will use the same browser testing becomes much easier. The disadvantage is that they typically do not support the DVB-HTML extensions, but this may not be much of a problem. It may even be an advantage, due to reduced code size and complexity in the browser. This does cause a few extra headaches for application developers because two networks may use different browsers, but overall this approach seems to work best for providing HTML content. Network operators have to license a browser, of course, but the cost of this can often be lower than the cost of building and updating information services written in Java.

Application developers should validate their content to avoid any potential problems moving between browsers, but this is usually a minor issue in any case. The advantage of working in an MHP environment is that legacy browsers are much less common, and thus many of the browsers in use will have reasonable support for XHTML and CSS.

At the time of writing, network operators in Germany and Finland already use downloaded browsers to display digital teletext pages. For this type of content, HTML will probably remain an important technology. The benefits of it are simply too significant to ignore. The only downside to this approach is that it potentially delays the introduction of MHP 1.1 receivers that include full support for DVB-HTML applications. This may not be a big problem, though, given the likely future direction of DVB-HTML. Chapter 8 of the MHP 1.1 specification, which defines DVB-HTML, opens with the following note.

The contents of this chapter may be substantially replaced in a later release of this specification. Readers are encouraged to check for the existence of such a later release before proceeding.

This is a fairly clear statement that we should expect HTML support in MHP and OCAP to change substantially in future versions.

The Future of DVB-HTML

DVB recognizes that DVB-HTML is in an early stage of development, and that no one who is deploying MHP really cares about the extra features added by DVB-HTML right now. Because HTML support is far more important in markets outside Europe, DVB is happy to follow the lead of those organizations with a bigger stake in getting it right.

Many of the features of DVB-HTML need a powerful receiver to run them, and cost pressures mean that most receivers will not support this level of power in the immediate future. The companies that defined DVB-HTML were primarily approaching the DTV market from the PC side, and what is possible on a PC often is not possible on a DTV receiver. Although DVB-HTML could put those companies in a strong position if it were widely adopted, most broadcasters and receiver manufacturers have ignored it. The realities of the market mean that receivers with enough power to run a fully compliant DVB-HTML user agent are simply too expensive and complex, and PC-based DTV receivers are a niche product.

Adding DVB-HTML as an optional part of MHP 1.1 was very much a political move: it shows support for HTML and the wishes of the PC industry, but in a way that can be ignored if that support becomes inconvenient. Many of the parties standardizing MHP and OCAP have voted with their feet, and DVB-HTML has not been updated since it was introduced (although other elements of MHP 1.1 have, if only to bring them in line with changes in MHP 1.0.2 and 1.0.3). This is a partly a result of the focus on GEM in order to harmonize OCAP, ACAP, and MHP, and partly a result of the lack of commercial interest. No network operator has yet adopted it and no middleware vendor has built and deployed a full DVB-HTML user agent.

The United States is probably the most important market for HTML applications now. Like DVB-HTML, the ATSC DASE standard is substantially based on XHTML, and ACAP extends DASE to add support for MHP (in the form of GEM). HTML support is an integral part of ACAP, and thus more time has been spent making sure it includes all of the necessary features for building useful applications without being too big to fit in a DTV receiver.

To maintain compatibility, it is likely that MHP 1.1.x will replace DVB-HTML with the ACAP-X format (the XHTML format for HTML applications) once ACAP is formally ratified. At the time of writing, ACAP is a candidate standard and formal adoption is not too far away. Once ACAP is formally ratified and initial bugs have been resolved, expect MHP to rework substantially the current definition of DVB-HTML in order to harmonize HTML support with ACAP. It is not clear what would happen to OCAP-HTML in this case, but in all likelihood it would also move toward harmonization. Given the current direction of OCAP, this is not an unreasonable assumption.

For now, implementing HTML support using a downloadable application is probably the best way to approach this. Until the situation stabilizes, deploying a full DVB-HTML solution is a risky move. The current standardization work on the Portable Content Format (PCF) specification for declarative applications only complicates things further, although in many cases HTML may be an ideal format for implementing PCF support. In that this work is still in the early stages, only time will tell.

16 MHP 1.1

MHP 1.1 introduces several new features to MHP receivers, such as APIs for controlling Internet clients and communicating with smart cards, and the ability to store applications in the receiver. This chapter introduces these features of MHP 1.1 and several more, looking at the opportunities they offer to application developers and network operators. There are some risks to using MHP 1.1 that must be considered, and we will also take a look at how these affect the deployment of MHP 1.1 applications and receivers.

About 15 months after finishing MHP 1.0, DVB published the MHP 1.1 specification. MHP 1.1 is about 500 pages longer than the original MHP specification, and thus there is a lot of new material for middleware implementers and application developers to come to grips with. DVB-HTML applications are the biggest addition to MHP 1.1, as we saw in the last chapter. In this chapter, we will look at some of the other additions and at how these new features extend the functionality of MHP and how applications can use them.

The Internet Access Profile

As we have seen in earlier chapters, MHP 1.0.x defines the Enhanced Broadcast Profile and Interactive Broadcast Profile so that receiver manufacturers can choose a specific subset of the MHP functionality that meets the price point for a given product. These profiles also provide application developers with a way of guaranteeing that a known subset of the functionality will be present in an MHP receiver, even though it may not include all of the features the specification defines.

MHP 1.1 adds an extra profile to provide additional capabilities for high-end receivers. The Internet Access Profile adds support for a web browser, e-mail clients, and other Internet technologies commonly found in DTV receivers with a return channel, but which are too complex to be a standard part of the middleware.

Although many implementations that support the Internet Access Profile will also include support for DVB-HTML, these are actually two different things. The Internet Access Profile is much more flexible in the level of HTML support that receivers must implement, and thus a receiver could choose to support a limited subset of DVB-HTML and still be a valid implementation of the Internet Access Profile. This is deliberate. The requirements for supporting DVB-HTML are quite high, and thus this gives developers an alternative way of using HTML in their applications. Similarly, we can use DVB-HTML support in the Enhanced Broadcast Profile of MHP 1.1 without having to support the Internet access APIs.

As well as adding support for Internet client software, the Internet Access Profile defines an API that lets normal MHP applications manipulate those clients so that an MHP application can browse the Web or send an e-mail.

The Philosophy of the Internet Client API

Before we look at the API in detail, we need to understand a little bit more about the philosophy behind it. The Internet client API is the first of the MHP APIs to consider explicitly the concept of resident applications, wherein MHP-compatible applications are built in to the MHP receiver instead of being downloaded from the broadcast channel. Unlike OCAP, this was something MHP did not have to deal with in the first versions of the specification.

In practical terms, the final solution was pretty close to that used by OCAP for supporting unbound applications. OCAP uses the concept of abstract services to define which unbound applications can coexist, and this implies that unbound applications and service-bound applications cannot run in the same service context. The Internet client API takes the same basic approach, for the same reasons OCAP did. This gives the best combination of flexibility and reliability. Network operators always know which applications will run at the same time, whereas the middleware manufacturer can choose to let resident applications run at the same time if the receiver is powerful enough.

There are a couple of differences from OCAP's approach, however. The Internet client API does not rely on abstract services to decide which applications can run together. Instead, the middleware treats every client application as a separate service. An MHP application can start an Internet client application by selecting the appropriate service. If the receiver has enough resources, an application can create a new service context and select the service representing that client in the new service context, as we see in Figure 16.1. If the receiver does not have enough free resources to run the client and the original MHP application at the same time, the MHP application can choose to select the client's service in its own service context. This will allow the client application to run, but will kill any other applications that may be running in that service context. Sometimes, however, this will be an acceptable choice for the network operator and application developer.

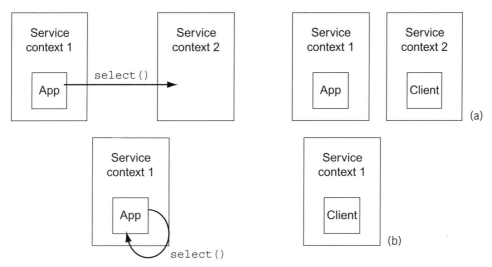

Figure 16.1. Launching an Internet client acts like selecting a new service: (a) launching an Internet client in a different service context, and (b) launching an Internet client in our own service context will kill our application.

Using the Internet Client API

The Internet client API is located in the `org.dvb.internet` package, which contains two class hierarchies of interest to application developers. The first of these consists of `InternetClientService` and its subclasses. These represent the abstract services the client applications will run in, but they also define operations that are common to all instances of a client application, whether it is running or not. This includes operations such as adding an entry to an e-mail address book or to a web browser's bookmarks list and querying the capabilities of the browser.

The second hierarchy consists of `InternetClient` and its subclasses. An instance of one of these classes represents a running copy of a client application, and thus these classes define operations that act on a particular instance of a client. For instance, this can include telling a web browser to go to a particular URL.

If you look at the definitions of these classes, you will notice that MHP defines both hierarchies as interfaces rather than classes. Why do we have two hierarchies for handling these functions? The answer is really quite simple. Because all of these are interfaces, there is no way of defining static methods. If we could do this in Java, all of the methods defined by `InternetClientService` and its subinterfaces would be static.

General Operations on Internet Clients

The `InternetClientService` interface is a subclass of `javax.tv.service.Service`, and thus it has all of the methods one would expect from a `Service` object. Applications can call the `getLocator()` method to get a locator they can use to start that client. There is a lot more to this class than simply acting as a way of starting an application, however. Because an Internet client may not always be able to run, the `canRunApplication()` method lets a downloaded application find out if a given Internet client can run at the same time. This gives downloaded applications a way to find out whether a receiver can support some functions without having to actually try it, thus letting applications change their user interface to account for the capabilities of the receiver.

Sometimes it is useful to find out the state of a specific client application. An application can register as a listener for events from a specific instance of an Internet client using the `InternetClient.addInternetClientListener()` method. These events, which are subclasses of the `InternetClientEvent` class, allow the application to find out whether a particular operation succeeded or failed. Because the client may not be able to respond immediately (after all, the term *world-wide wait* does have a certain amount of truth to it), this allows the applications to get some status without blocking.

The `InternetClient` interface is a subinterface of `javax.tv.service.ServiceContentHandler`, just like any other class that presents content from a service. This provides a couple of methods that are not terribly useful in this case. The `getService()` method returns a reference to the `InternetClientService` object that corresponds to the type of client represented by this object (e.g., a `WWWBrowser` object would return a `WWWBrowserService` object), whereas a call to `getServiceContentLocators()` returns an array containing the locator for the service. This has the same effect as calling `InternetClientService.getLocator()` for the appropriate type of service, except that the locator is returned in an array with one element.

E-mail Clients

E-mail clients are represented by the `EmailClient` and `EmailClientService` interfaces. The `EmailClientService` interface, as follows, adds two methods to those we have already seen.

```
public interface EmailClientService
   extends InternetClientService {

   public java.lang.String getUserEmailAddress();

   public void addToAddressBook(
      java.lang.String address,
      java.lang.String name)
      throws EntryExistsException, java.io.IOException;
}
```

As you would expect, these let the application get the e-mail address of the user or add an entry to the user's address book. These are both fairly obvious, although it is worth noting that the application will need the appropriate permission before it can read the user's e-mail address. If it does not have this permission, calls to `getUserEmailAddress()` will throw a `SecurityException`. Similarly, applications may not have permission to add an entry to the address book, and in this case an `IOException` will be thrown.

Receivers will typically only have a small amount of space for storing address book entries, and thus it may not be possible to add a new entry. If the storage space is full, any attempt to add a new entry will also throw a `java.io.IOException`. Other methods in the Internet client API that add entries to bookmarks or newsgroup lists also have this behavior. The `EmailClient` interface, as follows, lets an application actually send an e-mail.

```
public interface EmailClient extends InternetClient {

    public void createMessage(
        java.lang.String to,
        java.lang.String subject,
        java.lang.String messageBody,
        java.lang.String sender)
        throws ClientNotRunningException;

}
```

By allowing the application to specify the sender for a given e-mail, the API allows more secure use of the e-mail functionality. The user must explicitly send any message that uses the user's default e-mail address (e.g., by clicking on the Send button in the e-mail client). This gives them a chance to review any messages an application sends in their name and to make sure that applications only use their e-mail address for purposes they agree to. Applications can use other sender addresses to send e-mail automatically because there is less room for abuse if the customer's real e-mail address is not used.

Web Browsers

The MHP middleware splits support for web browsers between the `WWWBrowser` and `WWW-BrowserService` interfaces. As with e-mail clients, the `WWWBrowserService` class, which has the following interface, lets an application carry out more general operations on a web browser.

```
public interface WWWBrowserService
    extends InternetClientService
{
    public java.lang.String[] getAcceptedMediaTypes();
    public java.lang.String[] getSupportedPlugins();
    public java.lang.String getUserAgent();
    public boolean areFramesSupported();

    public void setHomepage(java.net.URL defaultUrl);
```

```
public void addBookmark(
   java.net.URL bookmarkUrl,
   java.lang.String name)
   throws EntryExistsException, java.io.IOException;

public void addBookmark(
   javax.tv.locator.Locator locator,
   java.lang.String name)
   throws EntryExistsException, java.io.IOException;
}
```

An application can use the addBookmark() method to add an entry to the browser's bookmarks list. Bookmarks in a DTV receiver can refer to TV channels as well as web pages, and thus one version of this method takes a java.tv.locator.Locator instead of a URL. It may not always be possible to add a bookmark, however, due to the limited storage space that may be available in the receiver. In this case, the addBookmark() method will throw an IOException.

If an application has the appropriate permissions, it can even change the browser's home page using the setHomepage() method. This cannot be a dvb:// URL, and thus the act of opening the browser will not select a new service. If the application does not have permission to do this, this method will throw a SecurityException. The browser may not support some URL types or protocols, and thus an IllegalArgumentException may be thrown to tell the application that a particular URL is not supported.

Different browsers have different capabilities, and thus the WWWBrowserService class defines a set of methods an application can use to query the browser capabilities. This is useful for directing the browser to a page that includes a specific set of features, and in the case of a DTV receiver it is probably more accurate than using a browser detection script in the target page. This way, the application can get accurate answers rather than having to guess from the browser's user agent string.

The WWWBrowser interface, as follows, gives the application a way of directing the browser to a specific URL, using the goToURL() method. This causes the browser to open that URL in its active window. Any web page the browser is currently displaying will be lost.

```
public interface WWWBrowser extends InternetClient {

   public void goToURL(java.net.URL url)
      throws ClientNotRunningException;

}
```

Like setting the home page, there is no guarantee that the browser will support a particular protocol, and calling goToURL() with an unsupported URL type will throw an IllegalArgumentException.

News Readers

To represent a Usenet news reader application, MHP defines the `UsenetClient` and `UsenetClientService` interfaces. `UsenetClientService` defines a single method, as follows, that lets an application subscribe to a news group.

```
public interface UsenetClientService
  extends InternetClientService {

  public void subscribe(java.lang.String newsgroup)
    throws java.io.IOException;
}
```

The argument to `subscribe()` gives the name of the news group that should be subscribed to, such as `rec.arts.movies.reviews`. We have already seen that receivers may only be able to subscribe to a limited number of news groups, but the behavior in this case is slightly different from the behavior for other Internet clients.

If the list of news groups is full when the application calls this method, the middleware may remove an entry from the list to make room for it. Only entries that were added with this method will be affected, and thus any news groups the user has manually added will not be removed. If the application cannot subscribe to that news group (for instance, the list of news groups is full and it is not possible to delete one), an `IOException` will be thrown. The `UsenetClient` interface, as follows, allows the application to direct the news reader application to a specific group or message using the `selectGroup()` or `selectMessage()` method.

```
public interface UsenetClient extends InternetClient {

  public void selectMessage(java.net.URL message)
    throws ClientNotRunningException;

  public void selectGroup(java.net.URL group)
    throws ClientNotRunningException;
}
```

Both of these will throw an `IllegalArgumentException` if the URL is not a `news://` URL. `selectMessage()` will also throw this exception if the URL does not include a message ID. For `news://` URLs, it is especially important to remember that a particular group or message may not be carried on the user's news server and thus applications should not rely on the availability of any group or message.

A Practical Example

Now that we have seen the various elements of the Internet client API, let's take a look at how it is used in practice. The following example relies heavily on the JavaTV service selection and service information APIs we saw in Chapters 5 and 9. If you are not familiar with those APIs, now would be a good time to review them.

The first thing we have to do is to get a reference to a `WWWBrowserService` object that represents the web browser. To do this, we use the JavaTV service information API, because we can use an `org.dvb.internet.InternetServiceFilter` to find the appropriate object easily.

```
// Create an SIManager instance that we can use to do the
// query
siDatabase = SIManager.createInstance();

// Create an InternetServiceFilter that filters on web
// browsers only. We will use this with the JavaTV
// service navigation API to get the service representing
// the web browser
InternetServiceFilter myWebFilter;
myWebFilter = new InternetServiceFilter(
   InternetServiceFilter.WWW_CLIENT);

// Get the list of services that match our filter
ServiceList services;
services = siDatabase.filterServices(myWebFilter);

// Now that we've got the list of services, we need to
// find the first (and probably only) instance of a
// WWWBrowserService that has been found by the filter.

// This will store the WWWBrowserService instance once
// we've found it
Service browserService;

// Create an iterator to navigate the service list
ServiceIterator iterator;
iterator = services.createServiceIterator();

// Iterate over the service list until we find the entry
// we want.
while (iterator.hasNext()) {
  Service currentService;
  currentService = iterator.nextService();
  if (currentService instanceof WWWBrowserService) {
    browserService = currentService;
    break;
  }
}
if (browserService = = null)
  return;
```

At this point, the `browserService` variable will contain a reference to an instance of the `WWWBrowserService` class if the middleware supports a browser. Next, we need to start the browser by selecting the service we have just found, as follows. In this case, we check to see if we can run the browser and our application at the same time, and fail if we cannot. Alternatively, we could choose to start the browser in our own service context and kill our own application.

```
// This represents the service context in which we'll run
// the browser
ServiceContext context;

// Now we establish which service context we will use.
// First we check to see if we can run the browser
// without killing ourselves.
if (browserService.canRunApplication()) {

  // We can, so try to create a new service context.
  ServiceContextFactory contextFactory
  try {
    context = contextFactory.createServiceContext();
  }
  catch (InsufficientResourcesException ire) {
    // If we can't create a new service context, we have
    // to make a decision whether we use our own service
    // context to start the browser, which will kill this
    // application and any others running as part of this
    // service. In this case, we will just exit
    return;
  }
  // start the web browser.
  context.select(browserService);

}
else {
  // We can't, so we could choose to kill ourselves and
  // run the browser in our own service context. In this
  // case, we will just return.
  return;
}
```

The browser will now be starting. This may take a little time, and thus applications should register a `ServiceContextListener` with the service context they used to run the browser and then wait for a `javax.tv.service.selection.NormalContentEvent` to indicate that the browser has started normally. In that this is sample code, we will not show that here.

Once the browser has started, we can select a new URL we want to show using the `WWW-Browser.goToURL()` method. Before we can do this, however, we need to get a reference to the `WWWBrowser` instance that represents the running browser, as follows.

```
// In order to get a reference to the instance of the
// browser that we've just started, we need to get the
// ServiceContentHandlers for the new service and find
// the WWWBrowser instance
WWWBrowser browser;

ServiceContentHandler[] handlers;
handlers = context.getServiceContentHandlers();

int i;
for (i = 0; i < handlers.length; i++) {
  if (handlers[i] instanceof WWWBrowser) {
    browser = handlers[i];
    break;
  }
}

// Now that we've got a reference to the browser, we can
// do stuff.

// Send the browser to a web site. Let's go visit the
// offical MHP site today. . .
browser.goToURL(new URL("http://www.mhp.org"));
```

Although this looks complicated, it is not really any more complex than selecting any other service. Most of the work here is involved in selecting the service and finding the browser, rather than in using the browser when we have it. For a typical application in which the focus will be on using the browser rather than merely opening it, we can easily hide this complexity in one or two methods.

If we started the browser in our own service context, our application will not be able to carry out the last step we showed, because it will already have terminated. This makes it more difficult for applications to do this and still manipulate the Internet client they launched. Depending on the needs of the application, this may be acceptable, but application developers need to think carefully about how they handle this situation. Similarly, middleware developers and receiver manufacturers need to think about the needs of their customers and decide whether their products will support running Internet clients and downloaded applications at the same time. Sometimes, the ability to do this will be worth a slight increase in the cost of the receiver.

Inner Applications

The next new feature in MHP 1.1 is support for inner applications. An inner application is one that is embedded within another MHP application, such as a DVB-J application running within a DVB-HTML application or a DVB-HTML application running within another DVB-J application. Figure 16.2 shows how this works. Inner applications are an easy way of sup-

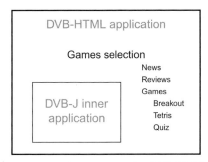

Figure 16.2. A DVB-J inner application running as part of a DVB-HTML application.

porting this type of feature without having to author the inner application specifically for that purpose. It is possible to do this without inner applications, of course, but issues such as use of screen real estate and application priority can become a problem.

There are some limitations on what inner applications can do. Most importantly, an application cannot have an inner application of the same type. This means that DVB-HTML applications can only use DVB-J inner applications, whereas DVB-J applications can only use DVB-HTML inner applications. Although this may seem arbitrary, there are other ways of starting applications of the same type, and thus adding support for this to inner applications may weaken the existing application models.

The API for inner applications is contained in two packages. The `org.dvb.application.inner` package provides support for embedding DVB-J applications within DVB-HTML applications, whereas `org.dvb.dom.inner` contains the classes needed to support embedding DVB-HTML applications within DVB-J applications. In both cases, the structure is similar. The middleware uses an instance of the `org.dvb.application.inner.InnerApplication` class to represent a DVB-J inner application, and instances of `org.dvb.dom.inner.HTMLApplication` to represent DVB-HTML inner applications.

Creating an Inner Application

Readers who look at the `InnerApplication` class will see that its only method is a protected constructor that cannot be called by applications. Because DVB-J applications are not allowed to create an instance of this, there is no need for a complex interface. The parent DVB-HTML application provides all of the information the middleware needs to launch a particular inner application in the `object` element that defines the application. The HTML fragment following shows how a DVB-HTML application can include an inner application and pass runtime arguments to it.

```
<object id= "myInnerApp"

  width = "320"
  height = "240"
```

```
codetype = "application/dvbj"
classid = "myInnerApplication.class"
codebase = "inner_apps/app1" >

<param name = "appid" value = "0x0123"/ >
<param name = "arg_0" value = "AnInnerApp"/ >
<param name = "arg_1" value = "42"/ >
<param name = "arg_2" value = "ATestString" / >

< /object>
```

By adding the `declare` attribute to the `object` element, a DVB-HTML application can declare an inner application without instantiating it or running it. In this case, application developers must use a separate `object` tag to actually run the application.

For DVB-HTML inner applications, we must use the `HTMLApplication` class. This has the following interface.

```
public class HTMLApplication
   extends org.dvb.application.inner.InnerApplication {

   public HTMLApplication(
      java.net.URL physicalRoot,
      java.lang.String initialPathBytes,
      java.lang.String parameters);

   public HTMLApplication(
      java.net.URL physicalRoot,
      java.lang.String initialPathBytes,
      java.lang.String parameters,
      java.lang.String[] label,
      java.lang.String[] regex);
}
```

These constructors let an application define the parameters needed to start a DVB-HTML inner application. The first version of the constructor lets a DVB-J application specify the directory containing the HTML application, the first page of that application, and any arguments the middleware should pass to it. The second version also defines a set of labels for files within the application, and a set of regular expressions that indicate to which files those labels apply. This defines the boundary of the application, and this information can be used to prefetch data the application will need. Exactly how this is done will vary between implementations and thus we cannot give too many details here.

In both cases, the information contained in the constructor arguments is identical to the information that can be included in the DVB-HTML application descriptor or application boundary descriptor that would be included in the application signaling. To the inner application, it will be the same as if the application manager started it like any other application.

Drawing an Inner Application

Creating an application is not much use if we cannot see what it does, and thus we need some way of giving it some screen real estate within the area used by its parent application. For a DVB-J inner application, we have seen in the previous example that its parent application can define a width and height for it using attributes in the object element. Similarly, its position will be defined in its parent XHTML document, just like any other XHTML element. Its size may be zero in either dimension (or even negative), in which case the inner application does not have a user interface. This is useful when we want to use the application as an interface to other middleware components, such as SI or section filtering, that are normally not available to HTML applications.

If we do want to display a user interface, though, the inner application needs some type of container to which it can add components. Obviously, the DVB-HTML application has already defined its size and position in the associated object element, and thus we cannot simply create an HScene for it as we would with a normal DVB-J application. Instead, we use another method to get our top-level container. The javax.tv.graphics.TVContainer class gives us a way to get our root container (if one has been assigned) using the static getRootContainer() method, as follows.

```
public class TVContainer {

  public static java.awt.Container getRootContainer(
    javax.tv.xlet.XletContext ctx);
}
```

Calling this from our inner application will return an object that implements the org.dvb.application.inner.DVBScene interface. DVBScene is a cut-down version of HScene, which provides an inner application with all of the methods it needs in its top-level container. This object must also implement the org.havi.ui.HComponentOrdering interface in order to allow applications to change the Z order of components as they do with normal HScenes. Although this may look like an HScene and behave like an HScene, it may not actually be an HScene and thus application developers should take care not to use any methods from the HScene class that may not be implemented on a given platform. As we saw in Chapter 7, we can also use this method to get the HScene for a normal DVB-J application.

This means that we can use the behavior of the TVContainer class to find out whether our application is running as an inner application or as a normal DVB-J application. For a normal application, getRootContainer() will return an HScene object. If we have already created an HScene for that application, getRootContainer() will return that HScene. If we have not yet created an HScene, this method will create a default HScene (if possible) and return it to us. In some cases, this may not be possible, and thus we may get a null reference. In this case, we unfortunately cannot tell what type of application we have. Either the application is a normal DVB-J application that cannot get an HScene or it is an inner application with its height or width set to zero. The following code shows how we can exploit this behavior.

```
Container myContainer;
myContainer = TVContainer.getRootContainer(myXletContext);

// Now that we have our container, we can find out what
// type of container it is
// check to see if our container is null. This may
// happen if we can't create a container for the
/ application, and this usually means that we can't
// create an HScene for a normalDVB-J application
if (myContainer = = null) return;

if (myContainer instanceof HScene) {
      // we are not running as an inner application

      // if we have not created an HScene earlier in the
      // application we should try to resize and position
      // the HScene as we want to, since this will be a
      // default HScene.
}
else if (myContainer instanceof DVBScene) {
      // we are an inner application
}
```

For DVB-HTML applications embedded in DVB-J applications, we have a similar problem. This time, we need some way of positioning the HTML application on the screen and changing its size. To solve this problem, MHP defines the `org.dvb.application.inner.InnerApplicationContainer` class. This abstract class is a subclass of `HContainer` and implements the `HNavigable` interface. DVB-J applications use instances of this class to define the area of the screen the DVB-HTML user agent will use to draw the inner application. This is similar to the `HVideoComponent` class used to present video in an AWT component (see Chapter 11).

Because this is an abstract class designed to support any type of inner application, applications must use the concrete `org.dvb.dom.inner.HTMLInnerApplicationContainer` class when they embed a DVB-HTML application. As well as defining a public constructor, this class adds the `performAction()` method, which allows the parent application to send DOM actions to the inner application.

The Life Cycle of Inner Applications

Because inner applications are not signaled in the same way as normal MHP applications, their life cycle is a little different from what you might expect. In general, inner applications have the same behavior, but the way we initiate that behavior changes.

DVB-J inner applications will start when the DVB-HTML engine first renders the element that defines them. As we saw earlier, applications can be declared separately from their definition by using the `declare` attribute. It is only when an application is actually defined that it will be initialized and started. For instance, we can use this behavior to declare an application that we then display after the user agent has rendered the document. This could happen in response to an external trigger such as a stream event or in response to a DOM event. An embedded DVB-J application will only terminate when it kills itself or when the `object` tag that defined it is no longer part of the displayed document (either because a DOM operation has removed it or because a different page is being displayed).

A DVB-HTML inner application will be started when its parent application creates an `HTML-InnerApplicationContainer` to display it. The middleware treats this component like any other AWT component, and thus there is no explicit way of destroying it or stopping the associated application. When the parent application removes all references to the embedded DVB-HTML application and its container, the receiver will automatically stop the application and reclaim any memory used by it or by classes within the application.

The application ID of an inner application is supplied by its parent application and this may conflict with other applications being signaled in the AIT. An MHP or OCAP receiver will not start an application if another application with the same application ID is already running, be it an inner application or a normal application. The only possible exception to this is when the two applications are running in different service contexts.

Inner applications will inherit the application priority and permissions from their parents. This makes sure that there is no way for inner applications to subvert the security mechanisms of the middleware stack, because an inner application can only do what its parent can do.

Stored Applications

One of the more useful changes to MHP with version 1.1 is the ability to store applications in the receiver instead of having to download them every time. In general, MHP takes an approach similar to that used by OCAP, but there are a few differences that application developers and network operators must consider.

The most obvious of these is that an OCAP receiver can only store unbound applications. MHP allows unbound applications (known in MHP as standalone applications) or applications bound to a particular service to be stored. In the former case, the application is completely standalone and the AIT entry for the application is stored along with the application files. In the latter case, however, the receiver will cache files belonging to the application, but its life cycle is still controlled by the broadcast AIT. This lets a network operator store commonly used applications such as a news ticker locally, while still associating them with a particular service or event.

To indicate that an application is storable, the network operator includes an additional descriptor in its AIT entry. This is similar to the application storage descriptor used by OCAP (and even has the same name), but it contains slightly more information that is targeted at MHP receivers. The application storage descriptor in MHP systems has the format shown in Table 16.1, and it tells the receiver the priority for that application to be stored with respect

Table 16.1. Format of the application storage descriptor.

Syntax	No. of Bits	Identifier
application_storage_descriptor() {		
descriptor_tag	8	uimsbf
descriptor_length	8	uimsbf
storage_property	8	uimsbf
not_launchable_from_broadcast	1	bslbf
reserved	7	bslbf
version	32	uimsbf
priority	8	uimsbf
}		

Source: ETSI TS 102 812:2003 (MHP 1.1.1 specification).

to other storable applications. The higher the priority the more likely it is that the application will be stored.

For some applications, it may not be possible to start them directly from the broadcast stream. This usually applies to applications transmitted using very low bandwidth, or to other cases where it would take too long to download the application. The not_launchable_from_broadcast flag tells the receiver that it should not let the user start the application until it has been completely downloaded and stored.

The storage_property field tells the receiver how the life cycle of an application is controlled. As of MHP 1.1.1, this can take one of two values. A value of 0 means that the application is broadcast related, and that the life cycle of the application is controlled by the application signaling in the AIT. In this case, the application is stored locally in order to reduce start-up time. Other applications may be signaled with a storage_property value of 1. This indicates a standalone application that can be started independently of the application signaling. This does not mean that we cannot control it via the AIT, however. If the AIT includes an entry for that application, it will behave like any other application that is signaled in the AIT.

The application storage descriptor tells the middleware that it should store an application, but unless we know which files to store this is not very useful. For this reason, every application that has an application storage descriptor signaled in the AIT must have an application description file associated with it.

The application description file is an XML file that lists the files needed for that application. For the receiver to know which application description file belongs to which application, broadcasters should transmit the application description file in the root directory of the application the file describes. Also for this purpose, the file should be named according to the following format.

    dvb.storage.ooooooo.aaaa

The characters oooooooo in this file name should be replaced with the organization ID of the application, in hexadecimal format (padded to eight characters by leading zeroes, if necessary), and the characters aaaa replaced with the application ID, also in hexadecimal format and padded to four characters with leading zeroes if necessary.

The exact format of the application description file is defined in the MHP specification and is too complex to cover here in detail. In general terms, the application description file lists the files and directories that should be stored with some extra information such as the size of a file and the priority of a directory. The following example is an application description file for a simple application.

```
<?xml version = "1.0"?>
<!DOCTYPE html PUBLIC
  "-//DVB//DTD Application Description File 1.0//EN"
  "http://www.dvb.org/mhp/dtd/applicationdescriptionfile-1-0.dtd"

<applicationdescription>
  <dir name = "classes" priority = "1">
    <file name = "MyApp.class" size = "2468">
    <file name = "Utils.class" size = "19343">
    <dir name = "package">
      <file name = "MyPackageClass.class" size = "5693">
    </dir>
  </dir>

  <file name = "background.png" priority = "2" size = "25847">
  <file name = "data.dat" priority = "3" size = "124">

</applicationdescription>
```

Plug-ins

Many digital networks already have some interactive content, and converting this content to MHP can be difficult and expensive. To help solve this problem, MHP 1.1 introduces the concept of plug-ins. These work just like plug-ins in a web browser, providing the middleware with a way of handling content formats MHP does not directly support. This gives MHP receivers a way of handling applications written in MHEG, BML, or another content format such as the Portable Content Format currently under development within MHP.

Plug-ins fall into two categories: interoperable and non-interoperable. Non-interoperable plug-ins are written by the middleware developer, and the way these plug-ins are used is not defined by MHP. A non-interoperable plug-in can be used in any way that meets the needs of the network operator.

Interoperable plug-ins are written in Java, and the way they are used and signaled is defined in the MHP 1.1 specification. Before we see what a plug-in looks like, we will look at how we signal a plug-in. The most important thing to remember is that plug-ins are only used for running applications in a non-MHP format—typically a declarative content format such as MHEG-5 or HTML. Plug-ins cannot be used to add support for MP3 files to an existing DVB-J application (unless we treat each MP3 file as a separate application, which is not usually what we want).

Plug-ins and Application Signaling

Applications executed by a plug-in are signaled in the AIT just like other MHP applications. Although they use the same descriptors for this signaling, some of the fields in those descriptors (e.g., the profile and the version fields) may contain values that are not defined by MHP and that are specific to the type of application being signaled.

Interoperable plug-ins are also signaled like normal MHP applications, although they must also include an extra descriptor (the plug-in application descriptor) in their AIT entry. This descriptor has the format outlined in Table 16.2.

The `application type` field tells the receiver what content format the plug-in can present. Application types must be registered with DVB in order to avoid any clashes and to make sure that each value will be unique. Just like MHP applications, other content formats may also include several profiles and versions. The plug-in application descriptor also lets the middleware know which versions of that content format the plug-in supports.

In some cases, more than one plug-in may support a given content type. In this case, the receiver may choose which one to use by including a delegated application descriptor in the signaling of the application. This tells the receiver which plug-in it should use to handle that application, and has the format outlined in Table 16.3.

When more than one plug-in for that content type is available, a network operator may prefer that a specific plug-in be used to present it. The delegated application descriptor can specify

Table 16.2. Format of the plug-in application descriptor.

Syntax	No. of Bits	Identifier
plugin_application_descriptor() {		
descriptor_tag	8	uimsbf
descriptor_length	8	uimsbf
application_type	16	uimsbf
for(i=0; i<N; i++) {		
application_profile	16	uimsbf
version.major	8	uimsbf
version.minor	8	uimsbf
version.micro	8	uimsbf
}		
}		

Source: ETSI TS 102 812:2003 (MHP 1.1.1 specification).

Table 16.3. Format of the delegated application descriptor.

Syntax	No. of Bits	Identifier
delegated_application_descriptor() {		
descriptor_tag	8	uimsbf
descriptor_length	8	uimsbf
for(i=0; i<N; i++){		
application_identifier	48	uimsbf
}		
}		

Source: ETSI TS 102 812:2003 (MHP 1.1.1 specification).

a list of application identifiers, each corresponding to one plug-in, and the receiver will check these application IDs in the order they are specified to see whether that plug-in is available. If the receiver cannot run any of the plug-ins listed in the delegated application descriptor, or if the descriptor is not included in the application signaling, the middleware will choose a plug-in based on its own algorithm.

Building a Plug-in

Now that we have seen how to signal a plug-in, we can turn our attention to how we actually write one. Plug-ins use a small API defined in the `org.dvb.application.plugins` package in order to create a suitable instance of a plug-in and to start, stop, and terminate that plug-in.

All interoperable plug-ins must implement the `Plugin` interface, as follows. This gives the application manager in the middleware a way of creating and starting a specific plug-in.

```
public interface Plugin {

    public boolean initPlugin();
    public void terminatePlugin();

    public boolean isSupported(
        org.dvb.application.AppAttributes app);

    public javax.tv.xlet.Xlet initApplication(
        org.dvb.application.AppAttributes app)
        throws InvalidApplicationException;

    public Xlet initApplication(InnerApplication app)
        throws InvalidApplicationException;
}
```

Starting a plug-in for a piece of content is a three-stage process. First, the application manager will initialize the plug-in using the `initPlugin()` method. This lets the plug-in set up any general data structures it needs and perform any time-consuming processing needed to initialize the plug-in itself.

After the plug-in is initialized, the application manager will create a specific instance of the plug-in that will be used for presenting the content by calling the initApplication() method. We have already mentioned that the middleware will treat any content presented using an interoperable plug-in as if it were a normal application, and thus the plug-in instance we create should implement the javax.tv.xlet.Xlet interface. Calls to init-Application() will take either the AppAttributes object representing the information that was signaled for the piece of content or an InnerApplication object. This allows the receiver to display the content as part of a DVB-HTML application rather than as a stand-alone DVB-J application.

initApplication() will return either an Xlet, which can be manipulated like any other Xlet and which will follow the same life cycle, or a null reference if the plug-in cannot handle that content at that time (for instance, because not enough memory is available). To actually start presenting the content, the receiver has to call initXlet() and startXlet() on the Xlet that is returned by initApplication().

Once the middleware has finished presenting the content and has destroyed any running instances of the plug-in (by calling destroyXlet() on the Xlet objects representing those instances), it can unload the plug-in by calling the Plugin.terminatePlugin() method. This will free any resources used by the plug-in, and will tell the middleware that it can unload the plug-in. Figure 16.3 shows the life cycle of a plug-in and any applications that run in it.

In some cases, the content a plug-in is presenting may itself include Xlets. The most obvious case of this (and one of the most obvious cases for plug-ins) is when a receiver presents DVB-HTML applications using a plug-in. In this case, the DVB-HTML application may include DVB-J inner applications the plug-in will want to isolate from the rest of the middleware.

MHP 1.1 lets a plug-in intercept some system calls from an Xlet that is running in that plug-in. This lets the plug-in provide a "sandbox" for the Xlet so that it can run safely within the plug-in and without knowing anything about the rest of the system. To do this, plug-ins can provide an object that subclasses the XletSystemCall class, as follows. This abstract class lets plug-ins override certain system calls.

```
public abstract class XletSystemCall {

  public final void register(
     Plugin p,
     javax.tv.xlet.XletContext ctx);

  public final void unregister(
     Plugin p,
     javax.tv.xlet.XletContext ctx);

  public abstract java.awt.Container getRootContainer(
     javax.tv.xlet.XletContext ctx);
}
```

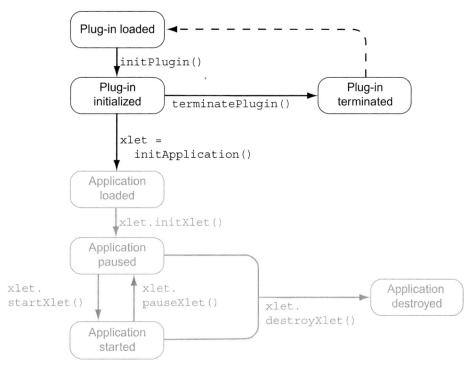

Figure 16.3. The life cycle of a plug-in.

Plug-ins can register an `XletSystemCall` implementation with the middleware to inter-cept requests from specific instances of the plug-in. The `register()` method takes two arguments: the plug-in for which that `XletSystemCall` is valid and the Xlet context for a specific Xlet these implementations should be used for. Note that in this case the Xlet context belongs to the inner application that is contained within the content presented by the plug-in, and not to the plug-in itself.

It is not possible to register an `XletSystemCall` implementation for all inner applications presented within a plug-in (see Figure 16.4). We have to do this for every application, every time. Although this may be a nuisance to developers, the semantics of registering several Xlet system calls (some global and some specific to certain inner applications) could get very complex. This avoids that problem by forcing the plug-in to specify explicitly which system call implementation it wants to use for every application.

As of MHP 1.1.1, we can only override one system call. Any calls to the `javax.tv.graph-ics.TVContainer.getRootContainer()` method from an Xlet will be redirected by the middleware to the `getRootContainer()` method of the registered `XletSystemCall` implementation. This can create and return a root container for the Xlet to use.

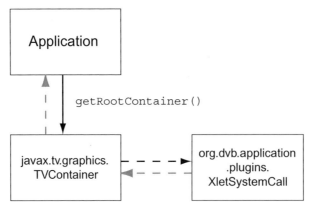

Figure 16.4. We can use the XletSystemCall class to redirect system calls for particular applications.

In some cases, we may want to limit the parts of the AWT hierarchy the Xlet can see, and to do this we can use the org.dvb.application.plugins.XletContainer class. This is a subclass of java.awt.Container, which plug-ins can use to specify which container should act as the parent for the XletContainer. Using this class, a plug-in can completely isolate any Xlets embedded in the content it is presenting from the rest of the AWT hierarchy in the receiver.

The Smart Card API

Most DTV receivers will use a smart card for subscriber management and as part of the CA system. Many new receivers are adding extra smart card slots that application developers can use for other purposes, such as banking cards for online payment or electronic banking, memory cards for storing pay-per-view information, or other types of cards.

To support these types of applications in an MHP receiver, MHP 1.1 includes an API for communicating with smart cards. For now, this is limited to smart cards that are not related to the CA system, partly in order to avoid any potential security problems and partly because it is not really necessary. CA smart cards do not usually have any other interesting services an MHP application would use and we can use the CA API for communicating with these smart cards.

Instead of defining their own smart card API, DVB chose to use the embedded version of the OpenCard Framework (OCF) for smart cards. This defines a set of Java classes that can be used by MHP application developers, receiver manufacturers, and smart card issuers to build portable applications that use smart cards.

Even though OCF is the current choice for smart card APIs, this may not be the case for very long. At the time of writing, the DVB Project is in the process of integrating Sun's Security and Trust Services API from JSR 177 into MHP 1.1 as a replacement for OCF. This is based on the J2ME Generic Connection Framework, and features a number of optional packages. Only the SATSA-APDU package will be required for MHP 1.1. This is a more basic API than OCF, but in practice this is likely to mean less overhead for users.

DTT receivers in Italy (including MHP 1.0.x receivers) must support the APIs from JSR 117, and thus it is likely that this will become the standard API for smart card access in MHP. Because this is not yet a formal part of MHP, we will not discuss it in any detail in this chapter because it is not yet clear exactly how it will be integrated with the existing MHP APIs. More information about this API is available in the JSR 117 documentation, available from the Java Community Process web site.

The OCF Architecture

Many of the features of the full OCF specification are not needed in an embedded environment, and thus the embedded OCF specification is a subset of that found in other OCF implementations. The architecture of the OCF may look pretty complex at first, but MHP application developers can safely ignore many of the complex elements. We can split the framework itself into two basic parts: the part that deals with the device containing the smart card reader and the part that deals with the smart card itself and the services it offers.

In the OCF, each smart card reader is known as a card terminal (represented by the `open-card.core.terminal` package). The `CardTerminal` class represents a specific smart card reader, which may have more than one smart card slot. Applications can use the `CardTerminal` class to find out how many slots are available in a given reader, and to receive notification when the user inserts or removes a card. Embedded OCF assumes that there is only one card terminal available to the application, unlike other OCF implementations. This makes the OCF architecture a little simpler in our case.

The other main part of the API is contained in the `opencard.core.services` package. This defines the `SmartCard` class, which represents a specific smart card inserted into a reader. Smart cards on their own are not very interesting to an application, because applications are usually more interested in the functionality a smart card can provide. This functionality is represented by instances of the `CardService` class. Different types of cards will support different card services, each providing a high-level API that lets an application use the functions of the card. One smart card may implement more than one type of function, and thus a single `SmartCard` object may be associated with several `CardService` objects.

Usually, smart cards in a DTV receiver will be tied to a specific application (or set of applications), and thus the application authors can define their own common card services for their cards. Applications can also use a lower-level API to communicate with a smart card, which we will see in material to follow. Sometimes, this may be easier than defining a complete card service if the smart card offers a limited set of features. Figure 16.5 shows the architecture of the OCF.

Querying the Smart Card Reader

We have already seen that the CardTerminal class represents any smart card readers in the receiver. The interface to this class follows. Although it may seem rather complicated, we can ignore many of these methods for now.

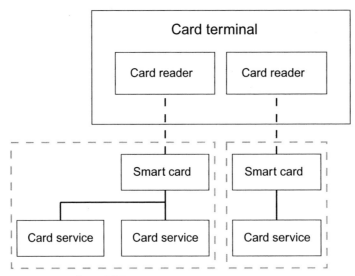

Figure 16.5. The architecture of the OCF.

```
public abstract class CardTerminal {

  public static final CardTerminal getCardTerminal();

  public abstract void open()
    throws CardTerminalException;

  public abstract void close()
    throws CardTerminalException;

  public final int getSlots();

  public abstract boolean isCardPresent(int slotID)
    throws CardTerminalException;

  public abstract CardID getCardID(int slotID)
    throws CardTerminalException;

  public final boolean isSlotChannelAvailable(int slotID);

  public final SlotChannel openSlotChannel(int slotID)
    throws java.lang.IndexOutOfBoundsException,
        CardTerminalException;

  public final SlotChannel openSlotChannel(
    int slotID, java.lang.Object lockHandle)
    throws java.lang.IndexOutOfBoundsException,
        CardTerminalException;
```

```
    public final void closeSlotChannel(
        SlotChannel slotChannel);

    public final CardID reset(SlotChannel slotChannel)
        throws CardTerminalException;

    public final ResponseAPDU sendAPDU(
        SlotChannel slotChannel,
        CommandAPDU commandAPDU)
        throws CardTerminalException;

    public abstract int poll()
        throws CardTerminalException;
}
```

An MHP receiver will only have one card terminal, and so we can get a reference to this using the getCardTerminal() method. (Other OCF implementations can support more than one smart card reader, and thus they use a different method to get references to available card terminals. Embedded OCF only allows one card terminal, however.) Before a terminal can be used, it should be reset using the open() method. This reinitializes the smart card reader so that it is in a known state when the application tries to use it.

Once we have opened the CardTerminal, we can use it to query the available hardware and to find out how many slots are present and which ones currently contain smart cards. We can also use this to find the ID of the smart card in a particular slot. This is represented by a CardID object, which gives us some information about the type of smart card.

This only tells us about cards currently inserted, however. An application can also receive events that notify it when the user has inserted or removed a card, using the classes contained in the opencard.core.event package. To receive particular events, an application should register an opencard.core.event.CTListener with the OCF implementation. This listener has the following interface.

```
    public interface CTListener {

    public void cardInserted(CardTerminalEvent event)
        throws CardTerminalException;

    public void cardRemoved(CardTerminalEvent event)
        throws CardTerminalException
}
```

Each CardTerminalEvent contains the ID of the slot where a card was inserted or removed. Events are generated by an instance of the opencard.core.event.EventGenerator class. OCF uses this singleton object to dispatch all card terminal events. Applications can add a listener to this class using the addCTListener() method, as follows.

```
public final class EventGenerator implements Runnable {

   public static EventGenerator getGenerator();
   public void addCTListener(CTListener listener);
   public void removeCTListener(CTListener ctListener);

   public void createEventsForPresentCards(
      CTListener listener)
      throws CardTerminalException;
}
```

In some cases, an application may wish to find out what cards are already available. Calling the `createEventsForPresentCards()` method will generate `CardTerminalEvents` for any cards currently inserted.

Using Card Services

The `CardTerminal` class and card terminal events do not tell us anything about the services supported by any available smart cards, however. To find and use a particular smart card, we need to use the classes contained in the `opencard.core.services` package. The two most important classes in this package are the `SmartCard` class and the `CardService` interface. As their names imply, these represent a physical smart card and a service offered by a smart card. We have already said that an application is usually not interested in the physical smart cards that are available, and thus the `SmartCard` class mainly provides a way of finding which card supports a particular service, as follows.

```
public final class SmartCard {

   public SmartCard(
      CardServiceScheduler scheduler,
      CardID cid);

   public void close() throws CardTerminalException;

   public CardID getCardID();

   public CardService getCardService(
      java.lang.Class clazz, boolean block)
      throws java.lang.ClassNotFoundException,
            CardServiceException;

   public static SmartCard getSmartCard(
      CardTerminalEvent event, CardRequest request)
      throws CardTerminalException;

   public static SmartCard getSmartCard(
      CardTerminalEvent event,
      CardRequest request,
      java.lang.Object lockHandle)
```

```
        throws CardTerminalException;
    }
```

The static `getSmartCard()` method queries a smart card to see if it supports a specific card service. This takes an event that identifies the slot containing the card, and a `CardRequest` object that identifies (among other things) the type of card service needed and whether only newly inserted cards should be considered. This last element may be ignored in an Embedded OCF implementation, but the specification is not very clear on this and thus application developers and implementation developers should be careful to do the right thing.

Every service offered by a smart card will extend the `CardService` class. From an application perspective, this interface does not offer very much because most of the methods it defines are aimed at allowing a card service to interoperate with the rest of the OCF implementation. This is important, because card services will typically be written by the card supplier rather than by the middleware vendor.

A full OCF implementation defines a number of standard card service interfaces that may be used for common types of card services. Unfortunately, none of these is part of the embedded OCF specification, and thus they are not available by default to an MHP application. This is not a big problem, however. In a DTV receiver, specific smart cards will typically be used with specific applications, and thus the application can implement these card services itself.

Because no card services are standardized, we cannot describe exactly how an application should use them. In an MHP receiver, application authors may need to know how to implement a card service, though, and thus we will look more closely at this later in the chapter.

A Practical Example

Now that we have seen how applications can use the OCF, let's look at a slightly more concrete example. The following code shows how an application can query a card terminal and find a specific card service.

```
public class MySmartCardUser
    implements opencard.core.event.CTListener {

    public void findSmartCard() {

        EventGenerator gen;
        gen = EventGenerator.getGenerator();

        // add a listener for card terminal events
        gen.addCTListener(this);

        // now tell the middleware to generate events for
        // any cards that are already inserted
        try {
```

```
      gen.createEventsForPresentCards(this);
   }
   catch (CardTerminalException e) {
      // ignore exceptions in this example
   }
}

// Most of the work gets done here. Since this method
// can throw a CardTerminalException, we don't catch
// any of these exceptions here. Ideally, we should do
// in order to avoid re-throwing the exceptions.
public void cardInserted(CardTerminalEvent event)
   throws CardTerminalException {

   // get a reference to the card terminal
   CardTerminal terminal;
   terminal = CardTerminal.getCardTerminal();

   // MySmartCardService is the class implementing the
   // card service that we want to use.
   CardRequest request;
   request = new CardRequest(
      CardRequest.ANYCARD,
      terminal,
      MySmartCardService.class);

   // Does the smart card support the card service we
   // want?
   SmartCard card;
   card = SmartCard.getSmartCard(event, request);
   if (card = = null) {
      // return if this card doesn't support the service
      return;
   }

   // Now that we've got a smart card that supports that
   // card service, we can get a reference to the
   // service itself
   CardService myService;
   try {
      myService = card.getCardService(
         MySmartCardService.class, false);
   }
      catch (ClassNotFoundException cnfe) {
         return;
      }
```

```
    catch (CardServiceException cse) {
      return;
    }

    // We can now use the card service. Any calls to the
    // service would go here

    // Close the card service when we're done so that the
    // middleware can free any associated resources
    card.close();
  }

  public void cardRemoved(CardTerminalEvent event)
    throws CardTerminalException {
    // ignore
  }
}
```

Implementing a Card Service

A card issuer will often implement the card services for a particular smart card, although sometimes application developers may want to write their own services to handle a specific card. Doing this is not difficult, although there may be a lot to learn at first. For the sake of brevity, we will only give an outline of what is involved. Readers who are interested in the gory details can find more information at the OCF web site (*www.opencard.org*).

Any card service in an embedded OCF implementation must be a subclass of the `CardService` class. The interface for this class looks as follows.

```
public abstract class CardService {

  protected void initialize(
    CardServiceScheduler scheduler,
    SmartCard smartcard,
    boolean blocking)
    throws CardServiceException;

  public final SmartCard getCard();

  protected void allocateCardChannel()
    throws InvalidCardChannelException;

  public void setCardChannel(CardChannel channel);

  public final CardChannel getCardChannel();
  protected void releaseCardChannel();
    throws InvalidCardChannelException;

  public final CHVDialog getCHVDialog();
```

```
    public void setCHVDialog(CHVDialog dialog);
}
```

The card service will use a `CardChannel` object to communicate with the card terminal. Card channels are scarce resources, and thus card services should allocate card channels only when they need them and release them as soon as they have finished communicating with the card. Calls to `allocateCardChannel()` will block until a card channel is available, and thus card services need to take care in order to avoid potential deadlocks.

Card services will communicate with the card itself using the `CommandADPU` and `Response-ADPU` classes. These represent APDU (application protocol data unit) messages that a card service sends to the card itself, and the responses to those messages. An APDU is a string of bytes that follows the basic format shown in Figure 16.6. We will not examine APDUs in any detail, because every card will use different data in their command and response APDUs.

An OCF implementation cannot create a card service directly, because it does not know which services are available at any given time. To solve this, OCF uses a card service registry and one or more card service factories. A card service factory is responsible for creating a specific CardService object for a smart card or a family of smart cards. Card service factories may support several different card services, depending on the smart cards they support. Like the card services themselves, card service factories will usually be implemented by the card issuer or the application provider. This means that we need some way of telling the middleware about a given card service factory, and that way is the card service registry.

The `CardServiceRegistry` class acts as a central point where applications can register a card service factory. When an application requests a card service, the card service registry will query the card service factories that applications have registered with it until it finds one that supports the card in question. It will then ask that card service factory to create a card service object of the appropriate type. More than one card service factory may support a given type of card, and thus if a card service factory cannot create the desired card service the card service registry will try any other card service factories that support that card.

We have glossed over this process slightly, but this should give you enough information to understand the concepts you need in order to implement a card service. For a more in-depth look at how to use the OCF for developing smart card applications, the documentation on the OpenCard web site is a good place to start.

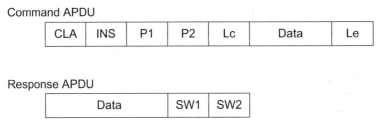

Figure 16.6. The format of command and response APDUs.

MHP 1.1 in the Real World

Although MHP 1.1 has been available for quite a long time, very few companies have built an MHP 1.1 implementation yet, and even fewer have actually deployed it. This is partly a consequence of the greater size of MHP 1.1, and partly a consequence of its immaturity. MHP 1.0.x is complete enough for most people, and thus now everyone is concentrating on getting MHP 1.0.x receivers into the market and making that successful before adding any more complexity.

Even though DVB updated it to include bug fixes and clarifications from MHP 1.0.2 and 1.0.3, there is still a lot of work needed before MHP 1.1.x is truly ready for the market. MHP 1.1.1 included very few bug fixes to the MHP 1.1 APIs, and it appears that fixing the outstanding issues with MHP 1.1 is relatively low on the priority list. Given the other work that has been carried out by DVB since the publication of MHP 1.1, this is is not entirely surprising. The GEM specification has been far more important in terms of MHP as a whole by defining MHP as the basis for other standards such as OCAP and ACAP.

MHP 1.1 adds some useful APIs and functionality, but it also adds a great deal of complexity and thus companies are currently focusing on deploying MHP 1.0.x solutions before moving to MHP 1.1.x. Right now, getting a basic platform deployed and working is more useful than getting the benefit of a few extra APIs. It is therefore difficult to blame anyone involved for taking this approach. The biggest thing to be added in MHP 1.1.x is HTML support, and as we saw in the last chapter that has a number of problems in its current form.

Because there are no test suites for the MHP 1.1.x APIs, and a number of bugs and ambiguities in the new APIs, companies are avoiding it for now. Taking advantage of all of the features of MHP 1.1 means risking interoperability problems, and few companies are willing to do this when MHP 1.0.3 provides enough for most network operators and application developers.

There are a few areas in which the features of MHP 1.0.3 are not enough—most notably HTML support and smart card access. Even here, where MHP 1.1 already adds these features, not many companies are currently looking to it as a solution. In the last chapter, we saw that many network operators are using a downloaded browser written in Java to handle HTML content. Something similar is happening with smart card access, whereby most of the activity is currently happening outside the MHP specification. In that using the MHP 1.1 smart card API involves using the rest of MHP 1.1.x as well, many companies are turning to third-party suppliers for solutions.

Two companies have shown an interest in this new market for smart cards: Modirum of Finland and TVCard of France. Modirum is an early player in this market that has not promoted their technology too heavily in the MHP market. On the other hand, TVCard have been promoting their solution at various exhibitions, and have shown several applications that used their TVCard solution with MHP. Smart card solution providers are beginning to believe that the time is right to start the real marketing effort and show the value of transactional TV to broadcasters and operators.

As DTV solutions are deployed more widely, more and more network operators are finding that applications using smart cards offer a number of new services and revenue streams. Although these solutions are ideal for pay-TV networks, they are much more difficult to deploy in a horizontal market unless there is agreement from all parties on the use of a specific technology. Both Modirum and TVCard are proprietary solutions, and thus these types of solutions can only be deployed in a horizontal network when receiver manufacturers, application developers, and network operators all agree to use the same solution. With the adoption of JSR 177, using a proprietary system may become less attractive to network operators.

No one needs to move to MHP 1.1 now, and thus there is no real driver for getting these issues resolved. Now that work on GEM is nearing completion, the MHP standardization team is turning its attention back to MHP 1.1.x and looking at resolving the outstanding issues. As receivers and applications become more advanced, and as bugs are fixed in future versions of MHP 1.1.x, more receiver manufacturers and network operators will deploy it. For the moment, however, there is no "killer application" that needs enough of the MHP 1.1 APIs to justify a full-scale move to MHP 1.1. In the future, this spot may be taken by home shopping or home banking applications that use the smart card API (either the current API or an alternative) and DVB-HTML or the Internet client API, but only time will tell. Some countries (such as Italy) are experimenting with e-government services using DTV that will require the use of smart cards, but it is not clear whether this will be a driver toward MHP 1.1.

17 Advanced Topics

There are many features of MHP and OCAP we have not covered elsewhere in this book. This chapter looks at three features useful to application developers and network operators: managing the return channel and connecting to external servers, tuning to a new transport stream, and controlling one application from another. We will see how developers can use these features to extend the capabilities of their applications.

In this chapter, we will look at some of the more advanced APIs in MHP and OCAP. These provide functionality many applications will not need to use, but some applications may find them useful.

Out of the APIs we cover here, the return channel API is probably the most commonly used because many applications will need to communicate with a server at the network operator's head-end or on the Internet. We will look at the issues involved in doing this, as well as some of the other APIs that offer functionality beyond the basic features an application will need.

Using the Return Channel

The Interactive Broadcast and Internet Access profiles of MHP include support for using a return channel in the MHP receiver to talk to the outside world. Similarly, OCAP and ACAP both support some type of return channel connection.

Depending on which standard we are using, this return channel can take many forms. OCAP receivers will typically use the cable TV infrastructure for the return channel, whereas MHP or ACAP receivers will support just about any type of return channel, such as a PSTN modem, a cable modem, ADSL, Ethernet, or something more exotic such as DVB-RCS (return channel via satellite).

Past the basic interface of the return channel, life gets slightly easier for application developers. In general, using the return channel in MHP, ACAP, or OCAP is just like using an IP connection in any other Java application: all of the middleware stacks use the standard `java.net` API, with only a couple of major changes. The first of these is not a big problem, but it can have an influence on the rest of the system. When we say that the return channel provides us with an IP connection, this is exactly what we mean: the MHP 1.0.x specification requires support for HTTP 1.0 and DNS over the return channel on top of the basic TCP and UDP protocols, but everything else is optional. MHP 1.1 and OCAP 2.0 add support for HTTPS, but they add no other mandatory protocols such as SMTP or FTP.

The second difference is in the area of session management. The `java.net` API assumes that a permanent network connection is available, and this may not be the case depending on the type of return channel used by the receiver. Given this, MHP defines some extensions that let applications set up a modem and connect to a service provider. These extensions are defined in the `org.dvb.net.rc` package, and OCAP and ACAP receivers will also support them because they are part of GEM.

In many cases, we do not care about the details of session management. Connecting a `java.net.Socket` or `java.net.URLConnection` object that refers to a remote host will make the receiver automatically connect to its default service provider if the return channel is available and if the application has permission to use it. Similarly, after a certain period of inactivity, the receiver will automatically disconnect the return channel.

If this is enough for our application, we do not need to look at the return channel API in any more detail. Sometimes, though, we want to connect to a different service provider, or we want more control over the session management. To do this, we need to use the return channel API.

Return Channel Interfaces

Each interface to a return channel in an MHP or OCAP receiver is represented by an instance of the `RCInterface` class, as follows. This is a simple class that models a couple of the basic properties of the interface; namely, its type (e.g., cable modem, PSTN, ISDN, or others) and its data rate.

```
public class RCInterface {
   public int getType();
   public int getDataRate();

}
```

This class also defines a number of constants that represent the different types of interfaces an MHP or OCAP receiver can support. These are outlined in Table 17.1.

To support the different types of cable modems that are supported by OCAP systems, OCAP defines the `org.ocap.net.rc.OCRCInterface` that extends `org.dvb.net.rc.RCInterface`. This adds a `getSubType()` method that returns the type of cable modem in use if the `getType()` method returns the `TYPE_CATV` constant. The supported subtypes are defined by two constants in the `org.ocap.net.rc.OCRCInterface` class, as outlined in Table 17.2.

Table 17.1. Return channel types in MHP and OCAP.

Constant	Return Channel Type
TYPE_PSTN	PSTN (standard modem)
TYPE_ISDN	ISDN
TYPE_DECT	DECT telephone
TYPE_CATV	Cable modem
TYPE_LMDS	Local Multipoint Distribution System (LMDS) wireless return channel
TYPE_MATV	Master antenna TV return channel
TYPE_RCS	DVB-RCS (return channel via satellite) return channel

Table 17.2. Subtypes for cable return channels in OCAP.

Constant	Return Channel Type
TYPE_CATV_DOCSIS	A DOCSIS cable modem
TYPE_CATV_OOB	An SCTE-55 mode A or mode B return channel using an out-of-band channel

In other cases, or those cases where the return channel is not a cable modem, the value returned by getSubType() is invalid and implementation dependent.

Getting Access to a Return Channel Interface

Because a receiver may support more than one return channel interface, we need some way of getting access to the correct one for our application. The RCInterfaceManager class (as follows) lets us do this, and gives us some way of managing return channel resources.

```
public class RCInterfaceManager
    implements org.davic.resources.ResourceServer {

    public static RCInterfaceManager getInstance();
    public RCInterface[] getInterfaces();

    public RCInterface getInterface(
        java.net.InetAddress addr);

    public RCInterface getInterface(
        java.net.Socket s);

    public RCInterface getInterface(
        java.net.URLConnection u);

    public void addResourceStatusEventListener(
        org.davic.resources.ResourceStatusListener listener);
```

```
    public void removeResourceStatusEventListener(
       org.davic.resources.ResourceStatusListener listener);
}
```

The `RCInterfaceManager` is a singleton class, and applications can get a reference to it using the `getInstance()` method. Once the application has a reference to the `RCInterfaceManager`, it can get an interface to an appropriate return channel in several ways. `getInterfaces()` returns references to all of the return channel interfaces the application can use. Depending on the permissions assigned to the application, it may not be able to access any return channels, and thus this method may return an empty array even if the receiver does support a return channel.

`getInterface()` has several variants, which can take either a `java.net.URLConnection`, a `java.net.Socket`, or a `java.net.InetAddress`. In the first two cases, the MHP middleware assumes that a connection already exists (for instance, because the receiver automatically connected the return channel when a socket or `URLConnection` was connected) and returns the interface used for that connection. In the last case, the middleware returns the interface that will be used for that connection or a null reference if it cannot tell which interface should be used. Return channel interfaces may also be scarce resources, and thus the `RCInterfaceManager` lets us register listeners for `ResourceStatusEvents` concerning the return channel interfaces in the receiver.

Connection-based Return Channels

Many of the return channel types the `RCInterface` class defines represent an always-on interface, but some do not. The receiver middleware will not treat an always-on interface like a cable modem or ADSL return channel as a scarce resource, simply because the application has no say in what provider the return channel will connect to. That is entirely up to the middleware, and thus to the application it is just an IP connection it can use.

Other return channel interfaces may be a scarce resource. A PSTN modem, for instance, must dial an ISP and connect before it can be used. Depending on the service provider it connects to, an application may only be able to access a specific set of remote servers and other applications may not be able to use that connection to connect to their servers.

The `ConnectionRCInterface` class extends `RCInterface` to add support for return channel interfaces that must be connected before they are used. It adds a number of methods to handle connecting to and disconnecting from a service provider and for reserving access to the modem, as demonstrated in the following.

```
    public class ConnectionRCInterface
       extends RCInterface
       implements org.davic.resources.ResourceProxy {

       public boolean isConnected();

       public float getSetupTimeEstimate();
```

```
public void reserve(
    org.davic.resources.ResourceClient c,
    java.lang.Object requestData)
    throws PermissionDeniedException;

public void release();

public void connect()
    throws java.io.IOException, PermissionDeniedException;

public void disconnect()
    throws PermissionDeniedException;

public ConnectionParameters getCurrentTarget()
    throws IncompleteTargetException;

public void setTarget(ConnectionParameters target)
    throws IncompleteTargetException,
        PermissionDeniedException;

public void setTargetToDefault()
    throws PermissionDeniedException;

public int getConnectedTime();

public org.davic.resources.ResourceClient getClient();

public void addConnectionListener(
    ConnectionListener l);

public void removeConnectionListener(
    ConnectionListener l);
}
```

Many modems that do not have a permanent connection will need to connect to a specific service provider before applications can use them. This service provider is known as the target for the interface, and it is set using the `setTarget()` method.

Before we can set the target or make a connection using this return channel interface, we need to reserve the interface (using the `reserve()` method) so that no other application can use it. Once we have successfully reserved the interface, we can set the target and make a connection.

Every return channel interface will usually have a default target an application can connect to if it has the necessary permissions. This allows applications to access servers via a common service provider, such as one that is a partner of the broadcaster or the receiver manufacturer. To reset the interface to its default state, the `setTargetToDefault()` method removes other target settings and uses the default target the middleware manufacturer has defined. If our application does not have permission to reset the target, this method will throw a `SecurityException`.

Each target is defined by a number of different settings, and these are encapsulated in a `ConnectionParameters` object. The `ConnectionParameters` class defines (as follows) the most important parameters needed to set up a connection, such as the phone number to dial and the user name and password the receiver should use to authenticate the connection.

```
public class ConnectionParameters {

    public ConnectionParameters(
        java.lang.String number,
        java.lang.String username,
        java.lang.String password);

    public ConnectionParameters(
        java.lang.String number,
        java.lang.String username,
        java.lang.String password,
        java.net.InetAddress[] dns);

    public java.lang.String getTarget();
    public java.lang.String getUsername();
    public java.lang.String getPassword();
    public java.net.InetAddress[] getDNSServer();
}
```

Any changes to the `ConnectionParameters` object that describes our target will be reflected when we connect to that target, even if those changes are made after we have set the target. After we have connected, any changes will not have an effect until we disconnect and reconnect.

Once we have set the target for the interface, we are ready to connect to the service provider. The `connect()` method makes a connection to the target specified for that interface, and once the connection is made we can use the normal `java.net` operations to make an IP connection to the host we want to connect to.

In the case of PSTN return channels, the middleware will ask the user before it dials any number. If the user allows the application to dial that number, it may be stored in a white list to avoid asking the user every time, but this is a choice for middleware implementers. This happens when the application calls `connect()`, and thus if the user does not allow the call we will get a `ConnectionFailedEvent`. If the user allows the call and the middleware connects successfully, the application will receive a `ConnectionSucceededEvent`. After we have finished communicating, we can disconnect the interface using the `disconnect()` method, and then call `release()` to allow other applications to use that interface.

As you can see, this is a slightly different use of the resource notification API than we have seen in some other APIs. The `RCInterfaceManager` acts as the resource server, but we have to use `ConnectionRCInterface` objects to reserve or release an interface. Specific implementations may choose to implement the resource management in the `RCInterfaceManager` class (or another central class), but this is not visible to the application.

Another difference from other uses of the resource notification API is that applications cannot create new resource proxies. These are implemented by the `ConnectionRCInterface` class, and instances of this can only be created by the middleware. This allows the middleware to return the appropriate return channel interface to a given application, depending on the host it wants to connect to.

To cite an example of this, consider a receiver with a cable modem for general Internet connectivity and a PSTN modem for secure home banking or home shopping. In this case, the PSTN modem may only be able to connect to one or two servers (i.e., it may only be allowed to dial one or two toll-free numbers) and thus it cannot be used as a general-purpose Internet connection.

In this case, the middleware will assign the interface that should be used depending on the host the user wants to connect to. Connections to the appropriate server at the bank or network operator are made over the PSTN modem, whereas all other connections would use the cable modem. If the application could create new resource proxies, there is no guarantee it would create one of the correct type for the interface it would use. Removing this responsibility from the application means there is one less thing that can go wrong with making a connection.

Making a connection to a service provider may take some time (especially for a PSTN modem), and thus applications should register a `ConnectionListener` with the interface they are using. This will receive `ConnectionEvents` from the middleware that tell it the current state of a connection. When a connection is successfully established, the middleware will send a `ConnectionEstablishedEvent` to any listeners registered. Of course, a connection may not always succeed (for instance, someone else may be using the telephone), and thus a `ConnectionFailedEvent` will be generated should this happen. Similarly, a `ConnectionTerminatedEvent` will be sent to any listeners if the connection is broken for some reason or when the application disconnects the return channel.

Using a Return Channel

To use this API to connect to a remote server, we could use the following code.

```
// First, we get a reference to the RCInterfaceManager
RCInterfaceManager rcm = RCInterfaceManager.getInstance();

boolean connectionMade = false;

// Now, we get the list of return channel interfaces that
// are available to our application. This returns all
// the available interfaces
RCInterface[] interfaces = rcm.getInterfaces();

// Now choose which interface we use.
```

```
// Since we get all the interfaces, we need to check that
// we get the right type. So, we check if the first
// interface is a ConnectionRCInterface. If it was not,
// we could do the same thing using another interface,
// but in this example we won't for the sake of
// simplicity.
if (interfaces[0] instanceof ConnectionRCInterface) {
  // If it's a ConnectionRCInterface, it's not
  // permanently connected, so we need to connect it
  // first.
  ConnectionRCInterface myInterface;
  myInterface = (ConnectionRCInterface)interfaces[0];
  // Now that we've got a reference to the interface, we
  // can start to use it
  try {
    // First,we reserve the connection
    myInterface.reserve(this, null);
  } catch (PermissionDeniedException e) {
      // we can't reserve the interface, so return
    return;
  }

  // Set up the connection parameters
  ConnectionParameters myConnectionParameters;
  myConnectionParameters = new ConnectionParameters
    ("+ 0191", "username", "password");

  // Then we set the target to point to our phone
  // number
  try {
    myInterface.setTarget(myConnectionParameters);
  }
  catch (IncompleteTargetException ite) {
    return;
  }
  catch (PermissionDeniedException e) {
    return;
  }

  // Now that we've done that, we can actually connect
  try {
    myInterface.connect();
  }
  catch (IOException ioe) {
    return;
  }
```

```
      catch (PermissionDeniedException e) {
        // we can't reserve the interface, so return
        return;
      }

        // set the flag that tells us we need to disconnect
        // after we are done
        connectionMade = true;
    }
  // do whatever we want to now that we've got a
  // connection
  if (connectionMade) {
    // Once we're done, we disconnect the interface and
    // release the resource
    try {
      myInterface.disconnect();
    }
    catch (PermissionDeniedException c) {
      return;
    }
    myInterface.release();
  }
```

Once our application has an IP connection, it can use that for any purpose it sees fit. MHP and OCAP make no restrictions on the type of data that can flow over this connection. Applications may need to implement higher-level protocols themselves, however, because MHP or OCAP receivers are only required to implement support for HTTP and DNS.

Advanced Application Management

In Chapter 4 we saw how the network operator can control the life cycle of applications through the application signaling in the AIT and XAIT. This may not be enough in some cases, and thus MHP and OCAP both include APIs that let applications query the application database and control the life cycle of other applications directly.

The application listing-and-launching API is contained in the org.dvb.application package, although OCAP adds some extensions in the org.ocap.application package (which we will examine later).

Before we can control any applications, we need to know which are currently available. The AppsDatabase class gives us an interface to the underlying application database the middleware has built up from the application signaling, and thus this can give us information about applications that have been signaled either in the AIT or the XAIT, or that are built into the receiver. The interface looks as follows.

```
public interface AppsDatabase {

  public static AppsDatabase getAppsDatabase();

  public int size();

  public java.util.Enumeration gettAppIDs(
    AppsDatabaseFilter filter);

  public java.util.Enumeration getAppAttributes(
    AppsDatabaseFilter filter);

  public AppAttributes getAppAttributes(AppID key);
  public AppProxy getAppProxy(AppID key);

  public void addListener(
    AppsDatabaseEventListener listener);

  public void removeListener(
    AppsDatabaseEventListener listener)
}
```

The `AppsDatabase` is a singleton object, and thus an application can get a reference to it by calling the `getAppsDatabase()` method. It may not be a true singleton object, however: each application will only see those entries in the application database that are associated with its own service, or that are signaled as unbound applications. For this reason, there may be more than one instance of the `AppsDatabase` class in the middleware, although an application will always see one (and only one) instance.

The `size()` method tells the application how many applications are currently available in the application database for that service. This may not remain constant, because the content of the database can change with time depending on actions by the broadcaster or the user. For instance, the broadcaster may choose to stop signaling one application, while signaling another in its place (e.g., when one TV show finishes and another starts). Alternatively, the user may switch to a different channel, which is signaling a completely new set of applications. In OCAP systems, the monitor application may also affect the content of the database by registering or unregistering unbound applications.

An application can receive notification of any changes to the content of the application database by listening to `AppsDatabaseEvents`. The `addListener()` and `removeListener()` methods on the `AppsDatabase` let applications register to receive events when the content of the AIT changes, as well as to remove a previously registered listener. The receiver will generate an `AppsDatabaseEvent` whenever an application is added or removed from the AIT, when the AIT entry for an application is changed, or when the entire AIT is changed (e.g., when the user changes channel or when an application selects a new service).

Although it is useful to know when the application database is updated, most of the time we are more interested in specific applications. The philosophy behind the design of the application listing-and-launching API is that there are two types of interactions between applications and the application database. Some applications will only want (or only be allowed) to get information about other applications that are being signaled. Others will go a step further and will actually want to control the life cycle of other applications. These two sets of functionality are represented by two different classes. The `AppAttributes` interface provides access to the information about an application that is contained in the AIT or XAIT. To control an application, the `AppProxy` interface and its subclasses define a set of methods that let us start and stop a particular application.

By separating these functions into two different objects, the design of the API makes it much easier for the application manager to check that applications have the right permissions for the actions they want to take. Any application (including unsigned applications) can access information about other applications, and thus no security checking is necessary in the implementation of `AppAttributes`. On the other hand, the middleware will check any request an application makes to an object implementing `AppProxy` in order to make sure that the application has permission to control applications.

You might have noticed that both `AppAttributes` and `AppProxy` are interfaces. There was much argument during the design of this API as to whether separate classes were needed to provide these two sets of functionality, or whether they could be merged into one. In the end, the decision was that for the security reasons we mentioned earlier separate classes were more secure and easier to implement.

By defining these as interfaces, however, the MHP specification lets middleware implementers implement both interfaces in a single class if it suits the middleware architecture better. This is a compromise, but does not significantly affect security because applications cannot assume that these two interfaces will be implemented by the same class.

Getting Information About an Application

Applications can get an `AppAttributes` object for a particular application using the `AppsDatabase.getAppAttributes()` method. There are two versions of this, one taking an instance of the `AppID` class that represents an application ID/organization ID pair and one taking an `AppsDatabaseFilter`. This allows the application to select only those applications that match a set of conditions. The `AppID` class (as follows) encapsulates the organization ID and application ID of an MHP or OCAP application, and gives us a way of uniquely identifying an application in the application database.

```
public class AppID {

   public AppID(int oid, int aid);

   public int getOID();
   public int getAID();

   public java.lang.String toString();

   public boolean equals(java.lang.Object obj);
```

```
    public int hashCode() ;
}
```

MHP defines two `AppsDatabaseFilters` that applications can use. The first of these is the `CurrentServiceFilter`, which selects only those applications signaled as part of the current service. At first, this may not seem very useful in an MHP implementation, but it has its uses. The `AppsDatabase` contains all applications signaled as part of the current service, or that are allowed to run even if they are not signaled (i.e., if the application signaling includes an external application authorization descriptor for that application). The `CurrentServiceFilter` will select only those applications that are actually signaled on the service, and thus any applications running because of an external application authorization descriptor will not be included. Similarly, any unbound applications on an OCAP receiver will not be included.

The second filter is the `RunningApplicationsFilter`, which selects those applications currently running in the same service context as the current application. This may include applications running because of an external application authorization descriptor. This is one of the few ways an application can find out whether a specific application is running, and thus it may be useful for purposes other than controlling the life cycle of that application.

Other standards may define more filters, or applications can choose to override these filters and change their functionality. By default, OCAP uses the same basic set of filters as MHP, although some versions of it add the `org.ocap.application.AppFilter` for use by the monitor application. Each filter has an `accept()` method that takes an `AppID` object as an argument. Any `AppIDs` for which the `accept()` method returns `true` are passed by the filter, whereas all others are rejected.

The `AppsDatabaseFilters` we have seen so far can narrow down the list of applications we are interested in, but we need to use the `AppAttributes` class to get details about the individual applications. The full interface of the `AppAttributes` class follows.

```
public interface AppAttributes{

    public final int DVB_J_application;
    public final int DVB_HTML_application;

    public int getType();
    public String getName();

    public String getName(java.lang.String iso639Code)
        throws LanguageNotAvailableException
;
    public java.lang.String[][] getNames();

    public java.lang.String[] getProfiles();

    public int[] getVersions(java.lang.String profile)
        throws IllegalProfileParameterException;
```

```
public boolean getIsServiceBound();
public boolean isStartable();
public AppID getIdentifier ();
public AppIcon getAppIcon ();
public int getPriority();
public org.davic.net.Locator getServiceLocator();

public java.lang.Object getProperty(
   java.lang.String index);
}
```

Table 17.3. Properties available via the `AppAttributes` class.

Property Name	Description
dvb.j.location.base	The application base directory
dvb.j.location.cpath.extension	Class path extension
dvb.transport.oc.component.tag	Component tag of the elementary stream carrying the root of the object carousel (MHP only, and only valid for applications carried in object carousels)
dvb.ait.descriptors	The binary data forming the descriptor loop for that application's AIT entry (MHP 1.1 only)

There is nothing too surprising here, and any of the values these methods return will be those defined in the AIT or XAIT. The `getType()` method returns the type of the application (DVB-J, OCAP-J, DVB-HTML, or OCAP-HTML), depending on the platform and the type of signaling. Other standards, such as ACAP, may return other values.

Most of the other methods are fairly obvious, except for the `getProperty()` method. This allows us to get the value of fields in the application signaling that are not available in any other way. MHP defines the properties outlined in Table 17.3, and the first two of these are included in the GEM specification.

Because the XAIT can include some extra application-signaling information, OCAP defines the `org.ocap.application.OcapAppAttributes` interface. This extends the `AppAttributes` interface, and gives applications a way of finding out about additional OCAP-specific signaling. The full interface to this class follows.

```
public interface OcapAppAttributes
   extends org.dvb.application.AppAttributes {

   public static final int AUTOSTART;

   public static final int DESTROY;
   public static final int KILL;
   public static final int PREFETCH;
```

```
    public static final int PRESENT;
    public static final int REMOTE;

    public static final int OCAP_J;

    public int getApplicationControlCode();

    public int getStoragePriority();

    public boolean hasNewVersion();

    public void setApplicationPriority(int priority)
        throws java.lang.SecurityException,
                IllegalStateException;

}
```

Controlling Applications

To control an application, we have already seen that we must use an instance of the AppProxy class. The application can get an AppProxy object by calling the AppsDatabase.getAppProxy() method. Like the getAppAttributes() method, this takes an AppID object as a parameter, and returns an AppProxy object representing that application if the requesting application actually has permission to manipulate it. If it does not, getAppProxy() will throw a SecurityException. If no application is available that matches that AppID object, getAppProxy() will return a null reference.

The AppProxy class is also relatively simple. As we can see from the interface definition following, it provides a set of methods to control the state of the application.

```
public interface AppProxy {
    public int getState();

    public void start();

    public void start(java.lang.String[] args);

    public void stop(boolean forced);

    public void pause();

    public void addAppStateChangeEventListener(
        AppStateChangeEventListener listener);

    public void removeAppStateChangeEventListener(
        AppStateChangeEventListener listener);
}
```

As well as controlling the state of the application, we can find out its current state by calling the getState() method. This returns one of the constants defined in Table 17.4.

Table 17.4. Constants for application states used by the application listing-and-launching API.

Value	State
AppProxy.NOT_LOADED	The application has not been loaded.
AppProxy.PAUSED	The application has been loaded and initialized, but is currently paused. It may have been running at some point in the past.
AppProxy.STARTED	The application is currently running.
AppProxy.DESTROYED	The application has terminated.
AppProxy.INVALID	The application proxy is not valid on the currently selected service, or the application is no longer signaled.

The INVALID state tells an application that the AppProxy is no longer valid. An invalid AppProxy will never be valid again, even if we select a service on which that application is signaled. When a new AIT is received (e.g., when a new service is selected, but not when an application is added or removed from the current AIT), any AppAttributes or AppProxy objects will no longer be valid, unless they refer to an application that has an external application authorization descriptor in the new AIT. The receiver cannot know if an entry in the new AIT refers to the same application as one in the old AIT even if the organization ID and application ID are the same, and thus we have to get a new AppProxy or AppAttributes object for applications signaled on the new service.

We can monitor any changes in the state of an application by registering an AppState-ChangeEventListener with the AppProxy object that represents it. This will receive an AppStateChangeEvent whenever an application tries to change its state. Each App-StateChangeEvent will include information about the old and new states of the application, and will tell the listener whether the state change actually succeeded. This can be used, for example, to indicate when the middleware asked an application to terminate but the application refused, or whether a request to start an application succeeded. Because we can only start one copy of an application, attempting to start an application that is already running will generate an AppStateChangeEvent with both the old and the new states set to STARTED. Similarly, any attempts to start or stop an invalid application will generate an AppStateChangeEvent with both the source and destination states set to INVALID.

We have now seen the most important parts of the application listing-and-launching API. Let's take a look at an example of how it can be used. The following example starts all applications signaled on the current service.

```
// Get a reference to the applications database
AppsDatabase theDatabase;
theDatabase = AppsDatabase.getAppsDatabase();
```

```
// Use a CurrentServiceFilter to tell us what is
// being signaled on this service.
AppsDatabaseFilter filter;

filter = new CurrentServiceFilter();

Enumeration attributes;
attributes = theDatabase.getAppAttributes(filter);

// Since the version of the method that uses a
// filter returns an Enumeration to us, we need
// to iterate over the returned elements to find
// one that is interesting to us.
while(attributes.hasMoreElements()) {

    // First, get the attributes of the application
    // from the enumeration
    AppAttributes info;
    info = (AppAttributes)attributes.nextElement();

    // The attributes contain the application ID.
    AppID id = info.getIdentifier();

    // We then use this to get an AppProxy object
    // that represents the application.
    AppProxy proxy;
    proxy = (AppProxy)theDatabase.getAppProxy(id);

    // We don't register an AppStateChangeListener,
    // although we probably should

    // Now that we've got a reference to a proxy for
    // the application, we can start it.
    proxy.start();
}
```

Although this API lets applications ask the middleware to start or stop another application, there is no guarantee that this request will actually result in the target application changing its state. There are a number of reasons a request could fail, ranging from resource restrictions to security reasons to the target application itself not wanting to change its state in some cases. In an OCAP receiver, the monitor application could also deny a request for an application to be started, as we will see in material to follow. As with so many other places in the MHP or OCAP APIs, an application should be flexible enough to handle situations in which the middleware denies a particular request.

Managing Applications in an OCAP Receiver

Given the power of the monitor application to control other aspects of the receiver's behavior, it is probably no surprise to hear that it can also affect which applications are run. The `org.ocap.application` package defines a set of additional classes the monitor application can use to manage applications. The most important class is the `AppManagerProxy` class (with the following interface), which allows the monitor application to interact with the application manager in the receiver.

```
public class AppManagerProxy {

  public static AppManagerProxy getInstance();

  public static int[] getSupportedApplicationTypes();

  public void registerUnboundApp(java.io.InputStream xait)
      throws java.lang.SecurityException,
            java.io.IOException;

  public void unregisterUnboundApp(
      int serviceId, org.dvb.application.AppID appid)
      throws java.lang.SecurityException;

  public void setAppFilter(AppsDatabaseFilter filter)
      throws java.lang.SecurityException;

  public void setAppSignalHandler(
      AppSignalHandler handler)
      throws java.lang.SecurityException;

  public void setSecurityPolicyHandler(
      SecurityPolicyHandler handler)
      throws java.lang.SecurityException;

}
```

This interaction usually involves setting a filter or handler for a specific action in the application manager. For example, a network operator may only want a receiver to run a certain set of applications, depending on who wrote those applications.

To do this, the monitor application can register an `AppFilter` with the `AppManagerProxy` to check whether an application is allowed to run. Each `AppFilter` can contain a list of rules, where each rule consists of a pattern for the organization and application ID, a priority, and an action. The middleware will apply these rules in the order of their priority, and thus rules with a higher priority will be applied first. The action for a rule will be executed when the rule is matched, and these can tell the application manager to allow the application to run, or to stop it from running.

These rules can also tell the application manager to ask the monitor application whether an application is allowed to run. To use this feature, the monitor application should define a

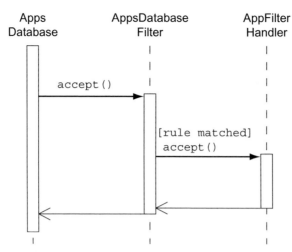

Figure 17.1. The operation of `AppFilter`s and `AppFilterHandler`s.

class that implements the `org.ocap.application.AppFilterHandler` interface and should register an instance of it with the `AppManagerProxy`.

Each `AppFilterHandler` includes the `accept()` method, which takes an `AppID` and an `AppPattern` as arguments. This method returns a `boolean` value that tells the middleware whether the application with the given application ID should be allowed to run. Whenever an `AppFilter` matches a rule whose action tells it to ask the monitor application, the application manager will check whether an `AppFilterHandler` has been registered with the `AppManagerProxy`. If an `AppFilterHandler` has been registered, the middleware will call the `accept()` method for that `AppFilterHandler` to determine whether the application should run. The application will be allowed to run only if the `accept()` method returns `true`. Figure 17.1 shows the operation of `AppFilter`s and `AppFilterHandler`s.

We have already seen most of the other methods on the `AppManagerProxy` class in Chapter 4, and thus we will not go into them in any more detail here.

Tuning to a Different Transport Stream

We have seen in previous chapters that most APIs only work on the transport stream that is currently connected, and that they will not implicitly tune to the transport stream they need if it is not already being received. Out of all of the APIs MHP and OCAP support, only the tuning API and the service selection API will cause the receiver to tune to a new transport stream.

The main reason for forcing explicit tuning is the way transport streams are broadcast. Every transport stream is broadcast on a different frequency, and thus switching transport streams is a much more complex task than switching between services within the same transport stream. In addition, the receiver can only get data from the transport stream to which it is currently tuned. Any applications that rely on data from the original transport stream will suffer as the result of a tuning operation, and we will take a more detailed look at what this means in material to follow.

Now that we have seen the dangers involved in tuning, let's assume we still want to do this. How do we actually do it? To tune to a new transport stream, we have to use the tuner control API, which can be found in the `org.davic.net.tuning` package. This was originally part of the DAVIC specification, and has since been included in MHP and OCAP.

The first class in the API we need to look at is the `StreamTable` class. This database lists all transport streams the receiver can access. Although this is all that is visible to applications, the stream table may also store the network interface each transport stream is associated with and any parameters that are needed to tune to it. `StreamTable` only exposes two methods to applications, which can get a list of the transport streams represented by a given locator or a list of all available transport streams, as follows.

```
public class StreamTable {

  public static org.davic.mpeg.TransportStream[]
    getTransportStreams(
      org.davic.net.Locator locator);
      throws NetworkInterfaceException;

  public static org.davic.net.Locator[]
    listTransportStreams();

}
```

This is about as much as application developers need to know about this class. Middleware developers care a little more, and there are two things they should think about. The first of these is what information is actually stored in the class, which may include tuning parameters for each transport stream. The second thing we need to consider is how the middleware actually gets this list of available transport streams.

Typically, the receiver can search the SI to find out what streams are available for a given network (by searching the NIT on a DVB system, or the VCT on an ATSC system), although this may only find transport streams belonging to a single network. Alternatively, it can find which transport streams are available by scanning the entire frequency range a receiver supports. Although this will not find analog signals in an ATSC network, this is not a problem because the `StreamTable` class explicitly lists digital signals only.

Network Interfaces

The second class we care about is the `NetworkInterface` class, which has the following interface. This represents a broadcast network interface (a tuner) we can use to access a transport stream. There will be one `NetworkInterface` instance for every physical tuner in the receiver.

```
public class NetworkInterface {

  public org.davic.mpeg.TransportStream[]
    listAccessibleTransportStreams();

  public org.davic.mpeg.TransportStream
    getCurrentTransportStream();

  public int getDeliverySystemType();

  public org.davic.net.Locator getLocator();

  public boolean isLocal();
  public boolean isReserved();

  public void addNetworkInterfaceListener(
    NetworkInterfaceListener listener);

  public void removeNetworkInterfaceListener
    NetworkInterfaceListener listener);
}
```

This class does not let an application change any of the settings on the tuner, but it does let us see the status of the tuner and find out more information about the tuner itself. Because this gives us read-only access to the state of the tuner, more than one application can use an instance of this class at a time for any given tuner.

Most of the methods here are fairly obvious. The getCurrentTransportStream() and listAccessibleTransportStreams() methods return the current transport stream and a list of all transport streams accessible from this network interface, respectively. The getDeliverySystemType() method returns an integer value that identifies the type of network this interface supports. For instance, the delivery system could be a cable, satellite or terrestrial network, or even a DSL network for IP-based systems. getLocator() will return a network-bound locator representing the transport stream to which that interface is currently tuned.

The two most interesting methods are the isLocal() and isReserved() methods. The isLocal() method is a holdover from the era in which carrying audio or video digitally over a home network looked much closer than it currently does. Because few operators have historically been willing to trust CA systems other than those they use, this feature is not very common at present. This method was intended to allow a unified interface to tuners both in the receiver and in other devices, while still allowing the application to be able to tell if it was using a local or remote interface. Although this type of functionality is not much in demand now, it is becoming more common.

The isReserved() method tells the applications whether the network interface has been reserved by an application for tuning. The model here will be familiar to most developers: several applications or threads can have read access to the network interface (i.e., they can read its settings via the NetworkInterface class), but only one can reserve it for read-write access (i.e., tuning).

Finding the Right Network Interface

Before we can tune to a transport stream, we must find which network interface we need to use. Because some transport streams will only be available on some networks, we cannot simply use any available network interface to access any transport stream.

The `NetworkInterfaceManager` class is responsible for keeping track of the `Network-Interface` instances in the receiver, and for providing applications with a means of getting a reference to these network interfaces. This is a singleton object, and we can get a reference to it by calling `NetworkInterfaceManager.getInstance()`. The interface to the `NetworkInterfaceManager` class follows.

```
public class NetworkInterfaceManager
   implements org.davic.resources.ResourceServer {

   public static NetworkInterfaceManager getInstance();

   public NetworkInterface[] getNetworkInterfaces();
   public NetworkInterface getNetworkInterface(
      org.davic.mpeg.TransportStream ts);

   public void addResourceStatusEventListener(
      org.davic.resources.ResourceStatusListener listener);

   public void removeResourceStatusEventListener(
      org.davic.resources.ResourceStatusListener listener);
}
```

Once we have a reference to the `NetworkInterfaceManager`, we can use it to find which `NetworkInterface` instance can access a given transport stream or to list all of the network interfaces in the system.

This class also lets us find out about when an application reserves or releases a network interface by registering as a listener for `ResourceStatusEvents`. We will take a closer look at how the tuning API manages resources in material to follow.

Tuning to a New Transport Stream

So far, we have seen how we can get a network interface that can access a specific transport stream, and we have seen how we can use the `NetworkInterface` class to read the status of that network interface. On its own, this is not much use because we cannot use that to actually tune to a new transport stream.

To do this, we need to use the last major class in the tuning API. The `NetworkInterface-Controller` class, as follows, provides a mechanism for applications to control a network interface and use it to tune to a new transport stream.

```
public class NetworkInterfaceController
   implements org.davic.resources.ResourceProxy {
```

```
    public NetworkInterfaceController(
        org.davic.resources.ResourceClient rc);

    public void reserve(
        NetworkInterface ni,
        java.lang.Object requestData)
        throws NetworkInterfaceException;

    public void reserveFor(
        org.davic.net.Locator locator,
        java.lang.Object requestData)
        throws NetworkInterfaceException;

    public void tune(
        org.davic.net.Locator locator)
        throws NetworkInterfaceException;

    public void tune(
        org.davic.mpeg.TransportStream ts)
        throws NetworkInterfaceException;

    public void release()
        throws NetworkInterfaceException;

    public NetworkInterface getNetworkInterface();
    public org.davic.resources.ResourceClient getClient();

}
```

The `NetworkInterfaceController` implements the `ResourceProxy` interface, and like most other resource proxies it has a public constructor. This means that applications can create as many `NetworkInterfaceController` instances as they like.

Simply creating one of these objects does not mean that an application can control a network interface. The `NetworkInterfaceController` is not associated with any specific interface at this point. Before applications can use a `NetworkInterfaceController` to control a network interface, it must be bound to one of the network interfaces in the system. The `reserve()` method lets us bind a specific interface to that network interface controller and reserves that interface for tuning. Another approach is to use the `reserveFor()` method, which reserves any free network interface that can access the specified transport stream.

Once an application has successfully reserved a network interface, it can use the `Network-InterfaceController` to tune that interface to a new transport stream. There are two versions of the `tune()` method, one taking a locator as an argument and the other taking a `TransportStream` object. The only difference here is that a `TransportStream` object refers to a specific transport stream on a specific network interface, whereas the locator may not specify which network interface the middleware should use. Some transport streams or services may be available on more than one interface, and thus it may not matter which interface is used. Tuning using a `Locator` object and `reserveFor()` gives the middleware

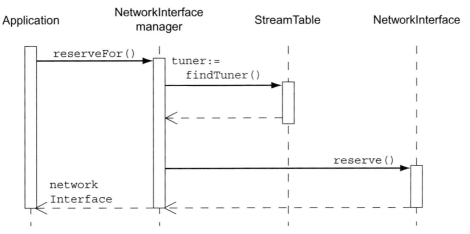

Figure 17.2. Using `NetworkInterface.reserveFor()` to reserve the correct network interface.

more flexibility to choose which network interface should be used, as indicated in Figure 17.2.

After the application has finished tuning, it should call the `release()` method so that the tuner can be used by other applications.

Tuning Events

Tuning to a new transport stream can take one or two seconds, due to the time required for the tuner to change frequencies and lock to the new signal. For this reason, tuning is an asynchronous operation in the tuning API. Calls to the `tune()` method on a `NetworkInterfaceController` start the tuning process, but it probably will not have finished by the time the method returns.

Applications can track the status of a given network interface by registering as a listener for `NetworkInterfaceEvents` using the `NetworkInterface.addNetworkInterfaceListener()` method, passing a `NetworkInterfaceListener` object as an argument. The `NetworkInterfaceListener` interface has one method, as follows.

```
public void receiveNIEvent(
    NetworkInterfaceEvent anEvent);
```

`NetworkInterfaceListeners` can receive two types of events. A `NetworkInterfaceTuningEvent` indicates that an interface has started to tune to a new transport stream, whereas a `NetworkInterfaceTuningOverEvent` tells the listener that tuning has finished. Because tuning may not succeed, the applications can use the `NetworkInterfaceTuningOverEvent.getStatus()` method to find out whether tuning was successful. If it was, this method will return the value `NetworkInterfaceTuningOverEvent.SUCCEEDED`.

Applications that want to carry out some processing once tuning has finished (which will be most applications that tune; otherwise, why is it tuning?) should register themselves as a `NetworkInterfaceListener` and wait for a `NetworkInterfaceTuningOverEvent` before they begin any processing that relies on the new transport stream. In material to follow we will see an example that shows one approach to doing this.

Resource Management in the Tuning API

We have already seen the basics of the resource management model in the tuning API, but there are a few points that are worth mentioning specifically. As you can see from the class definitions preceding, the resource server for this API (the `NetworkInterfaceManager` class) has no methods for reserving or releasing a resource. Instead, these are implemented on the resource proxy (the `NetworkInterfaceController` class).

In many middleware stacks, the implementation of `NetworkInterfaceManager` will handle the actual management of the underlying resources. The `NetworkInterfaceController` will forward any requests to reserve or release a network interface using private methods on the `NetworkInterfaceManager`. This lets the API take care of the resource management in one place, while providing the most flexible interface to the applications. Figure 17.3 shows the reservation and release of tuners.

Any applications that register as a listener for `ResourceStatusEvents` will receive two types of events from the `NetworkInterfaceManager`. A `NetworkInterfaceReservedEvent` is generated whenever an application reserves a network interface for tuning, whereas a `NetworkInterfaceReleasedEvent` is released when a network interface is released.

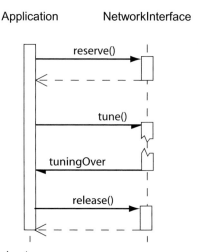

Figure 17.3. Reserving and releasing tuners.

An Example of Tuning

The following example shows how we can use the tuning API to tune to a new transport stream. In this case, our example class implements a `NetworkInterfaceListener` so that it can detect when the tuning operation finishes. The `tuneToNewTS()` method will block until tuning has finished, and thus this class also shows how we can convert an asynchronous tuning operation into a synchronous one. When we are using the tuning API, we need at least one object that implements the `org.davic.resources.ResourceClient` interface to handle resource notification. Here, we declare our example class as the one that implements that interface, although for reasons of clarity we have not shown the implementations of any of the `ResourceClient` methods.

```
public class TuningExample
   implements org.davic.resources.ResourceClient,
            NetworkInterfaceListener {

NetworkInterfaceManager manager;
NetworkInterfaceController controller;
NetworkInterface networkInterface;

// We use these fields for synchronization during the
// tuning operation
Object tuningFinished = new Object();
boolean tuningHasFinished = false;

// Methods inherited from ResourceClient are not shown
// here for clarity

/**
 * Tune to a new transport stream
 */
public void tuneToNewTS(
  org.davic.net.Locator targetTS) {

  // convert the locator to an appropriate
  // TransportStream object using the stream table. We
  // take the first suitable transport stream that we
  // get
  org .davic.mpeg.TransportStream ts;
  try {
      ts = StreamTable.getTransportStreams(targetTS)[0];
  }
  catch (NetworkInterfaceException nie) {
    return;
  }

  if (ts == null)
   return;
```

```
// Before we can do any tuning, we need to get a
// network interface that can tune to the correct
// transport stream. First, we get a reference to
// the NetworkInterfaceManager
manager = NetworkInterfaceManager.getInstance();

// Now find a network interface that lets us tune to
// the appropriate transport stream
networkInterface = manager.getNetworkInterface(ts);

// We need to keep track of what the network
// interface is doing, so we add ourselves as an
// event listener for it.
networkInterface.addNetworkInterfaceListener(this);

// A NetworkInterface object gives read-only access
// to the network interface. To actually control it,
// we need a NetworkInterfaceController.
controller = new NetworkInterfaceController(this);

// Creating it does nothing – we have to reserve the
// network interface before we can do anything useful.
try {
  controller.reserve(networkInterface, null);

  // if this fails it will throw an exception and so
  // any code past here will not get called

  // Now we can tune to the transport stream we want.
  controller.tune(targetTS);
}
catch (NetworkInterfaceException nie) {
  // There's not much we can do if an exception is
  // thrown, so we will just return
  return;
}

// Before this method exits, we wait for the tuning
// operation to complete. We do this by using the
// synchronization object that we declared earlier.
synchronized(tuningFinished) {
  try {
    // The 'notify' that corresponds to this 'wait'
    // takes place in the event handler method below.
    tuningFinished.wait();
  }
  catch (InterruptedException ie) {
```

```
        // Ignore the exception, since there's not much
        // we can do about it.
      }
    }
  }
  /**
   * This method is inherited from
   * NetworkInterfaceListener, and gets called when the
   * tuning API generates an event for the
   * NetworkInterface object that we have registered
   * ourselves as a listener for.
   */
  public void receiveNIEvent(
    NetworkInterfaceEvent event) {

    // If the event indicates that the tuning operation
    // is over, we release the resources that we claimed.
    if (event instanceof NetworkInterfaceTuningOverEvent)
    {

      // Release the network interface because we don't
      // need it any more
      try {
        controller.release();
      }
      catch (NetworkInterfaceException nie) {
        // Ignore the exception for now
      }

      // Set the flag that says tuning has finished.
      tuningHasFinished = true;

      // We also need to notify the main method (and
      // anything else that is waiting for tuning to
      // finish) that tuning has finished.
      synchronized(tuningFinished) {
        tuningFinished.notify();
      }
    }
  }
}
```

Tuning and Other Components

We have already mentioned a little bit about the effect tuning has on the rest of the system, but this is such an important topic that we will take a closer look at the effect on other middleware components. Given that so many parts of the middleware rely on having access to the right transport stream, tuning has a serious effect on other components. This is also influenced by the way MHP and OCAP treat a tuning operation. Other components in the middleware are not told that the receiver has tuned to a new transport stream, and thus they will just act as if the previous transport stream had disappeared. They will not look for any information from the new transport stream, because that implies a much deeper change.

The relationship between tuning and service selection is a delicate one, but it is very important. Selecting a new service may cause the receiver to tune to a new transport stream, but the effect is far more fundamental than that. Selecting a new service tells the receiver

Table 17.5. How tuning affects other middleware components.

Component	Effect
Section filtering	Section filters attached to the old transport stream will be disconnected, but the underlying resources may not be automatically released. If resources are released, the `notifyRelease()` method of the resource client for those section filters will be called.
SI	Cached information may still be available, but it will not be updated and no new information can be read from the stream. Any changes made to the SI after the receiver has tuned away from that transport stream will not be detected. SI that is read from an out-of-band channel in OCAP receivers will still be available. Attempts to read SI from the stream only will fail. Cached data may be flushed, depending on the middleware implementation and what new SI requests are made.
Broadcast file system access	Cached files may still be available, but any attempts to open a new file from an object carousel on the old transport stream may fail. Similarly, attempts to read data from cached files may also fail if only part of the file was cached. Similarly, NPT information will not be updated, and DSM-CC stream events in the old transport stream will not be detected. Cached data may be flushed, depending on the middleware implementation.
Media playback	Playback of any media components from the old transport stream will stop. If those components are available on the new transport stream, they will not be restarted automatically.
Service selection	No effect. The receiver will still assume that the old service is selected. Tuning alone does not select a new service.
Application management	Updates to the AIT will not be detected, although the AIT at the time of tuning will still be available. New applications may not be started, in that the broadcast file systems containing their class files and media may not be available.
Conditional access	New conditional access messages for the original service will not be received.

that it should start looking in the new transport stream for service information, that old broadcast file systems may not be available, and that old applications may be killed. Tuning on its own does none of these, and thus many components in the system will still be looking for data in the old transport stream (and because the data is not there, they will not find it). Table 17.5 outlines what happens to different components because of a tuning operation.

These effects do not just apply to the application that tunes to a new transport stream. Every application that is part of that service will suffer, as will any unbound applications that were using content from the original transport stream.

Applications can find out when the transport stream changes by registering a `NetworkInterfaceListener` for the network interfaces that are receiving a transport stream they care about. This is a lot of work for a situation that probably will not arise very often, however, and thus it is better for applications to avoid tuning unless they absolutely have to.

18 Building a Common Middleware Platform

The Globally Executable MHP (GEM) specification allows MHP to be used as the basis for other middleware standards. To help harmonize middleware markets around the world, specification bodies in the United States and Japan adopted GEM as the basis of their own middleware standards. In this chapter we will examine the relationship between those standards, and how we can use GEM to help us build applications and middleware stacks that can be targeted at any of the major digital television markets.

MHP comes from the DVB toolbox of specifications, and thus it makes use of many other DVB specifications such as DVB-SI and DVB subtitling. This allows for the integration into the DVB-T, -S, and -C world of receivers, but it does mean that some elements of MHP are specific to DVB networks.

Prior to MHP, other standards bodies tended to develop their own middleware solutions, such as DASE, MHEG, and BML. Following the publication of MHP, many organizations looked to the body of work produced by DVB to see if they could make use of it. There was a feeling at the various consortia that MHP could be the basis for their own flavor of middleware, and thus the concept of GEM was born. Following much discussion, both within DVB and between DVB and other specification bodies, GEM eventually became a way forward for the globalization of open standard ITV middleware. DVB has described the goals of GEM as follows.

- To maximize interoperability between GEM-based specifications from different organizations

- To maximize the presence of MHP components, enabling economies of scale for the entire interactive broadcast chain
- To take into account local business and technical constraints

This has a number of advantages for almost everyone involved in deploying the standards built around GEM. For application developers and network operators, it makes it much easier to move content between networks. For example, a network operator in Europe can reuse interactive content from a cable operator in the United States without having to invest too much time or money in porting the application, and application developers can sell applications around the world much more easily.

Although some localization and customization will still be necessary, we can remove many of the major technical hurdles. "The incorporation of the DVB's MHP standards into our OpenCable Application Platform (OCAP) was critical, I believe, because it greatly reduced risk for application developers. Having a worldwide agreement gives these developers, and service providers, confidence that an application written once will port to run anywhere in the world on set-top boxes," said CableLabs President and CEO Dr. Richard R. Green. "That, in turn, speeds worldwide deployment of middleware and therefore interactive applications."

For middleware developers, the main advantage of GEM is that it increases the potential for reusing software between the different open middleware standards. By building on a common foundation, middleware developers can build an MHP implementation based around GEM, and then make the changes needed to provide OCAP and ACAP middleware, then finally add support for ARIB as their business evolves and expands. Reusing many core elements shortens testing and debugging cycles, and makes it possible to get products to market much quicker. It also means that middleware developers have just one set of code in which to fix bugs, and thus fixing a bug in the MHP implementation may automatically fix the same bug in the OCAP and ACAP implementations (although we have to be careful doing this, as we will see later in the chapter). This advantage also extends to future middleware solutions based on GEM. By designing in expandability and reusability from the very beginning, it becomes easier to build, test, and deploy future middleware standards based on GEM or on the current crop of open standards.

GEM and Other Standards

Although GEM is based on MHP, it is important that we do not just treat it as "MHP lite." GEM is a complex specification on its own, and thinking of it as just a subset of MHP is not a good way to approach it. Although GEM is defined in terms of MHP, this is largely for historical reasons and to avoid rewriting the MHP specification in terms of GEM. Thinking of GEM in this way can lead to architects failing to spot components that should be reused, or components that cannot be reused. It is best to think of GEM as the basis for an MHP implementation, just as we may think of GEM as the basis for an OCAP implementation.

As other standards (such as ACAP and ARIB B23) harmonize with GEM, the picture gets even more complicated in terms of the relationships among the various standards. Although MHP, OCAP, ACAP, and ARIB B23 are all based on GEM, "based on" is not the same as "rigidly follows every aspect of the standard." Each of these standards uses GEM in its own way, adding extensions to and placing restrictions on the basic GEM specification. These differences are not surprising. After all, that is how GEM is designed to be used.

One thing we do need to consider is whether other standards refer to GEM 1.0 or GEM 1.0.1. The most obvious difference between the two versions is that GEM 1.0.1 provides cleaner access to the non-GEM parts of MHP. This makes it easier for standards that want to reuse other parts of MHP.

Replacement Mechanisms

If we look at the GEM specification, we can see that the GEM framework defines a replacement mechanism for parts of the specification that are market specific or that are tied too closely to DVB technology. Thus, specification bodies can replace pieces of GEM with components that meet their own needs, as long as the replacements are functionally equivalent to the original pieces. The functionally equivalent replacements have been negotiated among the various parties involved (DVB, CableLabs, ARIB, and others) and the GEM specification contains a list of the associated specifications that are fully compliant with the entire GEM framework. This list includes only MHP 1.0.x and OCAP 1.0 for now, but DVB approved SCTE standard 90–1 (*SCTE Applications Platform, OCAP 1.0 Profile*, which defines a basic profile of OCAP for use in U.S. cable systems) for addition to this list in early 2004, and thus this will be added in future versions of GEM. ACAP and ARIB B23 have not been approved for use with GEM at the time of writing, but they will probably get this approval in the not-too-distant future.

The functionally equivalent replacements for GEM components fall into two categories: business-level replacements and technically functional replacements. Business-level replacements are driven by differences in business models or market requirements. For example, the OCAP specification defines the monitor application, which is not present in MHP. This is considered a business-level replacement because the reasons for the replacement are commercial rather than technical. Other business replacements in the OCAP stack are unbound applications such as e-mail clients, video-on-demand clients, and so on.

Technically functional replacements are driven by technical requirements in a given market. An example of the technically functional replacement would be in the field of application transport protocols in the ARIB B23 specification, which uses an approach based on DSM-CC data carousels instead of object carousels as used by OCAP and MHP. Figure 18.1 shows some of these replacements.

In addition to replacing elements of GEM, specifications that use GEM may extend it, either with features from other standards or with completely new features. For example, the combination of the Java parts of OCAP and the HTML part from the original DASE-1 specification led to ACAP. DVB describes the ACAP specification as follows.

- A reference to GEM for the core of the specification, including the Java part; and a reference to OCAP for much of the remaining common infrastructure: security, application and graphics models, additional APIs, and so on.

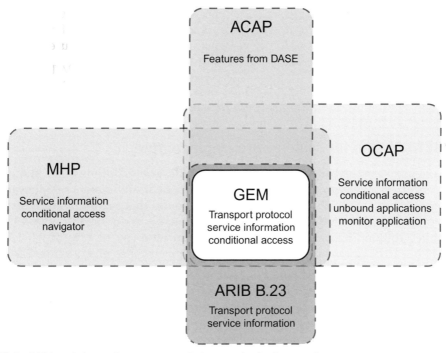

Figure 18.1. GEM and the replacements made by standards that use it.

- Definitions of the GEM functional equivalents—many of them being minor deviations from the DVB original, the major being the Service Information abstraction.
- For the ACAP-X declarative engine option, a reference to the HTML parts of DASE-1 and MHP 1.1 with the definitions of deviations, restrictions, and extensions. This is different from the HTML support described in MHP 1.1 and the OCAP 2.0 profile.
- Reference to both DASE-1 and OCAP for extensions beyond the scope of the MHP specification.

ACAP was previously known as DCAP, and was a standard created as a joint effort by Cable-Labs and the ATSC. The idea was to harmonize the cable and terrestrial environments, because in several cases (e.g., must-carry content) ITV-enhanced shows will be carried from terrestrial networks to cable networks. An ATSC group called S2 is now responsible for maintaining ACAP.

We have heard that the formal relationship between OCAP and ACAP is not completely decided so far, although most of the technical differences between the two standards are minor and work is under way to resolve these. A process is already in place to fix unintentional differences, but this will leave a few intentional differences between the standards that need further work.

The substantive differences between OCAP and ACAP are nontechnical: CableLabs and ATSC have different philosophies toward their standards, especially toward maintenance of the standards, the development of future versions, and conformance testing. CableLabs and ATSC must work out these differences and find a way of working together before OCAP and ACAP can truly be harmonized. At present, approximately 80% of TV viewers in the United States receive their signals over cable, and thus the relationship between OCAP and ACAP will be a key decision in the adoption of the ACAP specification.

What GEM Means for Middleware Implementers

We have already seen the benefits of a common middleware platform earlier in the chapter. The time, effort, and cost we can save by reusing components in different middleware implementations can be considerable. GEM provides a basis for this by specifying what many of the core components in a middleware stack are likely to be, as outlined in Table 18.1. This is not a full description of the differences, but it should give you an idea of what the three standards have in common and where they differ. This may also give you some idea of the strategy you should take for designing a middleware implementation that supports all three standards in a way that suits the needs of both middleware developers and receiver manufacturers.

Some of these differences are fairly superficial and easy to manage, but others are less so. Later in this chapter, we will take a look at techniques we can use to control and minimize the changes we have to make in using the various middleware stacks.

Before we can do this, however, we have to identify these changes in detail. Luckily, this is one area where we get a great deal of help from the standards themselves, because most standards that use GEM identify where they diverge from the basic GEM specification. At heart, the OCAP and ACAP specifications are largely a list of changes from GEM and other specifications, and GEM itself is largely a list of changes from the MHP 1.0.2 specification. This makes it very easy for middleware developers to compare and contrast the standards. The only complication here is later versions of MHP (both MHP 1.0.3 and MHP 1.1.1), which are not currently specified in terms of GEM.

Table 18.1. Differences among GEM, MHP, and OCAP.

Component	GEM 1.0	MHP 1.0.3	OCAP 1.0 ver. I11
Service information		DVB	ATSC/SCTE
Basic application life cycle	✓	✓	✓
SI API	✓ (JavaTV SI API)	Adds DVB SI API	Adds support for low-level access to SI tables and descriptors

Table 18.1. *Continued*

Component	GEM 1.0	MHP 1.0.3	OCAP 1.0 ver. I11
Section filtering	✓	✓	✓
Application signaling	✓	✓	Adds XAIT
Advanced application management	✓	✓	Adds additional classes for use by monitor application
Security	✓	✓	File names and some certificate certificate management techniques are changed from MHP Adds support for monitor application to assign permissions to downloaded applications
Graphics	✓ (Minimum device resolutions not specified)	✓ Specifies device resolutions	Specifies device resolutions; background device is optional
JMF	✓	Adds support for DVB subtitles	Adds support for closed captions
Basic MPEG concepts	✓	Adds classes for DVB-specific elements	Adds classes for out-of-band interfaces
Content referencing	✓	Adds DVB locator class	Adds OCAP locator class
Return channel API	✓	✓	✓
Tuning	✓	✓	✓
Service selection	✓	✓	Adds support for abstract services
Conditional access		✓	✓
Inter-Xlet communication	✓	✓	✓
Content formats	✓	✓	✓
Persistent storage	✓	✓	✓
DSM-CC API	✓	✓	✓
DSM-CC object carousel	✓ (Some elements abstracted away from specific SI standards)	Slightly more concrete, based on DVB signaling	Slightly more concrete, based on ATSC signaling
Resource management	✓	✓	Adds support for resource management by monitor application
Navigator		✓	
Monitor application			✓
System events			✓
Hardware control			✓
Stored applications		MHP 1.1 only	✓
Unbound applications		MHP 1.1 only	✓

OCAP is still not completely GEM compliant even though it is the only standard (apart from MHP) described as being fully compliant with the GEM framework. These differences are minor, however. OCAP needs to be updated to refer to the most recent version of GEM (version 1.0.1), and a few minor discrepancies between OCAP and GEM still need to be ironed out.

Design Issues

Having identified the differences among the various middleware platforms, we are in a better position to start looking at the design of our middleware. When we design our middleware stack, we need to consider three factors. These will influence our design in a number of ways, although the techniques we use to address these factors are quite similar in every case. Before we can really look at how we can define the best middleware architecture, we need to understand these factors and see how and where they influence our design. Based on our desire to create as much commonality as possible between our middleware implementations, we need to take the following into account.

- How easily we can port our middleware to new hardware platforms
- How easily we can modify our current middleware platform to add support for new middleware standards such as ACAP and ARIB B23
- How easily we can customize our implementation for different customers (in a vertical market) or different market segments (in a horizontal market)

In the next sections, we will look at each of these in a little more detail, before discussing the techniques we can use to achieve these goals.

Porting to a New Hardware Platform

Very few middleware stacks will exist on only one hardware platform, and even if the operating system is the same across platforms we will still face problems caused by different hardware capabilities. For companies providing middleware solutions to third parties, this is a big problem because they will almost certainly have to support more than one hardware platform. Even for companies that are only building solutions for their own products, it is unlikely that those receivers will use the same hardware platform throughout the life of the middleware. Thus, portability is a concern for all middleware developers.

The most common solution to this problem, and probably the best, is the use of an abstraction layer or porting layer to separate the hardware-specific parts of the middleware from the more platform-independent components. This porting layer will usually sit above the device drivers and the operating system, with the Java VM and other components that are largely platform independent sitting above it, as we see in Figure 18.2. These platform-independent components may be written in Java or in native code such as C or C++.

Although the Java VM may not be as portable as other elements of the middleware, basing it on top of the porting layer lets us avoid some of the porting problems we would other-

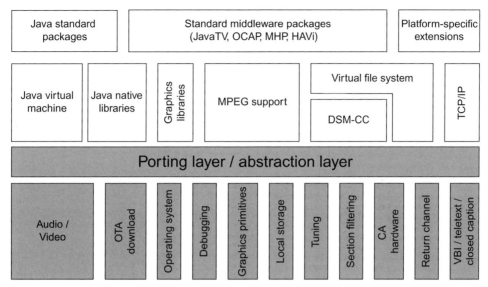

Figure 18.2. The porting layer is an important, but often overlooked, part of the middleware.

wise face, such as dealing with low-level graphics primitives, operating system calls, and memory management. At the same time, we should treat porting the VM as a separate activity to porting the other elements of the middleware. The JVM usually faces a separate set of compliance tests and licensing issues, and thus it is best to treat the Java VM as a separate component that just happens to use the same porting layer as the rest of the middleware stack.

As we can see in Figure 18.2, elements that will be hidden below the porting layer may include the interfaces to hardware MPEG section filters, the return channel, the tuner, and the MPEG decoder. We will not discuss these in detail here, but it is worth mentioning that each of these has its own set of dependencies, and that the order in which these are implemented during the porting process can affect the time taken to complete the entire process. For example, the middleware provider Osmosys suggests that the porting process be carried out in the stages outlined in Table 18.2.

This may not be the same for every middleware stack, and the details of the porting process will depend heavily on the elements present in the porting layer. In general, however, it is a pretty good guide to the order in which components should be implemented. Most of this porting layer will be native code, but code above the porting layer will be Java code or a mixture of Java and native code. The relative amounts of platform-dependent code, platform-independent native code, and Java code will vary from implementation to implementation, and will depend on other decisions made during the design process.

Table 18.2. Suggested porting process for a typical middleware stack.

Stage	Elements
1 – Basic OS interfaces	Threading, memory, debugging support, file system, synchronization primitives, and so on
2 – Graphics	Basic graphics primitives, drawing to TV display
3 – Java VM	Other elements needed to run the JVM
4 – MPEG filtering	Tuning, PES filtering, section filtering
5 – Audio/video support	MPEG decoding and other graphics or audio formats
6 – Security	MHP security, certificate management
7 – Other TV functionality	Remote control input, teletext decoding, subtitles/closed caption support, modem support
8 – Conformance testing	MHP/OCAP conformance testing and final debugging

Customizing Our Middleware

As well as the issues in porting the middleware to a new hardware platform, we have to consider customizability. In a vertical market, manufacturers will typically customize their receivers for each network operator, whereas in a horizontal market receivers with different feature sets (and different price points) may be aimed at different segments of the market. In each case, making our solution easy to customize makes life far simpler.

For instance, making it easy to customize the navigator's user interface is desirable in any implementation (and this may include support for character sets and locales other than the default Western European one). The ability to add or remove optional features that may not be part of MHP or OCAP is desirable in more advanced middleware implementations. These optional features may, for example, include web browsers, non-MHP smart card functionality, or PVR functions.

The challenge here is to define an appropriate set of interfaces between our middleware components so that we can easily modify or replace components as necessary, without affecting the rest of our middleware stack. In this case, we may be looking at components that sit entirely above the porting layer, or at components such as PVR functionality that cross that porting layer. The problems introduced by the other issues we describe in this chapter will depend heavily on the rest of our design, and in what other components we are providing. In general, however, we can solve them using the same techniques we use for other types of customizability. In each of these cases our goal is to maximize reuse without sacrificing performance or stability.

Part of this process involves keeping an eye on the future, and there is a certain amount of overlap between designing customizable middleware solutions and designing future-proof middleware solutions. CableLabs, for instance, has already defined a DVR extension to the

OpenCable specification, and thus it would make sense for anyone providing a PVR/DVR solution for MHP or OCAP receivers to ensure that their solution can easily support this extension.

Developing Other Middleware Solutions

As we have already seen, it is very rare for a middleware manufacturer to produce a single middleware platform. Given the similarities among the various GEM-based platforms, many companies are developing MHP and OCAP implementations, with most of them also planning to support ACAP and ARIB B23. This means that code needs to be reusable between standards, so that we gain all the advantages of a common code base that we discussed earlier in the chapter.

If we look at a typical middleware stack, such as that shown in Figure 18.3, we see that we can reuse many components (those shaded components in the diagram) between implementations of MHP and OCAP. Although we have not discussed them here for the sake of simplicity, the ability to reuse components is roughly the same for other standards.

Even for the components we can reuse, we may still need to make some changes. In some cases, these changes are minor (such as a class that must be renamed or moved to a different package, classes that have been added or removed, or possibly some constants that have different values). The shaded components shown in Figure 18.3 may need these types of changes, and we will look at ways of managing these changes later in the chapter. For other

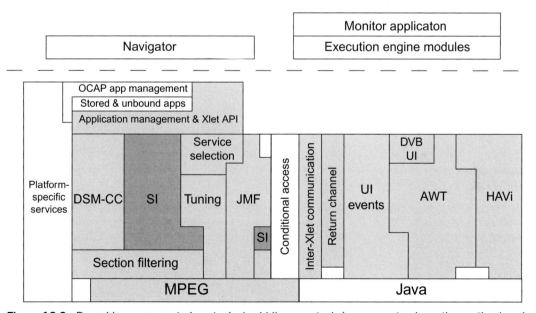

Figure 18.3. Reusable components in a typical middleware stack (components above the porting layer).

components, the changes will be much more significant, and we will either have to rewrite that component or at least make a much larger number of changes.

Even for those components we cannot reuse completely, we can often reuse a large number of them if we design them properly. In particular, the SI component should be designed carefully in order to maximize reusability between middleware standards. We will look more closely at how we can do this later in the chapter, after we have discussed the basic techniques we can use to maximize code reuse.

Reusability is a very desirable goal, but we must also be realistic and accept that we will not always be able to achieve this. This may happen because the cost of making our code reusable (either the effort in the design or coding process, or the performance impact of the changes needed) is so high that it is not practical. It may also be that some components simply are not useful in a given middleware stack. For example, the use of DSM-CC data carousels in the ARIB B23 standard means that there is no need for the object carousel implementations used in MHP and OCAP.

In this last case, we may still be able to improve reusability by using a common API for our data carousel and object carousel implementations. Depending on the nature of the middleware architecture and the requirements of the components, we must make this type of decision on a case-by-case basis.

Techniques for Improving Reusability

Now that we have seen the issues that affect the design of our software, we can look at ways of improving reusability within those constraints and requirements. There are many different techniques we can use, which fall into two broad categories: software development techniques and design techniques. We will not discuss software development techniques in too much detail, because these are already familiar to most developers. There are already many books available on the software development process, and everyone has their own favorite methodology.

The most obvious thing we can do to make porting easier is to follow some of the standard software engineering practices, such as avoiding "magic numbers" in our code and using named constants for any values that could conceivably change. This could be used for specifying graphics resolutions, for instance, in order to handle the differences in display resolution between MHP and OCAP systems.

Another common technique is the use of conditional compilation to maintain code for different middleware platforms. Although using this technique too often in our native code can make it difficult to read and maintain, we can apply a similar technique in Java by using abstract base classes and different concrete implementations for each platform. This can be very readable while still being extremely powerful, as we can see in Figure 18.4. Using project management tools such as integrated development environments or the programmer's favorite, `make`, we can compile and use only those classes that implement a specific middleware platform, while still sharing as much code as possible in a readable way.

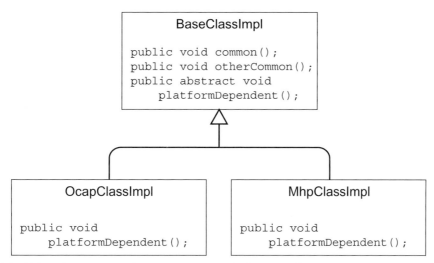

Figure 18.4. Using abstract base classes to separate platform-specific Java code from platform-independent code.

One choice that faces both developers and architects is the separation of native code and Java classes. Sharing code between platforms becomes much simpler if we think carefully about which components will be implemented primarily in Java and which will be implemented primarily in native code. Java code has a number of portability and maintainability advantages that are difficult to ignore, especially because many of the components in our middleware stack will have a Java API. Writing these components entirely in Java can avoid having to implement a potentially messy interface between Java and native code.

At the same time, Java is not the answer to all of our problems. Although Java code is generally more portable, ANSI C or C++ code can also be extremely easy to port, and careful use of conditional compilation techniques may actually result in native code that is more maintainable than the Java equivalent would be. The performance penalties for Java may force developers to spend more time optimizing than would be necessary if the same component were written in C++, and this should be taken into account when deciding whether a component should be written in Java or native code. It is likely that a lot of the code above the porting layer will still be written in C or C++.

Designing Reusable Components

These techniques are useful, but a more powerful approach is to design our middleware components so that we can easily reuse them on different hardware platforms and in different middleware stacks. There are several ways of doing this, which tend to complement one another fairly well in terms of GEM-based middleware stacks.

The first thing to do is to ensure that the interfaces between components are appropriate for all of the tasks they need to do across all of the middleware implementations we may use them for. If a component already has an API defined by MHP or GEM, we can often use this for communicating with other components in the middleware as well as with applications. For example, components can use the DAVIC tuning API to carry out any tuning operations. Reusing APIs where we can means that we are not putting effort into an already-solved problem. This is especially valuable in that many different people have reviewed the MHP and OCAP APIs at many different times, thus reducing the risks of unforeseen problems in the design of the API.

We need to remember, though, that the designers of the MHP APIs may have had a different set of requirements than middleware developers. Many of the MHP APIs take great care to avoid potential security problems or to avoid giving applications access to internal data, which can mean that they are not efficient enough for communication between middleware components. Because we can generally trust middleware components more than we can trust a downloaded application, we can use more efficient mechanisms to get access to some of the data we need if existing APIs are too cumbersome. As long as both components are aware of the limitations of the interface between them, we can use relatively less safe approaches that save time and memory. This is especially valuable when the components in question share a lot of data.

One example of where we could use this is the interface between the SI component and the section-filtering component. An API written in Java may not be suitable if both components are largely written in native code. In this case, we probably want to define our own API at a more appropriate level.

If we want to use Java interfaces but still improve performance, we can provide a simpler API to the section-filtering component that bypasses some of the features designed to restrict malicious or badly written applications. Because we can trust both of the components, this is not necessarily a dangerous thing to do provided we stay within the bounds of good software engineering practice. Depending on the design of our middleware stack, these types of changes could greatly improve performance, and we will look at this in more detail later in the chapter.

So far, we have concentrated on the obvious middleware components to help improve reuse. When we discuss this, most developers will probably think about the standard components shown in Figure 18.2, which are split roughly along the lines of the APIs provided by MHP. In some cases, however, it may be helpful to split those components into several smaller components to help improve usability. The advantage of this approach is that we can concentrate the platform-dependent parts into a few of these smaller components and reuse more code than is possible if we rely on having a few monolithic components.

To give an example of this, let's consider the code that supports application signaling. A GEM implementation will include code that parses the AIT, monitors it for changes, and makes any changes to the application database. An OCAP implementation needs additional code to monitor the XAIT and parse OCAP-specific descriptors. Because the XAIT uses the same format as the AIT, we can keep the XAIT-specific code from any shared functionality if we define appropriate interfaces between the following five main elements.

- AIT monitoring code
- XAIT monitoring code
- AIT parser
- Parser for XAIT descriptors
- Code for managing the application database

Using this approach, we can make the XAIT monitoring functionality an extension of AIT monitoring while keeping the OCAP-specific code in separate components. A final technique we can use is related to this. By putting hooks in our implementation that are used in the appropriate standard but ignored in others, we can easily add functionality that is only needed by one standard. Some of the functionality needed by the monitor application in OCAP is an example of this. The monitor application can register external code with the security manager and the application manager, whereas MHP implementations do not need this feature. In that it is not harmful to an MHP implementation, however, we could provide hooks for this functionality in the core middleware, and simply not use these hooks in our MHP implementation.

One example of this is the resource management component in an OCAP receiver. This includes the `ResourceContentionHandler` class, which allows the monitor application to influence the resource management strategy of the receiver so that a network operator can guarantee the same behavior across all of the receivers in their network. We can add this behavior to a common middleware platform very easily, needing no changes to be compatible with either MHP or OCAP. To give an example of how we could do this, let's consider a part of our resource management class, as follows.

```
public MyResource allocateResource(
  ResourceUsage requestingApp,
  ResourceUsage[] currentOwners) {

  MyResource resource;

  resource = resourceAvailable();
  if (resource != null)
    addToResourceUsage(requestingApp, resource);
    return resource;
  }
  resource = getResourceFromOtherApps();
  if (resource != null) {
    addToResourceUsage(requestingApp, resource);
    return resource;
  }

  // Now try to get an OCAP resource contention handler.
  // If this is an MHP platform, then we need some stub
  // classes to handle this
```

```
ResourceContentionHandler contentionHandler;
contentionHandler = getResourceContentionHandler();

if (contentionHandler != null) {
   ResourceUsage[] newResourceAllocation;
   newResourceAllocation =
      contentionHandler.resolveResourceContention(
      requestingApp,
      currentOwners

   if (newResourceAllocation.length > 0)
      resource = allocateResourcesToApps(
         newResourceAllocation, resources);
      return resource;
   }
}

// Otherwise, we remove the resource from the
// application with the lowest priority (assuming that the
// requesting
// application has a higher priority) and return that
// resource or null.

}
```

In this case, we are using some OCAP-specific classes, and thus we need to provide an implementation of the missing classes if we use this approach in an MHP implementation. Although this means a bit more work for middleware developers, it does mean that we have fewer changes between our middleware platforms that can cause problems. Components such as the application management component and OCAP application API, the OCAP system API, or some of the changes to the OCAP security model can all benefit from this type of approach.

Reusability Outside GEM

So far, we have concentrated on reusing components within the GEM framework. This may be the area in which we can gain most, but it is not the only way we can reuse parts of our design. Some standards-specific elements may actually be used in more than one standard, and it is worth making sure that these components can also be reused in your middleware design. Table 18.3 outlines how these middleware standards reuse some of the same components.

Even if these components are not explicitly reused, they may be present but described in a slightly different way with different terminology, such as the use of unbound applications (for supporting the client applications in the Internet access API) and stored applications in MHP 1.1. Although some of the names are different from those we find in OCAP, and are

Table 18.3. Many standards will reuse elements from other standards, even if they have different names.

Standard	Uses
OCAP 1.0 profile	• Text-rendering behavior from MHP • Application signaling from MHP • Object carousel from MHP • Stored applications from MHP 1.1
OCAP 2.0 profile	• HTML support from MHP 1.1
ACAP	• Text-rendering behavior from MHP • Application signaling from MHP • Object carousel from MHP • SI format and SI API from OCAP • OCAP locator classes • Closed-caption support from OCAP • UI event extensions from OCAP • HTML support from MHP 1.1/DASE • Monitor application support from OCAP

not used in exactly the same way, the functionality is very similar. Identifying these elements is not always easy, and exactly which components fall into this category will change depending on the architecture of the middleware. Not only should we make sure that we design these components to use platform-independent APIs where possible but we should look at the features offered by those components in the different standards, and at how these features are used.

Reusing code can offer a number of advantages, but we need to be careful not to reuse too much. Although there are a large number of similarities between the standards we have discussed here, there are also some important differences. Sometimes these are deliberate, but sometimes it is merely the result of the standardization process. We will not look at all of the differences between the standards, but we will point out one of the main reasons for unintentional differences between the standards. Each of the standards is developed by a different group, and each of these groups follows its own schedule for publishing standards (which is decided partly by market pressures). This makes it difficult to coordinate the activities of the various standards bodies, and can often lead to the situation in which one standard refers to an old version of another standard.

The result of this complex relationship among the standards is that some standards will contain bugs that have been fixed in other standards, and thus any code reuse must take account of this. In some cases, it is possible to carry some of these bug fixes between middleware stacks, even when the standards in question do not explicitly mention the similarities. Every standard has some ambiguity, and it is possible to use this to your advantage. In other cases, different standards will make different assumptions and have different de facto behavior for these ambiguities, and middleware developers need to handle this in order to be truly interoperable.

As the standards evolve and stabilize, this problem is slowly being fixed. Most standards now refer to GEM 1.0.1 and MHP 1.0.3. Now that a test suite has been released for OCAP, this will also help by providing one version of OCAP that all of the players will standardize on, by solving the problems of implementations based around old versions of the GEM or MHP specifications, and by helping to stabilize future versions of ACAP.

An Example: The SI Component

Now that we have seen some of the reusability issues we face and some techniques for solving these, let's look at how we can put these into practice. The SI component is a good example of the choices middleware developers face. Applications will use the SI component, but other middleware components will also use it to get SI for their own purposes. In turn, the SI component will use the section-filtering component to get the private sections that contain the SI tables, which results in a very close relationship between these two components.

This results in two very different demands on this component. When handling requests from applications, the SI component must be secure and must avoid any problems caused by malicious or badly written applications. This is not a problem when other middleware components make a request, but performance may be much more important in these cases. These two demands are often mutually exclusive, and thus we need to find a design that balances these while still being flexible enough that we can reuse it in an MHP, OCAP, or ACAP implementation.

The other problem we face when designing a reusable SI component is the need to provide several APIs to it. OCAP or ACAP implementations primarily use the JavaTV SI API, but OCAP also adds a small API for lower-level access to the PMT and PAT. MHP implements both the JavaTV API and the DVB SI API. We looked at how we could design the SI component to support the different APIs in Chapter 9, and thus we will not cover the same ground here. Instead, we will focus on the factors from outside the SI component that affect our design with respect to maximizing code reuse.

The most obvious way for the different components to talk to each other is through the existing APIs. This approach has several big advantages. First, it means we do not need to spend the time and effort designing a new API. Second, many people have reviewed and implemented the existing APIs, which means that they are well understood and that most of the bugs and limitations have already been identified and/or fixed. Third, it means that we only have one set of entry points into each component, which makes the implementation simpler and thus less error prone as well as simplifying resource management.

Taking this approach may cause us some other problems, however. The existing GEM and JavaTV APIs are designed to meet the needs of applications, and to concentrate on security rather than performance. In particular, the SI APIs and section-filtering API are all asynchronous, which may mean that more threads are needed within these components to handle event dispatching and to handle the extra complexity of scheduling the increased number of requests. Heavy use of asynchronous requests and callbacks may be too inefficient, and thus designers may choose to share data structures between the components instead.

If we used the normal section-filtering API, the SI component would have to copy data from each `org.davic.mpeg.sections.Section` object and then set the object to be empty. This means that the implementation will have to copy a lot of data between components, which is not very efficient. By defining our own API, the section-filtering implementation may simply pass the SI component a reference to the section data and then create a new `Section` object to replace the one it passed to the SI API. Because we can trust the middleware components to behave correctly and free any resources they use, this type of approach may be much simpler.

Some implementations may bypass the existing section-filtering API completely in favor of a lower-level section-filtering implementation that is used by both the SI API and the DAVIC section-filtering API. This has the advantage of avoiding much of the overhead that would come from using the DAVIC section-filtering API for getting SI, while still ensuring a good level of separation between the different components.

Depending on the data needed and the trust we place in our middleware components, an even more direct approach is possible. If we know the caching strategy the SI component is using, other components (such as the DSM-CC implementation) may benefit from being able to read data directly from the SI cache. This provides an even faster interface for cases in which the implementer of the component knows that the SI data will always be in the cache and that this information may change frequently. Giving other components this level of access is usually not a good idea, because it may limit portability between different SI implementations and caching strategies, but it does show how much we could improve performance by removing unnecessary complexity from the APIs between components. We have to make sure the complexity really is unnecessary, however.

A final reason for not using the existing APIs may be the language in which our middleware components are implemented. If most of our section-filtering and SI components are written in native code, it would be foolish to use a Java API for communicating between them. This may mean that middleware designers may need to implement an API in C or C++ instead, but this should have no more implications for reusability than defining a new Java API between the two components.

Limits to Reusability

With all of these issues facing middleware developers, there may be a temptation to make your middleware stack as general as possible to solve the problems that arise. This may not be a good idea, however. Overgeneralizing our design can result in poor performance, and this can cause other problems. A slow product, for example, will attract fewer customers. Balancing flexibility and performance is one of the biggest challenges that will face companies developing multiple middleware stacks, and this usually means finding the correct balance between reusing code and writing optimized code to improve performance.

Because this balance will vary depending on a large number of factors resulting from other design decisions, we cannot make too many suggestions about what is right for an individual case. All we can say is that middleware developers should check to see how much code within a component will really be reused, and evaluate whether the return on the investment needed to make that component reusable is worth it, given the target middleware platforms. In the case of Sun's "On ramp to OCAP" specification, for instance (JSR 242), the amount of code shared with a GEM platform is likely to be very small, just because the capabilities of a JSR 242 implementation are very different from those of an OCAP receiver. Because of this, it probably does not make sense to spend much time and effort building components that are reusable between the two platforms.

19 Deploying MHP and OCAP

Although deploying an open standard middleware solution is similar to deploying other DTV solutions, there are a number of differences we need to take into account. In this chapter we will look at the challenges facing operators deploying MHP, both from the commercial and technical sides. We discuss the differences in business models, in interoperability testing, and in the broadcast chain itself, and also look at the practical issues with moving to MHP or OCAP.

So far in this book, we have looked at the technical details of MHP and OCAP and how we can build high-quality applications and middleware implementations. Building a product is a good start, but the story does not end there. We have to deploy it successfully and start making money before we can really think of it as a success.

In this chapter we will look at the issues we face in deploying products based on the current crop of open standards. Some of these are marketing issues, some are technical, and some are political. Some of these issues may fundamentally change the business models of companies that move to open standards. For a typical middleware solution today, the usual process of deploying it is as follows.

1. Get the specification
2. Build the software
3. Implement it on a given hardware platform and operating system
4. Purchase the test suite
5. Complete the lab trial
6. Build samples of the product
7. Start the field trial
8. Debug the software
9. Build the final product
10. Deploy the system

For proprietary solutions, steps 2 and 3 usually involve a relatively simple porting process. However, with complex software such as MHP or OCAP we have a different and more complex set of dynamics because middleware implementers must each build their own solution (or license a third-party solution). This, coupled with the need to make sure that different solutions are interoperable, makes the process much more complex. It has taken several years to get MHP up and running in the market.

Some people criticize the time it has taken, but we have been with the process all the way through and are more positive. This is not a bad track record for deploying such complex technology in a traditionally conservative industry. MHP is being deployed in large volumes in several countries and it is now a market reality. OCAP is where MHP was approximately three years ago, waiting for the first broadcaster to "switch it on" before we can make any real progress. With all the enthusiasm in the world, we still need broadcast signals before we can successfully deploy set-top boxes using the new middleware. This is a fact of life, and we will see why later in this chapter.

From Vertical Markets to Horizontal Markets

A receiver manufacturer's market has traditionally been in the business-to-business (B2B) sector, where the network operator purchases receivers and then issues them to customers. Network operators may provide the receivers to subscribers free of charge as part of the subscription package, rent the receivers to subscribers at an additional cost, or even sell them outright. Each of these cases is an example of a vertical market, where the receiver manufacturer sells the receivers to the network operator (typically customizing them for each operator) and the network operator then controls distribution of the receivers to consumers.

A horizontal market is one in which the network operator no longer distributes or owns the receivers. Instead, receiver manufacturers sell their boxes through the retail sector, and consumers purchase a DTV receiver just as they would any other CE product (such as a DVD player). The rationale for moving toward horizontal markets is not difficult to understand. The vertical market is expensive for network operators, and the benefit of digital broadcasting in a vertical market is low for broadcasters who do not follow the subscription-based business model. Horizontal markets, on the other hand, offer the potential for a wider audience and different business models that can benefit public broadcasters, advertising-driven operators, and subscription-based operators alike.

One of the most obvious savings for network operators in a horizontal market is the cost of the receivers. This is a huge capital investment, especially in the move from analog to digital. Imagine that you are a large or medium-size network operator in the process of deploying DTV in a vertical market. You may have several hundred thousand to several million viewers, all of whom must be migrated to the new system and all of whom need you to supply them with a receiver. An initial order of 100,000 receivers from an STB vendor is enough to get started. It does not take a rocket scientist to see that $100,000 \times \$150$ (example price) = A Great Deal of Money, and you would typically need to purchase many more receivers than this.

Presently this financial burden is quite squarely on the shoulders of the network operator, and it does not stop with the purchase of the receivers. The network operator must then get those receivers out to the customers, manage a backup stock, and provide maintenance, repair, and call centers for customer support, marketing, and all other aspects of business as a hardware provider. Selling the receivers through a retail market allows the network operator to get away from all of this. It allows them to concentrate on their core businesses of content, services, and (if they so desire) ITV applications. At a recent U.S. conference in early 2004, the CEO of a large MSO (multiple system operator) stated the following: "The most important technology for the future of cable is plug-and-play and OCAP. We have, for too long, missed out on content creativity and content innovation, concentrating far too much on STB technology. This is not our job!"

Unfortunately, the negative side of retail is that the cost of the receiver reflects the extra margins added by the various players, and thus the end price of a receiver may be significantly higher than in a vertical market. This often leads to the argument that it will not be possible to grow a retail market, because the price points for receivers need to be the same as those in the vertical market. However, in many cases it is the available services that drive subscriptions in a vertical market, rather than the cost of the receiver in the vertical market. In addition, a horizontal market means that the subscription costs can be lowered once the network operator does not have to bear the cost of the receivers.

Despite the high initial cost of receivers, it is not all bad news for consumers. Horizontal markets can offer more choice for the customer, from low-level just-watch-TV products to more advanced receivers supporting PVR, Dolby sound, and media center applications that can share content with a PC or that may even be based on a PC platform. Network operators usually do not purchase advanced receivers because of the high cost (look at the installed base of the Motorola DCT 2500 in the United States), but customers purchasing receivers in the retail market may want these extra features. This, combined with the well-known benefits of competition in the retail market, offers some clear benefits to the end user.

Apart from the lack of choice in terms of the features in the receiver, vertical markets have one other disadvantage for the customer. Vertical markets typically lead to market fragmentation, with different operators choosing different middleware stacks and CA systems. In Western Europe alone, over 25 different combinations of middleware and CA systems are in use today, without considering whether those systems are deployed in cable, satellite, or terrestrial networks. Obviously, this level of fragmentation offers few economies of scale to receiver manufacturers or to application developers, and this in turn leads to increased costs for the network operator. Unless this level of fragmentation is reversed, it can only bring problems for countries that wish to completely move to DTV.

As with any new software technology of this nature (e.g., WAP), MHP initially suffered from the "chicken and egg" scenario, with broadcasters pointing the finger at receiver manufacturers and receiver manufacturers pointing at broadcasters for the time taken to launch products and services. If there are no MHP broadcasts, there is no market for MHP receivers, and without receivers there is no way to work on, understand, and deploy MHP services successfully (the same thing applies to OCAP and ACAP). Fortunately, Finland pioneered MHP broadcasting and MHP services were being transmitted almost one year before the first receivers were available to the public. Many other countries have run trials since then, and this has kick-started the availability of receivers so that it is now possible to purchase receivers off-the-shelf and use them for testing MHP broadcasts almost anywhere in the world. Now that services are available, many manufacturers are providing receivers. This in turn makes it easier for network operators and application developers to produce and deploy other services.

The place of MHP and OCAP in the horizontal market is important, but we would like to alter the message that many organizations (including the DVB and CableLabs) use today to promote MHP and OCAP. The message should not be about the devices, it should be about the fact that we can create a horizontal market for ITV *content*. If DTV receivers become more standardized in the same way the PC has, this will allow content created for those devices to be written once and run anywhere. That is what is important about open-standard middleware.

The Web has gone from simple hypertext to a rich (and profitable) media experience because of the wide adoption of open standards, and TV can do the same thanks to the liberating effects of a common platform for interactive content. Horizontal markets are beneficial for consumers because producing and deploying content gets easier, not because they save money for network operators or because they give a wider market to a small number of middleware providers.

The Fight for Eyeballs: Cable, Satellite, and Terrestrial

The entire relationship among satellite, cable, and terrestrial broadcasting is a fight for eyeballs. Issues abound, ranging from complicated technical problems to old-fashioned politics (political issues may in fact be some of the most difficult to solve, as we will see later in the chapter). One of the most important commercial differences between the types of networks is the business models. Satellite and cable networks are largely pay-TV systems in which viewers must subscribe to receive the services, although some content on certain networks may be free. Terrestrial markets are approximately 90% free-to-air (public service broadcasting and/or advertising funded channels). There are also a few new pay-TV terrestrial services starting even after the earlier failures of ONdigital and QuieroTV. Italian broadcasters are going to offer terrestrial pay-TV coverage of major sporting events. Whereas Freeview in the United Kingdom is largely a free-to-air service, it does include some pay-TV services.

As we have already said, pay-TV broadcasting was predominantly a vertical market, even before the introduction of digital services. Changing this mind-set and moving these broadcasters and network operators toward a horizontal market is a difficult proposition and not one that will happen overnight, and this is why it appears that MHP is growing faster in the free-to-air market. The main reason for this is that the free-to-air market is already a horizontal market. The consumer is used to purchasing a reception device from an electronics store or other retail outlet, and they expect that their reception devices will work when they take them home. Because of this, there is much less of an issue in the sales process of inter-active receivers. It is a bonus to the basic ability to watch DTV.

MHP is further helped by the fact that these DTT systems are often "green field" markets, where there is no previous middleware to oust. Vertical market broadcasters and network operators have a very different business model, and they face a completely new set of problems moving toward a horizontal market. They need to give up some of their control over the receiver, as well as take care of retail distribution, advertising, and all manner of other retail marketing incentives they have not had to worry about in the past.

One of the other major commercial issues for pay-TV operators is churn: the loss of customers or inability to retain a customer for a long period. Interactivity seems to be overcoming (or at least addressing) some of the churn issues. Digital channels in the United Kingdom have already proven that clever use of interactive applications can drastically increase consumer interest, and customer "stickiness." This can in turn lead to "channel brand" loyalty and ulti-mately to a reduction in churn. At the fifth annual Interactive TV Show in 2003, German content provider RTL New Media stated that one show with an associated interactive appli-cation increased viewer numbers by 27% over seven weeks. Another show increased viewer numbers by 30%, with interactive competitions playing a large part in this. Statistics such as these offer concrete proof that ITV applications really can affect the profitability of a network, and these examples are not isolated incidents.

A Mandatory Middleware Platform?

Free-to-air markets often have regulatory restrictions that do not apply to pay-TV operators in the same way. Typically, these restrictions are imposed to make sure that public broad-casting meets certain requirements for content and for accessibility to all viewers. Among the decisions facing regulators is whether a common middleware platform should be manda-tory for all DTV systems. This debate is particularly fierce in Europe, where the European Union is carrying out public hearings on the matter. Over 40 companies have entered the debate on both sides and it is unlikely that the companies will reach a consensus.

We will not cover the details of this debate here, but we will say that the disagreement is largely between two groups. In one camp are those companies with commercial interests in the present fragmented market (largely proprietary middleware manufacturers or operators with a large investment in proprietary middleware deployments). The other camp (open standards) consists of those companies that see the benefits of moving to a common mid-dleware platform. Fundamentally, some parts of the industry are in favor of the EC man-dating MHP in order to help move from a fragmented market into one where economies of scale are more achievable. Other players oppose this, and they would like to see other mid-dleware or alternative "common content type systems" as a way of sorting out the issue. One initiative is the PCF (Portable Content Format) for declarative middleware solutions such as MHEG, which has become a work item for the DVB Project.

The consensus in the broadcast industry seems to be that mandating a middleware solution may not really be required, but that something must be done to help reduce fragmentation in the market for the good of the industry (in fact, this was why MHP was built). Trying to please all of the players all of the time is likely to hurt the market rather than help it, but no one agrees as to what steps should be taken to rectify this. Nowhere is this more apparent than the process of moving toward full digital broadcasting and the end of analog services, the so-called "analog switch-off."

Switching Off Analog

Analog switch-off is a subject that provokes much debate, both within the industry and in politics. Selling off the frequencies used by analog TV can bring huge amounts of money to governments, in the same way the sell-off of 3G mobile phone licenses can. In the United Kingdom alone, the sale of mobile phone licenses raised 30 billion euros in 2000, and the sale of the analog TV spectrum in the United Kingdom will probably raise a further 2 to 3 billion euros.

At the same time, analog switch-off cannot happen until a critical mass of the population already receives DTV, or the political consequences will be too much to bear. For this reason, politicians are careful not to push this process too quickly, especially because predicted revenues from the sell-off of the analog spectrum is much lower than it was before the dot-com crash. Switch-off targets differ between countries, and those dates change depending on advances in the technology and the political climate. The United Kingdom was originally planning analog switch-off in 2006, but this has been pushed back to 2010, and the Hong Kong government has also delayed the analog switch-off. The two terrestrial broadcasters in Hong Kong, Asia Television (ATV) and Television Broadcasts (TVB), are required to simulcast digital and analog terrestrial TV services through the end of 2007. What is more interesting is that the two companies will choose the DVB-T standard if the Chinese government has not introduced its national standard by 2006 (full switch-off in China is planned for 2015). Mexico is even further behind, with analog switch-off not scheduled until 2021. It is not all bad news, however. Germany, has proposed an accelerated migration to digital, with switch-off now proposed in 2006 instead of 2010.

Spain highlights the political issue quite succinctly. A recent change in government has meant a review of DTT plans and the possible scrapping of all previous plans that were created by the previous government for the allocation of digital multiplexes. Catalunya and Andalucia were not happy with the distribution of the 266 multiplexes (the regions were only allocated 20 each) and have lobbied successfully for changes to reflect the actual TV landscape in Spain. This probably means that the current plans will be scrapped completely, leading to a longer delay in the Spanish progression toward DTT.

Subsidies and compensation have their problems, however. At the time of writing, DTV deployments in both Sweden and Germany have come under investigation by the EU to check whether compensation for the costs of switching to digital amounts to "state aid," which is not permissible under EU rules. These investigations have only just started, and it is not yet clear whether there is a problem or not, but it does illustrate that governments must be careful when taking this approach.

Although analog switch-off is a tricky proposition for politicians, it is possible to do it without too much upset. Germany began its move to full digital broadcasting in 2003, and the government simply turned off some of the analog terrestrial channels in the Berlin region. We are oversimplifying this slightly, but the result was that some of the analog channels were switched off and only their digital replacements were available. Digital zapper receivers (set-top boxes or intergrated digital TVs with no support for interactive applications or return channel) were available on the shelf and in general the process was relatively painless. Some analog cable companies suddenly lost their must-carry content, but that problem was soon rectified.

The very low penetration of terrestrial TV compared to cable and satellite networks helped the process in Berlin, although it remains to be seen whether this approach can work in the rest of Germany. In the United Kingdom, a similar trial will be conducted in the very near future, albeit on a much smaller scale. Countries such as Singapore and Taiwan, which also have low penetration rates for terrestrial TV, may be in a similarly easy position, but in many European countries and other parts of the world it will not be as straightforward.

Italy has created a different landscape in order to encourage the move to digital by subsidizing the purchase by consumers of the new digital decoders (which also support MHP). In 2004, the 150€ subsidy has driven the market so successfully that it will probably be continued, but with a reduced subsidy of around 80€. You have to prove you have paid your TV license to get the subsidy, and thus the Italian government kills two birds with one stone by offering an incentive to also pay the license fee. Unsubsidized boxes are still being sold, however, because some consumers feel that the benefit of not paying the license fee outweighs the subsidy.

As well as the technical and political aspects of the move to digital, governments must avoid promoting the public services and free-to-air services too heavily in order to ensure fair competition between terrestrial operators and existing satellite or cable TV operators. In markets where digital cable or satellite services are already established, this can be a delicate balancing act. Promote DTV too little, and no one will migrate to the new services; promote it too heavily, and pay-TV operators will complain about unfair competition. This is already becoming a problem in Europe. As previously mentioned, after formal complaints, the EC announced in July of 2004 that it is investigating the financing of the Swedish DVB-T network and the funding of the German DVB-T switchover. Both cases are considered potential illegal "state aid" under articles 87–89 of the EC Treaty.

The main conclusion we can draw from this is that analog switch-off needs clear government backing, with a consistent policy, in order to reduce the risks and the delays. The companies in the receiver industry need to encourage broadcasters to look at "future-proofing" their networks with reasonable cost-effective yet capable gamut of receivers and not simply create a dumbed-down cheap digital legacy. The example of Italy shows that this is possible, but there is still a lot more work to do before full analog switch-off is possible in any country. In order to make this a success, both the industry and governments need to play their part in a positive manner. The status of digital terrestrial deployments in Europe, and switch-off plans for the various countries, is available on the web site of the European Radiocommunications Office at *www.ero.dk*.

Making Money from ITV

As we have previously stated, the horizontal market is an easy market in some respects for middleware. For most operators, it is a "green field" market in which there is no legacy technology apart from the analog equipment. In this case, the temptation is to add some form of additional services such as interactivity to increase the perceived added value in the transition from analog to digital. Interactivity does not have to mean open standards, of course, but because that is the focus of the book we will examine the opportunities and the pitfalls from the perspective of open standards and consequently a new horizontal market. Many of the lessons have come from vertical market operators, and thus they can apply just as easily to other segments of the market.

We would be lying if we said that the news was entirely good, and that would not be helpful to anyone reading this book who actually wants to make some money in this market. We have tried to be balanced, though, and thus there is both good news and bad news for companies wanting to deploy MHP or OCAP in a horizontal market.

The Good News

The move to digital can offer a number of new revenue streams for those companies that choose to make the most of interactive services. Although many of these do not require MHP, horizontal markets can bring these revenue streams to operators that otherwise could not afford the cost of deploying interactive services.

Finland's national lottery is an example of this, whereby an MHP lottery application will run on any MHP-capable receiver in the market (presently cable and terrestrial). The use of a common platform and a horizontal market mean that deploying this type of service across more than one broadcaster is much easier than it would otherwise be, especially for applications that have nationwide appeal (such as a lottery). The ability to deploy applications across networks in this way may provide a huge boost to the number and complexity of applications that are deployed in both pay-TV and public networks.

To a certain extent, public broadcasters are in a tough position when it comes to making money from MHP, because they were never intended to become commercial channels. We see in the United Kingdom that the BBC has some soul-searching to do in this area. Its opportunities are greater than most due to their successful and much admired content capabilities, especially BBC Interactive, and the challenge is to exploit this content in the new digital markets. There is a need to balance revenue generation against the mission of public broadcasters, but any income may help to pay for the migration to digital through receiver subsidies or through offsetting the cost of upgrading transmission equipment.

For commercial channels, the situation is more clear-cut and they must examine any opportunity to make money. The Finnish commercial channel MTV3 probably has the most experience in generating revenue with open middleware, because it was one of the first to launch MHP services. For MTV3, SMS revenue sharing, the Finnish national lottery, and other forms of "services for payment" (such as chat and games) have all proved lucrative. T-commerce is also a winner, and it can be tied to individual shows or broadcast as a standalone application. We will look at T-commerce in more detail later in this chapter.

The types of services deployed by MTV3, Mediaset (Italy), Skylife (Korea), and Astra (mainland Europe) to date are only the tip of the iceberg, however. Commercial operators using other middleware already have a great deal of experience in this area, and many of the lessons from existing closed systems are also applicable to open systems. The U.K. broadcaster Channel 4 presented a number of case studies at the Fifth Annual Interactive TV Show in 2003 regarding revenue creation in ITV. The following presents a key example of how interactivity can influence viewer participation.

For those readers not familiar with the show, "Big Brother" is a reality TV show where contestants share a house for several weeks, with TV cameras showing life inside the house 24/7 and no contact allowed with the outside world. Viewers can vote each week on who should be expelled from the house, as well as voting on other issues related to the contestants and life within the house.

Big Brother: Fun Voting (no impact on the outcome)

If you were on the rich side of the house, would you:

- Flaunt it
- Be humble
- Feel guilty

Outcome = 5,013 votes at 25 pence (0.38€) per vote

Big Brother: Impact Voting (something that affects the outcome)

What should the housemates get for breakfast?

- Full English
- Continental
- Kippers
- Fruit

Outcome = 68,819 votes at 25 pence (0.38€) per vote (an increase of 1,372%)

This is a clear-cut example that there is money to be made from interactivity if the viewers actually get something from it. In this case, they had the ability to affect the show directly. While this may be a simple example, it shows how using interactive applications to increase the participation of the viewers can be a money spinner. Using an interactive application, the viewers do not even need to leave their armchair to register a telephone vote or log on to their ISP, thus making impulse participation more likely.

At the same event, RTL New Media (Germany) stated that during an interactive Formula One broadcast in Germany, over a million viewers participated when prompted by the host and that SMS pinboards were receiving 19,000 messages per day at 0.49€ per SMS. This is considerably attractive enough to excite broadcasters who wish to create a new viewer experience and increase customer loyalty. We have already mentioned that RTL New Media also saw viewer numbers for some shows increase by as much as 30%, largely due to interactivity and participation in ITV-based competitions.

Finally, another example from this conference highlighted an ITV "Angling Times" show on the Discovery Channel. This ran an interactive competition in association with the show in which 52,000 people participated. The broadcaster also experimented with T-commerce (E-commerce via TV), and sold approximately 190 fishing sets at 100€ per set. This is not a massive sum of money, but it is a start in the revenue generation process, and this money went a long way toward paying for the interactive part of the show.

None of these cases were MHP or OCAP applications, but they were interactive applications that could easily be made available on open middleware. The message is clear: interactivity can be profitable even for broadcasters and network operators who do not offer pay-TV content. Use of SMS in interactive applications is about to start in U.S. NASCAR broadcasts, and has one or two ITV gurus scoffing at its merit. This is mostly because SMS is highly underrated (largely due to a lack of experience and knowledge) and less common in the United States, unlike Europe and Asia where tie-ins between SMS and TV shows are a well-known way of making money.

The examples we have seen so far are all related to individual TV shows (bound applications), but these are not the only types of services we can deploy. Digital broadcasting opens up new types of applications that are not tied to existing TV shows (unbound applications). Information services such as super teletext and EPGs (electronic program guides are the most common examples of this, but it can also cover games, lottery applications, news and stock tickers, traffic reports, and other types of services). As before, these are not limited to MHP or OCAP systems, but using open standards for middleware enables broadcasters to deploy these services to a wider audience than may otherwise be possible.

With the types of statistics we have shown here, the misguided notion that ITV is not an important part of the puzzle is an overstated and particularly narrow-minded view of the modern DTV industry. It is possible to make money, but only if the service is available to be exploited by curious consumers. Companies need to make an investment in infrastructure before they can really exploit interactivity, just like telephone companies had to make similar investments. No telephone service means no calls, and exactly the same forces apply in ITV: no compelling content means no takers.

Some typical applications are shown in Figure 19.1 below. As you can see from these screenshots, there is a wide range of applications currently deployed in various MHP networks, and many of these offer some way for a network operator to make money. These may not be obvious at first, but through partnerships with other companies, product tie-ins, and other revenue streams (such as SMS voting) it is possible to make money from most types of applications. For instance, sports tickers such as that shown in Figure 19.1a could be branded by the sponsor of the event or of the TV show. Similarly, information services such as those shown in Figure 19.1b could include advertisements or sponsorship. Although this is risky, as seen by web users' opinions of advertisements in web pages, it is one potential way of making money.

(a)

(b)

Figure 19.1. A few of the MHP applications currently being broadcast around the world: (a) ticker application showing hockey scores during prime time matches (Copyright MTV Oy); (b) super teletext information services (Copyright MTV Oy);

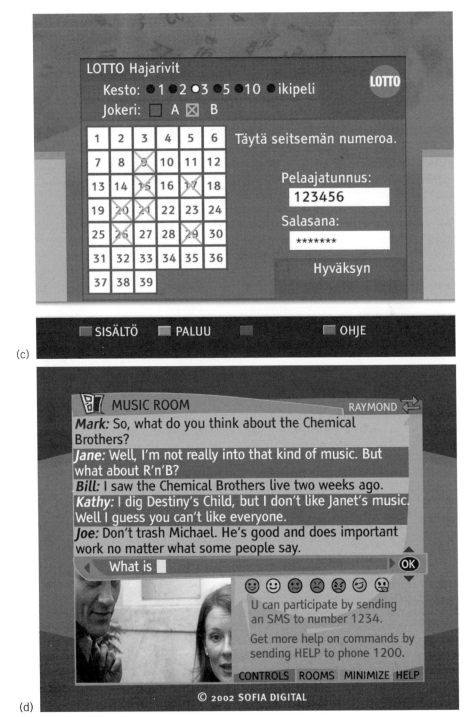

(c)

(d)

Figure 19.1. *Continued* (c) ITV lotto application for the Finnish national lottery (Copyright Sofia Digital); (d) ITV chat applications let you discuss your favorite TV shows while you watch them (Copyright Sofia Digital);

(e)

Figure 19.1. *Continued* (e) a portal application can provide a wide variety of features while still allowing viewers to watch TV shows (Copyright TVC Multimedia).

Applications such as ITV lotto offer a more direct way of making money, either by taking a percentage of ticket sales or by tying the application in with a lottery show. Finally, there are applications that do not directly make money, but which contribute to viewer "stickiness." A chat application such as that seen in Figure 19.1d is a good example of this, where viewers may stay on line to chat about a show even after it has finished. Again, this also offers the potential for sponsorship to be tied into the show in question. Another example of this is a portal application that may be targeted at a specific audience segment. Figure 19.1e shows TVC Multimedia's 3xl.net portal, which is aimed at teenagers and young adults. This can reduce viewer churn in specific groups of viewers by providing them with extra information, while still showing the TV show in part of the screen. This lets viewers see more information while not missing their favorite TV show.

Deploying applications is one thing, but how does a network operator know which applications are helping to generate revenue? Many of the discussions related to ITV revolve around the network operator's ability to maximize the return on investment and capitalize on revenue-generating applications. Many people talk about the infamous "killer application," but these are very difficult to predict. SMS was one of the killer applications in the telecommunications sector, and that came about by accident. The SMS phenomenon has already contributed to growth in ITV revenues for broadcasters and network operators due

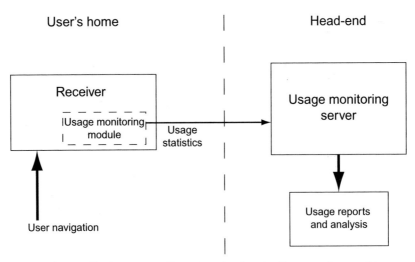

Figure 19.2. Measuring application usage allows broadcasters to discover when and how consumers use interactive applications.

to voting applications such as Big Brother, but ITV does not yet have a killer application of its own.

To find the killer application, operators need to measure application usage and analyze how the popularity of ITV applications corresponds to the investment made in building those applications. How do we understand what goes on at the application level? Is there a way to monitor application usage, or is it up to companies such as Nielsen Media Research to gather this information in the traditional way?

Companies such as Digisoft (*www.digisoft.tv*) have developed ITV measurement and monitoring technology that diminishes the need for traditional ratings research, and places the control of statistics firmly in the hands of the network operator. Like link tracking in a web site, this technology allows network operators to see which applications, and even which parts of an application, are used most by viewers. This is not a completely passive technology, and must be included either in applications or in the receiver middleware, as shown in Figure 19.2. Typically, this technology will contact a server at the network operator's head-end to transmit usage information, although the type of network and return channel will dictate exactly how we do this. In receivers with an always-on return channel, this information can be sent back in real time, but for receivers with a PSTN modem this information will typically be cached and transmitted back in one burst. In this case, of course, this application-tracking feature must be built into the middleware.

To give some examples of how we can use this information, let's consider two cases.

- **Application co-branding and sponsorship**
  1. The application developer (or the network operator's ITV manager) measures when an application is started and stopped.

2. This information is aggregated into a report detailing when that application is used and how often.

3. Sponsorship of applications can then be priced based on usage statistics, just like the price of an advertising slot depends on the ratings of the shows around it.

4. The look and feel of the application can be branded to the sponsor's look and feel.

- **Bandwidth optimization**

  1. The application developer (or the network operator's ITV manager) records which categories are selected in a news application.

  2. A report is created listing the most popular categories for each day (e.g., sport on Sunday, weather on Friday).

  3. The amount of content generated for each category can be streamlined to suit the number of readers (possibly even on a daily basis), optimizing bandwidth usage for content.

  4. By optimizing bandwidth in this way, the broadcaster can save money or broadcast more applications.

This type of measurement has a number of privacy issues associated with it, and consumers will often be concerned about the effect this will have on their privacy (especially in that network operators could track application usage without the consumer's knowledge). The traditional statistics-gathering process is a voluntary scheme, and the automatic system should be run in an identical fashion in order to reassure consumers and avoid privacy concerns. Italian operators have used this technique in field trial systems, and the first Italian MHP users have been using and testing applications while also helping broadcasters assess usage of the initial services they deployed. In Italy, this tracking is built into the STB middleware and can be turned on or off by the broadcaster.

For future tracking schemes, subscribers could pay reduced subscriptions if they are part of the ITV analysis program, or they could receive some other benefits from taking part in this type of activity. The type of information these activities generate will be very valuable to network operators, and will help to concentrate development activity on the types of applications people actually want to use. Some pay-TV networks are already carrying out this type of measurement, but moves toward a horizontal market change how this information is gathered and used.

T-commerce

Network operators have always understood that the TV shopping phenomena is something ITV services can exploit. Millions of people spend money with the home shopping channels, even though it means logging on to a web site or calling the broadcaster before they can actually make a purchase. Home shopping channels are poised to make even more money now that ITV is possible, but other channels and shows can exploit this. Imagine being able to purchase materials from your favorite home makeover show at the click of a button, or placing an order for the ingredients used in a recipe shown on a cooking show. Customers will be able to shop at the click of a button via the remote control, and fully automated back-end systems can offer savings not currently possible. Back-end systems used for e-commerce via web sites can be reused in many cases, allowing operators to extend their services for only a small extra cost.

We have already discussed some cases of T-commerce via the TV, and although these services may not make a great deal of money at the moment it is likely that both standalone T-commerce and T-commerce associated with specific shows will be a growing part of the revenue stream for many operators.

Reduced Piracy

For any pay-TV operator, theft of services can have an important effect on revenues. Estimates show that European operators lost up to 1 billion euros in 2003, with the United States losing between 1 and 4 billion dollars a year due to piracy. DTV systems do not eradicate piracy, as these numbers show, but it does become more difficult due to the more sophisticated CA systems network operators can deploy. The MHP and OCAP specifications also cover pay-TV services, and even some terrestrial operators are considering some types of pay-TV content. Conditional access companies such as Philips, Irdeto, Nagra, Conax, Viaccess, and NDS are all involved in protecting premium content, and all of these have integrated their CA solutions with MHP middleware. Similar moves are under way in the United States as more companies provide CableCARD and POD modules for their CA systems.

As we move forward, and as receiver performance increases, software encryption becomes more feasible. A recent announcement by Latens has shown that there is a place for pure software encryption after they secured their first satellite deal in Turkey. Latens claims that its system can reduce the cost of conditional access by about 50%. The system is certainly more flexible, because scrambling algorithms can be changed and the network resecured almost as soon as the system is hacked. Whether these systems produce all of the benefits they claim remains to be seen, however, but it does provide a new weapon in the war against pirates.

Mobile Services

Because data-only services do not have to be tied to a particular TV show, they can be used in mobile applications in which analog broadcasting cannot easily compete. Trains, buses, or even private cars could use a DVB-T system for traffic reports, customer information, advertising, and other services. Even audio/video services are feasible. Singapore's TVMobile is already broadcasting mobile services such as advertising and lifestyle programming to commuters on public transport, as we mentioned in Chapter 2.

Open standards make data-only services much easier to deploy, due to the availability of off-the-shelf receivers and a well-known set of APIs. The EU Multimedia Car Platform (MCP) project began looking at information services in mobile environments in 2000 and 2001, and this work has continued in the MHP-Automotive group within DVB. Despite their names, these activities focus on the use of MHP in all mobile applications. Typically, these applications fall into two main categories, including travel guides and navigation services for front-seat passengers and entertainment services similar to those deployed in other DTV networks for rear-seat passengers.

Figure 19.3. The DVB-T Mobile World on Wheels, a modified VW Caravelle, at Expo 2000, Hanover.

During the Hanover Expo 2000, Volkswagen minibuses from the World on Wheels project (part of the MCP work; see Figure 19.3) demonstrated the use of data services over DVB-T and the DVB-T Mobile framework as part of the MCP project. These minibuses transported businessmen and VIP guests to and from the Expo and acted as a rolling demonstrator for the types of mobile service possible in a DVB-T system. Many companies attended the latest meeting of the MHP-Automotive group, but only a single car manufacturer was present (Volkswagen, who have been involved in several mobile MHP projects in the past as well as the Hanover Expo project). The risk is that there may be insufficient traction to maintain the work of the group.

The Bad News

Although we are seeing a number of digital deployments using MHP, revenues are still small at the moment. Many of the broadcasters that have launched MHP services are not pay-TV operators, and this limits the revenue streams available to them. Commercial operators in the United States are moving toward OCAP, but at the time of writing no projects have moved past the field trial stage. Even when interactive applications are deployed, there is no guarantee that they will make money.

This is not to say that broadcasters cannot make money from open middleware, however. Finland is at the forefront in MHP, and Finnish broadcasters have had the chance to understand what works and what does not in a market based on the open middleware standard. Already, MTV3 Oy has claimed that it is making money through ITV applications linked to SMS via revenue sharing, and it is not alone. There are still a number of problems facing broadcasters, however. Some of these are general issues for any DTV network, whereas others are more specific to networks using open middleware standards.

Low Penetration

It is inherently difficult to transform an entire technology base overnight, and this is especially true in the DTV market. Analog TV systems are thoroughly entrenched, and the deployment of proprietary digital systems complicates moves toward MHP and OCAP. Network operators have committed to these proprietary technologies and invested heavily in them, and they need to amortize these investments. Furthermore, the entire process of deploying these open middleware standards has required that systems and products implementing those standards become more mature. For those companies that have made the move to open standards, the selection of partners by everyone involved has been made more difficult by the number of different companies offering MHP/OCAP products (or quirky hybrids of MHP/OCAP and other systems) in an immature market.

Finally, the required education process for the industry has added to the overall delay of its launch. There still are, as can be imagined, companies with proprietary and legacy systems who still do not wish to see MHP become successful. These negative forces are still at play, and MHP developers have had to counter this and prove that MHP really can work. This is less of an issue for OCAP. Many elements of the OCAP infrastructure are already in place, and OpenCable receivers are already shipping in a horizontal market (without OCAP middleware, so far). Despite this, OCAP still faces stiff competition from proprietary middleware solutions.

It is true that we need further deployments and commitment in order to build up the market. However, if we consider that it has only been two years since the real launch of the horizontal market the present status of MHP is very encouraging. Table 19.1 shows which countries have already launched MHP services and which are testing MHP in a trial phase (some of which are due to launch in 2004–2005). Already, about 14 broadcasters in five countries have launched MHP. Similarly for OCAP, many U.S. network operators, as well as operators in other countries, are exploring it with an eye to deployment.

Table 19.2 lists the companies that are offering MHP receivers as of March 2004, although not all of these have passed the MHP conformance tests. In addition to these manufacturers, a number of other companies have products under development, or provide middleware stacks that can be incorporated into MHP or OCAP products.

Viewers Resisting the Move to Digital

One of the reasons behind the current low penetration of free-to-air DTV is the resistance of viewers. This may not be resistance to DTV as such (after all, the success of digital pay-TV services shows that this is not the case). Instead, the resistance mostly falls into two categories: those viewers who are not willing to pay for a receiver, and those viewers who have no reason to move.

Table 19.1. Countries that have launched MHP, or which are testing it.

Country	Network Type
Launched	
Finland	Terrestrial/cable
Germany	Satellite, terrestrial
Spain	Terrestrial
South Korea	Satellite
Italy	Terrestrial
Testing, trials carried out or due for launch in 2004–2005	
Australia	Terrestrial
Austria	Terrestrial/satellite
Belgium	Terrestrial/cable
Czech Republic	Terrestrial
Denmark	Terrestrial
Finland	Cable
France	Terrestrial
Germany	Cable
Hungary	Terrestrial
Norway	Terrestrial
Poland	Terrestrial
Portugal	Terrestrial
Spain	Terrestrial
Singapore	Terrestrial
Saudi Arabia	Terrestrial
Sweden	Terrestrial
Switzerland	Terrestrial/cable
Taiwan	Terrestrial/cable

The first of these is the most difficult to address, but the cheap free-to-air receivers and subsidies for more advanced products are slowly making an impression. In particular, the latter issue is the one that will most help the success of open middleware standards. Cheap zapper receivers will not support any type of interactivity, and this will in turn slow the progress of MHP and OCAP.

The second issue is one of viewer education. In general, consumers polled in surveys on the street and at home will talk negatively about things they do not really know of or understand, especially if those things may cost them money (this is human nature). However, put someone in front of an EPG for a couple of weeks and then put them back in front of traditional analog TV. Ask them if DTV and ITV is a good idea once they have used it for a while (and especially if they are forced to go back to the old system) and you may get some very different answers.

Table 19.2. Companies offering MHP or OCAP receivers at the time of writing.

Access Media	Advanced Digital Broadcast	Co-ship
Digit All World	Digital Multimedia Technology (DMT)	DTVIA
Echostar	Homecast	Finlux
Force	Fujitsu-Siemens	Galaxis
Handan	Hi-Top	Humax
Hyundai Digital Technology	IBM	Kaon Media
KISS	LG	Motorola
Nokia	OpenTech	Pace Micro Technology
Panasonic	Philips Electronics	Samsung
Skardin	Seodu Inchip	Sony
Techmate	Thomson/Canal Satellite	WisPlus
Zentek		

We have a question for all of the BSkyB or TiVo users reading this who travel regularly on business: how many times have you picked up the TV remote in your hotel and tried to find the button (the "i" button) that tells you which shows are on, what they are about, and when they started? The difference between "old TV" and "new TV" must be experienced, and we have seen that this user experience is what really counts in driving the adoption of new TV-related technologies. Michael Powell, Chairman of the FCC, stated in public that his TiVo personal video recorder was "God's Machine" (he could not live without it). Although this statement was not popular with some segments of the broadcasting industry, it does illustrate the importance of getting customers to try the product and see the advantages for themselves.

Freeview in the United Kingdom has shown that free-to-air digital services can be successful, with over 2.5 million receivers sold. With average sales of 100,000 receivers per week, this has proven to be a very successful venture. Indeed, it has been so successful that BSkyB, the U.K. pay-TV satellite operator, has recently announced its own free-to-air satellite service. Considering that Freeview is partly owned by BSkyB, this is a powerful statement about the current state of the free-to-air market in the United Kingdom. Whereas many consumers are cynical about the poor quality of content that is on offer on many of the digital channels, they are still purchasing the product. Nearly 55% of U.K. homes (13.7 million households) now watch DTV in one form or another, and although pay-TV is still the largest segment of the market free-to-air TV is catching up fast.

Datamonitor recently predicted that by 2007 over 18 million households in the United Kingdom would be watching DTV, an increase of 30% over the current viewer figures. The bad news for us is that not all of these households will be using MHP, but the good news is that a larger DTV market almost inevitably means a larger market for MHP, especially as other countries in Europe move to MHP for public broadcasting. A small piece of a large pie is often better than a large piece of a very small pie.

Competition from Pay-TV Operators

Many times in this book, we have stated that content is king for DTV, just like it is for so many other markets. Pay-TV operators have a well-defined business model that gives them a steady income, and this helps them to purchase or develop high-quality content. Interactive applications alone are not enough to get viewers to move to digital, and deploying those applications is no more difficult in a vertical market. The availability of premium content such as sporting events and movies, as well as a wider range of other content, is what motivates viewers to move to these services.

The common element to each of these is money. Obtaining content costs money, and without a steady flow of income network operators cannot afford the best content. Broadcasters funded by central government or by advertising revenues alone may not be able to compete with pay-TV operators in terms of the variety of content they offer, and thus they need to find ways of encouraging viewers to switch to digital without switching to a pay-TV service.

Interactive applications may be part of this, but they are unlikely to be the only one. Freeview in the United Kingdom has managed this by offering a wider range of services than is available to analog viewers (including some interactive services), but with the advantage of a one-time cost for the receiver. Many viewers are reluctant to pay subscription fees, and thus the ability to buy a receiver in the retail market is more attractive to them. BSkyB have recently introduced their free-to-air service partly because of this reason. Free-to-air services in the United Kingdom are expected to reach viewer numbers close to those of the pay-TV services in the next few years, simply because a large segment of the market does not want pay-TV and is not willing to pay subscription fees for a basic DTV service, no matter how low those fees may be.

Other Types of Services

So far, we have examined revenue-generating services we can deploy on open middleware platforms. These are not the only types of services that can be offered, especially on public networks. In Italy, for instance, the government is planning to use DTV as an important part of its e-government initiative.

The Italian government plans to issue a National Services Card (*Carte Nazionale dei Servizi*, or CNS) for use with e-government applications. Over 10 million of these cards are due to be issued before the end of the current parliament in 2006, and they will be used to identify and authenticate holders in any e-government related applications that will be deployed in Italy.

Each card will contain identification data for the holder, such as name, date of birth, and place of residence, as well as some authentication information such as a unique ID number for the card and the name of the issuing administration. Cards will also contain a basic digital signature function and some storage for other digital certificates, allowing these cards to be used for authentication as well as identification. Similar cards are already in use in the Lombardy region for access to health services, but these cards can be used for a much wider range of services. By offering these services via DTV networks, the Italian government intends to provide much wider access to e-government services than would be possible via the Web.

There are problems with this type of system, however. Because the Italian solution is based around a smart card, this means that receivers must be able to read those smart cards in order for them to be part of the system. In that only MHP 1.1 receivers can communicate with non-CA smart cards, this means that any MHP 1.0.x receivers must use technology outside the MHP standard. In this case, they have chosen to use the elements of the Java Security and Trust Services API (JSR 177), which DVB has adopted for future versions of MHP 1.1.

Conditional Access and Horizontal Markets

Conditional access is an important part of a pay-TV system, as we have already seen, but it can cause several problems when we try to move to a horizontal market. Using our definition of a horizontal market, customers should be able to purchase a receiver from an electronics store and use that to connect to a DTV service. Although this is possible, many pay-TV services use proprietary CA systems for content protection. Many of these rely on smart cards to contain encryption keys, and in most DVB systems the CA system must be integrated with the middleware or even the receiver hardware itself.

This means that any receiver must be built for a specific CA system, and as you can imagine this causes problems in a horizontal market. Even if two networks use the same middleware, consumers may still have to purchase different receivers for each network, and manufacturers may have to build receivers for two networks. This is a big obstacle for horizontal markets: consumers would not purchase a DVD player that only works with DVDs they rent from one store, and the same is true for DTV receivers. This restriction vastly limits the number of viewers who will purchase a receiver outright, because they are then tied to a single network operator unless they purchase another receiver. A recent survey, which made no claims to be comprehensive, counted 10 different CA systems in use across Europe. In some countries, such as Germany and the Netherlands, at least four different CA systems are in use.

For OpenCable systems, this is less of an issue because OpenCable receivers use a Cable-CARD module for conditional access. The CableCARD interface is standardized and thus consumers can purchase a receiver with a CableCARD slot and then plug in a CA module supplied by their network operator. Should the customer wish to move to a different network (e.g., because they are moving), they simply return their current CableCARD module and get a new one from their new network operator. This lets equipment manufacturers sell receivers through retail channels while maintaining interoperability, and this is likely to be an important factor in the adoption of DTV in U.S. cable markets. Already, a wide range of CableCARD-ready receivers are available to consumers from electronics stores and other outlets.

Although DVB defines the DVB Common Interface (DVB-CI) for conditional access systems, this is less popular. Partly, this is a cost issue, because DVB-CI modules can be as expensive as an entire receiver that has the CA system built in. Partly, it is a security issue. Companies such as NDS have expressed concerns about the security of DVB-CI because of the architecture of the interface, and thus they are reluctant to produce DVB-CI modules for their scrambling systems.

Without a pluggable solution for conditional access, horizontal markets become much less feasible for pay-TV operators in DVB countries. This can only harm MHP (and indirectly OCAP, because fewer deployments of this family of standards means less content will be produced for those standards). Public networks in some countries are taking steps to resolve this by choosing a common CA system, as Finland did with Conax, but this may not be enough if the DTV market as a whole remains fragmented.

"MHP Lite" and Low-end Solutions

There have been calls for a low-footprint solution such as "MHP Lite" and "OCAP Lite." Although these may seem like good technical ideas for backward compatibility, they will slow the uptake of the existing open middleware platforms. These provide a short-term solution for low-end receivers that are already deployed on pay-TV networks. The CPU power and memory in these receivers is very small, and they were never really intended to provide full interactive capabilities.

Many people in the TV industry look at backward compatibility from the perspective of ITV, instead of from the perspective of the fundamental services (TV and audio). This makes things more difficult for middleware developers, because they are being pressured into shoehorning interactive solutions into inadequate and elderly equipment. Although Sun is making some efforts in this area with the "On ramp to OCAP" specification (JSR 242), CableLabs and DVB currently have no intention of producing low-end versions of their specifications.

Companies with existing solutions for low-end platforms tried to generate interest in MHP Lite, but only a few network operators and receiver manufacturers were interested and thus this initiative was abandoned. It could be argued that this was one of the reasons MHP has not been that successful to date. Each of these discussions led to a delay in the launch of MHP, and this stopped people committing to MHP as they watched the debate roll on and on.

In the United States there are a large number of low-end receivers such as the Motorola DCT 2500, for which network operators would like to support interactive services. The TV industry faces some difficult choices with the move toward horizontal markets and open standards. Although consumers do not upgrade their CE equipment as often as they upgrade their PC, it is in the best interest of the industry to remove older DTV receivers so that more advanced solutions can be rolled out more quickly. At the same time, network operators have a huge investment in receivers such as the DCT 2500, and we cannot simply ignore these.

However, the risk is that these "lite" versions of the technology become the norm and (like the "digital zapper" phenomenon) create another low-level platform that will not allow the growth of ITV applications and services (i.e., a sort of dumbing-down of ITV). As long as network operators treat this as a stepping-stone to OCAP for those receivers that are already deployed, and do not simply deploy this on all of their future receivers, the situation should help the growth of ITV services and revenue.

This is one area in which a horizontal market may help deployment of receivers. Over the last few years, consumers have been willing to upgrade some of their devices, provided that there are obvious extra features that are worth the money. People may not see this with traditional TV sets, but the mobile phone industry has seen an explosive growth in the features available (and in the number of times consumers will upgrade). Although this is partly due to the use of mobile phones as a fashion accessory, it is also partly due to the new features modern phones offer (such as cameras, games, and 3G services). If the receiver manufacturers can offer new features consumers actually want, this could help speed the move away from these legacy solutions into a market in which receivers supporting the full OCAP or MHP specification are the norm.

Interoperability

For a horizontal market to succeed, we have to make sure that every piece of equipment in the broadcast chain, from the MPEG encoder to the receiver, works together no matter which company builds it. Interoperability issues are something we have to consider seriously when we discuss middleware, simply because of the complexity of this particular piece of the chain and the fact that this is the component consumers will interact with directly. Consumers will likely blame the receiver for any problems because that is the part of the system they are most familiar with. It is simplistic to expect that there will not be any issues and that a homogenous platform will solve all problems. Experience has shown that this is an unrealistic attitude that can only lead to problems. The entire market has to realign hardware vendors with open-standard software suppliers in order to eventually simplify the entire system, and this will undoubtedly take time and effort.

In the MHP world you would think that there is a greater interoperability risk because of the self-certification system implemented for conformance testing. This is not in fact the case, given the strict nature of the tests and the efforts under way outside the formal testing regime to ensure conformance. In the OCAP world, CableLabs will handle this much more formally with a Wave certification system that requires products to be tested in the CableLabs laboratories. For companies deploying MHP and OCAP in some markets, bodies other than DVB and CableLabs will also carry out interoperability testing to help guarantee that deployments of MHP- or OCAP-enabled receivers will be successful. Italy has put a mechanism in place at the national level, with the DTV and DGTVi laboratories ensuring interoperability between receivers.

Furthermore, individual cable or satellite network operators (Premiere in Germany is an example) will expect vendors to pass a network-specific interoperability test before any product aimed at those networks can go into the retail market. Network operators in vertical markets already use tests such as these when they deploy receivers from more than one manufacturer, but in this case the MHP test suite and pressure from other operators to be interoperable can offer help in ensuring interoperability. Many DTV watchdogs (such as the

DGTVi in Italy and NorDig in Scandinavia) also publish their own implementation guidelines for middleware and receivers.

Satellite and cable operators may well choose to remain in the vertical market while using open middleware standards to help reduce their costs and increase the availability of applications. In the end, viewers care only about content, and as more companies deploy MHP, OCAP, and related middleware standards the amount of content produced for that middleware will rise. With MHP and OCAP providing a worldwide market for content, more and more content developers will be attracted to this new common platform.

At present, a common platform usually means an additional middleware solution on top of proprietary middleware, or declarative content that is pre-processed and then displayed using a middleware-specific presentation engine. MHP, OCAP, and GEM bring a true common platform to a wide range of markets, and as content providers such as the Discovery Channel, MTV, and HBO become worldwide brands, common platforms become more important if interactive applications are to play a part in the future.

MHP Interoperability Events and Plug-fests

In addition to formal interoperability testing, MHP vendors also carry out slightly less formal but highly practical testing at interoperability workshops in Europe. The first MHP interoperability workshop took place at the Institut für Rundfunktechnik (the joint research arm of a number of German broadcasters) in Munich in January of 2002. The idea was to have a "plug-fest" in which companies involved in MHP could come together in the pre-competitive phase of the market to make sure that different MHP solutions really were interoperable (see Figure 19.4). At this time, there was no test suite available and thus this was the only way to make sure that products worked together as they should.

The only criticism that could be leveled at this first session was that as much time was spent fixing problems as testing implementations. This may not seem like a problem, but it made it more difficult to work out whether a particular problem was caused by a middleware implementation, an application, or a piece of head-end equipment. Several more interoperability workshops have been held since then, roughly every three to four months, and as products have matured and more companies have become involved the focus has shifted from bug fixing to interoperability testing.

Table 19.3 outlines the companies that were present at the most recent (at the time of writing) interoperability event in 2004. This level of participation from all aspects of the market shows how companies are taking real steps to achieve interoperability and not just relying on test suites.

In addition to these interoperability workshops, a more political plug-fest took place in Brussels in 2002 to show the European parliament how work was progressing on open middleware standards. A group of 20 companies demonstrated MHP receivers and services to EU Commissioner Erkki Liikanen, members of the European Parliament, and representatives of the member states. This allowed politicians to see that many companies were adopting

Figure 19.4. The first MHP interoperability "plug-fest" at IRT in Munich.

Table 19.3. Companies participating in the most recent MHP interoperability workshop.

ADB	Alticast	Canal+ Technologies
Convergence	DR Dënmark	FH Salzburg
Fraunhofer-Institut für Medienkommunikation	Gist Communications	ICT embedded B.V.
Institut für Rundfunktechnik	MHPeople	MIT xperts
Motorola	Nionex	Nokia
Ortikon	Osmosys	Panasonic
Philips Electronics	RTL New Media	Salzburg AG
Sofia Digital	Softel	Sony Belgium
Tektronix	TVC Multimedia	Universität Duisburg-Essen

MHP and working together to make sure that truly horizontal markets were possible through interoperability, and was generally regarded as a great success.

It has become apparent at the various interoperability workshops that the common platform is well and truly on its way. By gathering different players in a single room to hash out MHP applications, receivers, and head-end equipment, these events show that no network operator or receiver manufacturer needs to be locked into a single middleware supplier any longer, and this is now being shown in the marketplace. Today, there are multiple receiver manufacturers selling their products in Germany, Finland, and Italy with different MHP implementations. For the broadcaster and the customer, the support for different middleware implementations is transparent, and thus it is becoming more obvious every day that open standards for middleware can work in the real world.

Conformance Testing

Interoperability testing is not enough, however. Products must also conform to the MHP specification in order to avoid the type of bug-compatibility issues that have plagued the World Wide Web. Conformance testing is a crucial element to the success of MHP, as it will be for OCAP and other open standards, and test suites are necessary to verify that products coming to market are indeed MHP compatible.

Philips, Nokia, Panasonic, Canal + Technologies, Sony, Alticast, and IRT formed a test suite consortium outside DVB, and these companies together with Sun Microsystems provided the first input to the MHP. Following this, the consortium members paid the princely sum of 183,000€ each for an external contractor to produce an automated test environment (ATE) and a variety of test cases written as Java Xlets. The initial test suite for MHP was initially due in November of 2001, but the complexity of the work delayed its release until June of 2002. This version of the test suite was aimed at version 1.0.2 of MHP, and in December of 2002 the DVB Steering Board approved version 1.0.2b of the test suite (which added more tests to improve coverage of the MHP 1.0.2 standard).

At present, there are no test suites available for MHP 1.1.x, and this is one of the reasons behind the slow adoption of MHP 1.1.x by broadcasters and receiver manufacturers. Without a test suite, it is impossible to test conformance, and with no implementations in the market it is difficult to test interoperability. We have discussed the issues surrounding interoperability and conformance testing of MHP 1.1 in Chapters 15 and 16, and thus there is no need to discuss these issues again here.

In 2000, at the same time early versions of the test suite were under development, the MHP Experts Group (MEG) was set up within DVB to maintain the test suite. This group is led by Dr. Rainer Schäfer of the Institut für Rundfunktechnik. The work of the MEG includes verifying the tests being produced by the test consortium for their relevance and validity, approving the addition of tests to the test suite, and to receiving any feedback about tests that are considered incorrect. In its role as maintainer of the test suite, the MEG is also responsible for issuing new versions of the test suite to ETSI.

Work continues to extend the test suite to cover other versions of MHP, but work on GEM has delayed this because many of the participants in the MEG were also involved in defining GEM. Politically, GEM is extremely important for harmonization of middleware platforms around the world, as we have already seen, and this had to take priority over extending the test suites. Now that work on GEM is complete, work can resume on the test suites.

Anomalies in the Conformance Testing Program

So far, the test suites have only covered receivers. These are not the only piece of the MHP puzzle, of course, and it is equally important that applications and head-end equipment also conform to the MHP specification. DVB recognizes the need for some conformance testing of applications, or at least a set of guidelines for application developers, but so far no company has committed to producing this.

The need for MHP certification of head-end equipment is also a contentious issue within DVB. Some companies believe that some head-end products will not require the types of stringent compliance tests defined for MHP receivers. Because head-end equipment must typically conform to other standards (such as DSM-CC), some manufacturers believe that testing many types of head-end equipment is outside the scope of MHP. This remains an unresolved point in DVB, and achieving a consensus will not be easy.

We have seen over the years that many companies, even large CE vendors, do not have the expertise to locate (let alone fix) all of the software bugs and issues involved in MHP. A good testing regime is essential to the success of MHP, for receivers, applications, and head-end equipment. Although DVB may only define some elements of this testing regime, it is vital that all parts of the chain conform to the appropriate standards.

The MHP Conformance Testing Process

If a company wants to sell an MHP-enabled device, they must pursue the following process outlined to obtain the MHP logo.

1. The company requests the MHP test suite from the custodian (ETSI). This is subject to an administration fee of 1,000 euros.
2. The custodian issues the test suite to the company under a nondisclosure agreement.
3. The company tests its device (self-certification) and lodges a test certificate with the custodian if the device passes the tests.
4. The custodian informs DVB that the company has completed and passed the tests.
5. The company pays a fee of €10,000 for rights to use the MHP logo.
6. DVB issues the DVB MHP logo to the company and registers the compliant product in their database.

Things may not be quite this straightforward in practice. There are several scenarios in the testing process, and these are outlined in the DVB Blue Book. A full description of the procedure, and all necessary documents, are available from the ETSI web site at *http://portal.etsi.org/dvbandca/MHP/mhp_conformance.htm*.

From this point, the company has the right to use the MHP logo in association with that product, either on the receiver itself or on any packaging and marketing material. After the first year, the company must pay a maintenance fee of €5,000 to continue using the logo for each subsequent year.

Black 30%
White 100%

Figure 19.5. The MHP logo may be used on products that conform to the MHP specification and that have followed the process described in the text. Color versions of the logo, and different black-and-white versions, are also available.

Use of the MHP logo is subject to strict rules, and manufacturers must adhere to a strict set of guidelines dictating how the logo should look and how it should be displayed. One version of the MHP logo is shown in Figure 19.5, and other versions are available in several color schemes to fit the different needs of receiver manufacturers. DVB and MHP are registered trademarks of the DVB Project, and thus the use of these names or logos without the permission of DVB is not allowed.

Testing MHP: A Case Study

This section gives an overview of the testing process from the point of view of a leading MHP vendor, documented in part in the following. By including this, we hope to highlight the complexity involved, and hopefully make the reader understand that conformance testing is not a trivial task but is probably one of the most crucial parts of creating an MHP implementation suitable for the market.

We received the test suite from ETSI at the beginning of September 2002. We fully completed all testing and passed the test suite on 23 December 2002 using build #402 of our MHP implementation on our standard terrestrial retail STB.

This test suite contains 10,840 tests, of which 235 were excluded as bad tests. These were highlighted a result of appeals to the MEG (MHP Expert Group). During the whole process our own company submitted 44 appeals which related to around 90 tests.

In January 2003 we received 1.0.2b version of the test suite containing an additional 41 tests, and we completed testing and passed 31 of those tests on 1/31/2003, 10 tests have been re-submitted to the MEG as incorrect.

Continued

We have been spending 30–40 man-months (120–160 man-weeks) on bug fixing and on preparing tools such as a test harness for automatic testing and status reporting. At its peak, we had 12 engineers working on the test suite.

Although the DVB test suite cannot give absolute guarantee or truly confirm that a build is fully functional and ready for the market, the test suite is a very effective regression test, confirming the reliability that gives the engineering group confidence with respect to the implementation. The test suite is a very important achievement and of course covers the largest part of the MHP implementation. It is however not sufficient in itself; we have to cover much, much more.

Quality assurance is and will be the key factor for the big, intricate software packages such as the MHP and OCAP middleware.

Testing OCAP

OCAP conformance testing is subject to a different set of rules from those of MHP. Cable-Labs runs Wave certification programs or their standards, and they have expanded this process to include OCAP. The first "practice run" certification wave for OCAP applications took place on January 17th 2005, with the first full certification wave starting on March 14th 2005. CableLabs has defined an Automated Test Environment (ATE), which currently consists of approximately 11,000 separate tests. This is still work in progress, and the final ATE is expected to comprise approximately 20,000 tests. Manufacturers can purchase the ATE for 25,000 US dollars, and they will be given access to the ATE approximately one month prior to the start of the certification wave. Before starting the certification wave itself, manufacturers must show that their product passes the ATE by providing CableLabs with test logs from five separate units of the product undergoing testing.

Besides the ATE, there will be a separate Conformance Test Package (CTP) available from CableLabs for a fee of $10,000 and a $5,000 annual maintenance fee.

As for MHP, there will be separate OCAP interoperability events, with four scheduled for 2005. These will give companies working with OCAP a chance to verify interoperability in a more "real-world" setting than the ATE can provide.

As with MHP, CableLabs strictly controls the use of the OpenCable logo and other Cable-Labs trademarks, and OCAP receivers must pass the conformance tests before they can use any OpenCable logos on the receiver or in marketing materials. CableLabs also produce a style guide that controls how the logo may be used.

Compliance and Quality

Building a middleware stack that meets the MHP, GEM, and OCAP specifications is only part of the problem facing middleware developers. For proprietary systems, porting the middleware to your platform and passing the conformance tests is enough. After that, an OpenTV stack (for instance) running on one receiver is the same as an OpenTV stack running on another, ignoring differences imposed by the hardware. This is not the case for open-standard middleware. Each implementation is more or less independent, based only on the published specifications. These specifications are sometimes unclear and ambiguous, and sometimes this ambiguity is deliberate. These ambiguous areas are one of the places where implementers can differentiate their middleware stack from their competitors.

A receiver that does only what the specification says will pass the test suites, but it will not be as attractive to consumers as one that takes more care with the details. Some features will help applications to run better on certain receivers, although the MHP specification says nothing about them. For instance, the MHP specification says that when a new service is selected any clipping or scaling of the video will be reset. Current best practice says that applications should register for `PresentationChangedEvents` for that player and immediately reapply any desired video transformations. Doing this may result in flickering as the video is scaled twice (first to its default size/position, and then back to what the application wants), but many platforms take steps to avoid this flickering problem. This is just one example of how we can improve the experience for the user and add value to our middleware stack.

Another UI-related example is the HAVi user interface classes. The MHP specification says that methods in these classes do not have to be thread safe, but this could cause major problems if more than one application attempts to change the screen configuration at the same time. (There are good reasons this is not required. The Java AWT implementation is also not required to be thread safe, and the HAVi GUI's reliance on AWT would make thread safety difficult to implement in every case.)

By enforcing thread safety even when it is explicitly not required, the middleware removes one possible source of problems users could see. Even limited thread safety is better than none in this case, if the effort is aimed at the right places.

To provide an example that is not related to the UI, but which most consumers will notice, think about the DSM-CC code. The caching strategy used by the middleware can strongly influence how quickly applications start, and this is one area in which attention to detail by the middleware manufacturer will pay off.

These types of issues will be what separate good receivers from those that are not so good. Although an application may run fine on both receivers, it will probably look and behave better on one that takes steps to avoid problems such as those just mentioned. To quote someone heavily involved in MHP standardization, "Applications will sometimes look worse on 'less good' MHP implementation than on good ones. This is not the fault of the application developer but either the fault of the manufacturer or the choice of the consumer to buy a cheaper MHP receiver . . . The MHP specification is not about preventing manufacturers making bad products." In a horizontal market, these types of differences are where middleware developers will compete.

Head-end Requirements

Deploying receivers is not enough if we cannot transmit MHP or OCAP content to those receivers. MHP and OCAP impose no special restrictions on most head-end equipment, and thus we can reuse existing DVB or OpenCable broadcast equipment for many parts of the broadcast chain. The move to MHP or OCAP will mainly affect those parts of the head-end that are related to applications, including the parts that produce the application's directory structure, convert that directory structure into an object carousel, and broadcast it.

Producing the directory structure involves checking all needed files, updating files, removing unused files, and general content management. It also involves taking the application files and adding any other necessary files, such as files related to security. This may all be part of the same tool set that generates the object carousels, especially in content management systems and object carousel generators that support MHP. Solutions such as Sony's MediaCaster, Thales' Coral, or TSBroadcaster from Strategy & Technology combine these features in a single tool that developers can use for both content management and object carousel generation. Some network operators may already have their own content management system, in which case separating these functions may suit their workflow better.

Many companies offer object carousel generators, and more are coming to market all the time, but some have better support for MHP than others. Object carousels in MHP are no different from those in other DVB systems, and their use in OCAP follows the DVB data broadcasting specification and thus there is no problem there.

Support for MHP in an object carousel generator usually means the ability to generate the application signaling. Combining this with object carousel generation is often extremely useful, both from a workflow perspective and from a technical perspective. At this stage of the broadcasting process, we will typically be thinking about applications and thus using one tool to manage both the content and signaling for those applications can make life easier. Similarly, this approach concentrates all of the MHP-specific functions in one part of the chain.

Although it is not necessary for the multiplexer to be MHP aware, it should be able to handle MHP- and OCAP-specific SI tables (such as the application information table, signaling the presence of MHP or OCAP applications) generated by other equipment. Typically, this means being able to add appropriate entries to the MPEG PMT for the service in question. Even in multiplexers that do not explicitly support MHP, there is usually a way to do this.

In some cases, there may also be a need to generate DSM-CC NPT and stream event descriptors and to embed them in a stream. Some object carousel generators can also do this, although it may be necessary to use a separate tool. Figure 19.6 shows the typical elements in the transmission system, and highlights which of these should be MHP aware. Shaded boxes show those components that need to generate or handle MHP-specific data.

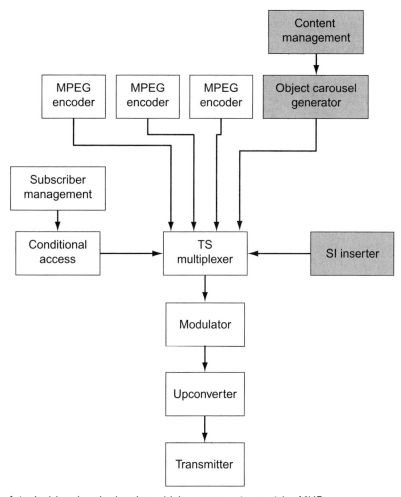

Figure 19.6. A typical head-end, showing which components must be MHP aware.

Remultiplexing Issues

As for standard DVB or OpenCable services, remultiplexing poses no great problems for MHP or OCAP applications. The biggest issue is the remapping of PIDs, and most multiplexers will support the features needed for interactive applications.

The other area that may pose problems is the use of DSM-CC stream events or NPT. This will not be a problem in most cases, but sometimes the PCR values of a stream will change when a stream crosses a network boundary if the network operator remaps the PCR values. Because NPT (and stream events that use it) uses this value to map from NPT time codes to actual points in the stream, remapping PCR values can corrupt the NPT time codes.

When dealing with streams that use NPT, network operators must take care to regenerate NPT time codes if the PCR is modified or remapped when a stream crosses a network boundary.

Application authors can reduce these problems by minimizing the use of NPT in their applications, and in particular they can use do-it-now stream events rather than scheduled stream events. Do-it-now events do not rely on NPT, and thus they are more resilient to changes in the PCR that may corrupt the NPT values. These are also easier to use in many cases, and are probably the best way to synchronize applications and broadcast content currently.

Conditional Access

The beauty of MHP and OCAP is that they work with any CA system. This is generally applicable to most middleware, of course, but with OCAP and MHP this includes support for CableCARD modules and DVB common interface modules that enable one receiver to support different conditional access systems. Just because open standards for CA modules are available, that does not mean we have to use them. MHP and OCAP will work equally well with proprietary CA systems, and most networks that have deployed MHP to date are using proprietary CA (or in the case of many public broadcasters, no CA at all).

Given the number of different conditional access systems that have been integrated with MHP receivers around the world, your choice of CA system is not likely to cause too many problems. There are still issues with using conditional access in a horizontal market, as we saw earlier in the chapter, but this is not a fundamental problem with MHP. Head-end equipment that deals with CA does not need any changes to support MHP or OCAP, and thus network operators can reuse existing equipment.

Using Object Carousels

Object carousels have been used for standards other than MHP (the MHEG-5 system in use in the United Kingdom is a good example), and data carousels have been used by still more operators for data broadcasting. There is nothing inherently different about the use of DSM-CC in MHP or OCAP, and thus we will not go into any detail in this chapter. We have already covered the basics of DSM-CC in Chapter 12.

Some network operators or application developers may want to share data between MHP or OCAP applications and legacy applications. There are good reasons for doing this. Bandwidth is often scarce, and if network operators have to simulcast different versions of an application it is easier to reuse assets where possible.

MHP or OCAP applications can mount other object carousels, and thus this is not a big problem for them. It is even possible to read data carousels with a little work. Although MHP and OCAP do not provide access to a data carousel through the MHP APIs, it is possible to write a data carousel parser in MHP using the SI and section-filtering APIs. We have seen this approach used to share data with OpenTV applications with remarkable success, although it obviously means more work for application developers unless they have a class library to support this.

Using the section-filtering API it is possible to parse most data formats from the transport stream if you try hard enough, and thus network operators can usually find some way to simulcast applications while using the least space possible. This means that we can only share data files, of course. You still need to transmit both sets of application files and the SI associated with them. Because of the nature of class loading in MHP, we can share Java class files if they are transmitted in object carousels or in IP streams using multiprotocol encapsulation, but not if they are transmitted using data carousels or any other data broadcasting techniques.

Sometimes, though, it is worth the extra bandwidth to retransmit data files, especially if they are not broadcast in object carousels. The effort required to parse other broadcast data formats can be high, and thus developing and testing the necessary code may take more time and effort than it is worth in saved bandwidth. We cannot offer a firm rule here, because the advantages and disadvantages of sharing data files between open and proprietary applications will change from case to case depending on the bandwidth available, and there is no "one size fits all" answer.

OTA Download and Engineering Channels

So far, we have concentrated on making sure that everything works together, but we have missed one important step in the process. Interoperability and conformance testing in the laboratory are one thing, but making them work in a real network can be much more difficult. Maintenance and the ability to upgrade software in the field are paramount to an efficient deployed system.

One of the most significant events in the early days of MHP deployment took place in 2002 in Northern Jutland, Denmark. With a transmitter and a local studio/laboratory, a small group of TV2Nord-Digital employees created a digital broadcast environment for testing DVB-T and MHP. Using some innovative authoring tools and a number of the first MHP receivers to hit the market, their work led to the creation of some 70 enhanced shows and interactive applications. Although it has gone relatively unnoticed in the MHP world, this real-world trial helped tremendously in the understanding of the issues involved in launching MHP both in Denmark and elsewhere due to close collaboration on debugging, testing, and implementation between the receiver manufacturer (Advanced Digital Broadcast in this case) and the broadcaster. In many ways, this was the prelude to the launch of MHP in the Finnish market.

One significant aspect of this trial was the testing of an over-the-air (OTA) download mechanism facility for upgrading the middleware in receivers that have already been deployed. This is an important component in almost any DTV deployment, in that logistics make it impossible to return every receiver to a central location in the event any bugs are found. In Finland, this feature was especially important. Deployment in Finland required going live with receivers that did not conform fully to the MHP specification, because Finnish broadcasters were in a hurry to launch services. This required parallel work on a fully MHP-compliant software stack, which would be downloaded OTA into the receivers that had already been deployed.

One of the STB manufacturers putting receivers in the market began discussions with Digita in Finland (who have 100% control of the terrestrial transmission network) regarding arrangements for an engineering channel for testing and for downloading new versions of middleware to receivers that had been deployed. Before negotiations regarding this crucial engineering requirement and methodology were complete, Digita declared that for them at least the downloading of software to MHP receivers came under the same contractual and financial terms and conditions as that of the other mainstream broadcasters in the Finnish market (such as MTV3 and YLE). Furthermore, there were no quality-of-service guarantees for downloading new middleware and no centralized STB farm for testing new products, applications, services, or software upgrades. More importantly from a commercial stand-point, they took no responsibility in the event of failure of their network to actually supply anything. This was not very encouraging.

This is just about the worst-case scenario for deploying a new digital system, no matter what middleware is used. Each network will make different choices about modulation, transport stream organization, and applications, and receiver manufacturers and broadcasters must collaborate to ensure that the best choices are made for that particular network. When we deploy interactive services to several different brands of receivers this type of testing and the ability to upgrade the receiver middleware become even more important. Despite the best efforts of all parties involved, there will always be misunderstandings, misinterpretations, and bugs in the software used at each stage. The only way to resolve these is collaboration, testing, and patience.

Although these problems affect everyone involved, they are generally worse for receiver manufacturers. Even in a field trial, receivers will be widely deployed, and thus fixing a bug in the middleware means getting new software to every receiver that is in someone's home. Returning those receivers for an upgrade is simply not feasible, and thus they have to be upgradeable OTA. Although OpenCable defines a way of doing this (the Common Download Specification), DVB does not. Thus, each receiver manufacturer and possibly even each network will have its own approach to this, and both the network and the receiver manufacturer must test this carefully before any boxes are deployed. Without a working upgrade mechanism, field trials become very difficult work indeed.

Despite these less-than-perfect conditions for deployment, it appears that MHP and digital broadcasting in Finland is still forging ahead, and luckily for everyone the deployment seems to have hit few major problems. For other countries, these particular issues have become less serious as more receiver manufacturers become fully MHP compliant and as more companies gain experience in deploying MHP.

By way of comparison, Italy has handled deployment in a far more discerning manner. Digital license holders in Italy have created the DGTi (the equivalent of the United Kingdom's Digital Terrestrial Group), which has arranged for a 64-Kb/s engineering channel. This is available to those companies that sign up to the Code of Conduct arrangements for compliance and interoperability of MHP receivers in Italy. In return for the use of this channel, manufacturers are expected to place samples of the hardware and software for pretesting at the DTGi control center to make sure that any new receivers or applications are good MHP citizens.

Both of these elements are important for every DTV system, but for networks where several broadcasters, content providers, and receiver manufacturers are all deploying products, this type of testing is vital. No matter how mature MHP or OCAP becomes, there will always be surprises. It is the job of all companies involved, and any regulatory authorities, to make sure that these surprises are discovered before the service goes live.

Convergence with the Internet: Fact or Fiction?

As digital entertainment media becomes ever more popular, both PC and consumer electronics companies are examining how they can profit from this move, both in terms of connecting existing devices digitally and developing new devices that exploit the nature of digital content. We are already seeing devices such as TiVo replace the VCR, and digital media adapters are beginning to let consumers watch PC content on their TV systems. Sony and Microsoft both have big plans for this particular segment of the market, with future versions of PlayStation and the XBox aiming to bring new levels of convergence.

PC component companies such as Intel and Dell have also been pushing convergence, with Intel announcing the Entertainment PC, which can watch TV, act as a personal video recorder, and play back other media clips. Sony has also launched the VAIO PCV-W500 media center PC in Japan, with styling and functionality that aims to bridge the gap between the office and the living room (and with PC, TV, DVD player, and CD player functionality). Microsoft is also aiming for this market with the introduction of Windows XP Media Center Edition, and many companies are now building media centers that can sit in the living room and be controlled with a remote control unit rather than a keyboard and mouse. This type of product appears to be increasingly popular in the United States and in Asia, and these types of products bring new competition to the traditional DTV receiver. For consumers with only a small amount of space for their entertainment center, this is an attractive solution (although they are still only a very small segment of the overall market).

Until now we have not seen ITV services on personal computers, but companies such as Osmosys and Zentek have developed software that enable users to watch DTV services with MHP applications on a Windows PC. Other projects are working on open-source solutions for Linux machines, although these are not currently as mature as their Windows counterparts. Some commercial receivers also use Linux, although there is no shrink-wrap MHP middleware for Linux at the moment. Consumers wanting to try MHP on their Linux systems must make do with one of the open-source packages.

Bringing DTV toward the PC has other challenges for broadcasters and middleware developers. We can be sure that there will be a limited number of set-top boxes and other receivers on the market, and we know that each of these will typically only have one or two configurations (or at least, only one or two configurations that affect us). Once the PC enters the scene, things start to change. No two PCs are the same, and thus manufacturers of MHP or OCAP solutions for PCs need to be even more careful about interoperability and conformance testing. We all know about compatibility problems in the PC world, and we cannot let those become problems for MHP or OCAP. In some ways, PCs have the advantage of being a more powerful and flexible problem, but we still have to ensure that these advantages do not outweigh any compatibility problems.

It is not just PC manufacturers who are getting into this market. In other moves toward convergence, demonstrations of the DVB-H (handheld) specification have taken place using a DVB-H enabled Nokia 7700 media device. DVB-H enables telecom providers to develop new and more advanced services that can be downloaded to mobile phones and other devices. Many PDAs are more powerful in terms of memory and CPU than the majority of DTV receivers, and the growth of 3G markets brings another potential application area for MHP. Although MHP in its present form is probably too big, many new handsets already run Java and a subset of the JavaTV or MHP APIs, such as the On ramp to OCAP (defined in Sun JSR 242), may prove an attractive proposition.

Many telecom providers are eager to deploy VOD and interactive services over broadband connections to both mobile devices and more traditional set-top boxes, and many of these companies are considering MHP as a middleware technology. As these IP-based video services take off, interactivity will become a differentiating feature (as it has in other DTV markets), and the need for a common middleware platform will become apparent in this market as well. Although MHP does not currently support pure IP-based services, there are no great barriers to this, and the DVB-IP specification should solve many of the outstanding issues.

Appendix A DVB Service Information

Service information (SI) is an important part of any DTV network. It carries a wide range of information that is useful to the receiver and to the viewer. This appendix provides an overview of SI in DVB networks, as well as information on the common elements of SI that are also found in other types of networks. The appendix also includes information on the format of SI tables, how the tables are related, and any restrictions these tables impose on receivers.

Transport streams contain two different types of SI. The first of these is defined by MPEG, and the receiver uses this to find what streams it should decode. This is called program-specific information (PSI). The second type is defined by another standards body, which is usually either DVB or ATSC. This provides more information about other transport streams in the network, and adds some user-oriented information about the content of the transport stream. DVB calls this DVB Service Information (DVB SI), whereas ATSC calls it the Program and System Information Protocol (PSIP). Both standards are based on the low-level formats specified by MPEG, but the information they carry is different.

The Organization of SI

Logically, we can think of PSI, DVB SI, and PSIP as relational databases. Each consists of a set of tables containing information about the transport stream and the programs within it. Each table contains one type of information about the transport stream, about the services and elementary streams it contains, or about its relationship to other transport streams that make up the network. Because each of these tables uses the basic structure defined in the

MPEG-2 specification, there are very few fundamental differences between the different SI standards. The differences are more in the structure of the tables, rather than the format of the tables themselves.

Descriptors

Much of the data contained in an SI table is transmitted as a set of descriptors. A descriptor is a basic unit of data that corresponds to part of a row in the table. There are many different types of descriptors, each providing a different piece of information that fits into the SI database. Using descriptors to transmit the tables has the following advantages when compared with other formats.

- *Space:* Some parts of the tables will be optional, and thus the broadcaster only has to send those descriptors that contain information that is appropriate for that situation.
- *Ease of parsing:* The use of descriptors makes the content much easier to parse than other approaches to making some elements optional. In this case, the receiver simply has to parse a set of descriptors, where each descriptor has a similar format.
- *Flexibility:* Network operators may send descriptors in any order without affecting their semantics.
- *Reusability:* If we find that we need to send the same type of information in more than one place or more than one table, we can reuse an existing descriptor if it is appropriate. This makes things easier for receiver manufacturers who have to write parsers, for head-end manufacturers who have to generate the SI, and for standards developers who can maintain consistency more easily.
- *Extensibility:* Need to add some data for a new situation, or to change the content of a certain part of the data? Just add another descriptor to handle it. Receivers that do not support it will simply ignore the new descriptor, whereas receivers that do can read it and understand it.

Descriptors are usually transmitted in sets known as loops. Typically, a table will contain one common loop, containing descriptors that apply to all of the elements in the table, and then another loop containing the descriptors that make up the rows of the table. For this loop, every iteration will contain the descriptors for a single row, and thus there may be a nested set of loops depending on which descriptors make up a row. Figure A.1 shows how this structure works for a typical SI table in any system.

Each descriptor is identified by a descriptor tag so that the receiver can parse it correctly. The only other field common to all descriptors gives the length of the descriptor so that a receiver can skip any descriptors that have an unrecognized descriptor tag. The standards body that defines a particular descriptor will define any other fields it contains, and receivers that do not recognize a given descriptor will just ignore it.

The more common descriptors that can appear in the various SI tables are examined further later in the appendix. Although the DVB SI specification defines the intended location of many descriptors, it does not restrict a particular descriptor to these locations, and in many cases does not specify which descriptors are mandatory. The DVB implementation guide-

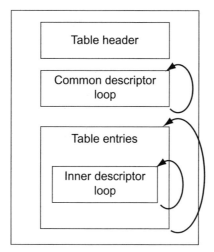

Figure A.1. Descriptor loops in an SI table.

lines for SI (ETSI document number TR 101 211) has more information about this, and so this is an important source of information for anyone who is interested in DVB SI. We recommend that you check both this document and the DVB SI specification (ETSI document number EN 300 468) to make sure that you have the latest information, and to get a detailed description of the restrictions on DVB-SI.

Transmitting an SI Table

For transmission, each SI table is broken down into a number of packets called sections. Most packets containing SI data use the MPEG-2 private section format. This is discussed in Chapter 10, and part 1 of the MPEG-2 specification contains some more details. Once we have split the table into sections, these are then transmitted as part of an MPEG elementary stream, and the entire table is broadcast at regular intervals to make sure the receiver will receive an entire copy of the table.

To make sure the receiver knows where to find these sections, some tables are broadcast on well-known PIDs. These tables will differ between standards, but they will tell the receiver where to find the other tables that it needs. The receiver will monitor these well-known PIDs, retrieve the tables from them, and then examine these to see which other PIDs it needs to monitor in order to load all of the SI being broadcast.

The receiver must continue monitoring these well-known PIDs, however, because the network operator may update the content of a table at any time. Each table contains a version number, which will change when the network operator updates that table. When a receiver detects that the version number of a table has changed, it will flush any cached data from the original table and replace it with the new copy.

Program-specific Information

This section looks at table content. Because PSI is common to both DVB and ATSC systems, we will look at that first. A transport stream can carry more than one program (that is, more than one group of associated audio, video, and data streams). In DTV terms, a program is also known as a service or a channel. To find out what programs are contained in a given transport stream, the receiver looks at the Program Association Table (PAT), which has the format outlined in Table A.1 (and a table ID value of 0x00). This lists all of the programs in the transport stream and tells the receiver where to find detailed information about each of those programs. The PAT is a table of contents, but nothing more.

Each entry in the PAT gives the PID containing a Program Map Table (PMT) for that program. The PMT is a slightly more detailed description of the program that tells the receiver what PIDs are part of that program and the type of data they contain. It provides no information that is useful for the user. The receiver needs this to work out which streams it should decode for a given program, but there is nothing user-oriented in this table. The format of the PMT is outlined in Table A.2, and each PMT has a table ID value of 0x02.

Table A.1. Format of the Program Association Table (PAT).

Syntax	No. of Bits	Identifier
program_association_section(){		
table_id	8	uimsbf
section_syntax_indicator	1	bslbf
'0'	1	bslbf
reserved	2	bslbf
section_length	12	uimsbf
transport_stream_id	16	uimsbf
reserved	2	bslbf
version_number	5	uimsbf
current_next_indicator	1	bslbf
section_number	8	uimsbf
last_section_number	8	uimsbf
for(i=0; i<N; i++){		
program_number	16	uimsbf
reserved	3	bslbf
if(program_number=='0') {		
network_PID	13	uimsbf
}		
else{		
program_map_PID	13	uimsbf
}		
}		
CRC_32	32	rpchof
}		

Source: ISO 13818-1:2000 (MPEG-2 systems specification).

Table A.2. Format of the Program Map Table (PMT).

Syntax	No. of Bits	Identifier
TS_program_map_section(){		
table_id	8	uimsbf
section_syntax_indicator	1	bslbf
'0'	1	bslbf
reserved	2	bslbf
section_length	12	uimsbf
program_number	16	uimsbf
reserved	2	bslbf
version_number	5	uimsbf
current_next_indicator	1	bslbf
section_number	8	uimsbf
last_section_number	8	uimsbf
reserved	3	bslbf
PCR_PID	13	uimsbf
reserved	4	bslbf
program_info_length	12	uimsbf
for(i=0; i<N; i++){		
descriptor()		
}		
for(i=0; i<N1; i++){		
stream_type	8	uimsbf
reserved	3	bslbf
elementary_PID	13	uimsnf
reserved	4	bslbf
ES_info_length	12	uimsbf
for(i=0; i<N2; i++){		
descriptor()		
}		
}		
CRC_32	32	rpchof
}		

Source: ISO 13818-1:2000 (MPEG-2 systems specification).

The PID belonging to an elementary stream can appear in more than one PMT, and in this case it will be shared between all of the programs that refer to that stream. No matter how many programs refer to it, we only need one instance of that elementary stream in the transport stream.

Conditional Access Information

In any transport stream, one or more elementary streams may be encrypted so that they are available only to a network's subscribers. There are many different conditional access systems in use, and thus the receiver needs some way of knowing which CA system is used

Table A.3. Format of the Conditional Access Table (CAT).

Syntax	No. of Bits	Identifier
A_section(){		
table_id	8	uimsbf
section_syntax_indicator	1	bslbf
'0'	1	bslbf
reserved	2	bslbf
section_length	12	uimsbf
reserved	18	bslbf
version_number	5	uimsbf
current_next_indicator	1	bslbf
section_number	8	uimsbf
last_section_number	8	uimsbf
for(i=0; i<N; i++){		
descriptor()		
}		
CRC_32	32	rpchof
}		

Source: ISO 13818-1:2000 (MPEG-2 systems specification).

for a given stream. This information is contained in the Conditional Access Table (CAT), outlined in Table A.3. This table has a table ID value of 0x01.

The descriptor loop in a CAT will contain zero or more conditional access descriptors. The conditional access descriptor is defined as part of the MPEG-2 specification, and each conditional access descriptor tells the receiver about one CA system used in that transport stream and identifies which PID is used for carrying CA messages relating to that CA system. CA descriptors in the CAT identify the PIDs containing messages that apply to the entire system, such as entitlement management messages, whereas CA descriptors in a PMT entry identify messages such as entitlement control messages that apply only to that program.

DVB SI

The following examines DVB SI (see Appendix B for information on ATSC PSIP). Readers who need a complete description should read the appropriate standards documents. Table A.4 lists the standards that contain some information about SI in a DVB network.

Finding Information About the Network

PSI tables are fine for describing the content of a single transport stream, but networks usually consist of several transport streams. This means that a receiver needs some way of knowing which transport streams are part of the same network and how it can get access to them. In DVB systems, this information is part of the Network Information Table (NIT).

Table A.4. SI standards for DVB systems.

Standard	Description
TR 101 154	Implementation guidelines for MPEG in broadcasting applications.
EN 300 468	DVB Service Information (DVB-SI) specification.
TR 101 211	Implementation guidelines for DVB-SI.
EN 301 192	DVB Data Broadcasting specification. Two versions of this specification are of interest to MHP developers. Version 1.2.1 applies to version 1.0.2 and earlier of MHP, as well as MHP 1.1 and GEM. Version 1.3.1 applies to MHP version 1.0.3 and MHP 1.1.1.
TR 101 202	Implementation guidelines for DVB data broadcasting.
TR 101 162 (also known as ETR 162)	Allocation of SI codes for DVB systems.

The NIT gives a list of the transport streams in the network and the parameters the receiver needs in order to tune to them. In some cases, it may be split into two separate subtables: the NIT for the current network (known as the NIT-actual) and the NIT for any other networks that a receiver may be able to receive (known as the NIT-other). Every network must carry an NIT-actual table, but the NIT-other is optional and is only needed in some circumstances.

Because the NIT is so important to the receiver, it is always carried on the same PID (0x10) within a transport stream. The NIT-actual and NIT-other are both carried on the same PID, but they have different table IDs so that the receiver can tell them apart. The table ID value for sections containing the NIT-actual will be 0x40, and sections containing the NIT-other will use a table ID value of 0x41. Every section containing the NIT has the structure outlined in Table A.5.

Each network has a unique network ID that identifies it to the receiver. Within each network, every transport stream also has a unique transport stream ID, although one network may allow other networks to retransmit one or more of its transport streams. If this happens, we may end up with the case in which two networks are transmitting the same transport stream with the same transport stream ID but different network IDs. To help a DTV receiver identify when this happens, the NIT also carries the network ID of the network that originally produced the content. This is known as the original network ID. Together, the original network ID and the transport stream ID will uniquely identify a given transport stream no matter which network is currently transmitting it, and the NIT entry for a transport stream will include both of these values.

The first descriptor loop of the NIT must include a network name descriptor, but any other descriptors are optional under most conditions. The inner descriptor loop, which describes individual transport streams, must include a descriptor that identifies the delivery system (either a satellite delivery system descriptor, a cable delivery system descriptor, or a

Table A.5. Format of the Network Information Table (NIT).

Syntax	No. of Bits	Identifier
network_information_section(){		
table_id	8	uimsbf
section_syntax_indicator	1	bslbf
reserved_future_use	1	bslbf
reserved	2	bslbf
section_length	12	uimsbf
network_id	16	uimsbf
reserved	2	bslbf
version_number	5	uimsbf
current_next_indicator	1	bslbf
section_number	8	uimsbf
last_section_number	8	uimsbf
reserved_future_use	4	bslbf
network_descriptors_length	12	uimsbf
for(i=0; i<N; i++){		
descriptor()		
}		
reserved_future_use	4	bslbf
transport_stream_loop_length	12	uimsbf
for(i=0; i<N; i++){		
transport_stream_id	16	uimsbf
original_network_id	16	uimsbf
reserved_future_use	4	bslbf
transport_descriptors_length	12	uimsbf
for(j=0; j<N; j++){		
descriptor()		
}		
}		
CRC_32	32	rpchof
}		

Source: ETSI EN 300 468:2000 (DVB SI specification).

terrestrial delivery system descriptor). Again, several other descriptors can be included, but their presence is not mandatory.

Bouquets

Transport streams provide the physical organization of streams in a network. Each service will belong to a single transport stream, and each transport stream is broadcast on a different frequency with a specific set of modulation parameters.

Sometimes, though, network operators want to organize their services in a different way. This may be to group all of the same types of services (e.g., one group for sports channels,

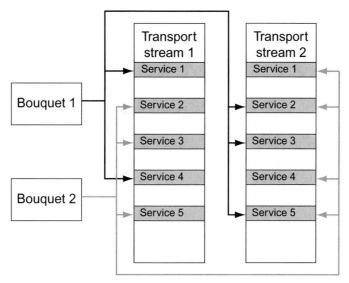

Figure A.2. Bouquets are logical groups of services that may come from different transport streams.

another for news channels, and a third for movie channels), or it may be to group services in subscription packages. For instance, all of the services in the basic package may be in one group, whereas another group may contain the services that make up the movies package, while yet another set of services are grouped together for the children's package. To complicate things further, the network operator may want some channels to appear in more than one group.

We need a way of grouping services beyond the concept of transport streams. DVB uses the concept of bouquets to group services such as this. A bouquet (see Figure A.2) is just a set of services that are logically grouped, no matter what transport stream they are part of. Bouquets can even group services across different networks.

Bouquets are described in an optional table called the Bouquet Association Table (BAT). Network operators can use this to identify which bouquets are present in a transport stream, and which services from that transport stream are part of those bouquets. Table A.6 outlines the structure of the BAT.

BAT sections have the table ID value 0x4A. The BAT is split into several subtables, each describing one bouquet. Each entry in a subtable refers to a single transport stream, and contains a set of descriptors that must include a service list descriptor that identifies all of the services from that transport stream that are included in the bouquet. A service can belong to more than one bouquet, and thus the same service may appear in several different BAT subtables.

Each bouquet has a name, which is defined by a bouquet name descriptor or a multilingual bouquet name descriptor carried in the outer descriptor loop. The outer descriptor loop may

Table A.6. Format of the Bouquet Association Table (BAT).

Syntax	No. of Bits	Identifier
`bouquet_association_section(){`		
`  table_id`	8	uimsbf
`  section_syntax_indicator`	1	bslbf
`  reserved_future_use`	1	bslbf
`  Reserved`	2	bslbf
`  section_length`	12	uimsbf
`  bouquet_id`	16	uimsbf
`  reserved`	2	bslbf
`  version_number`	5	uimsbf
`  current_next_indicator`	1	bslbf
`  section_number`	8	uimsbf
`  last_section_number`	8	uimsbf
`  reserved_future_use`	4	bslbf
`  bouquet_descriptors_length`	12	uimsbf
`  for(i=0; i<N; i++){`		
`    descriptor()`		
`  }`		
`  reserved_future_use`	4	bslbf
`  transport_stream_loop_length`	12	uimsbf
`  for(i=0; i<N; i++){`		
`    transport_stream_id`	16	uimsbf
`    original_network_id`	16	uimsbf
`    reserved_future_use`	4	bslbf
`    transport_descriptors_length`	12	uimsbf
`    for(j=0; j<N; j++){`		
`      descriptor()`		
`    }`		
`  }`		
`  CRC_32`	32	rpchof
`}`		

Source: ETSI EN 300 468:2000 (DVB SI specification).

also contain up to two country-availability descriptors for each BAT subtable. This lets the network operator identify specific countries that may receive the bouquet and specific countries that may not receive it. The second descriptor loop must include a service list descriptor. A DVB receiver will ignore any other descriptors in this loop.

Describing Services in DVB

Using the NIT and BAT, our receiver knows about the way transport streams are organized in a network, but that is not much use to a viewer who cares more about what channels he

Table A.7. Format of the Service Description Table (SDT).

Syntax	No. of Bits	Identifier
service_description_section() {		
table_id	8	uimsbf
section_syntax_indicator	1	bslbf
reserved_future_use	1	bslbf
reserved	2	bslbf
section_length	12	uimsbf
transport_stream_id	16	uimsbf
reserved	2	bslbf
version_number	5	uimsbf
current_next_indicator	1	bslbf
section_number	8	uimsbf
last_section_number	8	uimsbf
original_network_id	16	uimsbf
reserved_future_use	8	bslbf
for(i=0; i<N; i++){		
service_id	16	uimsbf
reserved_future_use	6	bslbf
EIT_schedule_flag	1	bslbf
EIT_present_following_flag	1	bslbf
running_status	3	uimsbf
free_CA_mode	1	bslbf
descriptors_loop_length	12	uimsbf
for (j=0; j<N; j++){		
descriptor()		
}		
}		
CRC_32	32	rpchof
}		

Source: ETSI EN 300 468:2000 (DVB SI specification).

or she can watch. The PAT and PMT tell the receiver what is inside a transport stream, but that is not very helpful to the viewer either, in that they focus more on what the receiver needs to know than on what the viewer wants to know.

To provide more information to the user, DVB systems use the Service Description Table (SDT). This contains more user-oriented information about the stream and the programs within it, including the names of the various services, parental rating information, and a textual description of the service. The structure of the SDT is outlined in Table A.7.

The SDT may be split into two subtables in order to make it easier for the receiver and the user to find out what services are available. The first of these subtables describes the services that are available on the current transport stream. This subtable is known as the SDT-actual and has a table ID of 0x42. The second subtable is called the SDT-other and describes what

Table A.8. Possible values for the running status of an event.

Value	Meaning
0	Undefined
1	Not running
2	Starts in a few seconds (e.g., for video recording)
3	Pausing
4	Running
5 to 7	Reserved for future use

Source: ETSI EN 300 468:2000 (DVB SI specification).

services are available on other transport streams in the network. In this case, the table ID is 0x46.

The SDT-other is an optional table, and thus not every transport stream will contain it. Network operators can use this table to let the receiver know about all of the services in the network without forcing the receiver to tune to every transport stream and read its SDT-actual. Although receivers only need to do this at startup to build their list of available services, it can be a slow process. Using an STD-other means that the receiver does not have to cache this information between reboots or search for it every time.

Each SDT entry has a pair of flags (the EIT_schedule_flag and the EIT_present_following_flag) that tell the receiver what event information is present so that it may avoid searching for nonexistent tables. Services also have a running status that indicates whether a given service is currently being broadcast. This can take one of the values outlined in Table A.8.

Each SDT entry contains a descriptor loop that must include a service descriptor. This gives some basic information about the service, such as its name and the type of service (for instance, a DTV service, a digital radio service, or a data-only service). An electronic program guide can use this to give the viewer more information about the services available.

Another important descriptor that may be present is the data broadcast descriptor. This is more useful to the middleware in the receiver, and it identifies data streams in the service. It tells the receiver what types of data streams are included, and lists any parameters it needs to access that stream. It may also include a textual description of the stream so that the receiver can give a little more information about it to the user. Depending on the type of data-broadcasting service, this descriptor may be included in the EIT instead of in the SDT. If a service contains more than one data stream, the SDT entry for that service will contain one data broadcast descriptor for each data stream. To make it easy for receivers to find the SDT, every transport stream will carry the SDT (both SDT-actual and SDT-other) on PID 0x11.

Describing Events

The SDT gives the user a way to find the channels they want, but it does not tell them anything about what shows are going to be on those channels. The Event Information Table (EIT) describes the shows that will be shown on a particular service.

The EIT can be split into several subtables. Every transport stream must carry information about the current and next event for every service in that transport stream. This is carried in the EIT present/following table. Optionally, we can also add information about events in the future, so that an EPG can display a program schedule to the user. This is carried in the EIT schedule table.

As with the SDT, we can also choose to carry information about events on other transport streams. Both EIT present/following and EIT schedule information from other transport streams can be included as separate subtables so that an EPG does not have to tune to a new transport stream to get all schedule information. Complete schedule information for all services on the network can take a lot of space, and thus broadcasters will often choose to broadcast only present/following information for other transport streams.

Because of the amount of space schedule information can consume, DVB has defined how it should be carried within the EIT. Each block of eight sections in an EIT schedule table is called a segment. Every segment represents a three-hour period in the schedule, and that segment will contain entries for any events that start within that period. These entries will be ordered chronologically within the segment, and segments will be ordered chronologically within the table. To make it easier for the receiver to find the entries for a specific time period, the first segment in the table will contain schedule information starting at midnight on the current day. Receivers can use this to calculate which segment they need to load.

Every table may contain up to 256 sections, which corresponds to four days of schedule information. Because this may not be enough for some applications, EIT schedule information may be split across several subtables. These subtables will use table IDs 0x50 to 0x5F for the actual transport stream, and 0x60 to 0x6F for other transport streams, and these table IDs will also be used in chronological order. Thus, table ID 0x50 will be used for the first four days of schedule information for the current transport stream, table is 0x51 for the next four, and so on.

Broadcasting complete schedule information would use a lot of bandwidth, and thus broadcasters do not usually broadcast all of the subtables and will often choose to include only present/following information for other transport streams. They may also choose not to broadcast schedule information for past segments. In this case, the segments must still be broadcast, but they may contain no entries. All sections carrying EIT information follow the format outlined in Table A.9.

Each EIT section describes events for only one service. Different sections may hold event information for different services within the same transport stream. Every event has an event ID that uniquely identifies it within the service (although event IDs can be reused within the same transport stream). As well as giving the start time and duration of any events, the EIT

Table A.9. Format of the Event Information Table (EIT).

Syntax	No. of Bits	Identifier
event_information_section(){		
table_id	8	uimsbf
section_syntax_indicator	1	bslbf
reserved_future_use	1	bslbf
reserved	2	bslbf
section_length	12	uimsbf
service_id	16	uimsbf
reserved	2	bslbf
version_number	5	uimsbf
current_next_indicator	1	bslbf
section_number	8	uimsbf
last_section_number	8	uimsbf
transport_stream_id	16	uimsbf
original_network_id	16	uimsbf
segment_last_section_number	8	uimsbf
last_table_id	8	uimsbf
for(i=0;i<N;i++){		
event_id	16	uimsbf
start_time	40	bslbf
duration	24	uimsbf
running_status	3	uimsbf
free_CA_mode	1	bslbf
descriptors_loop_length	12	uimsbf
for(i=0;i<N;i++){		
descriptor()		
}		
}		
CRC_32	32	rpchof
}		

Source: ETSI EN 300 468:2000 (DVB SI specification).

tells the receiver whether or not any parts of the event are scrambled (using the free_CA_mode flag).

The start time for every event is represented as a time/date pair. This field is coded using two different representations, and thus it may be easier to think of it as two separate fields that have been compressed into one. The most significant 16 bits of the field contain the date (using the Modified Julian Date representation, given by the number of days since November 17, 1858). The least significant 24 bits of the field represent the time of day the event starts. This is coded using 4-bit BCD, so that each byte represents the hours, minutes, or seconds of the starting time (on a 24-hour clock). Thus, a starting time of 10.45 p.m. would be stored in the least significant 24 bits as 0x224500. The duration is also coded with 4-bit BCD. In this case, we do not need any date information because events rarely last longer than one complete day.

Events do not always start when they should, and thus each EIT entry includes the `running_status` field. This tells the receiver about the status of events in the EIT present/following table, and can be used to cover situations in which the nominal start time of an event has passed but the event has not actually started yet (or has started and been interrupted). This field can take one of the values shown in Table A.8. For entries in the EIT-schedule table, the running status will always be undefined.

Although most of the values in Table A.8 are clear, the pausing state needs a little explanation. This state means that an event has been started, but is not currently running. This could happen when one show interrupts another, such as a news flash about a breaking news story that interrupts normal programming. Generally, this implies that the event will resume and that any pause is only temporary.

Each event description must include one or more component descriptors in its descriptor loop. These describe the audio or video streams that are part of that event. Data streams will use the data broadcast descriptor instead, and in this case one data broadcast descriptor should be present for every data stream in the event. If every event in that service uses the same data streams, the SDT can carry a data broadcast descriptor for those streams instead of the EIT.

As well as the component descriptors, every event description will include a short event descriptor. This contains the name of the event and possibly a short textual description. In case this is not enough, a network operator can also choose to include an extended event descriptor, which as its name suggests will contain a longer description of the event.

If many events have their running status updated at the same time, the network operator may choose to broadcast this information separately in a Running Status Table (RST). This lets the network operator update the running status of one or more events without having to rebroadcast the entire EIT containing that event. The running status defined in the RST overrides that which was broadcast in the EIT for that event, but the EIT must be updated with the correct running status before it is retransmitted. Sections containing the RST have the table ID 0x71, and they will follow the format outlined in Table A.10.

Telling the Time

Knowing when an event starts is only useful if the time at the receiver is synchronized to the time the network operator is using. To make sure this happens, network operators broadcast the Time and Date Table (TDT) that tells the receiver the current time. Because the TDT is so small, it is carried in one MPEG-2 section that has the structure outlined in Table A.11. The table ID for the TDT is 0x70.

The time is the current UTC time, coded in the same format used for start times in the EIT tables. Of course, not every country that uses DVB systems uses UTC time, and thus DVB defines a second table that indicates what offset the receiver should apply to a UTC time

Table A.10. Format of the Running Status Table (RST).

Syntax	No. of Bits	Identifier
running_status_section(){		
table_id	8	uimsbf
section_syntax_indicator	1	bslbf
reserved_future_use	1	bslbf
reserved	2	bslbf
section_length	12	uimsbf
for (i=0; i<N; i++){		
transport_stream_id	16	uimsbf
original_network_id	16	uimsbf
service_id	16	uimsbf
event_id	16	uimsbf
reserved_future_use	5	bslbf
running_status	3	uimsbf
}		
}		

Source: ETSI EN 300 468:2000 (DVB SI specification).

Table A.11. Format of the Time and Date Table (TDT).

Syntax	No. of Bits	Identifier
time_date_section(){		
table_id	8	uimsbf
section_syntax_indicator	1	bslbf
reserved_future_use	1	bslbf
reserved	2	bslbf
section_length	12	uimsbf
UTC_time	40	bslbf
}		

Source: ETSI EN 300 468:2000 (DVB SI specification).

value in order to get the correct local time. This is called the Time Offset Table (TOT). Like the TDT, this is contained in a single MPEG-2 section as outlined in Table A.12, and it has the table ID 0x73. Both the TDT and TOT are optional.

As with the TDT, this table includes a UTC time value. However, it can also include a local time offset descriptor in the descriptor loop. This descriptor defines the offsets that apply in various countries between UTC and the local time, and has the format outlined in Table A.13.

Table A.12. Format of the Time Offset Table (TOT).

Syntax	No. of Bits	Identifier
time_offset_section(){		
table_id	8	uimsbf
section_syntax_indicator	1	bslbf
reserved_future_use	1	bslbf
reserved	2	bslbf
section_length	12	uimsbf
UTC_time	40	bslbf
reserved	4	bslbf
descriptors_loop_length	12	uimsbf
for(i=0; i<N; i++){		
descriptor()		
}		
CRC_32	32	rpchof
}		

Source: ETSI EN 300 468:2000 (DVB SI specification).

Table A.13. Format of the local time offset descriptor.

Syntax	No. of Bits	Identifier
local_time_offset_descriptor(){		
descriptor_tag	8	uimsbf
descriptor_length	8	uimsbf
for(i=0; i<N; i++){		
country_code	24	bslbf
country_region_id	6	bslbf
reserved	1	bslbf
local_time_offset_polarity	1	bslbf
local_time_offset	16	bslbf
time_of_change	40	bslbf
next_time_offset	16	bslbf
}		
}		

Source: ETSI EN 300 468:2000 (DVB SI specification).

The local time offset descriptor can describe a time zone offset for several countries, where each country is identified by its ISO standard three-letter code (e.g., FRA for France). Each country also has a region ID that is used to distinguish different time zones in the same country. For instance, all entries for the United States would have the country code USA, but different region IDs for EST, CST, and other time zones.

For countries with only one time zone, the region ID will always be zero, but for countries with more than one time zone the region ID starts at 0x01 for the easternmost time zone and increases by one for every time zone west of that. Taking the United States as our example again, this means that the Atlantic time zone will have a region ID of 0x01, the Eastern time zone will have a region ID of 0x02, the Central time zone will have a region ID of 0x03, and so on.

The local offset is defined by two fields: a polarity, which indicates whether the offset is ahead of UTC or behind UTC, and the offset itself (which indicates the time difference). The offset is stored as four 4-bit BCD-coded digits in the format HHMM, so that an offset of +8 hours would be coded as 0x0800. Offsets will always be in the range −12 to +13 hours.

This offset may change over time, thanks to daylight savings time beginning or ending. To handle this, the local time offset descriptor also tells the receiver when daylight savings time starts or ends, and what the new time offset will be. All times (as opposed to time offsets) in the local time offset descriptor are stored using the same format that is used for the start time of events in the EIT.

Putting It All Together

The previous represent most of the tables that make up DVB-SI. There are a few not included here, but these are typically less important for understanding the concepts of DVB-SI. Figure A.3 shows how the various SI tables relate to one another.

Every table is transmitted repeatedly in order to make sure that the receiver has received the whole of the table. Exactly how often these tables are repeated is up to the network operator, but DVB does define a minimum rate for each table.

Table A.14 outlines which PIDs are used to broadcast the various SI tables, and how often those tables should be repeated. Note that these are the minimum allowed repetition rates, and thus network operators can choose to repeat tables more often if it improves performance on their network. Although we list the minimum repetition rates for each table, there is more to broadcasting SI than that. DVB specifies that there should be a delay of less than 25 milliseconds between sending sections of the same table, and thus network operators may need to increase the repetition rate, depending on the size of SI tables.

Repetition rates for the EIT vary dramatically depending on the type of network involved and the event information the network operator is transmitting. Table A.15 outlines the different repetition rates that can apply.

As well as those PIDs listed in Table A.14, DVB reserves several other PIDs. Network operators should also avoid PIDs used by ATSC PSIP (see Appendix B). Doing this makes it easier to share content between ATSC and DVB networks, and although this may not be a consideration in many cases it is easier to avoid problems before they arise.

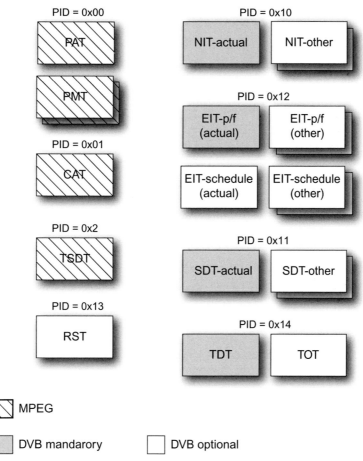

Figure A.3. The relationships among the various SI tables. *Source:* ETSI EN 300 468:2000 (DVB SI specification).

Table A.14. Minimum repetition rates for DVB SI tables.

Table	PID	Min. Repetition Rate
PAT	0x0000	100 milliseconds (recommended)
PMT	defined in PAT	100 milliseconds (recommended)
CAT	0x0001	Not specified
NIT	0x0010	10 seconds
BAT	0x0011	10 seconds
SDT	0x0011	2 seconds (SDT-actual)
		10 seconds (SDT-other)
EIT	0x0012	See Table A.15
Stuffing Table (ST)	0x0010	Not specified
	0x0011	
	0x0012	
	0x0013	
	0x0014	
RST	0x0013	Not specified
TOT	0x0014	30 seconds
TDT	0x0014	30 seconds
TSDT	0x0002	10 seconds (recommended)
Discontinuity Information Table (DIT)	0x001E	Not specified
Selection Information Table (SIT)	0x001F	Not specified

Table A.15. Repetition rates for EIT subtables.

EIT Subtable	Repetition Rate (sec)
Satellite or Cable Networks	
EIT present/following (actual)	2
EIT present/following (other)	10
EIT schedule (events in the next 8 days)	10
EIT schedule (other events)	30
Terrestrial Networks	
EIT present/following (actual)	2
EIT present/following (other)	20
EIT schedule (actual, events in the first full day)	10
EIT schedule (other, events in the first full day)	60
EIT schedule (actual, all events)	30
EIT schedule (other, all events)	300

Optimizing Bandwidth Usage: The Transport Stream Description Table

In some cases, we may have a set of descriptors that apply to every service within a transport stream. Although we could include these in the appropriate BAT or SDT, this would waste bandwidth, and thus we can include these descriptors in a separate table called the Transport Stream Description Table (TSDT). Table A.16 outlines the format of sections carrying the TSDT.

Not every descriptor can be carried in the TSDT, and thus some descriptors may have to be carried separately in every STD or BAT entry. Usually, though, receivers should check the TSDT first before parsing the other tables in order to make SI parsing as simple as possible.

Table A.16. Format of the Transport Stream Description Table (TSDT).

Syntax	No. of Bits	Identifier
`TS_description_section(){`		
`table_id`	8	uimsbf
`section_syntax_indicator`	1	bslbf
`"0"`	1	bslbf
`reserved`	2	bslbf
`section_length`	12	uimsbf
`reserved`	18	bslbf
`version_number`	5	uimsbf
`current_next_indicator`	1	bslbf
`section_number`	8	uimsbf
`last_section_number`	8	uimsbf
`for(i=0; i<N; i++) {`		
`descriptor()`		
`}`		
`CRC_32`	32	rpchof
`}`		

Source: ISO 13818-1:2000 (MPEG-2 systems specification).

Appendix B ATSC Service Information

OCAP networks will use the ATSC Program and Service Information Protocol (PSIP) to carry information about the services and events on the network. This appendix provides an introduction to PSIP, and examines the format of the various PSIP tables we may find in an OCAP network. We will also examine how the use of PSIP in an OCAP system differs from PSIP in other types of networks.

Like SI in DVB, PSIP is an extension of the PSI found in all MPEG transport streams. In North America, the way signals are broadcast is slightly different from the way they are broadcast in countries that use the DVB system, and thus ATSC made some different choices in the format of this data. The result of this is that the two systems are not compatible, although SI in ATSC networks contains much the same information as that found in the DVB network, and someone who understands DVB SI should be able to understand ATSC SI pretty easily. SI in ATSC systems is known as the PSIP.

PSIP is a complex specification, mainly because of the number of different documents that actually describe it. Although ATSC is the driving force behind PSIP, the Society of Cable Telecommunications Engineers (SCTE) has also been defining elements of it, and the Electronic Industries Alliance (EIA) and OpenCable also have something to say on the matter. Table B.1 outlines the documents that say something about how PSIP is used.

This appendix is not exhaustive regarding ATSC SI. We refer the reader to the extant voluminous literature. This appendix focuses on specific aspects of the SI that are important to OCAP and ACAP, including the major elements of SI you will find in a terrestrial, cable, or satellite system. Some systems may deviate from the ATSC standards. In particular, systems outside the United States and Canada may not follow these standards exactly.

Table B.1. Standards that discuss ATSC PSIP.

Standard Name/Number	Standards Body	Description
A/55	ATSC	Program guide for DTV
A/56	ATSC	SI for DTV
A/57	ATSC	Content identification and labeling for ATSC transport
A/65b (also known as DVS 097)	ATSC	PSIP standard for cable and terrestrial systems
A/68	ATSC	PSIP standard for Taiwan
A/69	ATSC	PSIP implementation guidelines for broadcasters
ANSI/SCTE 08 (previously known as DVS 11)	SCTE	Cable and satellite extensions to the ATSC System Information Standard
ANSI/SCTE 18 (previously known as DVS 208)	SCTE	Emergency alert message for cable TV systems
ANSI/SCTE 54 (previously known as DVS 241)	SCTE	Digital video service multiplex and transport system standard for cable TV
ANSI/SCTE 65	SCTE	SI delivered out-of-band for digital cable TV
EIA-766	EIA	U.S. and Canadian region rating tables and content advisory descriptors for PSIP

PSIP uses the same format as DVB SI, using tables and descriptors that have the same basic structure (see Appendix A). This appendix examines the changes made by ATSC and the other standards bodies to meet the needs of the North American markets.

One of the major changes that will be obvious to those readers familiar with DVB SI is that far fewer PIDs are used to carry the data. ATSC SI often uses fewer tables, with a single PID carrying several tables. This is important in those systems that use POD (point of deployment) CA modules. The POD system requires all streams to be transmitted to the POD so that it can manipulate the data, and only a limited number of streams can be sent to a given POD module concurrently. The result of this is that SI must be more efficient in the way it uses PIDs. To do this, tables are combined and some tables that are specific to a service in DVB SI become shared across the entire network in ATSC SI.

The result of this for an ATSC SI implementation is that the complexity of the internal processing may be a little higher. To be more efficient, an implementation may choose to use a different internal representation for the SI data than that which is broadcast. Event information is an excellent example of this. ATSC event information for satellite systems is broadcast in tables that may describe the entire network, but some implementations may find it more efficient to split this table internally into a number of smaller tables, each covering a single channel.

This appendix examines the format of many of the important tables in PSIP. Although the format of these is not likely to change drastically in later versions, we recommend that you check the latest versions of the PSIP standards on the ATSC web site at *www.atsc.org*. These will give you the latest information about the PSIP standards, and will cover any restrictions not covered in this appendix.

Describing Available Channels

ATSC systems use the basic MPEG PSI information to describe to the receiver how a transport stream is organized, and to tell it how many programs are actually in the transport stream and which programs contain which elementary streams. Although this is very useful to the receiver, it does not say anything about other transport streams that are out there and what they contain.

The viewer does not care about any of this, however. They are interested in what TV channels are available, and they should not have to know anything about transport streams and elementary streams or any other part of the broadcasting system.

The Virtual Channel Table

PSIP uses the Virtual Channel Table (VCT) to solve both of these problems. The VCT tells the receiver about all channels available on the network, and provides enough information for the receiver to tell the user something about those channels. In DVB systems, this is spread across three different tables: the NIT, SDT-actual, and SDT-other. Having all of this information in one place makes it far easier for the receiver to know how to access every channel on the network at the expense of a little extra bandwidth for SI.

Different transmission systems use slightly different versions of the VCT to carry information that is specific to the transmission system in use. Table B.2 outlines the format of the VCT for terrestrial systems.

For terrestrial and cable networks, only one VCT will be carried in each transport stream, on PID number 0x1FFB (this is known as the base PID). On the other hand, transport streams on satellite networks may contain several VCT instances carried on any PID (and one PID may carry more than one VCT instance). This is because a single satellite may carry signals from several different network operators. In this case, different instances of the VCT will have a different VCT ID in order to identify which VCT the receiver should use. The PSIP specification does not say how the receiver knows which VCT to use in this case. This can be tied to the CA system, or it can be hardcoded into the receiver.

Cable networks use a VCT format that is almost identical to that used for terrestrial networks. Apart from minor differences in the channel descriptions, the most notable difference is the table ID. Terrestrial VCTs use a table ID of 0xC8, whereas cable VCTs use a table ID of 0xC9.

Some cable systems (including OpenCable systems) can receive PSIP data via an out-of-band interface as well (e.g., the extended channel on the CableCARD interface). These systems will use a format slightly different from that of PSIP defined in ANSI/SCTE standard 65, and

Table B.2. The format of the VCT in terrestrial networks.

Syntax	No. of Bits	Identifier/Value
terrestrial_virtual_channel_table_section() {		
table_id	8	0xC8
section_syntax_indicator	1	'1'
private_indicator	1	'1'
reserved	2	'11'
section_length	12	uimsbf
transport_stream_id	16	uimsbf
reserved	2	'11'
version_number	5	uimsbf
current_next_indicator	1	bslbf
section_number	8	uimsbf
last_section_number	8	uimsbf
protocol_version	8	uimsbf
num_channels_in_section	8	uimsbf
for(i=0;i<num_channels_in_section;i++){		
short_name	7 × 16	uimsbf
reserved	4	'1111'
major_channel_number	10	uimsbf
minor_channel_number	10	uimsbf
modulation_mode	8	uimsbf
carrier_frequency	32	uimsbf
channel_TSID	16	uimsbf
program_number	16	uimsbf
ETM_location	2	uimsbf
access_controlled	1	bslbf
hidden	1	bslbf
reserved	2	'11'
hide_guide	1	bslbf
reserved	3	'111'
service_type	6	uimsbf
source_id	16	uimsbf
reserved	6	'111111'
descriptors_length	10	uimsbf
for(i=0;i<N;i++){		
descriptor()		
}		
}		
reserved	6	'111111'
additional_descriptors_length	10	uimsbf
for(j=0;j<N;j++){		
additional_descriptor()		
}		
CRC_32	32	rpchof
}		

Source: ATSC document number A/65b.

which includes two different formats for the VCT. The format discussed here is known as the long-form VCT. ANSI/SCTE 65 also defines a short-form VCT, which carries a little less information and restructures the VCT into three subtables. Readers interested in the short form should read ANSI/SCTE 65 for more information.

Satellite systems use a slightly different format of the VCT, which reflects the different parameters satellite receivers need before they can correctly tune to a transport stream. Table B.3 outlines the format of the VCT in a satellite network.

Most of Table B.3 is similar to that used for terrestrial and cable VCTs. Channel descriptions have a slightly different format (examined in material to follow), but all other changes are related to areas of the table noted "reserved in table B.2."

Describing Individual Channels

For all types of VCTs, each entry in the VCT describes a single channel. As well as telling the receiver how to access that channel (the frequency and modulation parameters of the transport stream, and the program number within that transport stream), it includes information such as the channel number and a reference to an extended text description.

Defining a channel number in the VCT lets a network operator assign a consistent set of channel numbers to all of their channels, no matter which transport stream or program number upon which they are actually transmitted. The network operator can then make changes to the way services are carried in the network, without the viewer actually noticing. A network operator can reorganize their transport streams, change the frequency of them, or even move services from one satellite transponder to another without the user having to change any settings in their receiver. When hundreds of services might be available to a user, this is an important advantage.

The channel number cannot be used as a unique way of identifying a channel, however, and thus each channel entry includes a source ID that acts as a unique identifier for that channel. There are two possible levels to this uniqueness. Source IDs between 0x0001 and 0x0FFF have to be unique in the transport stream carrying the VCT, whereas source IDs between 0x1000 and 0xFFFF must be unique at the regional level. This lets us handle cases in which the receiver can see two or more networks, each transmitting their own VCT. In this case, channel numbers may not be unique (and they probably will not be). By having unique source IDs, we always know which channel on which network we are referring to, even when we can see more than one network. In cases in which only one VCT will be available (for instance, a cable network), the network operator has no restrictions on the source IDs that may be used.

In all channel descriptions, the `ETM_location` field tells the receiver where it can find an Extended Text Message (ETM) structure that gives an extended description of the virtual channel. This field can take one of the values outlined in Table B.4.

These extended descriptions can contain several languages, and thus it is possible to provide multilingual names for channels using this mechanism. (Extended description functionality is further described later in the appendix.)

Table B.3. The format of the VCT in satellite networks.

Syntax	No. of Bits	Identifier/Value
satellite_virtual_channel_table_section() {		
table_id	8	0xDA
section_syntax_indicator	1	'1'
private_indicator	1	'1'
reserved	2	'11'
section_length	12	uimsbf
SVCT_subtype	8	uimsbf
SVCT_id	8	uimsbf
reserved	2	'11'
version_number	5	uimsbf
current_next_indicator	1	bslbf
section_number	8	uimsbf
last_section_number	8	uimsbf
protocol_version	8	uimsbf
num_channels_in_section	8	uimsbf
for(i=0; i<num_channels_in_section; i++) {		
short_name	8*16	uimsbf
reserved	4	'1111'
major_channel_number	10	uimsbf
minor_channel_number	10	uimsbf
modulation_mode	6	uimsbf
carrier_frequency	32	uimsbf
carrier_symbol_rate	32	uimsbf
polarization	2	uimsbf
FEC_Inner	8	uimsbf
channel_TSID	16	uimsbf
program_number	16	uimsbf
ETM_location	2	bslbf
reserved	1	'1'
hidden	1	bslbf
reserved	2	'11'
hide_guide	1	bslbf
reserved	3	'111'
service_type	6	uimsbf
source_id	16	uimsbf
feed_id	8	uimsbf
reserved	6	'111111'
descriptors_length	10	uimsbf
for(i=0; i<N; i++) {		
descriptors()		
}		
}		
reserved	6	'111111'
additional_descriptors_length	10	uimsbf
for(j=0; j<N; j++) {		
additional_descriptors()		
}		
CRC_32	32	rpchof
}		

Source: ATSC document number A/81.

Table B.4. Possible values of the `ETM_location` field in the VCT.

ETM Location	Meaning
0x00	No ETM
0x01	ETM located in the PTC carrying this PSIP
0x02	ETM located in the PTC specified by the channel TSID
0x03	[Reserved for future ATSC use]

Source: ATSC document number A/65b.

Satellite signals have a much greater number of parameters that control the way the signal is encoded, and satellite VCTs need to carry this information so that the receiver can tune to a signal correctly. As well as the modulation mode and frequency, channel descriptions in a satellite VCT tell the receiver the symbol rate, FEC coding, and feed ID it should use to tune to the transport stream that contains that channel.

Although the first two of these are obvious, the feed ID may not be. Network operators can use this to identify the satellite containing the signal, or to carry some additional information about the frequency band or polarization the receiver needs to know. The exact use of the feed ID is not standardized, and thus network operators can use this in any way they see fit.

One thing to notice when looking at the differences among satellite, terrestrial, and cable VCTs is the length of the `short_name` field. For satellite VCTs, this may be up to eight characters long, whereas for other network types it can only be seven characters. We are not aware of any good reason they are different, but it is something middleware developers and network operators need to be aware of.

Event Information

For cable and terrestrial networks, we can use the Event Information Table (EIT) to obtain information on events broadcast on various services. The EIT in an ATSC network is used in basically the same way as an EIT in a DVB network. EITs are split into 3-hour blocks of the schedule, which are broadcast as separate subtables. More than one subtable can be broadcast at any time, and thus the subtables being broadcast at any given time are referred to as EIT-0 (the EIT for the current 3-hour period) to EIT-N (the EIT covering the period Nx3 hours from now). EIT blocks can only start at midnight, 03:00, 06:00, 09:00, 12:00, 15:00, 18:00, or 21:00. All of the times used by the EIT are UTC times, and the receiver will adjust these to local times based on its current time zone.

One difference from a DVB system is that each 3-hour period is a separate subtable. In DVB systems, each 3-hour block (see Figure B.1) is a segment of a subtable that consists of eight sections. EITs in DVB systems still use subtables, but data is organized slightly differently. The second difference is the time period for which data is broadcast. In a DVB system, the

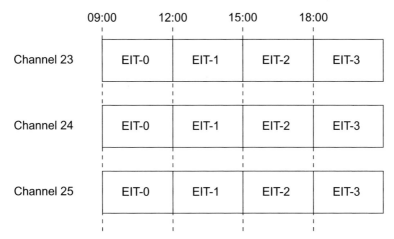

Figure B.1. Each EIT is split into 3-hour blocks for transmission.

first segment of the first subtable will always contain data for events starting at midnight on the current day. Thus, there may be several empty segments for periods that have passed and for which the network operator does not want to broadcast event information. In an ATSC system, the first subtable will always correspond to the current period, and the network operator will not broadcast subtables for previous periods.

It is possible to transmit up to 128 EIT subtables for any virtual channel, but it is very rare that so many will be transmitted. Terrestrial systems only have to transmit EITs covering at least the current time slot and the next three time slots. An event that is shown in a time slot that covers more than one EIT will be listed in every EIT that covers the time that event is shown. If this happens, the event must have the same event ID in every EIT. Table B.5 outlines the format of a section containing EIT data.

Satellite systems have a slightly different structure for carrying event information (discussed further in material to follow). For every type of network, though, the EIT describes events in roughly the same way because every receiver needs to know the same things about the events on the network.

Each entry in an EIT has a 14-bit event ID that can be used with the source ID to uniquely identify that event. An entry also contains a start time and duration for that event, as well as a string giving the title. As for VCT entries, each event description includes an `ETM_ location` field that indicates whether an extended description of that event is available. This field can take the values outlined in Table B.6.

As with the extended descriptions for VCT entries, these descriptions take the form of multilingual strings, allowing the receiver to display extended descriptions in several languages.

Table B.5. The format of the Event Information Table (EIT) in a terrestrial or cable system.

Syntax	No. of Bits	Identifier/Value
event_information_table_section() {		
table_id	8	0xCB
section_syntax_indicator	1	"1"
private_indicator	1	"1"
reserved	2	"11"
section_length	12	uimsbf
source_id	16	uimsbf
reserved	2	"11"
version_number	5	uimsbf
current_next_indicator	1	"1"
section_number	8	uimsbf
last_section_number	8	uimsbf
protocol_version	8	uimsbf
num_events_in_section	8	uimsbf
for (j=0; j< num_events_in_section; j++){		
reserved	2	"11"
event_id	14	uimsbf
start_time	32	uimsbf
reserved	2	"11"
ETM_location	2	uimsbf
length_in_seconds	20	uimsbf
title_length	8	uimsbf
title_text()	var	
reserved	4	"1111"
descriptors_length	12	
for(i=0;i<N;i++){		
descriptor()		
}		
}		
CRC_32	32	rpchof
}		

Source: ATSC document number A/65b.

Table B.6. Possible values of the ETM location in the EIT.

ETM Location	Meaning
0x0	No ETM
0x1	ETM located in the PTC carrying this PSIP
0x2	ETM located in the PTC carrying this event
0x3	[Reserved for future ATSC use]

Source: ATSC document number A/65b.

Event Information in a Satellite Network

In a satellite network, or the out-of-band channel in a cable network, it can be inefficient to transmit a separate EIT for every channel because of the number of transport streams that may be used and the number of ways they may be modulated. To help make life easier for the receiver, the ATSC specification for satellite broadcasting combines all of the EITs into an Aggregate Event Information Table (AEIT). Satellite networks use AEITs instead of EITs, and have the structure outlined in Table B.7.

This structure is similar to the EIT. The main difference is that for an AEIT the loop describing the events is nested within an additional loop that defines the source those events are associated with. The other difference is that AEITs, other than the one for the current time-slot (AEIT-0), do not need to contain information about events that started in an earlier time slot.

For a satellite network, event descriptions do not include a pointer to the extended event description. This description is still present, but there is less need to tell the receiver where it is (see also the next section). Complete event information may be split across several MPEG-2 sections due to the size limitation that applies to each section. Each EIT or AEIT may be up to 256 sections long.

Cable systems may also use the AEIT to carry event information on out-of-band channels. If they do, the format will follow that defined in ANSI/SCTE standard 65, which is slightly different from the format used by satellite systems.

Extended Text Descriptions

A broadcaster can transmit an extended description of a channel or an event. To avoid making other PSIP tables too large, these descriptions are all contained in a separate table called the Extended Text Table (ETT). Every extended description has its own entry in the table called an extended text message, or ETM. ETMs are multilingual strings, and thus a single extended text message can carry all of the necessary descriptions for a given channel or event.

These textual descriptions can be quite long, and thus every ETM is carried in a separate section. Table B.8 outlines the format of sections that are part of the ETT.

Each ETM consists of a pair of values: an ETM identifier and a text message (there are other fields, but they are there to give a description of these two main fields). An ETM identifier is a 32-bit value, which takes a slightly different format depending on what the ETM describes. If it gives an extended text description of a channel, the most significant 16 bits contain the source ID, whereas the rest of the identifier is set to zero. Figure B.2 shows the format for the ETM identifier for an ETM describing an event.

Not every event may have a description, and searching the ETT for a given ETM identifier may not be very efficient. The `ETM_location` field in every VCT, EIT, or AEIT entry tells the receiver whether an ETM for that entry is available and where to find it, and thus the

Table B.7. Format of the Aggregate Event Information Table (AEIT).

Syntax	No. of Bits	Identifier/Value
aggregate_event_information_table_section() {		
table_id	8	0xD6
section_syntax_indicator	1	"1"
private_indicator	1	"1"
reserved	2	"11"
section_length	12	uimsbf
AEIT_subtype	8	uimsbf
MGT_tag	8	uimsbf
reserved	2	"11"
version_number	5	uimsbf
current_next_indicator	1	"1"
section_number	8	uimsbf
last_section_number	8	uimsbf
if(AEIT_subtype = = 0) {		
num_sources_in_section	8	uimsbf
for(j=0;j<num_sources_in_section;j++){		
source_id	16	uimsbf
num_events	8	uimsbf
for(j=0;j<num_events;j++){		
off_air	1	bslbf
reserved	1	"1"
event_id	14	uimsbf
start_time	32	uimsbf
reserved	4	"1111"
duration	20	uimsbf
title_length	8	uimsbf
title_text()	var	
reserved	4	"1111"
descriptors_length	12	
for(i=0;i<N;i++){		
descriptor()		
}		
}		
}		
}		
else		
reserved	n × 8	
CRC_32	32	rpchof
}		

Source: ATSC document number A/81.

Table B.8. Format of the Extended Text Table (ETT).

Syntax	No. of Bits	Identifier/Value
extended_text_table_section() {		
table_id	8	0xCC
section_syntax_indicator	1	"1"
private_indicator	1	"1"
reserved	2	"11"
section_length	12	uimsbf
ETT_table_id_extension	16	0x0000
reserved	2	"11"
version_number	5	uimsbf
current_next_indicator	1	"1"
section_number	8	0x00
last_section_number	8	0x00
protocol_version	8	uimsbf
ETM_id	32	uimsbf
extended_text_message()		var
CRC_32	32	rpchof
}		

Source: ATSC document number A/65b.

31	16 15	2 1 0
Source ID	Event ID	

Figure B.2. Format of an ETM identifier. *Source:* ATSC document number A/65b.

receiver can avoid searching for extended descriptions that do not exist. For more information about the ETM location field, see Tables B.5 and B.8.

Like the EIT, the ETT is split up into a number of separate subtables. These are organized in the same way as the EIT, so that every ETT subtable corresponds to one ETT subtable.

Extended Text Messages

ETMs are multilingual, and thus PSIP can carry text descriptions in more than one language simultaneously. For this reason, ETMs are encoded using multiple string structures, which can carry multi-byte Unicode characters in an efficient way. Table B.9 outlines the format of the multiple string structure. This format is used for carrying all of the extended text descriptions in PSIP.

Table B.9. Extended text message format.

Syntax	No. of Bits	Identifier/Value
`multiple_string_structure() {`		
`  number_strings`	8	uimsbf
`  for(i=0; i<number_strings; i++) {`		
`    ISO_639_language_code`	24	uimsbf
`    number_segments`	8	uimsbf
`    for(j=0; j<number_segments; j++) {`		
`      compression_type`	8	uimsbf
`      mode`	8	uimsbf
`      number_bytes`	8	uimsbf
`      for(k=0; k<number_bytes; k++)`		
`        compressed_string_byte[k]`	8	bslbf
`    }`		
`  }`		
`}`		

Source: ATSC document number A/65b.

Each string has a three-letter language code, as defined by the ISO 693 standard. The string itself is split into a number of segments in order to support Unicode text more efficiently. Broadcasters can use the `mode` field to provide a simple way of shortening Unicode strings by identifying a common high-order byte for all characters in the segment. For instance, the Unicode sequence [0x0930, 0x0912, 0x0948, 0x0921] could be represented as a segment with the value 0x09 for the `mode` field and [0x30, 0x12, 0x48, 0x21] for the string. This gives a simple way of run-length coding the string in addition to any other compression that is applied.

Extended Descriptions in a Satellite Network

As with the EIT and AEIT, in satellite systems a number of different sources will share a single table for their extended descriptions. This table is called the Aggregate Extended Text Table, or AETT. This table has the structure outlined in Table B.10.

One difference you will notice is that AETT sections can carry more than one extended text message, unlike sections in the ETT. This helps use space more efficiently in large AETT tables. As with event information, cable systems may use this aggregate table for carrying extended text descriptions on an out-of-band channel. The use of the AETT in these cases is defined in ANSI/SCTE standard 65.

Parental Ratings

ATSC SI includes information about parental ratings so that receivers can filter out in-appropriate content, based on user preferences and information included by the network

Table B.10. Format of the Aggregate Extended Text Table (AETT).

Syntax	No. of Bits	Identifier/Value
aggregate_extended_text_table_section() {		
table_id	8	0xD7
section_syntax_indicator	1	"1"
private_indicator	1	"1"
reserved	2	"11"
section_length	12	uimsbf
AETT_subtype	8	uimsbf
MGT_tag	8	uimsbf
reserved	2	"11"
version_number	5	uimsbf
current_next_indicator	1	"1"
section_number	8	uimsbf
last_section_number	8	uimsbf
if(AETT_subtype = = 0) {		
num_blocks_in_section	8	uimsbf
for(j=0;j<num_blocks_in_section;j++){		
ETM_id	32	uimsbf
reserved	4	"1111"
extended_text_length	12	uimsbf
extended_text_message()	var	
}		
}		
else		
reserved	n × 8	
CRC_32	32	rpchof
}		

Source: ATSC document number A/81.

operator. There are two parts to any type of rating system such as this: the rating itself and a description of what those ratings actually mean.

In practice, things get even more complicated than that, because there may be several different rating systems in use at the same time. This may happen because different types of content have different rating systems (e.g., one for movies and one for normal TV shows) or because content is transmitted to different countries that use different rating systems.

These cases are all handled by a PSIP table called the Rating Region Table (RRT). This defines a set of rating regions and a set of rating dimensions for each region. Each rating dimension is one method of rating content. For instance, the MPAA rating system is one rating dimension.

A piece of content may have more than one rating dimension, either for the same region or for different regions (discussed further in material following). Table B.11 outlines the format of sections carrying the RRT.

Table B.11. Format of the Rating Region Table (RRT).

Syntax	No. of Bits	Identifier/Value
rating_region_table_section() {		
table_id	8	0xCA
section_syntax_indicator	1	"1"
private_indicator	1	"1"
reserved	2	"11"
section_length	12	uimsbf
table_id_extension{		
reserved	8	0xFF
rating_region	8	uimsbf
}		
reserved	2	"11"
version_number	5	uimsbf
current_next_indicator	1	"1"
section_number	8	uimsbf
last_section_number	8	uimsbf
protocol_version	8	uimsbf
rating_region_name_length	8	uimsbf
rating_region_name_text()	var	
dimensions_defined	8	uimsbf
for(i=0; i<dimensions_defined; i++) {		
dimension_name_length	8	uimsbf
dimension_name_text()	var	
reserved	3	"111"
graduated_scale	1	bslbf
values_defined	4	uimsbf
for (j=0; j<values_defined; j++) {		
abbrev_rating_value_length	8	uimsbf
abbrev_rating_value_text()		var
rating_value_length	8	uimsbf
rating_value_text()		var
}		
}		
reserved	6	"111111"
descriptors_length	10	uimsbf
for (i=0;i<N;i++) {		
descriptor()		
}		
CRC_32	32	rpchof
}		

Source: ATSC document number A/65b.

Table B.12. Format of the content advisory descriptor.

Syntax	No. of Bits	Identifier/Value
content_advisory_descriptor() {		
descriptor_tag	8	0x87
descriptor_length	8	uimsbf
reserved	2	"11"
rating_region_count	6	
for(i=0; i<rating_region_count; i++) {		
rating_region	8	uimsbf
rated_dimensions	8	uimsbf
for(j=0; j<rated_dimensions; j++) {		
rating_dimension_j	8	uimsbf
reserved	4	"1111"
rating_value	4	uimsbf
}		
rating_description_length	8	uimsbf
rating_description_text()	var	
}		
}		

Source: ATSC document number A/65b.

This shows how the various rating systems are sent to the receiver, but it does not tell us how these ratings are applied to a channel or event. To do this, ATSC uses the content advisory descriptor. This descriptor has the format outlined in Table B.12, and it can be carried in the event loop of the EIT or the AEIT to give a rating for a single event, or it can be carried in the VCT to give a rating for an entire channel.

Each content advisory descriptor can specify which rating systems apply in a number of different regions. In addition to a set of ratings in the various dimensions, it can also carry a text description that provides more detail about the rating. Broadcasters could use this description to tell viewers why a piece of content has the rating it does (e.g., sex, violence, or language that may be offensive).

If a network only uses rating region 0x01 (the United States, U.S. possessions, and Canada), the RRT is optional. The full requirements for this rating region are covered in standard EIA-766, which must be implemented by every receiver in those regions, and thus network operators do not need to broadcast this information because the receiver already knows it.

Advanced Functions: Redirecting Channels

The tables we have seen so far provide the basic functions a network operator needs to make a DTV service work. In addition to these basic functions, it is possible to do some more advanced things that may be useful in some cases. The most interesting of these from an SI perspective is the ability to force a receiver to change channels, either unconditionally or based on a set of criteria specified by the network operator. This can enable a network operator to provide different functions based on language settings in the receiver, geographical location, demographics, or any one of a number of other features.

The table that makes this possible is the Directed Channel Change Table (DCCT), which has the format outlined in Table B.13. This is an optional table, and may not always be present in transmissions.

Each entry in the DCCT is a test that decides whether a receiver should change from the source channel to the destination channel. These tests can be time limited, so that a channel change may only be active for a fixed period (e.g., for the length of one event). This allows the network operator to target specific receivers during specific shows, and to show different content in those cases. One example could be a football game. Viewers in one area could watch one game featuring their local team, whereas the network operator could automatically redirect viewers in another area to a different channel showing a game featuring their local team instead. After the game is over, the redirection is automatically cancelled because it is timed to last the length of the game.

The `DCC_context` field tells the receiver how it should handle a channel change, should it need to make one. A value of 0 means that the redirect is temporary, and that if the DCCT tells it to return to the original channel at a later time (or if the user changes channel) the redirection is cancelled. In this case, the new channel has the same channel number as the original channel. A value of 1 for this field tells the receiver that the redirection is permanent, and the new channel is a completely different channel (with a completely different channel number) from the original one. The following should give you an idea of the types of tests available.

- Unconditional redirection
- Unconditional return to original channel
- Inclusion/exclusion based on postal code
- Inclusion/exclusion based on location information other than postal code
- Inclusion/exclusion based on genre
- Inclusion/exclusion based on demographic
- Rating blocked

More information about how these tests operate, and a detailed look at how the broadcasters can specify them, is available in Section 6.7 of the ATSC PSIP standard (ATSC document number A/65b). If the standard values for genres and locations defined in the ATSC PSIP standard are not enough, a broadcaster can extend these by broadcasting the DCC Selection Code Table (SCT). This carries additional data that the receiver can use in conjunction with the DCCT when checking to see whether a particular test condition is met. (More information on the SCT can be found in Section 6.8 of the ATSC PSIP specification.)

Table B.13. Format of the Directed Channel Change Table (DCCT).

Syntax	No. of Bits	Identifier/Value
directed_channel_change_table_section() {		
table_id	8	0xD3
section_syntax_indicator	1	"1"
private_indicator	1	"1"
reserved	2	"11"
section_length	12	uimsbf
dcc_subtype	8	0x00
dcc_id	8	uimsbf
reserved	2	"11"
version_number	5	uimsbf
current_next_indicator	1	"1"
section_number	8	0x00
last_section_number	8	0x00
protocol_version	8	uimsbf
dcc_test_count	8	uimsbf
for(i=0; i<dcc_test_count; i++) {		
dcc_context	1	uimsbf
reserved	3	"111"
dcc_from_major_channel_number	10	uimsbf
dcc_from_minor_channel_number	10	uimsbf
reserved	4	"1111"
dcc_to_major_channel_number	10	uimsbf
dcc_to_minor_channel_number	10	uimsbf
dcc_start_time	32	uimsbf
dcc_end_time	32	uimsbf
dcc_term_count	8	uimsbf
for(j=0; j< dcc_term_count; j++) {		
dcc_selection_type	8	uimsbf
dcc_selection_id	64	uimsbf
reserved	6	"111111"
dcc_term_descriptors_length	10	uimsbf
for(k=0; k<N; k++) {		
dcc_term_descriptor()		
}		
}		
Reserved	6	"111111"
dcc_test_descriptors_length	10	uimsbf
for(j=0; j<N; j++) {		
dcc_test_descriptor()		
}		
}		
Reserved	6	"111111"
dcc_additional_descriptors_length	10	uimsbf
for(i=0; i<N; i++){		
dcc_additional_descriptor()		
}		
CRC_32	32	rpchof
}		

Source: ATSC document number A/65b.

Telling the Time Correctly

Many of the elements of PSIP contain some information about the time something will happen, such as the start time of an event. This information is only useful if we have some way of knowing that the time in the receiver is correct. In this case, the correct time is the one the network operator is working to, and thus there needs to be some way of synchronizing the clock in the receiver to the clock used by the network operator.

As in DVB systems, the network operator broadcasts a time reference as part of the transport stream to make sure that all of the receivers have the correct time. This is contained in the System Time Table (STT), and this table is used to tell the receiver a number of things. The first and most obvious one is the time, because that is the whole point of the table, after all. This is broadcast as the number of GPS seconds since midnight on January 6, 1980, and is measured as UTC (GMT) time. Because this is not always useful, the GPS_UTC_offset field tells the receiver what time zone it is currently in.

The STT also contains daylight_savings structure telling the receiver whether daylight savings time should be applied. This tells the receiver three things: whether it is currently in daylight savings time or not, the day at which it should change its daylight savings setting, and the time at which it should do so. In the month that daylight savings time will start or end, the table contains this information to tell the receiver exactly when to change. At other times, the receiver should ignore this structure. The full structure of the STT is outlined in Table B.14.

Each STT is small enough to fit in a single MPEG-2 section, and thus more fields have predefined values than is the case for other tables.

Putting It All Together

To increase efficiency in the receiver, another table—the Master Guide Table (MGT) – contains references to all of the tables referenced previously. The MGT provides a detailed description of all of the other tables in the system: which tables are present, the PIDs on which they are transmitted, and the size in bytes of each table. This lets the receiver allocate memory appropriately before it actually starts parsing the tables (it also means that the receiver knows which tables to look for).

As any experienced programmer will tell you, programming an embedded system can sometimes be a complicated matter of juggling memory. This means not allocating too much (which wastes memory the rest of the system may need), but not allocating too little either, which results in problems when your table is too small. By providing this information, the MGT helps the receiver parse the SI data as efficiently as possible, which software developers always appreciate. Sections containing MGT data have the structure outlined in Table B.15.

The MGT does not provide any information the user will actually see, but it is essential for the receiver to show all of the SI that may be broadcast.

Table B.14. Format of the System Time Table (STT).

Syntax	No. of Bits	Identifier/Value
system_time_table_section() {		
table_id	8	0xCD
section_syntax_indicator	1	"1"
private_indicator	1	"1"
reserved	2	"11"
section_length	12	uimsbf
table_id_extension	16	0x0000
reserved	2	"11"
version_number	5	"00000"
current_next_indicator	1	"1"
section_number	8	0x00
last_section_number	8	0x00
protocol_version	8	uimsbf
system_time	32	uimsbf
GPS_UTC_offset	8	uimsbf
daylight_savings	16	uimsbf
for(i=0; I< N;i++){		
descriptor()		
}		
CRC_32	32	rpchof
}		

Source: ATSC document number A/65b.

PSIP Profiles in Cable Systems

For cable systems, we can transmit PSIP information via an out-of-band channel such as the POD module, using the format defined by ANSI/SCTE standard 65. One of the ways this standard differs from the ATSC specification for PSIP is that it defines six different profiles for the PSIP data that is carried on the out-of-band channel. Table B.16 outlines these six profiles.

Many of the differences among these profiles cover which tables should be broadcast and which descriptors are allowed in the different profiles. However, there are some deeper differences. In particular, the different profiles of PSIP can use one of two different forms for the VCT. Higher profiles use the long-form VCT. Other profiles use the short-form VCT, which reduces the size of the VCT by removing some fields and moving others to some new SI tables such as the NTT and the NIT. Similarly, some tables (such as the AEIT and AETT) may take a slightly different format from that shown here, in order to carry information that is needed by a cable system rather than by a satellite system. These tables, and the full requirements for each of the different profiles, are described in ANSI/SCTE 65. Table B.17 outlines the tables each profile requires.

Table B.15. Format of the Master Guide Table (MGT).

Syntax	No. of Bits	Identifier/Value
master_guide_table_section() {		
table_id	8	0xC7
section_syntax_indicator	1	"1"
private_indicator	1	"1"
reserved	2	"11"
section_length	12	uimsbf
table_id_extension	16	0x0000
reserved	2	"11"
version_number	5	uimsbf
current_next_indicator	1	"1"
section_number	8	0x00
last_section_number	8	0x00
protocol_version	8	uimsbf
tables_defined	16	uimsbf
for(i=0; i<tables_defined; i++) {		
table_type	16	uimsbf
reserved	3	"111"
table_type_PID	13	uimsbf
reserved	3	"111"
table_type_version_number	5	uimsbf
number_bytes	32	uimsbf
reserved	4	"1111"
table_type_descriptors_length	12	uimsbf
for(k=0; k<N; k++) {		
descriptor()		
}		
}		
reserved	4	"1111"
descriptors_length	12	uimsbf
for(i=0; i< N; i++){		
descriptor()		
}		
CRC_32	32	rpchof
}		

Source: ATSC document number A/65b.

Table B.16. Profiles of PSIP that may be used in different networks.

Profile Number	Description
1	Baseline
2	Revision detection
3	Parental ratings support
4	Standard EPG data
5	PSIP-compatible with OOB extensions
6	PSIP only

Table B.17. Tables supported in each PSIP profile.

Table Section	Table ID	Profile 1 Baseline	Profile 2 Revision Detection	Profile 3 Parental Advisory	Profile 4 Standard EPG Data	Profile 5 Combination	Profile 6 PSIP Only[a]
Network Information Table	0xC2						
Carrier Definition Subtable		M	M	M	M	M	—
Modulation Mode Subtable		M	M	M	M	M	—
Network Text Table	0xC3						
Source Name Subtable		O	O	O	M	M	—
Short-form Virtual Channel Table	0xC4						
Virtual Channels Map		M	M	M	M	M	—
Defined Channels Map		M	M	M	M	M	—
Inverse Channel Map		O	O	O	O	O	—
System Time Table	0xC5	M	M	M	M	M	M
Master Guide Table	0xC7	—	—	(b)	M	M	M

561

Table B.17. *Continued*

Table Section	Table ID	Profile 1 Baseline	Profile 2 Revision Detection	Profile 3 Parental Advisory	Profile 4 Standard EPG Data	Profile 5 Combination	Profile 6 PSIP Only[a]
Rating Region Table	0×CA	—	—	(c)	(c)	(c)	(c)
Long-form Virtual Channel Table	0×C9	—	—	—	—	M	M
Aggregate Event Information Table	0×D6	—	—	—	M	M	M
Aggregate Extended Text Table	0×D7	—	—	—	O	O	O

Legend:
M: Mandatory. O: Optional. –: Forbidden.
Notes:
a. Exception: System Time Table (table ID 0×C5 is used here instead of table ID 0×CD defined in PSIP) and other modifications.
b. Mandatory for outside of North America to describe any transmitted RRT. For region 0×01 (United States and possessions), delivery of an RRT is optional because this table is standardized in EIA-766.
c. Exception: Delivery of the RRT corresponding to region 0×01 (United States and possessions) is optional because this table is standardized in EIA-766.
Source: ANSI/SCTE 65, 2002.

For all cable systems, any PSIP data that is transmitted in-band (i.e., as part of the same transport stream as the audio and video data) should follow the ATSC PSIP standard rather than any of the profiles specified in ANSI/SCTE 65.

Broadcasting PSIP Data

Together, the VCT, RRT, MGT, and SST give the receiver all of the information it needs to find and decode the channels available on a network. These tables are known as the base tables, and they are all broadcast on the same PID, known as the base PID. Usually the base PID is numberx01FFB. On cable systems that support out-of-band SI transmission (such as OpenCable systems), the base PID for out-of-band data is PID number 0x1FFC.

Carrying all of these tables on the same PID makes it easier for the receiver to filter the PSIP tables it needs most efficiently. This in turn means that performance may be better on some receivers. From the network operator's point of view, it is no more difficult to carry all of

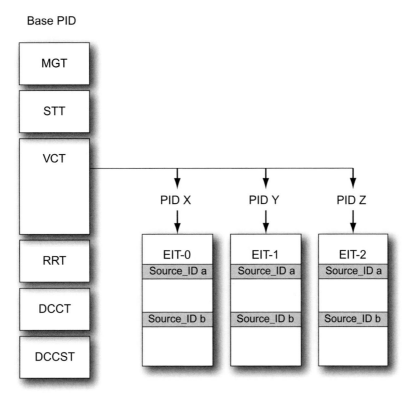

Figure B.3. Relationships among the PSIP tables. *Source:* ATSC document number A/65b.

these tables on one PID than it is on several, and thus ATSC took this opportunity to make life easier for receiver manufacturers.

Other tables, such as the EIT and ETT, do not have to be carried on any specific PID. Different versions of the EIT will often be carried on different PIDs, so that EIT-0 will be broadcast on one PID and EIT-1 on another. To avoid clashes when moving content to DVB systems, network operators should where possible avoid using PIDs that DVB reserves for its SI tables. The diagram shown in Figure B.3 shows the relationships among the various PSIP tables.

In addition to knowing how the various tables relate to one another, network operators need to be careful about how often tables are repeated. ATSC suggests repetition rates for the various tables in the PSIP implementation guidelines (ATSC standard A/69), but these repetition rates are not mandatory. In most cases, the suggested rates give a good balance between speed of access and bandwidth usage. Sometimes, however, there may be cases in which these suggested rates do not fit. Remember that the suggested values are only suggestions.

Index

Page numbers with "t" denote tables; those with "f" denote figures